22.50

PROCEEDINGS OF THE SIXTH BERKELEY SYMPOSIUM

VOLUME VI

PROCEEDINGS *of the* SIXTH BERKELEY SYMPOSIUM ON MATHEMATICAL STATISTICS AND PROBABILITY

Held at the Statistical Laboratory
University of California
April 9–12, 1971
June 16–21, 1971
and
July 19–22, 1971

with the support of
University of California
National Institutes of Health
Atomic Energy Commission

VOLUME VI

EFFECTS OF POLLUTION ON HEALTH

EDITED BY LUCIEN M. LE CAM
JERZY NEYMAN AND ELIZABETH L. SCOTT

UNIVERSITY OF CALIFORNIA PRESS
BERKELEY AND LOS ANGELES
1972

UNIVERSITY OF CALIFORNIA PRESS
BERKELEY AND LOS ANGELES
CALIFORNIA

CAMBRIDGE UNIVERSITY PRESS
LONDON, ENGLAND

ISBN: 0-520-02189-4

LIBRARY OF CONGRESS CATALOG CARD NUMBER: 49-8189

PRINTED IN THE UNITED STATES OF AMERICA

CONTENTS OF PROCEEDINGS
VOLUMES I, II, III, IV, V, AND VI

Volume I—Theory of Statistics

v

Asymptotic normality of sums of dependent random variables. B. V. GNEDENKO, Limit theorems for sums of a random number of positive independent random variables. M. ROSENBLATT, Central limit theorem for stationary processes. V. V. SAZONOV, On a bound for the rate of convergence in the multidimensional central limit theorem. C. STEIN, A bound for the error in the normal approximation to the distribution of a sum of dependent random variables.

Volume III—Probability Theory

Passage Problems

Yu. K. BELYAYEV, Point processes and first passage problems. A. A. BOROVKOV, Limit theorems for random walks with boundaries. N. C. JAIN and W. E. PRUITT, The range of random walk. H. ROBBINS and D. SIEGMUND, On the law of the iterated logarithm for maxima and minima. A. D. SOLOVIEV, Asymptotic distribution of the moment of first crossing of a high level by a birth and death process.

Markov Processes—Potential Theory

R. G. AZENCOTT and P. CARTIER, Martin boundaries of random walks on locally compact groups. J. L. DOOB, The structure of a Markov chain. S. PORT and C. STONE, Classical potential theory and Brownian motion. S. PORT and C. STONE, Logarithmic potentials and planar Brownian motion. K. SATO, Potential operators for Markov processes.

Markov Processes—Trajectories—Functionals

R. GETOOR, Approximations of continuous additive functionals. K. ITÔ, Poisson point processes attached to Markov processes. J. F. C. KINGMAN, Regenerative phenomena and the characterization of Markov transition probabilities. E. J. McSHANE, Stochastic differential equations and models of random processes. P. A. MEYER, R. SMYTHE, and J. WALSH, Birth and death of Markov processes. P. W. MILLAR, Stochastic integrals and processes with stationary independent increments. D. W. STROOCK and S. R. S. VARADHAN, On the support of diffusion processes with applications to the strong maximum principle. D. W. STROOCK and S. R. S. VARADHAN, Diffusion processes.

Point Processes, Branching Processes

R. V. AMBARTSUMIAN, On random fields of segments and random mosaic on a plane. H. SOLOMON and P. C. C. WANG, Nonhomogeneous Poisson fields of random lines with applications to traffic flow. D. R. COX and P. A. W. LEWIS, Multivariate point processes. M. R. LEADBETTER, On basic results of point process theory. W. J. BÜHLER, The distribution of generations and other aspects of the family structure of branching processes. P. S. PURI, A method for studying the integral functionals of stochastic processes with applications: III. W. A. O'N. WAUGH, Uses of the sojourn time series for the Markovian birth process. J. GANI, First emptiness problems in queueing, storage, and traffic theory. H. E. DANIELS, Kuhn-Grün type approximations for polymer chain distributions. L. KATZ and M. SOBEL, Coverage of generalized chess boards by randomly placed rooks. R. HOLLEY, Pressure and Helmholtz free energy in a dynamic model of a lattice gas. D. MOLLISON, The rate of spatial propagation of simple epidemics. W. H. OLSON and V. R. R. UPPULURI, Asymptotic distribution of eigenvalues or random matrices.

Information and Control

R. S. BUCY, A priori bounds for the Ricatti equation. T. FERGUSON, Lose a dollar or double your fortune. H. J. KUSHNER, Necessary conditions for discrete parameter stochastic optimization problems. P. VARAIYA, Differential games. E. C. POSNER and E. R. RODEMICH, Epsilon entropy of probability distributions.

Ecological Studies

Skeletal Plans for a Comprehensive Health-Pollution Study, Discussion and Epilogue

PREFACE

Berkeley Symposia on Mathematical Statistics and Probability have been held at five year intervals since 1945, with the Sixth Symposium marking a quarter of a century of this activity. The purpose of the Symposia is to promote research and to record in the *Proceedings* the contemporary trends in thought and effort. The subjects covered in the Berkeley statistical Symposia range from pure theory of probability through theory of statistics to a variety of fields of applications of these two mathematical disciplines. The fields selected are those that appear especially important either as a source of novel statistical and probabilistic problems or because of their broad interdisciplinary character combined with particular significance to the society at large. A wide field of application traditionally represented at the Berkeley Symposia is the field of biology and health problems. Physical sciences, including astronomy, physics, and meteorology are also frequently represented. Volume 5 of the *Proceedings* of the Fifth Symposium was entirely given to weather modification.

With the help of advisory committees and of particular scholars, the participants of the Berkeley Symposia are recruited from all countries of the world, hopefully to include representatives of all significant schools of thought. In order to stimulate fruitful crossfertilization of ideas, efforts are made for the symposia to last somewhat longer than ordinary scholarly meetings, up to six weeks during which days with scholarly sessions are combined with excursions to the mountains and other social events. The record shows that, not infrequently, novel ideas are born at just such occasions.

According to the original plans, the entire Sixth Berkeley Symposium was to be held during the summer of 1970, with the generous support of the University of California, through an allocation from the Russell S. Springer Memorial Foundation, of the National Science Foundation, of the National Institutes of Health, of the Office of Naval Research, of the Army Research Office, and of the Air Force Office of Scientific Research. This help is most gratefully acknowledged. Certain circumstances prevented the Biology-Health Section from being held in 1970 and the meeting held in that year, from June 21 to July 18, was concerned with mathematical domains of probability and statistics. The papers presented at that time, and also some that were sent in by the individuals who were not able to attend personally, fill the first three volumes of these *Proceedings*. Volume 1 is given to theory of statistics and Volumes 2 and 3 to the rapidly developing theory of probability.

The Biology-Health Section of the Sixth Symposium had to be postponed to 1971. Every postponement of a scholarly meeting involves a disruption of the plans and all kinds of difficulties. Such disruption and difficulties certainly occurred in the present case. As originally planned, the Biology-Health Section of the Sixth Symposium was to be comparable to that of the Fifth, the *Proceedings* of which extended close to 1,000 pages in print. This is much larger than

Volume 4 of the present *Proceedings* that summarizes the Biology-Health Section
held from June 16 to 21, 1971. However, the losses suffered in some respects
have been compensated by gains in others. Those gains are reflected in Volumes
5 and 6 of these *Proceedings*.

During the fall of 1970 we became much impressed by the development and
rapid growth of a new field of biological studies which includes the areas known as
"non-Darwinian" and "neo-Darwinian" studies of evolution. These are studies
based on the structure of macromolecules present in many now living species
and performing in them similar functions. One example is the hemoglobin mole-
cule, carried by all mammals as well as by fish. The differences among the
homologous macromolecules in different species are usually ascribed to mutations
that are in some sense inconsequential, and are supposed to occur more or less
at a uniform rate. The number of differences between any two species is indicative of the time that elapsed from the moment of separation from the presumed
common ancestor. The probabilistic-statistical problems involved in such studies
nclude the estimation of philogenetic trees of several species and, in particular,
the estimation of the time since two species separated from their ancestor.

It was found that, with only a few exceptions, mathematical statisticians are
not familiar with the new domain and that, at the same time, a great many
biologists make strong efforts to treat the statistical problems themselves. A
joint meeting of biologists and statisticians was clearly indicated and a separate
conference, especially given to novel studies of evolution, was held from April 9
to 12, as part of the Biology-Health Section of the Sixth Berkeley Symposium.
It is summarized in Volume 5 of these *Proceedings*. Somewhat unexpectedly, it
appeared that the new field of studies of evolution involves controversies that
are just as sharp as those that occasionally enliven the meetings of mathematical
statisticians . . .

We were introduced to problems of evolution treated on the level of macromolecules by Professor T. H. Jukes, V. N. Sarich, and A. C. Wilson. Their very
interesting seminar talks and later their advice on the organization of the conference on evolution are highly appreciated.

While studies of evolution involve observational research, particularly that
concerned with the relation between classical population genetics and novel
findings on the level of molecular biology, the whole domain is clearly conceptual.
Contrary to this, the third part of the Biology-Health Section of the Sixth
Symposium was totally given to observational studies in a domain of great
importance to society at large and of great public interest.

The domain in question, a highly controversial domain, is that of the relation
between environmental pollution and human health. The growing population in
the United States and in other countries needs more electric power, more automobiles, and other products. The relevant industries are eager to satisfy these
needs. However, the expanded industrial activity, unavoidably conducted with
an eye on costs, leads to pollution of the environment. The controversies at

public hearings, in the daily press, and in scholarly publications center around the question whether the currently adopted standards of safety are sufficient or not. The volume of research, largely statistical, surrounding this question is immense. The intention that the *Proceedings* provide a cross section of contemporary statistical work dictated the organization of a special conference entirely given to the problem of health and pollution. This conference, held from July 19 to 22, is summarized in Volume 6 of these *Proceedings*. In organizing the conference we benefitted greatly from the advice of Dr. S. W. Greenhouse of the National Institutes of Health, of Professor B. Greenberg of Chapel Hill, North Carolina, and of Drs. J. M. Hollander and H. W. Patterson of the Berkeley Lawrence Laboratory.

The first purpose of the Health-Pollution Conference was to take stock of the studies already performed. The second and the ultimate purpose was to see whether a novel statistical study is called for, hopefully more comprehensive and more reliable than those already completed. With this in mind, invitations to the conference were issued to Federal and State governmental agencies concerned with health and pollution, to authoritative scholarly institutions, and to a number of particular individual scholars known to have worked on one or another aspect of this problem.

As a special stimulus for thought on the entire problem of pollution and health, its present state and the future, the invitations to the conference were formulated to include a call for submission of skeletal plans for a fresh comprehensive statistical study, capable of separating the effects of particular pollutants. Four such plans were submitted and they are published in Volume 6.

All the participants had complete freedom of expression, both in their prepared papers and in their contributions to the discussion. Thus it is likely that the goal of providing a realistic cross section of contemporary statistical research on the problem is reasonably approached. Also it is not unlikely that the present state of knowledge on human health and pollution, and the scholarly level of the substantive studies prepared are fairly reflected in these *Proceedings*.

In addition to funds provided by the University of California and the National Institutes of Health, the Health-Pollution Conference was organized using a grant from the Atomic Energy Commission, Division of Biology and Medicine. This help is gratefully acknowledged.

The organization and the running of three distinct scholarly meetings, one in April, another in June, and the third in July 1971, each attended by some 100 to more than 300 participants, would not have been possible without the willing, efficient, and cheerful help and cooperation of the staff of the Department of Statistics and the Statistical Laboratory. Our most hearty thanks go to our successive "ministers of finance," Mrs. Barbara Gaugl and Mrs. Freddie Ruhl, who watched the sinking balances and surveyed the legality of proposed expenditures, some appropriate under one grant and not under another, etc. In addition to financial matters, Mrs. Gaugl supervised the local arrangements for scholarly sessions, for several social events and for servicing the participants. In this she was efficiently helped by Mrs. Dominique Cooke, by Miss Judy Whipple and by a number of volunteers from among the graduate students in the Department. Mrs. Cooke and Miss Whipple had their own very important domain of activities: to keep straight the correspondence and the files. Coming in addition to the ordinary university business, this was no mean job and the performance of the two ladies is highly appreciated.

All the above refers to the early part of the year 1971 and up to the end of the conferences. Then the manuscripts of the papers to be published in the *Proceedings* started to arrive, totalling 1849 typewritten pages, not counting figures and numerical tables. This marked a new phase of the job in which we enjoyed the cooperation of another group of persons, who prepared the material for the printers. At the time, the team of editors, Miss Carol Conti, Mrs. Margaret Darland, and Miss Jean Kettler, under the able guidance of Mrs. Virginia Thompson and supervised by Professor LeCam, Chairman of the Organizing Committee, worked assiduously on proofs of papers in Volumes 1, 2 and 3. The arrival of the material for Volumes 4, 5, and 6, unavoidably involving some correspondence with the authors and conferences at the University Press, created heavy burden. We are very grateful to the four ladies whose cooperation has been inspiring to us.

Last but not least, our hearty thanks to the University of California Press, Mr. August Frugé and his colleagues for their help, cooperation and also their patience when confronted with piles of manuscripts which we hoped to see published both excellently as in the past quarter of a century and "right away, yesterday!"

J. Neyman   E. L. Scott   L. Le Cam (Chm.)

CONTENTS

Problems of Monitoring

Pollutants in Food Chains

Ecological Studies

Skeletal Plans for a Comprehensive Health-Pollution Study, Discussion and Epilogue

STATISTICAL PROBLEMS AND STRATEGIES IN ENVIRONMENTAL EPIDEMIOLOGY

JOHN R. GOLDSMITH

CALIFORNIA STATE DEPARTMENT OF HEALTH, BERKELEY

1. Introduction and definitions

The purpose of this contribution is to outline a series of problems encountered mostly in work on air pollution health effects and to a lesser extent in studies on health effects of water pollution, and of noise, as well as in the field of chronic disease epidemiology. The presentation is in terms intended to help design statistically satisfactory studies of the association of environmental factors with the well being of human populations.

This report is in three parts, general problems in environmental epidemiology, prototypical problems, and statistical strategies. Emphasis is placed on problems and strategies, rather than on the critical evaluation of results. The critical evaluation of results in terms of well planned research is the essence of scientific analysis in the field, and not necessarily the introductory material which this article essentially treats.

"Environmental epidemiology" is a subdivision of epidemiology, deriving historically from chronic disease epidemiology. The derivation is traced in a previous paper [9]. The main fallout of significant contributions of chronic disease epidemiology to health appears to be in the identification and better management of environmental factors; it was this plus the experience with studying the components of chronic respiratory disease morbidity, including symptoms of cough, shortness of breath, and alterations in respiratory function, which led to the conviction that the association of environmental factors could be studied with such health parameters, even though the study of such parameters was not necessarily equivalent to the study of the epidemiology of disease. Accordingly environmental epidemiology does not depend only on the determination of the presence or absence of a disease, but may include the alteration of health associated with environmental exposures.

There are four essential requirements for the conduct of effective work in environmental epidemiology. These are: (1) a suitable set of statistical strategies; (2) the capacity to design, carry out, and report the necessary procedures for dependable research; (3) access to populations of sufficient size and appropriate characteristics for study; and (4) an adequate support base in resources and in personnel for carrying out the necessary work. This paper and this meeting are

1

primarily oriented to the problems of research capability and its dependence on statistical strategies.

It cannot be assumed that good study design is sufficient to assure good epidemiological studies because there are always factors of population accessibility and unestimated variables which influence results of work in the field.

Compared to experimental research on related problems, the epidemiologic studies of environmental effects tend to be expensive and hence there is a greater importance given to the planning and design of studies.

Most epidemiologic research deals with tests of association and almost never can a single epidemiologic study be interpreted with respect to causation. Thus, it is of great importance that precise language be used in order to make it clear that, for an identified problem, a statistical association is being sought and evaluated, and a necessary and sufficient cause is usually not approachable solely by a single epidemiologic study.

1.1. *Relation of epidemiological association to causation.* Causation is approached usually by convergence of many epidemiological studies or by convergence of evidence from experimental and epidemiological work. Convergence means here the tendency for different studies in different loci to support a similar association.

The steps which permit one to consider a verdict of causation when an epidemiologic association has been demonstrated have been reviewed by Sir A. B. Hill [15]. He describes the problem as follows. "Our observations reveal an association between two variables, perfectly clearcut and beyond what we would care to attribute to the play of chance. What aspects of that association should we especially consider before deciding that the most likely interpretation of it is causation?" He then lists nine yardsticks that are relevant. They are (with the author's interpretation) the following. (1) *Strength* of the association; primarily a statistical attribute. (2) *Consistency;* this is largely an epidemiologic attribute and related to the convergence mentioned above. (3) *Specificity;* a matter which is partly epidemiological and partly statistical in that if one has included all of the variables, an epidemiologic task, one could accept an inference of high specificity. (4) *Temporality;* how the possible exposure is related in time to the possible effect. This may be analyzed by statistical procedures and it may be the subject of epidemiologic strategies as well. (5) *Biological gradient;* this is predominantly an epidemiological observation couched in mathematical terms and may be analyzed for by statistical yardsticks, for example, the regression of blood lead on atmospheric lead or the increase in some effect with an increase in the number of cigarettes smoked. (6) *Plausibility;* particularly biological plausibility. This is neither an epidemiologic nor a statistical attribute but one based on knowledge of other data. (7) *Coherence;* the extent to which the proposed association fits with a large body of related information. (8) *Experiment;* its insight into causal mechanisms may be reflected in the convergence cited above. It is neither an epidemiological nor a statistical criterion. (9) *Analogy;* this yardstick does not require comment.

Hill makes the unqualified statement, "No formal tests of significance can answer those questions. Such tests can and should, remind us of the effects that the play of chance can create, and they will instruct us in the likely magnitude of those effects. Beyond that they contribute nothing to the 'proof' of our hypothesis."

2. Types of health effects and environmental exposures

Health effects. Environmental epidemiology deals with a broad array of disease and health states. Heading that array, of course, in terms of gravity, is mortality including case fatality rates; morbidity is estimated in a variety of ways, such as reported new cases of illness, aggravation of pre-existing illness, the occurrence of episodes of hospitalization or demands for medical service, the occurrence of accidents, the impairment of function or the production of symptoms.

Other less clearcut effects are also of importance, such as biochemical and physiological reactions which may not be easily understood in relation to long term implications [11], the occurrence of annoyance reactions [17], and the storage of potentially harmful material such as lead and pesticides in the human body [19], [12], [24].

Environmental exposures. Exposures to chemical or biological agents can occur to a given substance through inhalation or by oral ingestion or occasionally by absorption through the skin. Toxicity of the substance may vary considerably depending on the exposure mode. The location of exposures and the manner in which the given pollutant is introduced are also quite varied. In the case of radiation, exposures occur as a result of natural background levels which vary to some extent. There is the important contribution of medical diagnostic and therapeutic radiation, and there are occupational exposures and the possible community exposures in association with nuclear power development. Similarly in the case of exposure to pesticides, there is the incidental exposure of a few people who live adjacent to areas being sprayed, there is the residue in food, and occasionally a food chain gradient; there is the possibility of absorption through handling and occupational exposures of workers. Thus, the types of environmental exposures are quite varied, and they have a tendency to interact.

3. General logic and strategy for environmental epidemiology

In a review of the statistical aspects of air pollution medical research, Massey, Hopkins and Goldsmith reported [4] that the problems generally fell into three classes. There were (a) the problems of multiple factors influencing the single reaction and the corollary of multiple reaction from a single exposure *multivariate problems;* (b) the problems of complex interrelationships in time and space *time-space series;* and (c) problems which were best described as *systems problems* because they were so complex that existing single methods did not seem appropriate for them.

Beyond these, there is a further set of problems which have been recognized more recently. In the first place, there is the problem of *nonspecificity;* most environmental exposures do not produce reactions that are only produced by such exposures, but may change the probability of occurrence of other diseases (coronary heart disease) or reactions (increased air way resistance). In general, it is best to assume that an environmental exposure will only *unusually* be related to a single clearly defined clinical entity or alteration in health status. It is prudent to assume that environmental exposures may alter health status in nonspecific ways.

Secondly, there is the problem of the interaction of agents within biological systems and within the environment. For example, factors which impair oxygen transport function may include oxides of nitrogen, carbon monoxide, and altitude. The presence or absence of respiratory infection can affect the oxygen transport function. The interaction of cigarette smoking with exposure to radiation in uranium miners, produces a massive potentiation as judged by the joint association of these exposure with lung cancer.

Factors which interact in the environment include the complex, and not entirely understood photochemical reactions, the association of atmospheric radioisotopes to whatever the particulate matter may be in the atmosphere, and the importance of particulate matter in carrying sulfur oxides into the lungs in relationship to the long term respiratory disease reactions from sulfur oxide and particulate pollution. However, there are many other interactions in the environment, and they include actions which either remove pollutants, neutralize pollutants, or produce more harmful effects of pollutants.

There are a number of biologic processes which affect the strategy of environmental epidemiology. They include, for example, the processes of adaptation. With the first exposure of mice to a given level of ozone there may be very substantial toxicity, but if, at a lower level, there is pre-exposure then the toxicity to the given subsequent level is reduced by some mechanism as yet unknown. Whether this occurs in humans is not known [33]. The adaptation to heat by physiologic mechanisms is well known and well studied. Alterations in respiratory patterns as a result of inhaled irritants are well known. Changes in the rate and nature of red cell production as a result of exposure to altitude, to carbon monoxide, or to hemorrhage are also well known. Unfamiliar or sudden noises can produce vascular reactions but familiar or continuous noises do not. Many such reactions occur and there is always the question whether a reaction which is wholly adaptive is deleterious in some way. In general, experienced workers in this field assume that there is always a health cost to be paid for adaptation, but this may require years of follow up to demonstrate. Sensitization also can occur, for example, to inhaled allergens or to plants such as poison ivy or poison oak.

Another phenomenon in environmental epidemiology, is the phenomenon of avoidance. If living in a certain place produces uncomfortable symptoms people may leave; therefore, if one studies that population on a cross sectional basis alone, one may not find the unusually susceptible individual. Nevertheless, it is the unusually susceptible individuals, provided they can be defined in terms of age and medical status, which are of the greatest relevance, for example, to air quality standards. One must therefore take into account the potentiality for avoidance. This is of particular importance in occupational exposure to irritating materials or to materials which will produce allergic reactions (Ferris) [7].

In prospective study of mortality, there are competing risks; persons who are exposed to a dose which may in fact be carcinogenic, may not have the expression of cancer because the duration of time necessary for its manifestation is insufficient, and they die of other causes. For example, studies indicate that the exposure to cigarette smoking for 20 years is necessary to produce lung cancer [30]. Nevertheless, 20 years of observation is not always possible in a realistic population, and therefore one must not presume that the phenomena could not occur if a young population were exposed and studied for a sufficient period or if the rate of mortality were lower due to other causes which are "competing."

4. Prototypical problems

For purposes of brevity, the classes of prototypical problems are presented in schematic form in a series of tables. The tables include the identification of the particular health variable, and the environmental variables. In these entries, stress is placed on what has been published and not necessarily on what is conclusive, or what ought to be investigated. A crucial column deals with other variables, variables which are not thought to be directly relevant to environmental epidemiology but the control of which is essential for any reasonable interpretation of an association. Columns are also provided to reflect a judgment on the availability of data and a code is inserted as a crude estimate. The code ranges from 0 to 4 according to the relative availability of data.

0—Implies that a method needs to be developed and validated for obtaining the data.

1—Implies that data are accessible by available methods but are not compiled, organized, or collected.

2—Implies that data exist but will require considerable screening, review or reorganization before they are applicable to an epidemiologic study.

3—Implies that the data are in proper form for many purposes but need to be selected and extracted.

4—Implies that the data are well organized and are regularly published and widely available.

The references are not exhaustive and represent only a superficial selection. The question marks indicate a possible or doubtful relation.

Special comment is necessary for several problems. First, hospital morbidity, as well as requests for medical attention, represent a type of morbidity data which can be strongly biased by medical care practices which in turn may reflect economic factors which in turn may be interrelated with environmental factors.

Of importance is the fact that for study of some of the prototypical problems an extraordinary large volume of data is necessary. This is particularly the case for the analysis of mortality by day for which a population base of several millions is desirable and for the longitudinal study through survey methods of chronic respiratory disease morbidity, which may require a base population as great as a hundred thousand.

The references for each table appear at the end of the paper.

TABLE I

EFFECTS ON MORTALITY

Effects	Health data	Environmental data	Other variables	Availability of data H E		References
A. On mortality by type and area	deaths per year for cancer, infant mortality, heart disease, respiratory disease, leukemia— various locations	air pollutants, dust fall, SO_2, ozone per station —average per month radiation, mr per year (in food, water, soil, and air)	meteorological conditions demographic conditions diagnostic facilities socioeconomic factors cigarette smoking occupation	4	1	Winkelstein Sternglass ABCC (Bizzozero) Graham Reid Stewart McMahon Court-Brown and Doll Lillienfeld Setlser and Sartwell
B. On mortality by type, by trend	deaths per year for cancer, infant mortality, leukemia, chronic respiratory disease, for a series of years	same as A.	same as A.	4	1	Sternglass Deane Gilliam Hesse
C. On mortality by day	deaths per day in large metropolitan centers with age break or cause break	air pollutants, SO_2, particulate matter, NO_x, CO weather variables	"heat and cold spells" natural and technological disasters time of year	2	2	Bradley Greenburg Watanabe Hexter and Goldsmith Hechter and Goldsmith Cassell
D. On Case-fatality rate	deaths as proportion of ill persons (coronary disease)	pollutants (CO) indoors and outdoors weather smoking of cigarettes	day of week facilities and staff	1	2	Cohen, Deane and Goldsmith

TABLE II

MORBIDITY

Effects	Health data	Environmental data	Other variables	Availability of data H	E	References
A. Hospital morbidity	hospitalized illnesses per time period by broad cause	air pollution—SO₂, particulate, CO, oxidant	time of year temperature availability of beds, insurance survival of acute episode	3	2	Breslow State of California Sterling
B. Sickness absence	sickness absence rate per period per school or plant	suspended particulates sulfur oxides and black suspended matter	fog weather holidays socioeconomic factors time of year	3	3	Dohan Reid Ipsen et al. Anderson Paccagnella Gocke
C. Demand for medical care	request for medical service per day per type of chief complaint (asthma clinic visits)	oxidant, CO, SO₂, suspended particulates temperature	day of week time of year availability of alternative care socioeconomic factors	3	3	State of California Greenburg Weill et al.
D. Respiratory disease morbidity (cross sectional)	survey data on respiratory disease prevalence	black suspended matter, SO₂, oxidant	cigarette smoking socioeconomic factors occupation ?housing	1	3	Boudik et al. Wahdan Reid Holland Deane et al. Petrilli et al. Schoettlin Ferris Anderson Goldsmith McKerrow
E. Respiratory disease morbidity (prospective)	prospective incidence of respiratory disease in children or adults	air pollution indices by location (fuel consumption, and so forth)	socioeconomic factors occupation smoking of cigarettes crowding nutrition	1	1	Douglas and Waller Buell, Dunn and Breslow Lunn McCarroll

TABLE II, (continued)

MORBIDITY

Effects	Health data	Environmental data	Other variables	Availability of data H E	References
F. Disease aggravation	Aggravation of preexisting illness (asthma, bronchitis) by symptom by pulmonary function	oxidant, black suspended matter, SO₂, Nitrogen Oxides, CO	weather pollens time of year cigarette smoking	2 2	Lawther Schoettlin and Landau Schoettlin Remmers and Balchum Ury *et al.* Rokaw and Massey Shephard and Fair Carnow

TABLE III

IMPAIRMENT OF PERFORMANCE

Effects	Health data	Environmental data	Other variables	Availability of data H E	References
A. Diminution of pulmonary function in presumably normal subjects	$FEV_{1.0}$ airway resistance maximal expiratory flow diffusion capacity	suspended particulates SO₂ (Sulfation) oxidant	temperature and humidity time of day ?respiratory infection	1 3	Wright *et al.* Discher Toyama Lawther and Waller
B. Occurrence of motor vehicle accidents per hour per day of week	number of accidents	oxidant carbon monoxide nitrogen oxides	traffic density weather conditions of road and auto ?cigarette smoking ?drugs holidays alcohol	3 2	Ury Goldsmith, Perkins and Ury Clayton
C. Impairment of athletic performance by day	proportion of secondary school athletes whose performance improves	oxidant carbon monoxide	weather ?cigarette smoking ?motivation	2 2	Wayne and Wehrle McMillan

TABLE IV

PHYSIOLOGICAL AND BIOCHEMICAL EFFECTS INCLUDING STORAGE
(See also Table III,A, Diminution of Pulmonary Function)

Effects	Health data	Environmental data	Other variables	Availability of data H E		References
A. Growth and development (height, weight, skeletal age)	height by age, or increase in height per year by sex, by parents' height	sulfur oxide and suspended particulate ?sunshine ?oxidant ?CO ?water constituents	nutrition socioeconomic factors heredity	3	2	Kapalin and Novotna
B. Red cell indices	hemoglobin, hematocrit, RBC, WBC with differential methemoglobin (infants or ? adults)	suspended particulate SO₂ radiation nitrate in water or NO₂ in air	nutrition pregnancy parasitism ?hereditary hemoglobinopathy age of infant and formulae	2	2	Kapalin Petr and Schmidt Goldsmith, Shearer and Kearns
C. Biochemical indices	cholinesterase activity, ALA and heme metabolism indices	pesticide (Parathion, etc.) lead levels in paint, food, air, water	age nutrition hereditary factors	2	1	Selander and Cramer Nikkanen *et al.* California State Department of Public Health
D. Storage of potentially harmful pollutants	blood lead Carboxyhemoglobin Radio-isotope burden	atmospheric lead levels atmospheric CO levels radio-isotope levels, air, water, food	lead in food and water cigarette smoking occupational exposures	1	2	Goldsmith and Landau Goldsmith and Hexter Tipton and Schroeder Goldsmith, Ury and Perkins Morgan Curphey *et al.* Butt *et al.*

TABLE V

TERATOGENIC, MUTAGENIC, AND AGING EFFECTS

Effects	Health data	Environmental data	Other variables	Availability of data H	E	References
A. Fertility and sex ratio of newborn	relative number of successful pregnancies by sex	radiation exposures to gonads of parent or during gestation	age of parents nutrition ?occupation smoking complications of pregnancy	3	1	Jablon and Kato Meyer and Diamond
B. Birth defects physical and biochemical	birth defects	drugs (thalidomide) viruses (Rubella) ?parental radiation	time of year ?smoking	0–1	1	Macht and Lawrence
C. Leukemia (see also Table I,A)	leukemia onset by age, by cell type	radiation exposure (diagnostic, therapeutic, occupational) infectious agents	socioeconomic status other diseases competing risks	3	1	Graham et al. Stewart et al. McMahon Court-Brown and Doll
D. Spontaneous abortion	proportion of conceptions resulting in spontaneous abortion proportion of genetic defects in abortuses types of chromosome aberrations	drug use in parents radiation exposure heavy metal exposure infectious agents	nutrition age of parents other health factors ?socioeconomic factors	0	0–1	Bateman
E. Aging	age specific mortality	pollutants, CO, SO_2 and black suspended matter ozone and oxidants radiation	medical care nutrition infectious disease exercise cigarette smoking	4	0–1	Strehler and Mildvan Jones

TABLE VI

ANNOYANCE AND NONSPECIFIC HEALTH IMPAIRMENT

Effects	Health data	Environmental data	Other variables	Availability of data H E	References
A. Odor or noise annoyance	annoyance and response awareness of odor ?somatic reactions disturbed rest or enjoyment	olfactometry measured odorants or noise level	age, sex socioeconomic factors attitude or relationship to source air conditioning health status acoustic privacy	Odor 2 0–1 Noise 1 1	Jonsson Deane Goldsmith Lindvall
B. Residential instability	history of change of residence community dissatisfaction	oxidant odor visibility impairment noise level	socioeconomic status recency and basis for moving to area affected employment opportunities recreational opportunities	0–1 3	Hausknecht

5. Statistical strategies

Many of these prototypical problems have been tackled by a variety of strategies. This section outlines some of these strategies. In each case an example is cited and, if the results have been published the reference is given.

5.1. *Two-community strategy for maximizing one out of a set of environmental variables. The problem.* In Los Angeles there is an elevation of the photochemical oxidant level that is particularly marked during unusually hot weather. Thus in looking for the possible effects of oxidant on mortality the interaction of high temperatures is inevitable. In an attempt to maximize the contribution of oxidant while controlling for or stabilizing the effect of temperature we have developed the following plan.

Available data. The data available are for deaths by day by census tracts for Los Angeles County along with measured values of maximal temperature and measured values of photochemical oxidant and carbon monoxide. The number of monitoring stations is much smaller than the number of census tracts.

Strategic plan. The plan was to select artificial communities which had a common pattern of maximum temperatures for the month of September, but which had contrasting levels of photochemical smog and to carry out a difference analysis in order to maximize the difference between oxidant as an independent variable which might explain the difference in mortality for these two artificial communities. The work has been reported by Massey, Landau and Deane but has not been published.

Analytical process and results. The first problem in analysis was to allocate census tracts to monitoring stations. This was done by nearness, assuming the absence of physiographic barriers. The second problem was to obtain some idea of the homogeneity of the two communities which were being compared, that is whether they had similar demographic characteristics, for example; unfortunately they did not. Nevertheless, the analysis, when carried out, failed to show that the difference in oxidant was associated with a difference in mortality by day.

5.2. *Poisson strategy for testing for nonhomogeneity of the distributions of dependent variables with extreme value distribution of independent ones. The problem.* The problem dealt with whether or not extreme values of temperature and pollutants might contribute, and if so how much to the mortality of persons in nursing homes in Los Angeles County. For many years data was collected from nursing homes as a monitoring technique.

Available data. Data available were deaths per day from all nursing homes in Los Angeles County with 25 or more beds, maximum oxidant values, maximum daily temperature and values for other pollutants as needed from the available monitoring stations.

The strategic plan. This consisted of selecting days which had extreme conditions; for example, the days which had both temperature values in the upper 10th percentile and the oxidant values in the upper 10th percentile and to them adding the population of days which had extremely low values for both temperature and oxidant. The plan was then to test whether the daily mortality from the combined distribution for the extreme positions of environmental factors could be or could not be described by a simple Poisson distribution. If it is assumed that the number of deaths of people in nursing homes per day is a Poisson variable, the work of Tucker [29] made it possible to test whether for a given population of days the assumptions of Poissonnous were valid for a simple Poisson or whether there was more then one λ value. If this were so it would indicate a mixture of such distributions.

Analytic process and results. The analysis carried out failed to detect any effect of seasonality on nursing home mortality, an effect which could have been observed in general daily mortality as the work of Hechter and Goldsmith [13] has shown. It also failed to show any effect of extreme conditions on the distribution of daily mortality. Possibly this is due to the precarious state of health of these elderly people and the very high variance in daily mortality which reflects this.

5.3. *Fourier strategy for minimizing time of year effects in multiple regression analysis of mortality. The problem.* Mortality by day is a statistic which varies seasonally. Also environmental temperatures and pollutant values vary seasonally. We face a dilemma trying to see whether there is an association between pollutant values and daily mortality in that if the period of the approximately sinusoidal fluctuations is similar there will be a significant association using a sufficient set of data, regardless of whether there is any dependence of mortality on pollutants. For example Hechter and Goldsmith have shown that carbon monoxide levels are unusually high in the winter and mortality is unusually high in the winter. Both these are out of phase with maximal temperatures and maximal oxidant values (see Figure 1, [14]). So regardless of any biologic or physical processes, these data would produce a positive association between mortality and carbon monoxide and a negative association between mortality and oxidants.

Strategic plan. The plan was to add a sufficient number of Fourier terms according to the scheme of Bliss [2] to account for the seasonal variability of the underlying associations and then after adjusting for long term trend, test to see whether there were significant associations between pollutants and mortality among the residuals. The initial application [13] used simple cross correlation analysis with lagging and a weak association was observed, the statistical significance of which was not formally tested.

Analytic process and results. Hexter and Goldsmith [14] have applied a more complete Fourier strategy to a more recent set of daily mortality data for Los Angeles and have shown that there is a small contribution of carbon monoxide variation to the variation in total daily mortality, when trend, seasonal fluctuations and the effect of temperature are controlled for.

5.4. *Death clearance strategy for longitudinal study of cohorts. The problem.* The relationship of pollution to chronic disease problems has been a prominent political and public health problem but experience indicates that a longitudinal study is often necessary for a valid assessment of the effect of environmental factors. It was desired to study a cohort over a period of time but funds were not available for the repeated examination of a sufficiently large sample. Accordingly Buell, Dunn and Breslow [3] have developed the following approach.

Available data. Smoking, occupation, and residence history from a population of military veterans in California and from their spouses, a total of nearly two hundred thousand people, was obtained in 1957 and information is subsequently obtained as to when and from what cause they die, based on registration of deaths.

Strategic plan. To collect data on smoking, occupation and residence of a group of people in 1957 and then by checking the rosters of such individuals with those persons who die from lung cancer and from chronic pulmonary disease to find out whether these exposure variables are associated with the occurrence of these conditions.

Analytic process and results. These data have been analyzed in several reports and results indicate that there is no evidence of increased incidence of lung cancer occurring among persons in Los Angeles County, when compared to residents in other parts of California.

5.5. *Temporo-spatial strategy for reducing effects of location and time when environmental factors vary over both. The problem.* A large number of the studies referred to in Section 4 on Prototypical Problems are comparisons of morbidity or mortality between different locations with different environmental conditions or between different periods of time *in the same location,* when environmental conditions change. The weakness of spatial comparisons are that populations are often self-selected for living in a certain location and therefore two different locations almost invariably, have populations which differ in attributes which usually are important to the health reaction. The temporal comparison strategy has the weakness that a large number of factors may change over time including meteorological, cultural, and other environmental variables, and there may be changes in structure of the population as well so that these comparisons may be inappropriate because of inadequate control of these other variables.

The combined testing of the more complex hypothesis of the health reaction being more frequent or stronger *in the relatively more polluted place at the relatively more polluted time* seems intuitively to be a great deal more powerful. Therefore, following the example given by Toyama [28], and the two community strategy by implication, we prefer to cast each problem that is at all suitable into a combined temporo-spatial hypothesis testing framework rather than into a simple spatial or temporal hypothesis testing framework.

Data available. The data available for the example which we will cite were data on the myocardial infarction case fatality rate at hospitals in two parts of Los Angeles County, one of which was thought to have more pollution from carbon monoxide than the other on the basis of aerometric and monitoring data [6].

Strategic plan. The correlation of carbon monoxide values with case fatality rate within a given day of week for the total area was undertaken but has been treated as of lesser importance than the application of the temporo-spatial strategy. In this strategy, the question is asked whether the portion of the basin which has higher carbon monoxide compared to the other portion has a higher case fatality rate, during weeks of the year when carbon monoxide levels are high.

Analytical process and results. With the advice and suggestions of Hans Ury, the data were compared by a sign test in which four quartiles of weeks of the year were determined on the basis of a ranking of the average carbon monoxide value. We then scored a plus for each week when the higher area had a higher case fatality rate and a minus when the higher polluted area had a lower case fatality rate.

The sign tests showed that in the high quartile of the year there were twelve out of thirteen weeks in which the case fatality rate was higher in the more polluted area. On the other hand, during the other three quartiles, there were no differences between areas in the case fatality rate. We therefore feel that while a number of the variables were not controlled, for example the relatively polluted area is in the interior of the basin as opposed to the lesser polluted coastal area, and therefore the population may differ between these two areas, it seems unlikely that population variables would have their greatest effect during periods of high air pollution. However, these and other variables, for example cigarette smoking, should certainly be studied. We have drawn no causal conclusion from this study.

5.6. *Concordance sign strategy with classification for singling out contribution of dependent variables in a complex time series. The problem.* We wish to study whether or not there was an association between motor vehicular pollution and motor vehicle accidents. It is obvious that such pollution concentrations in Los Angeles are, in any event, highly dependent on motor vehicle traffic intensity in that the more the traffic the greater the emissions of primary pollutants. Further, the more the traffic the greater the population at risk of motor vehicle accidents. Thus an association of motor vehicle accident frequencies with pollution is greatly affected by the "hidden variable" of motor vehicle traffic. The fluctuation in traffic density is complex and difficult to describe by a simple mathematical relationship. We therefore look for a strategy for determining whether there was an excess of motor vehicle accidents during periods when the pollution concentration was higher.

Available data. Data were available for pollution measurements at the monitoring stations in Los Angeles and for the frequency of occurrence of motor vehicle accidents from the Los Angeles City Police Department. Other data such as whether visibility was normal, whether there was drug involvement or whether there was spillage of material on the roadway were also recorded.

Strategic plan. The plan was to compare adjacent weeks by hour of the day within day of week, with the assumption that for the same day-hour combination, traffic density at intervals of one week, would show only random differences. This plan would yield a large number of pairs of data sets for which both pollution concentrations could be defined and the number of accidents defined. We will test whether the differences between weeks go in the same direction (are concordant), tie, or go in opposite directions (are discordant). In the case of the pollutant carbon monoxide, the useful data extend over nearly all of the 24 hours. In the case of the pollutant oxidant, only the hours of daylight are used since the oxidant concentration drops to a very low value at night. Corrections are necessary for daylight saving time and a decision is needed as to whether the maximum value observed in the basin or the basin average value for a pollutant ought to be used. Ideally, one should use the pollution concentration nearest the place of occurrence of the accidents but the scarcity of the data and the small number of the monitoring stations makes such an analysis unattractive.

Analytic process and results. Ury [34], and Ury, Goldsmith and Perkins [35] have reported that, summing the sign test applied to the concordance or discordance of variation between pollutant and accidents in sets of pairs, there is a statistically significant association of photochemical oxidant and accidents. So far the results do not provide evidence of an association with carbon monoxide.

5.7. *Nonparametric correlation and trend strategy in the dual matrix trace metal problem. The problem.* The problems associated with the body burden of trace metals is one which presents some unusual problems we have worked with but on which we have not yet published. The analysis is based on a set of data made available through the courtesy of Dr. Edward Butt, formerly Director of Laboratories of the Los Angeles County hospital, on the levels of 14 metals in nine organs of 154 persons who had died at the hospital. The determination was made by emission spectroscopy. The analysis of the biologically important trace metals is usually carried out by dividing these metals into those which are called "essential" and those which are called "nonessential" and such an approach has been used, for example by Tipton and Schroeder [22], [27].

The distribution of any given metal concentration in any given organ is, in general, but not invariably, skewed. A log transformation would have been one possibility for analyzing the organ-metal combinations using normal distributional tests for associations. However we preferred a nonparametric approach through which we obtained medians and 95 per cent confidence intervals followed by a rank correlation analysis in which the concentration of metals in each organ was ranked for the 154 subjects.

Available data.   The data consist of, except for the occasional missing item, the concentration of 14 elements in each of nine organs for 154 persons. The data are represented in units of milligrams per 100 grams of dry tissue for soft tissue organs and micrograms per gram of ash for rib and calvarium.

Strategic plan.   The plan was to see how many inter organ correlations by metal could be explained by a small number of clusters. Rank correlation analysis yielded correlation coefficients for each pair of metals within an organ and for each pair of organs with respect to the same element. This therefore gave us two sets of correlation matrices, the interaction of the metal ranks within each of the organs and the interaction of organ ranking for each metal. From the analysis we selected those correlations which were statistically significantly different from zero by conventional yardsticks and then have described the clusters of correlations which occur both in the metal by metal matrix for organs and the organ by organ matrix for metals. The nonparametric trend test is then applied to see whether cancer patients or cigarette smokers have an unusual set of patterns.

Analytic process and results.   Work has not been completed on this problem. It is premature to give results. We were also able to carry out a nonparametric trend test for deaths due to cancer, deaths due to heart disease and to study the association of smoking with metal-organ distributions. There are, only four correlation clusters for organs and similarly there are apparently three for metals. We are able to define at least four general mechanisms by which the correlations might occur: (1) those due to measurement methods; (2) those due to common population attributes; (3) the differences in exposure of subgroups of the population to several elements (which can be studied by trend tests); and (4) common biological mechanisms of distribution of an element between organs.

5.8. *Day of week replication for reducing autocorrelation. The problem.*   In the study of daily mortality, one must start with the assumption that successive environmental measurements or successive mortality frequencies for adjacent days are not independent estimates of a random variate but are dependent to the extent that autocorrelation exists. Hechter and Goldsmith [13] have shown that there is significant autocorrelation and so the problem becomes one of treating this in carrying out a systematic analysis.

Available data. The data available are pollution measurements and mortality by age and cause, by each day in a period of over a year.

Strategic plan. The plan was to carry out calculations of correlations of pollutant measurements, including lags, trend, and Fourier terms with the daily mortality separately for the phenomena attributable to each day of week. This would include lagged pollution and weather measurements for the preceding days as well as for the day of mortality. By carrying out such calculations, one obtains presumably an independent set of seven replicates. However, the independence is also qualified in these replicates of interpenetrating data sets as it was in the original autocorrelation problem but the values for the individual correlations do not reflect significant autocorrelation. This strategic plan depends on the use of some correction for seasonal variation and the Fourier strategy mentioned above was used for that purpose.

Analytical process and results. Hexter and Goldsmith (unpublished results) have carried out the calculations and find consistency between the significance tests for each day in the mean square error explained by regression, and the importance of individual variables. By using replicates on a seven day basis or on a 28 day basis it is possible to obtain presumptively (but not truly) independent estimates of regression coefficients and to apply a t type of test to determine the consistency of these values. This was used in the early phase of the study reported by Hexter and Goldsmith [14].

5.9. *Combined variable strategy for relating monitoring data to possible morbidity and mortality. The problem.* A measuring system for the intensity of pollution which is dependent on a single compound or element or on the results of a single instrument may not reflect the burden of a pollution exposure which occurs. The same phenomena would exist in the case of a radiation estimate which did not include both beta and gamma radiation from a nuclear source. One therefore wishes to combine the impact of several pollutants in some way or other by weighting them with respect to their possible contribution to morbidity.

Available data. The data are, in general, restricted to the measurements made of pollutant concentration and include, of course, the meteorologic variables as well as pollutant data.

Strategic plan. The strategic plan outlined by Ipsen [16] consists of first removing trend and seasonal effects and then summing the levels of smoke shade, sulfate and particulate matter. Then they use the sum of these variables to study the association with respiratory infection prevalence. The sum is significantly positively correlated, while no single pollutant estimator is. A similar approach was suggested by the results of Spicer [23] who studied the effects of total variability in weather and pollution on lung function among a population in Baltimore.

Analytic process and result.   The analysis by Ipsen and Spicer does show that the total contribution of pollutants and meteorological factors is significantly associated with lung function and with prevalence of respiratory infection. Individual pollutants do not show such a consistent result.

5.10. *Convergence of epidemiological and toxicological data and its relevance to policy decisions. The problem.*   Epidemiological studies, even well planned ones, often suggest but do not prove causation. Because of the long lead time necessary for determining such effects as genetic and chronic disease associations, studies can hardly be justified if they lead to delaying the adoption of policies to minimize the risk. Yet if it happens that experimental and epidemiological data converge on even an approximate dose response relationship, policies to avoid a given "response" are scientifically supportable. We choose, as examples, effects of airborne lead on storage of lead in the body and on the enzyme delta-amino levulinic acid dehydratase, and effects of carbon monoxide exposures on coronary heart disease.

Available data.   "Three city" [33] "seven city" [25] and "freeway" [26] lead studies present data on blood lead in relationship to atmospheric exposure. Kehoe [18] has presented experiments with four subjects exposed an increasing number of hours per week to defined concentrations of lead in air. With increased blood lead biochemical effects on heme metabolism occur.

We have presented data on case fatality rate in myocardial infarction and on daily mortality in association with carbon monoxide exposures [6]. Ayres [1] has reported experimental and theoretical work on the impairment of oxygen extraction by the myocardium following human exposures to carbon monoxide. This is especially marked for persons with coronary artery disease. Forster [8], Coburn [5], and Permutt [21] have confirmed and expanded on the relatively greater effect of CO on organs and tissues with a high oxygen extraction ratio.

Strategic plan.   To compare dose response relationships between experimental and epidemiological studies, to see if the two sets of data appear to converge on a similar relationship. If so, the relationship can be depended upon as one basis for policy decisions.

Analytic process and results.   Goldsmith and Hexter [10] have presented the logarithmic regression for blood lead on estimated atmospheric exposure for a number of community studies and have plotted the experimental data of Kehoe on the same graph. A good agreement appears to be present. This relationship has been used by the California State Department of Health and the Air Resources Board of California as one basis for air quality standards.

The National Academy of Science-National Research Council Committee on Effects of Atmospheric Contaminants on Human Health and Welfare includes as one of eight tentative conclusions "a possible effect of increased ambient levels of carbon monoxide in coronary vascular disease" [20]. The Federal Air Quality Criteria for Carbon Monoxide [31] treat both epidemiological and toxicological results separately. However these citations [6], [1] are among the few cited in the Conclusions and Resume section of the report. We may thus conclude that the convergence was influential in the Federal policy decisions on air quality standards for carbon monoxide. It is doubtful if either toxicological or epidemiological results alone would have been of comparable importance for these policy considerations. By contrast, Federal Air Quality Criteria for oxides of Nitrogen [32] are based primarily on epidemiological studies. They are currently being contested by representatives of the motor vehicle industry.

$$\Diamond \quad \Diamond \quad \Diamond \quad \Diamond \quad \Diamond$$

The author wishes to thank the following statistical colleagues for their contributions to his understanding of this problem: Frank Massey, Carl Hopkins, Howard Tucker, Johannes Ipsen, H. H. Hechter, M. Deane, A. Hexter, N. Perkins and Dr. Hans Ury. Monroe Garnett and Louis Esclovon provided invaluable editorial assistance.

REFERENCES

[1] S. M. AYRES, S. GIANELLI, JR., and H. MUELLER, "Myocardial and systemic responses to carboxyhemoglobin," *Biological Effects of Carbon Monoxide, Ann. N.Y. Acad. Sci.*, Vol. 174 (1970), pp. 268–293.

[2] H. BLISS, "Periodic regression in biology and medicine," *Bull. Conn. Agric. Exper. Stat.*, Vol. 6151 (1958).

[3] P. BUELL, J. E. DUNN, JR., and L. BRESLOW, "Cancer of the lung and Los Angeles type air pollution," *Cancer*, Vol. 20 (1967), p. 2139.

[4] California State Department of Public Health Report to U.S. Public Health Service on Contract SAPH #73596, 1961 (contributions by F. Massey, C. Hopkins, and J. R. Goldsmith——not separately identified).

[5] R. F. COBURN, "The carbon monoxide body stores," *Op. cit.* [1], pp. 11–22.

[6] S. I. COHEN, M. DEANE, and J. R. GOLDSMITH, "Carbon monoxide and survival from myocardial infarction," *Arch. Environ. Health*, Vol. 19 (1969), pp. 510–517.

[7] B. J. FERRIS, W. A. BURGESS, and J. WORCESTER, "Prevalence of chronic respiratory disease in a pulp mill and a paper mill in the United States," *Brit. J. Industr. Med.*, Vol. 24 (1967), pp. 26–37.

[8] R. E. FORSTER, "Carbon monoxide and the partial pressure of oxygen in the tissues," *Op. cit.* [1], pp. 223–241.

[9] J. R. GOLDSMITH, "Environmental epidemiology and the metamorphasis of the human habitat," *Amer. J. Public Health*, Vol. 57 (1967), pp. 1532–1549.

[10] J. R. GOLDSMITH and A. C. HEXTER, "Respiratory exposure to lead: epidemiological and experimental dose-response relationships," *Science*, Vol. 159 (1968), p. 1000.

[11] J. R. GOLDSMITH, "Non-disease effects of air pollution," *Environ. Res.*, Vol. 2 (1969), pp. 93–101.

[12] ———, "Epidemiological bases for possible air quality criteria for lead," *J. Air Pollut. Contr. Assoc.*, Vol. 19 (1969), pp. 714–719.

[13] H. H. HECHTER and J. R. GOLDSMITH, "Air pollution and daily mortality," *Amer. J. Med. Sci.*, Vol. 241 (1961), pp. 581–588.

[14] A. C. HEXTER and J. R. GOLDSMITH, "Carbon monoxide: association of community air pollution to mortality," *Science*, Vol. 172 (1971), pp. 265–267.

[15] SIR A. B. HILL, "The environment and disease: association or causation," *Proc. Roy. Soc. Med.*, Vol. 58 (1965), pp. 295–300.

[16] J. IPSEN, M. DEANE, and F. E. INGENITO, "Relationships of acute respiratory disease to atmospheric pollution and meteorological conditions," *Arch. Environ. Health*, Vol. 18 (1969), pp. 462–472.

[17] E. JONSSON, "Annoyance reactions to external environmental factors in different sociological groups," *Acta Sociologica*, Vol. 7 (1964), pp. 229–263.

[18] R. A. KEHOE, "Criteria for human safety from contamination of ambient atmosphere with lead," *Proc. XV Congr. Int. Med. Travail*, Vienna, Vol. 3 (1966), p. 83.

[19] ———, "Toxicological appraisal of lead in relation to the tolerable concentration in the ambient air," *J. Air Pollut. Contr. Assoc.*, Vol. 19 (1969), pp. 690–700.

[20] National Academy of Sciences-National Academy Engineering, *Effects of Chronic Exposure to Low Levels of Carbon Monoxide on Human Health, Behavior, and Performance*, Washington, D.C., 1969, p. 66.

[21] S. PERMUTT and L. FAHRI, "Tissue hypoxia and carbon monoxide in effects of chronic exposure to low levels of carbon monoxide on human health, behavior, and performance," *Op. cit.* [20], pp. 18–24.

[22] H. A. SCHROEDER and I. H. TIPTON, "The human body burden of lead," *Arch. Environ. Health*, Vol. 17 (1968), pp. 965–997.

[23] W. S. SPICER, JR., P. B. STOREY, W. K. C. MORGAN, et al., "Variations in respiratory function in selected patients and its relation to air pollution," *Amer. Rev. Resp. Dis.*, Vol. 86 (1962), pp. 705–712.

[24] G. J. STOPPS, Discussion (of Dr. Goldsmith's paper), *Op. cit.* [12], pp. 719–721.

[25] L. TEPPER, "Seven city study of air and population lead levels: An Interim Report," supported by Contract PH 22-68-28 with the Air Pollution Control Office and by the American Petroleum Institute and the International Lead Zinc Research Organization, Inc., multilith.

[26] H. B. THOMAS, B. K. MILMORE, and G. A. HEIDBREDER, "Blood lead of persons living near freeways," *Arch. Environ. Health*, Vol. 15 (1967), p. 695.

[27] I. H. TIPTON and J. J. SHAFER, "Statistical analysis of lung trace element levels," *Arch. Environ. Health*, Vol. 8 (1964), pp. 58–67.

[28] T. TOYAMA, "Air pollution and its health effects in Japan," *Arch. Environ. Health*, Vol. 8 (1964), pp. 153–173.

[29] H. G. TUCKER, "Effects of air pollution and temperature on residents of nursing homes in the Los Angeles area," unpublished (available from Dr. Goldsmith or Dr. Tucker at University of California, Irvine).

[30] U.S. Dept. of HEW, "Smoking and health. Report on the advisory committee to the Surgeon-General of the Public Health Service," Publication 1103, U.S. Government Printing Office, 1964.

[31] U.S. Dept. of HEW, "Air quality criteria for carbon monoxide," NAPCA Publication #AP-62, 1970.

[32] U.S. Dept. of HEW, "Air quality criteria for nitrogen oxides," NAPCA Publication AP-84, 1971.

[33] U.S. Public Health Service, "Survey of lead in the atmosphere of three urban communities," USPHS Publication #999-AP-12, 1965.

[34] H. K. URY, "Photochemical air pollution and automobile accidents," *Arch. Environ. Health*, Vol. 17 (1968), pp. 334–342.

[35] H. K. URY, J. R. GOLDSMITH, and N. M. PERKINS, "Possible association of motor vehicle accidents with pollutant levels in Los Angeles," a report to U. C. project *Clean Air*, 1970 (available from Dr. Goldsmith).

Table I: Mortality

A. *On mortality by type and area*

[36] W. WINKELSTEIN, JR., S. KANTOR, F. W. DANS, et al., "The relationship of air pollution and economic status to total mortality and selected respiratory system mortality in men, I, suspended particulates," *Arch. Environ. Health*, Vol. 14 (1967), pp. 162–171.

[37] W. WINKELSTEIN, JR., and S. KANTOR, "Stomach cancer: Positive association with suspended particulate air pollution," *Arch. Environ. Health*, Vol. 18 (1969), pp. 544–547.

[38] E. J. STERNGLASS, *Bull. Atomic Sci. XXV*, Vol. 4 (1969), p. 18.

[39] E. J. STERNGLASS, *New Sci.*, Vol. 43 (1969), p. 178.

[40] O. J. BIZZOZERO, JR., K. D. JOHNSON, and A. COICCO, "Radiation-related leukemia in Hiroshima and Nagasaki," *New Eng. J. Med.*, Vol. 274 (1966), pp. 1095–1101.

[41] S. GRAHAM, M. L. LEVIN, A. M. LILLIENFELD, *et al.*, "Preconception, intrauterine, and post natal irradiation as related to leukemia," *Nat. Cancer Inst. Mon.*, Vol. 19 (1966), pp. 347–371.

[42] D. D. REID, "Air pollution as a cause of chronic bronchitis," *Proc. Roy. Soc. Med.*, Vol. 57 (1964), pp. 965–968.

[43] A. STEWART and R. BARBER, "The epidemiological importance of childhood cancers," *Brit. Med. Bull.*, Vol. 27 (1971), pp. 64–70.

[44] B. MACMAHON and M. A. LEVY, "Prenatal origin of childhood leukemia," *New Eng. J. Med.*, Vol. 270 (1964), pp. 1082–2085.

[45] W. M. COURT-BROWN and R. DOLL, "Mortality from cancer and other causes after radiotherapy for ankylosing spondylitis," *Brit. Med. J.*, Vol. 2 (1965), pp. 1327–1332.

[46] A. M. LILLIENFELD, "Epidemiological studies of the leukemogenic effects of radiation," *Yale J. Biol. Med.*, Vol. 39 (1966), pp. 143–164.

[47] R. SELTSER and P. E. SARTWELL, "The influence of occupational exposure to radiation on mortality of American radiologist and other medical specialists," *Amer. J. Epidem.*, Vol. 81 (1965), pp. 2–22.

B. *On mortality by type, by trend*

[48] E. J. STERNGLASS, *Op. cit.* [38], [39].

[49] M. DEANE, "Epidemiology of chronic bronchitis and emphysema in the U.S.: The interpretation of mortality data," (7th Conf. Res. in Emphysema, Aspen), *Med. Thorac.*, Vol. 22 (1965), pp. 24–37.

[50] A. G. GILLIAM, B. K. MILMORE, and J. W. LLOYD, "Trends of mortality attributed to carcinoma of the lung (The declining rate of increase)," *Cancer*, Vol. 14 (1961), p. 3.

[51] L. E. HESSE, "Mathematical models for predicting the trends of chronic diseases," presented at the Annual Meeting of the American Academy of Tuberculosis Physicians 7, 1964 (unpublished).

C. *On mortality by day*

[52] W. H. BRADLEY and A. E. MARTIN, "Mortality, fog and atmospheric pollution," *Monthly Bull. Minist. Health*, Vol. 19 (1960), pp. 56–72.

[53] L. GREENBURG, C. ERHARDT, F. FIELD, *et al.*, "Intermittent air pollution episode in New York City, 1962," *Publ. Health Rep.*, Vol. 78 (1963), pp. 1061–1064.

[54] H. WATANABE, "Air pollution and its health effects in Osaka," 58th Annual Meeting of Air Pollution Control Association, 1965, Toronto.

[55] A. HEXTER and J. R. GOLDSMITH, "Carbon monoxide: Association of community air pollution with mortality," *Science*, Vol. 172 (1971), pp. 265–267.

[56] H. H. HECHTER and J. R. GOLDSMITH, "Air pollution daily mortality," *Amer. Med. Sci.*, Vol. 241 (1961), pp. 581–588.

[57] E. S. CASSELL, D. W. WOLTERS, J. D. MOUNTAIN, *et al.*, "Reconsideration of mortality as a useful index of the relationship of environmental factors to health," *Amer. J. Publ. Health*, Vol. 58 (1957), pp. 1653–1657.

D. *On case-fatality rate*

[58] S. E. COHEN, M. DEANE, and J. R. GOLDSMITH, "Carbon monoxide and survival from myocardial infarction," *Arch. Environ. Health*, Vol. 19 (1969), pp. 510–517.

Table II: Morbidity

A. *Hospital morbidity*

[59] L. BRESLOW and J. R. GOLDSMITH, "Health effects of air pollution," *Amer. J. Public Health*, Vol. 48 (1958).

[60] STATE OF CALIFORNIA, "Clean air for California," report of the Air Pollution Study Project, Berkeley, 1955.

[61] T. D. STERLING, J. J. PHAIR, S. V. POLLACK, *et al.*, "Urban morbidity and air pollution," *Arch. Environ. Health*, Vol. 13 (1966), p. 158.

B. *Sickness absence*

[62] F. C. DOHAN and F. W. TAYLOR, "Air pollution and respiratory disease: A preliminary report," *Amer. J. Med. Sci.*, Vol. 240 (1960), pp. 337–339.

[63] F. C. DOHAN, "Air pollutants and incidence of respiratory disease," *Arch. Environ. Health*, Vol. 3 (1961), pp. 387–395.

[64] D. D. REID and P. H. LAMBERT, "Smoking, air pollution, and bronchitis in Britain," *Lancet*, Vol. 1 (1970), pp. 853–857.

[65] J. IPSEN, F. E. INGENITO, and M. DEANE, "Episodic, morbidity and mortality in relation to air pollution," *Arch. Environ. Health*, Vol. 18 (1969), pp. 458–461.

[66] D. O. ANDERSON and A. A. LARSEN, "The incidence of illness among young children in two communities of different air quality," *Canad. Med. Assoc. J.*, Vol. 95 (1966), pp. 893–904.

[67] B. PACCAGNELLA, R. PAVANELLO, F. PESARIN, *et al.*, "Immediate effects of air pollution on health of schoolchildren in some districts of Ferrara," *Arch. Environ. Health*, Vol. 18 (1969), pp. 495–502.

[68] T. M. GOCKE, P. McPHERSON, and N. C. WEBB, JR., "Predicting respiratory absenteeism," *Arch. Environ. Health*, Vol. 10 (1965), pp. 332–337.

C. *Demand for medical care*

[69] California State Department of Public Health, Report of Research Contract SAPH #73596 with the Department of Health, Education, and Welfare: "To investigate the applicability of suitable statistical methodology and to develop the accompanying data reduction program from examining the relationship between atmospheric pollution, meteorological conditions and mortality," 1961.

[70] L. GREENBURG, C. ERHARDT, F. FIELD, *et al.*, "Intermittent air pollution episode in New York, 1962," *Pub. Health Rep.*, Vol. 78 (1963), pp. 1061–1064.

[71] H. WEIL, M. M. ZISKIND, *et al.*, "Recent development in New Orleans asthma," *Arch. Environ. Health*, Vol. 10 (1965), pp. 148–151.

D. *Respiratory disease morbidity* (cross sectional)

[72] F. BOUDIK, J. R. GOLDSMITH, Y. TEICHMAN, *et al.*, "Epidemiology of chronic bronchitis in Prague," *Bull. WHO*, Vol. 42 (1970), pp. 711–722.

[73] M. B. WAHDAN, "Atmospheric pollution and other environmental factors in respiratory disease of children," Thesis, University of London, 1964.

[74] W. W. HOLLAND, D. D. REID, R. SELTSER, *et al.*, "Respiratory disease in England and the United States," *Arch. Environ. Health*, Vol. 10 (1965), pp. 338–343.

[75] M. DEANE, J. R. GOLDSMITH, and D. TUMA, "Respiratory conditions in outside workers," *Arch. Environ. Health*, Vol. 10 (1965), pp. 323–331.

[76] R. L. PETRILLI, G. AGNESE, and S. KANITZ, "Epidemiologic studies of air pollution effects in Genoa, Italy," *Arch. Environ. Health*, Vol. 12 (1966), pp. 733–740.

[77] C. E. SCHOETTLIN, "The health effect of air pollution on elderly males," *Amer. Rev. Resp. Dis.*, Vol. 86 (1962), pp. 878–911.

[78] B. G. FERRIS, "Effects of air pollution on school absences, and difference in lung function in first and second graders in Berlin, New Hampshire, Jan. 1966 to June 1967," *Amer. Rev. Resp. Dis.*, Vol. 102 (1970), pp. 591–606.

[79] B. G. FERRIS, I. T. T. HIGGINS, M. W. HIGGINS, et al., "Chronic nonspecific respiratory disease, Berlin, New Hampshire, 1961–67: a cross-sectional study," *Amer. Rev. Resp. Dis.*, Vol. 104 (1971), pp. 232–244.

[80] D. O. ANDERSON and A. A. LARSEN, "The incidence of illness among young children in two communities of different air quality," *Canad. Med. Assoc. J.*, Vol. 95 (1966), pp. 893–904.

[81] J. R. GOLDSMITH, "Epidemiology of bronchitis and emphysema (factors influencing prevalence and a criterion for testing their interaction)," *Med. Thorac.*, Vol. 22 (1965), pp. 1–23.

[82] C. B. McKERROW, "Chronic respiratory disease in Great Britain," *Arch. Environ. Health*, Vol. 8 (1964), pp. 174–179.

E. *Respiratory disease morbidity (prospective)*

[83] J. W. B. DOUGLAS and R. E. WALLER, "Air pollution and respiratory infection in children," *Brit. J. Soc. Med.*, Vol. 20 (1966), pp. 1–8.

[84] P. BUELL, J. E. DUNN, and L. BRESLOW, "Cancer of the lung and Los Angeles-type air pollution," *Cancer*, Vol. 20 (1967), pp. 2139–2147.

[85] J. E. LUNN, J. KNOWELDEN, and A. J. HANDYSIDE, "Patterns of respiratory illness in Sheffield infants and schoolchildren," *Brit. J. Soc. Med.*, Vol. 21 (1967), pp. 7–16.

[86] J. McCARROLL, "Measurements of morbidity and mortality related to air pollution," *J. Air Pollut. Contr. Assoc.*, Vol. 17 (1967), pp. 203–209.

F. *Disease aggravation*

[87] P. J. LAWTHER, "Climate, air pollution and chronic bronchitis," *Proc. Roy. Soc. Med.*, Vol. 51 (1958), pp. 262–264.

[88] C. E. SCHOETTLIN and E. LANDAU, "Air pollution and asthmatic attacks in Los Angeles area," *Pub. Health Rep.*, Vol. 76 (1961), pp. 545–548.

[89] C. E. SCHOETTLIN, *Op. cit.* [77].

[90] J. E. REMMERS and O. J. BALCHUM, "Effects of Los Angeles urban air pollution upon respiratory function of emphysematous patients (the effects of microenvironment on patients with chronic respiratory diseases)," presented at the meeting of the Air Pollution Control Association, 1965.

[91] H. K. URY and A. C. HEXTER, "Relating photochemical pollution to human physiological reactions under controlled conditions," *Arch. Environ. Health*, Vol. 18 (1969), pp. 473–480.

[92] S. N. ROKAW and F. MASSEY, "Air pollution and chronic respiratory diseases," *Amer. Rev. Resp. Dis.*, Vol. 86 (1962), pp. 703–704.

[93] R. J. SHEPHARD, J. J. PHAIR, and G. C. R. CAREY, "Effects of air pollution on human health," *Amer. Industr. Hyg. Assoc.*, Vol. 19 (1958), pp. 363–370.

[94] B. W. CARNOW, "Relationship of SO_2 levels to morbidity and mortality in "high risk" population," presented at the AMA's Air Pollution Medical Research Conference, 1970.

Table III: Impairment of performance

A. *Diminution of pulmonary function*

[95] G. WRIGHT, T. C. LLOYD, P. HAMILL, et al., "Epidemiological study of obstructive pulmonary disease in two small towns (A consideration of pulmonary function measurements)," *Amer. Rev. Resp. Dis.*, Vol. 86 (1962), pp. 713–715.

[96] D. DISCHER, cited by S. Rokaw, unpublished material, testimony for hearings about air

pollution submitted to Los Angeles County Environmental Quality Control Committee, June 28, 1971.

[97] T. TOYAMA, "Air pollution and its health effects in Japan," *Arch. Environ. Health*, Vol. 8 (1964), pp. 153–172.

[98] P. J. LAWTHER and R. E. WALLER, "Clean air and health in London," *Proc. Nat. Conf. Clean Air* (Great Britain) (1969), pp. 71–79.

B. *Occurrence of motor vehicle accidents*

[99] H. URY, "Photochemical air pollution and automobile accidents in Los Angeles (An investigation of oxidant and accidents, 1963 and 1965)," *Arch. Environ. Health*, Vol. 17 (1968), pp. 334–342.

[100] H. URY, J. R. GOLDSMITH, and N. M. PERKINS, "Possible association of motor vehicle accidents with pollutant levels in Los Angeles," Project Clean Air Research Report to the Regents of the University of California, Vol. 2 (1970), p. 1032.

[101] G. D. CLAYTON, "A study of the relationship of street level carbon monoxide concentrations to traffic accidents," *Industr. Hyg. J.*, Vol. 21 (1960), pp. 46–54.

C. *Impairment of athletic performance*

[102] W. S. WAYNE, P. F. WEHRLE, and R. E. CARROLL, "Oxidant air pollution and athletic performance," *J. Amer. Med. Assoc.*, Vol. 99 (1967), pp. 901–904.

[103] R. McMILLAN, unpublished paper presented at 9th Air Pollution Medical Research Conference.

Table IV: Physiological and biochemical effects including storage

A. *Growth and development*

[104] V. KAPALIN and M. NOVOTNA, "The physical development of children living in different social and housing conditions," reprint of a paper submitted to the Conference on the Influence of the Urban and Working Environment on the Health and Behaviour of Modern Man, Charles University, Prague, 1969, multilith.

B. *Red cell indices*

[105] V. L. KAPALIN, "The red blood picture in children from different environments," *Rev. Czech. Med.*, Vol. 9 (1963), pp. 65–81.

[106] B. PETR and P. SCHMIDT, "The influence of an atmosphere contaminated with sulfur dioxide and nitrous gases on the health of children," *Z. Ges. Hyg.*, Vol. 13 (1967), pp. 34–48.

[107] J. R. GOLDSMITH, L. A. SHEARER, and O. KEARNS, "Methemoglobin levels in infants in an area with high nitrate water supply," a paper for the American Public Health Association, Oct. 1971.

C. *Biochemical indices*

[108] J. SELANDER and K. CRAMER, "Interrelationships between lead in blood, lead in urine and ALA in urine during lead work," *Brit. J. Industr. Med.*, Vol. 27 (1970), p. 28.

[109] J. NIKKANEN, S. HERNBERG, G. MELLIN, *et al.*, "Delta-aminolevulinic acid dehydrase as a measure of lead exposure," *Arch. Environ. Health*, Vol. 21 (1970), pp. 140–145.

[110] CALIFORNIA STATE PUBLIC HEALTH SERVICE, "Lead in the environment and its effects on humans," prepared jointly by Bureau of Air Sanitation, Environmental Hazards Evaluation Unit, and Air and Industrial Hygiene Laboratory, Berkeley, 1967.

D. *Storage of potentially harmful pollutants*

[111] J. R. GOLDSMITH and S. A. LANDAU, "Carbon monoxide and human health," *Science*, Vol. 168 (1968), pp. 1352–1359.

[112] J. R. GOLDSMITH and A. C. HEXTER, "Respiratory exposure to lead: epidemiological and experimental dose-response relationships," *Science*, Vol. 158 (1967), pp. 132–134.

[113] I. TIPTON and J. J. SHAFER, "Statistical analysis of lung trace element levels," *Arch. Environ. Health*, Vol. 8 (1964), pp. 58–67.

[114] J. R. GOLDSMITH, H. URY, and N. M. PERKINS, "Statistical analysis of trace metal levels in human organs," unpublished.

[115] K. MORGAN, "The body burden of long-lived isotopes," *Arch. Environ. Health*, Vol. 8 (1964), pp. 86–99.

[116] T. J. CURPHEY, L. P. L. HOOD, and N. M. PERKINS, "Carboxyhemaglobin in relation to air pollution and smoking," *Arch. Environ. Health*, Vol. 10 (1965), pp. 179–185.

[117] E. M. BUTT, R. E. NUSBAUM, and T. L. GILMOUR, "Trace metal levels in human serum and blood," *Arch. Environ. Health*, Vol. 8 (1964), pp. 52–57.

Table V: Teratogenic, mutagenic aging effects

A. *Fertility and sex ratio of newborn*

[118] S. JABLON and H. KATO, "Sex ratio in offsprings of survivors exposed prenatally to the atomic bombs in Hiroshima and Nagasaki," *Amer. J. Epidem.*, Vol. 93 (1971), pp. 253–258.

[119] M. MEYER and E. DIAMOND, "Investigation of the effect of prenatal X-ray exposure of human oogonia and ooctyes as measured by later reproductive performance," *Amer. J Epidem.*, Vol. 89 (1969), pp. 619–635.

B. *Birth defects; physical and biochemical*

[120] S. H. MACHT and P. S. LAWRENCE, "National survey of congenital malformations resulting from exposure to roentgen radiation," *Amer. J. Roentgen*, Vol. 73 (1955), p. 442.

C. *Leukemia*

[121] S. GRAHAM, M. L. LEVIN, A. M. LILLIENFELD, *et al.*, *Op. cit.* [41].

[122] A. STEWART and R. BARBER, *Op. cit.* [43].

[123] B. McMAHON and M. A. LEVY, *Op. cit.* [44].

[124] W. M. COURT-BROWN and R. DOLL, *Op. cit.* [45].

D. *Spontaneous abortion*

[125] A. U. BATEMAN, "The use of spontaneous abortion in man for monitoring the mutation rate," Department of Cell Biology, New York University School of Medicine, multilith.

E. *Aging*

[126] B. L. STREHLER and A. S. MILDVAN, "General theory of mortality and ageing," *Science*, Vol. 132 (1960), pp. 14–21.

[127] H. B. JONES, "Some notes on ageing," *Symposium on Information Theory in Biology*, New York, Pergamon Press, 1958, pp. 341–346.

Table VI: Annoyance and nonspecific health impairment

A. *Odor or noise annoyance*

[128] E. JONSSON, M. DEANE, and G. SANDERS, "Community reactions to odors from pulp mills, a pilot study in Eureka, California," prepared for the Conference on Measurements and Evaluation of Odor in the Community in Relation to Odor Sources, Stockholm, 1970, multilith.

[129] M. DEANE and J. R. GOLDSMITH, "Health effects of pulp mill odor in Anderson, California for presentation at the Conference on the Dose-Response Relationships Affecting Human Reactions to Odorous Compounds, Cambridge, Mass., 1971, multilith.

[130] J. R. GOLDSMITH, "A suggested odor scaling system," prepared for the Conference on the Measurement and Evaluation of Odor in the Community in Relation to Odor Sources, Stockholm, 1970, multilith.

[131] T. LINDVALL, "On sensory evaluation of odorous air pollutant intensities," *Nord. Hyg. T.*, Suppl., Vol. 2 (1970).

B. *Residential instability*

[132] R. HAUSKNECHT, "Air pollution: effect reported by California residents," California State Department of Health, California Health Survey, 1962, multilith.

Discussion

Question: T. Sterling, Department of Applied Mathematics and Computer Science, Washington University

With respect to the statement that sometimes we formulate public policy in order to avoid harm even though the data are poor, let me point out that such a policy entails a serious loss that largely is overlooked. Presuming to know what causes a disease, we stop (or slow down) further research. The classical case is smoking. Much of the acceptance of cigarette smoking as a cause of lung cancer has been based on this type of thinking, namely, "what does it harm if we discourage smoking?" But, as a consequence, the progression of work investigating the effects of particular pollutants, namely, soot, dusts, and so forth, on lung cancer has been seriously hurt.

We have learned that pollutants that occur in trivial quantities in the environment may, nevertheless, be concentrated in the food chain. John, how does this problem fit into your thinking on strategies?

Reply: J. R. Goldsmith

When epidemiological data are used to formulate and support the need to enforce a health policy, there are many considerations other than scientific ones in the strict sense. They include not only the convergence of the data from epidemiological studies and experimental ones, but the 'cost' of the existing pattern which a policy is to modify as well as some evaluation of the health harm which we seek to prevent. To use an example familiar to me, the Lindane vaporizer was widely used on the basis of certification by the Department of Agriculture as to its effectiveness as a household insect control system. With no or trivial effort, our department heard of some eight cases of blood dyscrasia in California associated with the use of the vaporizer in the home. We published them along with our views that since there were other ways to control insects, the certification should be removed since some of the cases were fatal and all were serious. After some delay, they have been taken off the market. On the other hand, it is reasonable to expect that a number of studies controlling many of the relevant (or possibly relevant) variables should be undertaken before we ask to have major eating and smoking habits drastically affected by health policy decisions based on epidemiological research. I disagree with T. Sterling as to the comparative importance of data on particulate air pollutants and on cigarette smoking as regards their association with lung cancer (assuming I understand his question). The data on smoking and lung cancer is sufficient to convince me that there is a causal relationship. The data on particulate air pollution is much less impressive, and largely consists in the unequivocal and consistent demonstration that there is an urban factor in lung cancer, but the evidence that air pollution is that urban factor is not consistent, and my position is that the association of particulate air pollution and lung cancer is not causal because, among other reasons, it is not consistent. For example, there are no such associations in the British study of Buck and Brown (Tobacco Research Institute Report), nor in the study of W. Winkelstein, both of which did have other positive associations of a more convincing nature.

Decisions about health policy do have a massive impact on feasibility of further research, and this too should enter into the decision as to whether a policy is apt.

I have not included data on food chain effects on human health, since except for some radioactive compounds, there is little data of interest. The data on lead which were cited are associated with some food chain problems on which we are now working.

Question: R. J. Hickey, Institute for Environmental Studies, University of Pennsylvania, Philadelphia

(1) If I understood you correctly, regarding your reference to a report of Bradford Hill and the various statistical relationships mentioned (temporal, and so forth), and the association of environmental pollutant characteristics and health variables, you seemed to imply that if a number of related statistical relationships appear consistent, this increases confidence that a causal relationship exists. Was this implied?

(2) Since certain elements of public policy seem to have been determined on the basis of reported statistical associations, is this not an implicit acceptance of conclusion of causality from correlation? Further, is it not a commonplace of statistics that statistics can be used properly to reject a hypothesis, but never to establish that the hypothesis is true?

I find the nonrigorous use of statistics in determination of important matters of public policy quite disturbing. Perhaps in public policy matters it should be pointed out to the public and to legislators that statistics did not reject the hypothesis in question (if that is the case), and that public policy is, therefore, being determined on the basis of a hypothesis and not on the basis of a proven causality relationship.

Reply: J. Goldsmith

(1) Yes, this increases confidence, but it does not lead to a categorical conclusion. If the same relationship is found in different populations, using different

methods, this increases confidence more than a variety of statistical relationships. You should read Hill to obtain a more authoritative answer.

(2) Causality is a logical problem and public policy is a value problem. We often adopt policies in the face of uncertainty. We should participate in and encourage this, even though it precludes at times the opportunity for a definitive causal conclusion. This does *not* imply an acceptance of a causal relation on the basis of correlations. We agree on the use of statistics in hypothesis testing, and that the public and legislators should be informed as to the basis of recommended action.

Question: Alfred C. Hexter, California Department of Public Health

I agree with the previous discussants that these studies may establish association but the results do not show cause and effect.

Nevertheless, in many cases public policy depends upon such data. The health, or even the lives, of many people may depend upon decisions based upon such data. There may be many cases when we must *act* as though cause and effect is proven, because human safety requires that if we err, we err on the *conservative* side.

Reply: J. Goldsmith

I agree.

APPLICATION OF COMPUTER SIMULATION TECHNIQUES TO PROBLEMS IN AIR POLLUTION

JOSEPH V. BEHAR

UNIVERSITY OF CALIFORNIA, RIVERSIDE

The need for a rationale for deciding between alternative air pollution control strategies has been apparent for many years, yet to date, none exists. The utility and power of the techniques of systems analysis and numerical or computer simulation have been known for many years, yet, only recently, have they been applied to problems in air pollution.

In planning an epidemiological study of the effects of various pollutants it is reasonable to consider the potential contribution of simulation models of pollutant distributions. One such model has been utilized in this laboratory to simulate the distribution of total oxidants in the South Coast Air Basin in California. The advantages are clear in that knowledge of the spatial distribution of pollutants allows calculation of dose rates for the entire population within the mapped area. This in turn removes the important constraint, requiring that the sample population be located within a certain radius of an air monitoring station.

The computer program produces maps which graphically illustrate spatially disposed information, both qualitative and quantitative. As such, it is suited to a broad range of applications with only minor alterations. Raw data of any kind, be it physical, social, medical, economic, and so forth, may be related, manipulated, weighted and arrayed in any desired fashion. The concept, general design, and mathematical model for the program were first developed in 1963 by Howard T. Fisher at Northwestern Technological Institute. Since then, many others, too numerous to mention, have generously contributed ideas and improvements to the program.

The notion of mapping is that of graphically depicting spatially disposed quantitative information, in our case. concentration of total oxidants. The data consists of triplets (X, Y, Z) where X and Y are the abscissa and the ordinate data which define a point on the map surface and Z is the concentration of total oxidants at that point. The X, Y data must be continuous. Concentration data is also continuous, that is, is defined for every point X, Y on the map by a continuous mathematical function. Restricting ourselves in this way, the map which results is a contour map.

The program uses a "gravity" interpolation function, where, essentially, the concentration at an unknown point (X, Y) is directly proportional to the

29

weighted sum of all the known concentrations on the map. The weighting factor is the inverse square of the distance between the unknown concentration and the known concentration, hence the name "gravity" model.

In Figure 1 the shaded area depicts the region which was mapped. It is bounded on the south and west by the Pacific Ocean and on the north and east by the 1500 foot elevation contour. The area lies within the South Coast Air Basin, contains twenty air quality monitoring stations (represented by the filled circles), and has a population of about 9.7 million people. Figure 2 illustrates the ten concentration ranges represented by ten shades of gray. Owing to the uncertainty in printing and reproduction, these same ten ranges are also represented by a number between zero and nine. The levels of greatest interest, of course, are level 4 (0.05 to 0.09 ppm) which contains the health related federal ambient air quality standard for total oxidant (0.08 ppm); level 5 (0.09 to 0.15 ppm) which contains the State of California health related ambient air quality standard for total oxidant (0.1 ppm); level 7 (0.25 to 0.35 ppm) which contains the Riverside County school alert level (0.27 ppm)—0.35 ppm for Los Angeles County—at which point school children are brought indoors and restrained from strenuous physical activity; and last, level 9 (0.45 to 0.54 ppm) which contains the Los Angeles County first alert level (0.5 ppm).

The maps, Figures 3–36, are produced, utilizing actual oxidant measurements obtained through the cooperation of the Air Pollution Control Districts of Los Angeles, San Bernardino, Orange, and Riverside Counties. The data are hourly averages, hence are directly relatable to health effects based on dosage. There is one map for each hour of the day for thirty-four continuous hours on August 21, and 22, 1969, starting at 6:00 a.m. on the 21st.

The maps are self explanatory, however, certain features and trends are worth pointing out. At 6:00 a.m. (Figure 3), the basin in general is at essentially background levels of oxidant, except for a small region at the east end, bounded by Riverside, San Bernardino, Corona, Chino, and Pomona, where the total oxidant is slightly elevated. The reason for this elevated oxidant in this region is not completely clear as yet. This condition does recur later on, after the sun has and can be seen every hour between 7:00 p.m. (Figure 16) on the 21st and 6:00 a.m. (Figure 27) on the 22nd.

By 7:00 a.m. (Figure 4) almost right at sun-up, areas in Pasadena and southwestern Riverside are experiencing oxidant levels in excess of health related federal ambient air quality standards. At 8:00 a.m. (Figure 5) more than half of the area studied is engulfed by air which exceeds federal standards. If one assumes a uniform population distribution over the entire region, then by 9:00 a.m. (Figure 6) about 70 per cent of the population, slightly less than seven million people are breathing substandard air.

At 10:00 a.m. (Figure 7) the central portion of the basin as well as areas to the north around Burbank and Pasadena are experiencing levels in excess of the Riverside County school alert level. People in these areas should begin restraining strenuous physical activity. By 11:00 a.m. (Figure 8) the maximum oxidant readings are obtained, and if we again assume a uniform population distribution, everyone in the basin is breathing substandard air, and about three million of these persons should according to advice given by some physicians, curtail any form of vigorous activity. Note that between the hours of 11:00 a.m. to 1:00 p.m. (Figures 8 through 10) the entire air shed equals or exceeds the health related federal ambient air quality standard for total oxidant.

Starting at 1:00 p.m. (Figure 10) and going to 6:00 p.m. (Figure 15) one can observe two phenomena occurring simultaneously; first, a decrease in the rate of build up of oxidant, it apparently peaked between 11:00 a.m. and 12 noon; second, an increase in the effect of wind on the oxidant distribution. It is quite apparent that during the afternoon hours, the wind velocity has increased, and is moving the material from west to east. The sharp decrease in oxidant between 5:00 p.m. (Figure 14) and 7:00 p.m. (Figure 16) is consistent with the fact that the sun has set and the mechanism for oxidant production has been removed. By 7:00 p.m. (Figure 16) the basin in general has returned to near background levels of total oxidant.

Except for the apparent anomaly noted earlier, occurring in the eastern end of the basin, the oxidant levels remain at or near background throughout the night from 8:00 p.m. (Figure 17) to 6:00 a.m. (Figure 27) the following morning. Between 6:00 a.m. (Figure 27) and 7:00 a.m. (Figure 28) the sun rises and oxidant begins rising immediately. The pattern repeats itself, the following day being very much like the preceding one. The series continues to 3:00 p.m. (Figure 36) on August 22, 1969. One last input of information regarding this segment of data: the meteorology observed on August 21 and 22, 1969 was quite typical of the late summer season.

It should be noted that this series of maps, simulating actual concentration distributions of total oxidants, condenses into rapidly digestible form, an enormous amount of information. The implications are clear in terms of a much wider choice of sample populations. The specific method used here only begins to illustrate the enormous power of the more general technique of simulation modelling.

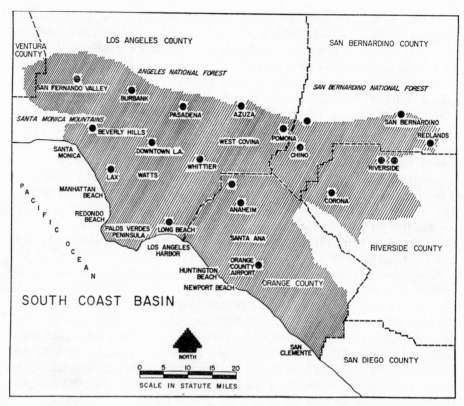

FIGURE 1

South Coast Air Basin study area, southern California.

OXIDANT CONTOURS
IN THE
SOUTH COAST BASIN
COMPUTER GENERATED

Levels		Total Oxidant (ppm)	Levels		Total Oxidant (ppm)
	0	0.00 - 0.01		5	0.09 - 0.15
	1	0.01 - 0.02		6	0.15 - 0.25
	2	0.02 - 0.03		7	0.25 - 0.35
	3	0.03 - 0.05		8	0.35 - 0.45
	4	0.05 - 0.06		9	0.45 - 0.54

FIGURE 2

Explanation of shadings for Figures 3 through 36.

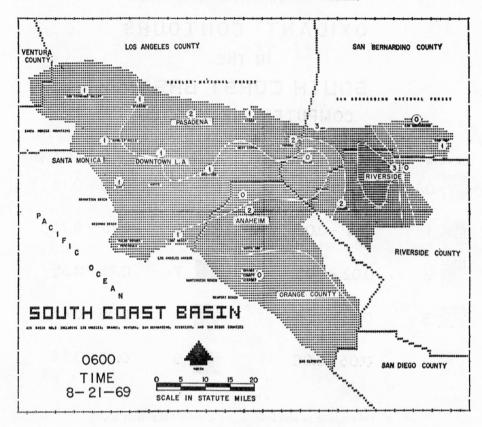

FIGURE 3

Extent of oxidant level at 0600 on August 21, 1969.

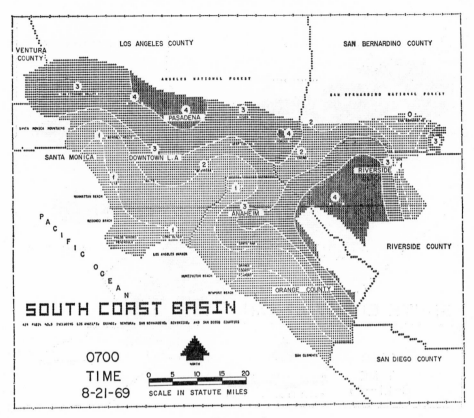

FIGURE 4

Extent of oxidant level at 0700 on August 21, 1969.

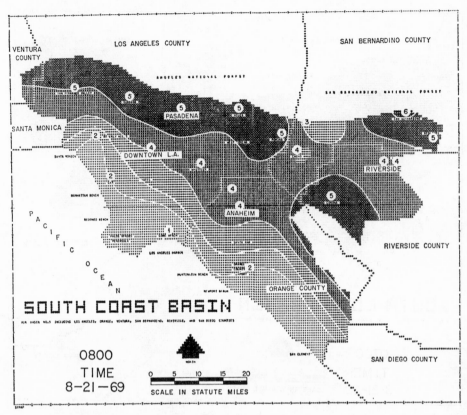

FIGURE 5

Extent of oxidant level at 0800 on August 21, 1969.

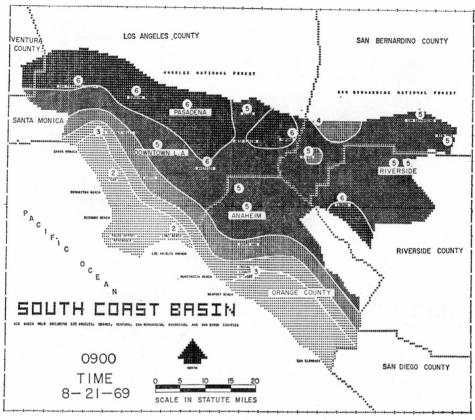

FIGURE 6

Extent of oxidant level at 0900 on August 21, 1969.

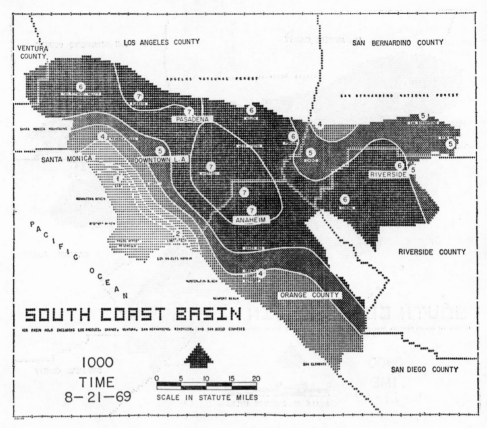

FIGURE 7

Extent of oxidant level at 1000 on August 21, 1969.

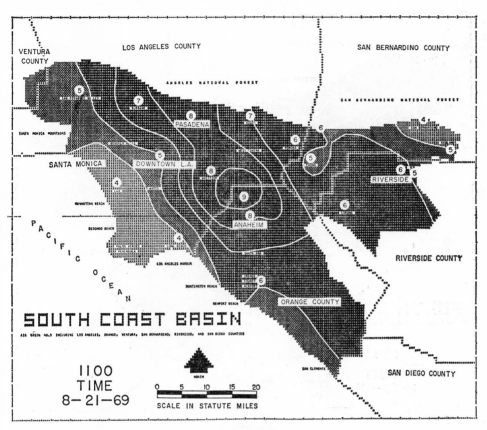

FIGURE 8

Extent of oxidant level at 1100 on August 21, 1969.

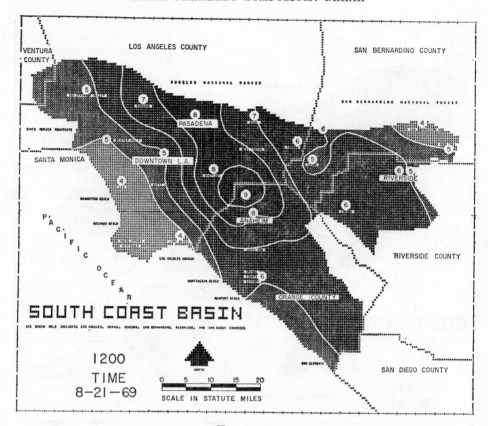

FIGURE 9

Extent of oxidant level at 1200 on August 21, 1969.

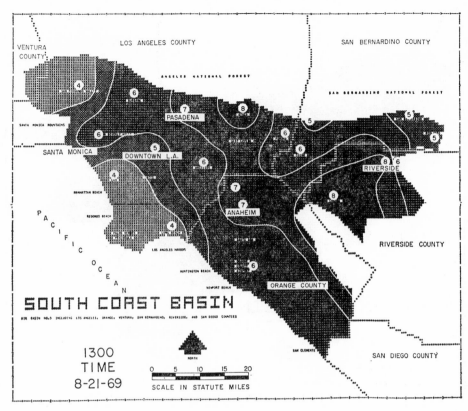

FIGURE 10

Extent of oxidant level at 1300 on August 21, 1969.

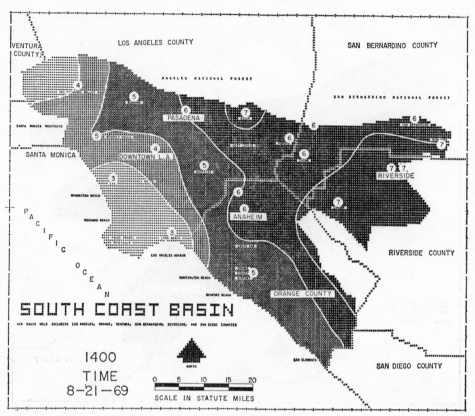

FIGURE 11

Extent of oxidant level at 1400 on August 21, 1969.

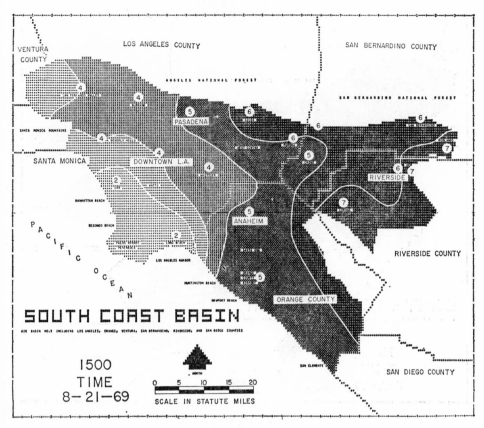

FIGURE 12

Extent of oxidant level at 1500 on August 21, 1969.

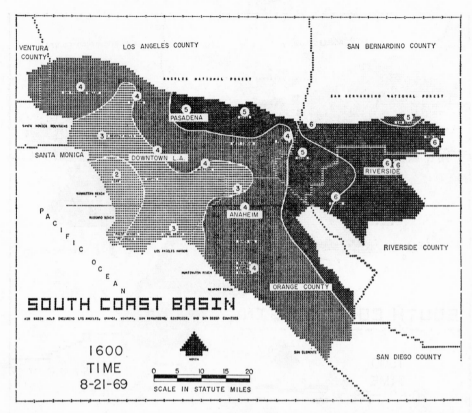

FIGURE 13

Extent of oxidant level at 1600 on August 21, 1969.

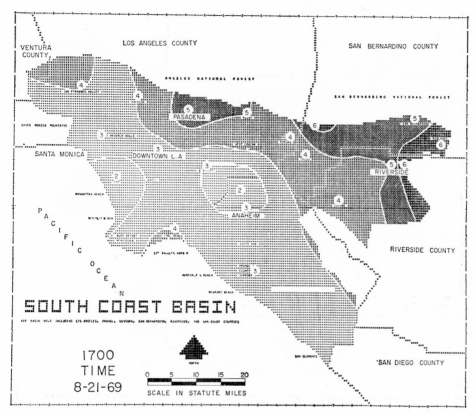

FIGURE 14

Extent of oxidant level at 1700 on August 21, 1969.

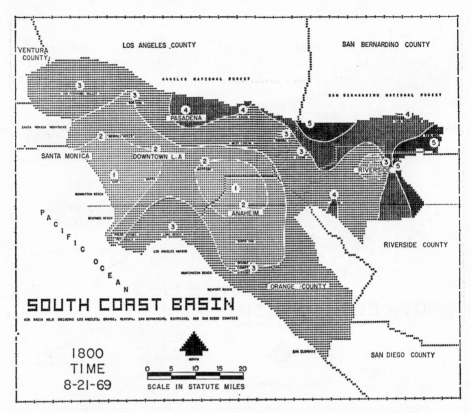

FIGURE 15

Extent of oxidant level at 1800 on August 21, 1969.

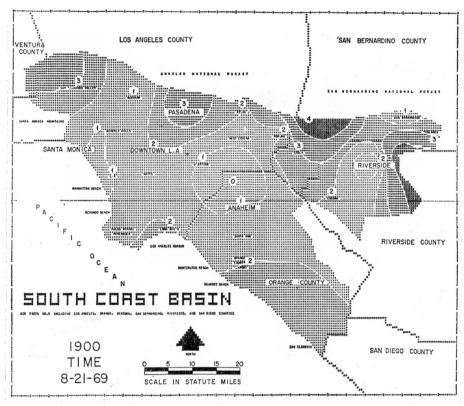

FIGURE 16

Extent of oxidant level at 1900 on August 21, 1969.

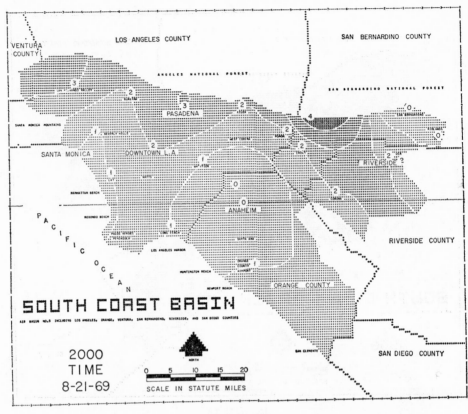

FIGURE 17

Extent of oxidant level at 2000 on August 21, 1969.

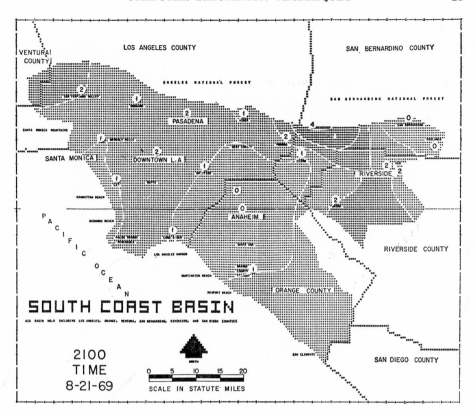

FIGURE 18

Extent of oxidant level at 2100 on August 21, 1969.

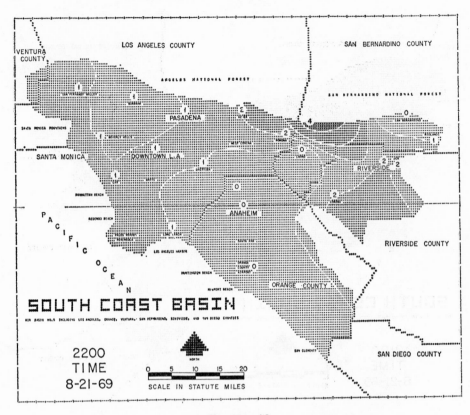

FIGURE 19

Extent of oxidant level at 2200 on August 21, 1969.

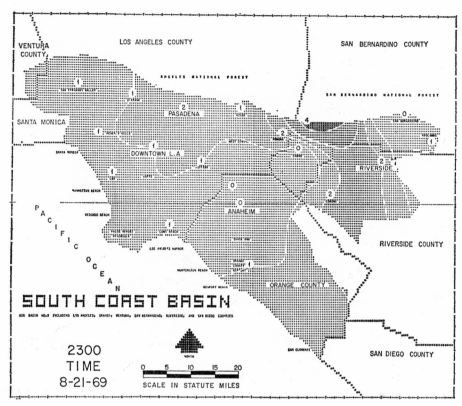

FIGURE 20

Extent of oxidant level at 2300 on August 21, 1969.

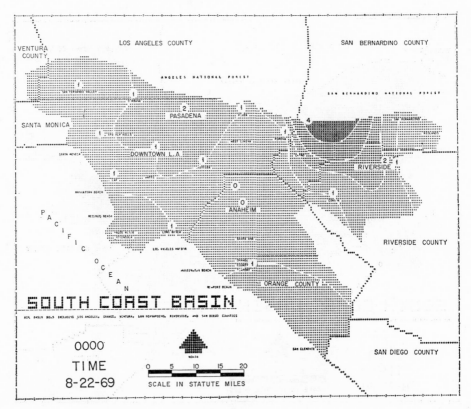

FIGURE 21

Extent of oxidant level at 0000 on August 22, 1969.

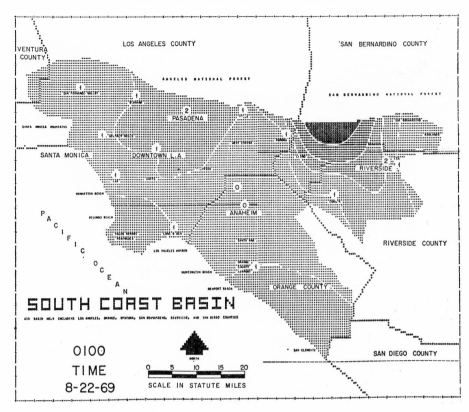

FIGURE 22

Extent of oxidant level at 0100 on August 22, 1969.

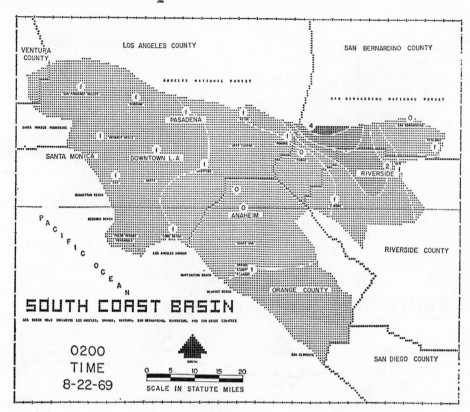

FIGURE 23

Extent of oxidant level at 0200 on August 22, 1969.

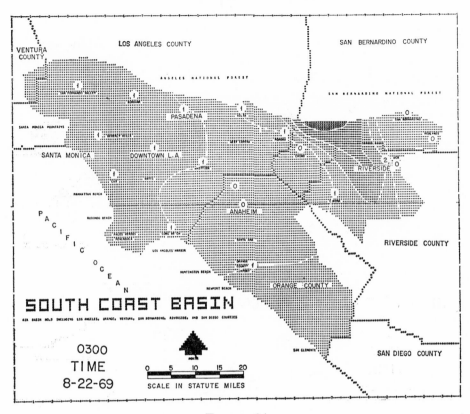

FIGURE 24

Extent of oxidant level at 0300 on August 22, 1969.

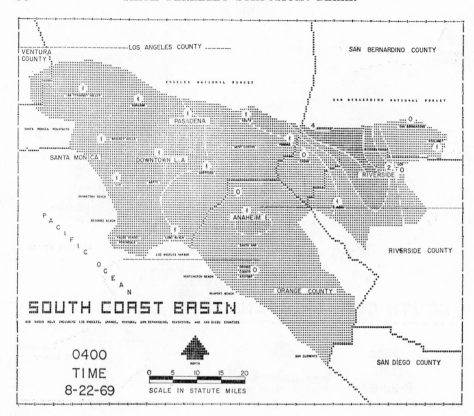

FIGURE 25

Extent of oxidant level at 0400 on August 22, 1969.

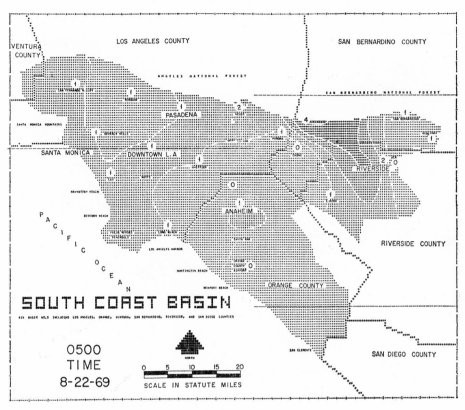

<figure>FIGURE 26</figure>
Extent of oxidant level at 0500 on August 22, 1969.

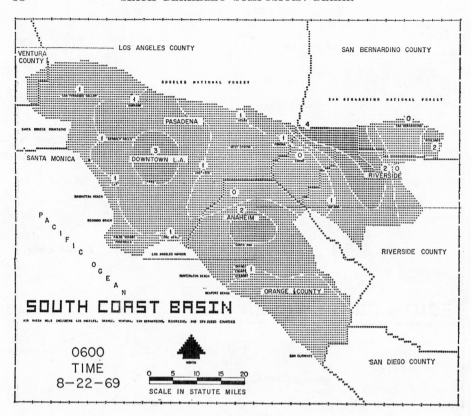

FIGURE 27

Extent of oxidant level at 0600 on August 22, 1969.

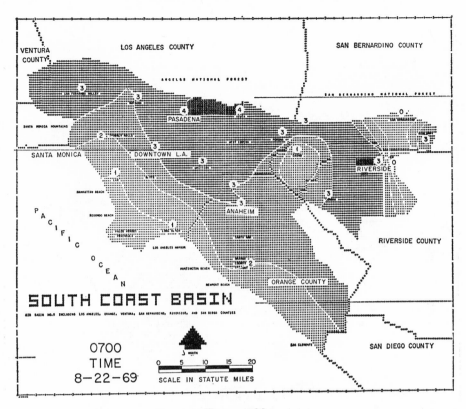

FIGURE 28

Extent of oxidant level at 0700 on August 22, 1969.

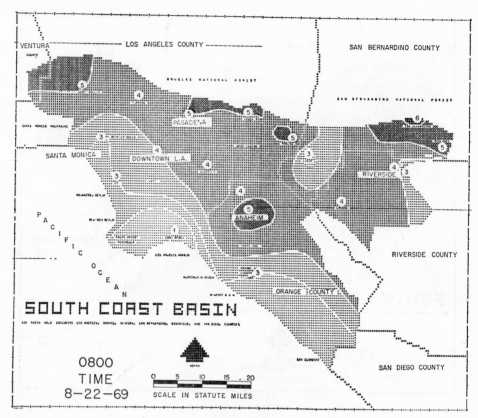

FIGURE 29

Extent of oxidant level at 0800 on August 22, 1969.

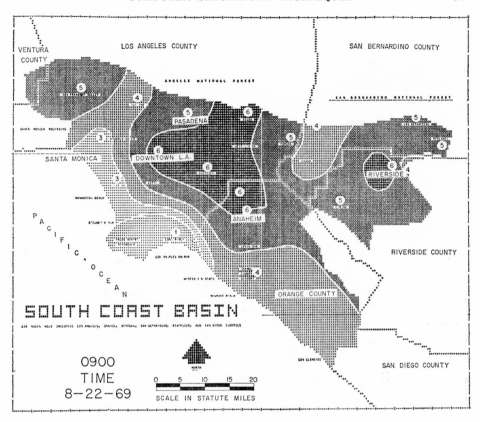

FIGURE 30

Extent of oxidant level at 0900 on August 22, 1969.

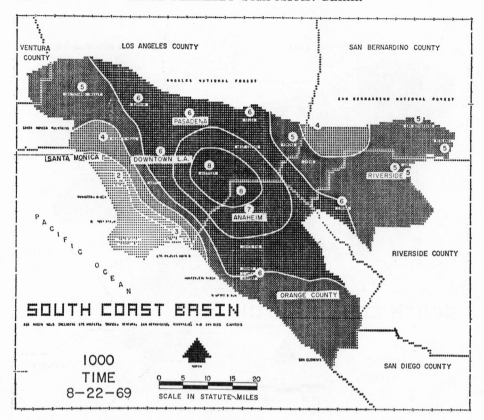

FIGURE 31

Extent of oxidant level at 1000 on August 22, 1969.

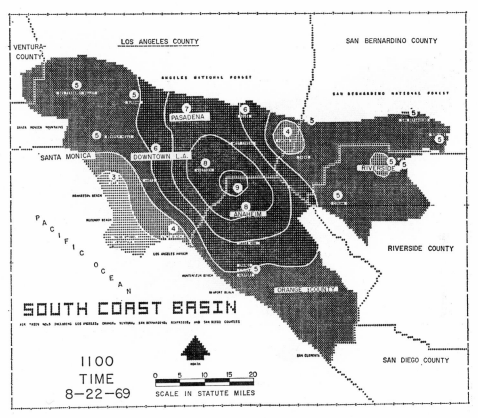

FIGURE 32

Extent of oxidant level at 1100 on August 22, 1969.

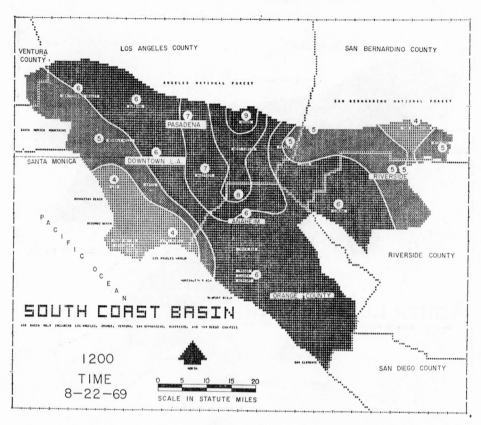

FIGURE 33

Extent of oxidant level at 1200 on August 22, 1969.

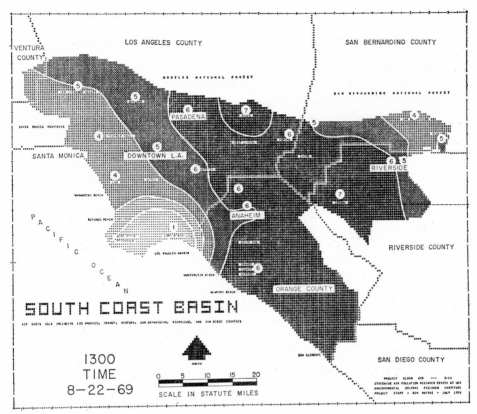

FIGURE 34

Extent of oxidant level at 1300 on August 22, 1969.

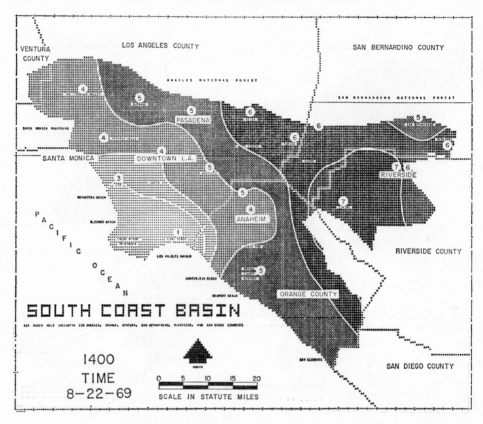

FIGURE 35

Extent of oxidant level at 1400 on August 22, 1969.

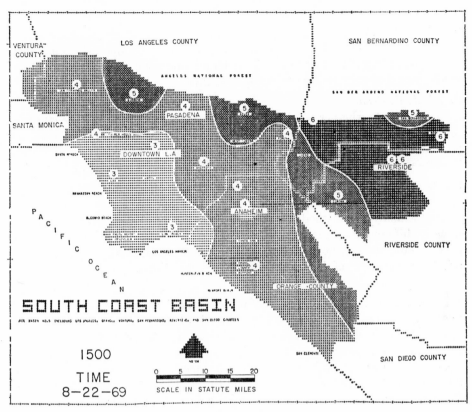

FIGURE 36

Extent of oxidant level at 1500 on August 22, 1969.

Discussion

Question: S. W. Greenhouse, National Institute of Mental Health

Do you have similar charts for a Sunday, particularly for the early hours, say, six a.m to ten a.m.? Presumably, this would provide some base line data of how much of the patterns are due to sunlight or rays.

Reply: J. V. Behar

In response to whether Sunday measurements might reflect only the effect of sunlight, note that the primary emissions of nitric oxide and hydrocarbons, as estimated by carbon monoxide, are only about ten per cent lower on weekends than on weekdays. The distribution by location might be different, however.

Question: R. J. Hickey, Institute for Environmental Studies, University of Pennsylvania, Philadelphia

Are the analytical methods for the determination of oxidant in California the same or different than the method of PHS? Has the method changed over the years? Consistency or nonconsistency of analytical methodology might possibly affect comparability of data in different places and times.

Reply: J. V. Behar

Both. There are several methods for the determination of oxidant. In most most cases, the colorimetric iodide technique is utilized.

Yes, the method has changed over the years. In the old days it used to be measured by rubber cracking. Presently, it is measured by the Mast meter.

Consistency or nonconsistency of analytical methodology will affect comparability of data taken at different places and times. However, for all of our studies, these factors are taken into account, and the data are normalized.

Question: J. F. Finklea, Environmental Protection Agency, Durham, N.C.

Scattered observations from eastern conifer forests indicate that background levels of oxidants may, on occasion, reach up to one half to two thirds of the national ambient air quality standard for photochemical oxidants (0.08 ppm).

Reply: J. V. Behar

I agree.

Question: V. L. Sailor, Brookhaven National Laboratory

Could you please explain briefly what you mean by oxidants?

Reply: J. V. Behar

Oxidants are generally thought to be composed by approximately 80–90% ozone, some NO_2, and some peroxyacetyl nitrate.

Question: B. E. Vaughan, Ecosystems Department, Battelle Memorial Institute, Richland, Wn.

In discussing toxic effects, it is important to realize that there is a question of environmental degradation and overall ecological effect. A datum for toxicity in any one animal or a datum for toxicity in a particular plant may not be an adequate measure of environmental degradation.

Reply: J. V. Behar

I agree. It ought to be pointed out that the technique here involves study of the dose rates and not the toxicities.

Question: H. L. Rosenthal, School of Dentistry, Washington University

Dr. Went has recently shown that the atmosphere in wooded areas contained large amounts of terpenoid and hydrocarbon emissions from plants. Do you have any information concerning the levels of oxidants in primeval areas due to such emissions? What can be considered an adequate background of oxidant acceptable for human life? In view of our evolutionary development, should a level of zero oxidants be advisable? (Dr. Goldsmith commented that zero levels are not really attempted.)

Reply: J. V. Behar

No, we do not have information concerning levels of oxidants in primeval areas. We do have measurements in forested areas located downwind from large urban centers, and measurements indicate that the levels of oxidants from the slopes of the wooded areas may be higher than in the general urban atmosphere. Reference: *Project Clean Air*, Vol. 2, University of California Task Force and Assessments, September 1970. See S-20, in the bibliography.

I do not feel qualified to answer Dr. Rosenthal's second question. As for the third, zero oxidant level is not attainable, in principle, since there is background level independent of human activity; precisely what that level is presumably in the range 0.02–0.05 ppm.

Question: R. W. Gill, Department of Biology, University of California, Riverside

What method was used to fit contour lines to data from only 20 stations?

Reply: J. V. Behar

This was stated in the paper.

RESEARCH PROGRAMS OF THE ATOMIC ENERGY COMMISSION'S DIVISION OF BIOLOGY AND MEDICINE RELEVANT TO PROBLEMS OF HEALTH AND POLLUTION

JOHN R. TOTTER

DIVISION OF BIOLOGY AND MEDICINE, ATOMIC ENERGY COMMISSION

1. Introduction

The program of the Division of Biology and Medicine has the major long term objective to measure and to evaluate the effects of radiations on man. Meeting this obligation is not a simple task. Many of the reasons for this will be obvious to you. Human experimentation is quite properly proscribed; we can make observations on man's responses to radiations only following exposure for reasons other than our need for radiobiological information, reasons such as accident, medical usage of radiations, and so on. We overcome this in part by resorting to the use of human cells in tissue culture. However, the responses of these may not give us a fair picture of what happens in a human being. Hence, we also use experimental animals and attempt to extrapolate the results of those studies to the human situation. Again, a mouse or a dog is not a man, and we need to find ways to improve confidence in our abilities to translate animal data into reliable estimates of hazards to man.

As with any presumably deleterious environmental contaminant, we have the problem of measuring dosages and of relating the magnitudes of observed effects to the dose. This involves not only measurements of external radiation, but the tracing of radionuclides through whatever environmental pathways they may follow in getting into man, determining localizations in human cells and tissues, measuring rates of turnover (biological half-life), and using this information to determine radiation dosages that can be ascribed to such internal emitters.

Except in case of accident, nuclear warfare, and so forth, the radiation doses that we are concerned with are quite small, and this leads to the problem that I do not need to detail to an audience of statisticians: when dosages are small, effects are small, hence difficult to measure because of the sheer magnitude of the observations that must be carried out to obtain samples large enough to permit establishing the statistical significance of differences that may be found. In addition, the effects produced by radiation are indistinguishable from those

71

arising spontaneously; consequently, small effects may be lost in the background noise.

We have a great need to know the effects of low, protracted doses of radiation to man, but we must depend on experiments with animals exposed to radiation doses that are higher by orders of magnitude, supplementing this by salvaging whatever bits of information can be gleaned from the very few human individuals or populations that have been exposed to sufficient radiation to produce observable changes. Thus, for example, the Commission and its contractors have cooperated in the studies of the Atomic Bomb Casualty Commission in Japan, a study of a Brazilian population living in a high radiation background area of monazite sands and studies of radium dial painters.

What kinds of effects are we concerned with? Effects on somatic (body) cells are of concern because these are the basis of damage to the exposed individuals; deleterious effects on germ (reproductive) cells may be transmitted to and result in damage to some subsequent generation. Most of these inherited changes (mutations) are known to be deleterious, and since they may be passed on from generation to generation, perhaps only occasionally resulting in a damaging effect when they occur in a suitable genetic combination, they are quite appropriately regarded as a threat not only to the individual who inherits them, but also to the population as a whole.

2. Genetic effects

With the advent of the nuclear age, it became mandatory that we learn more about the nature and magnitude of the genetic damage that could result from exposure to ionizing radiations. Initially this information came from studies on the spermatozoa (mature male germ cells) of the fruit fly, *Drosophila*. The knowledge of the mutagenicity of X-rays had been obtained by the pioneering work of the late Professor H. J. Muller [1], and the major genetic and cytogenetic effects (chromosomal aberrations, such as deletions, inversions, translocations) were well documented during the ensuing 10 to 12 years. Experiments by W. P. Spencer and C. Stern [2] under the auspices of the Manhattan District, showed that the earlier observations of proportionality of mutation yield to radiation dosage in *Drosophila* sperm that had been reported in the early 1930's held true even for doses as low as 25 Roentgens (R).

It was realized, however, that human germ cells might differ considerably from those of the fruit fly, and that attention should be centered on the immature germ cells, on the cells that are at risk for the major portion of the life cycle. Ideally, this information should come from a mammalian species. Hence, in 1956 an experiment was initiated by Dr. W. L. Russell at the Oak Ridge National Laboratory (ORNL) to measure the mutation rate in spermatogonia (immature germ cells of the male) of the mouse, using seven specific loci. The results of this experiment showed that mouse spermatogonia are some fifteen times more sensitive to radiation than similar cells in the fruit fly [3]. On the basis of these early mouse data, the various standards-setting bodies decided that the then current permissible occupational exposure levels were too high, and the maximum permissible annual dose for occupational exposure was lowered from 50 to 5 roentgens and for the first time a population permissible dose set at one-tenth this lower level. Subsequently, for a suitable sample 170 mR per year was given as a standard. This permissible exposure has not been changed up to the present. It gives a total genetically effective dose of 5 R per 30 year generation from man-made sources, exclusive of the medical usage and natural background radiation for the general population.

However, these results were for large doses administered acutely (that is, at high dose rates), whereas our major concern is for the low level, chronic exposures that might occur from various kinds of environmental radiation contamination. In view of the difficulty in carrying out low dose experiments as mentioned above, the question became what would be the effect of exposing the mice chronically (that is, at low dose rates), where the total radiation dose would be the same? In these subsequent experiments Dr. Russell [4] obtained the answer that at high doses, 300–600 R delivered at dose rates of 0.8 R per minute only one-third to one-fourth as many mutations were produced as compared to that induced by a similar dose administered at a rate of 90 R per minute. Lower dose rates down to 0.001 R per minute (normal background rate is approximately .001 R per 3.65 days) did not produce an additional sparing effect on mutation induction in irradiated spermatogonia.

When similar experiments were carried out in which oocytes (egg precursors) in female mice were irradiated, again a dose rate effect was found [5]. This effect was even more striking than that observed for irradiated spermatogonia. Mutation induction diminished as dose rate decreased from 90 R per minute down to 9×10^{-3}R per minute, such that there was a limiting dose rate below which few or no mutations were detected. Similarly, even with acute treatments, low doses of radiation (for example, 50 R) give fewer mutations than are expected on the basis of simple proportionality to dose ("linearity") [6]. In addition, there is every indication that as time from irradiation to conception increases, the premutational damage that is produced with acute exposures is completely eliminated [7].

Why is this so? We know of the existence of repair systems for the repair of damage to DNA, the genetic material, and, in a variety of organisms, can show the effects of repair of potential mutational damage, leading to a reduction in the yield of mutated genes. It is Dr. Russell's view that the effect of low level radiation exposure as well as time from exposure to conception is to permit much more effective repair of damage than is possible when the total damage is produced in a short period of time.

Again, why not get this information from man? You may be surprised to learn that to date no radiation induced mutations of any kind have been demonstrated to occur in man. You should not take this to mean that none have been produced by radiation, rather that genetic techniques have not yet been developed and used to detect such changes. The only rigorously treated genetic information available to date, obtained from human studies, is that reported on the progeny of the survivors of the atomic bombs at Hiroshima and Nagasaki. These studies indicated that genetic effects of the exposure to acutely delivered, high doses of radiation, which included both gamma rays and biologically more effective fast neutrons, could not be detected in these populations with the genetic endpoints available at that time (sex ratio, morbidity, mortality, presence of congenital abnormalities, and so forth, of children of the survivors) [8]. These negative results indicate that man is, in all probability, not much more sensitive to radiation than is the mouse (and he may be less sensitive).

There are many factors that must be known before we can be completely confident of our extrapolation from experimental animals to man. Recognizing this, the Atomic Energy Commission (AEC) has, for many years, maintained an active program in basic radiation genetics, including studies of somatic cell genetics, repair processes, the structure and function of the genetic material, and variations in response to radiation of different species.

Another important problem of concern is the behavior and expression of mutations in populations. A number of experiments have been carried out with mice, in which the spermatogonia of males and in some experiments both spermatogonia of the males and oocytes of the females have been irradiated with large acute doses of radiation in each generation, with the purpose of measuring accumulated genetic damage, where the damage was to be measured by the reduction in reproductive or Darwinian fitness—that is, effects on morbidity and mortality, fecundity (litter size), and so forth. A striking, common feature of these experiments is their failure to show evidence of significant cumulative damage [9].

2.1. *Early and late somatic effects of external radiation.* In a logical way the program of the Division of Biology and Medicine has evolved to assess three main potential exposures of individuals or large human populations; (1) possible exposure of one or few individuals to high doses as might result from an accident, (2) possible exposure of a small group of individuals to moderate doses in their occupation, and (3) possible exposure of a large population to minute amounts of radioactivity released to the environment by a nuclear industry.

The last possibility is the one of primary concern at present and differs very little from the overall problem of release of small amounts of any potentially toxic material by any industry to the environment.

Very early in the biological research program studies relevant to these considerations were being conducted since only the potential sources of radioactivity have changed from nuclear weaponry to nuclear power for peacetime application.

The three modes of potential exposure require the understanding of several variables, among which are total dose, dose rate, and quality of the radiation, which are potential determinants of both the qualitative and quantitative aspects of possible biological effects.

It was decided quite early in the Division of Biology and Medicine (DBM) program that prospective or retrospective epidemiological studies on accidentally or purposely exposed human populations would not be sufficient for evaluation of human hazards. Moreover, it was concluded that such studies would contribute little to a basic understanding of how the biological effects were produced or how the human body reacts to the damage produced by normal recovery and repair processes.

Experimental animals, chiefly small rodents and beagle dogs have comprised the major organisms used for evaluation of both acute and delayed somatic effects of ionizing radiation. An important aspect of the radiobiology program sponsored by the DBM has been to provide a strong basic research program on cellular and molecular radiobiology to provide fundamental biological principles which strengthen the more applied work and facilitate the often difficult extrapolation of radiation hazards to man. A good example is the case of a collaborative effort between a group at the University of California, Laboratory of Radiobiology, and one at ORNL. Studies at ORNL established that in microorganisms extreme sensitivity to ultraviolet radiation can be attributed to genetic defects which preclude the synthesis of enzymes which are involved in repair of chemical defects produced in the DNA by the radiation. In humans the genetically controlled cancerous disease, known as xeroderma pigmentosum, is characterized by abnormal sensitivity of the skin to sunlight and the irradiated cells become cancerous. A prompt application of the finding on bacteria using similar techniques has led to the conclusion that the cells of the affected humans lack one of the enzymes contained in normal cells that repairs radiation damage to the DNA. While this study does not imply that this is an important aspect of radiation carcinogenesis it does show that normal cells have enzymes for repair of radiation damage.

It will not be possible in this presentation to describe in detail the many important studies which have been carried out over the year as part of the radiobiology program sponsored by the AEC. Results of many individual studies have produced some general principles which will be highlighted, and an effort will be made to describe important new research programs to fill in gaps in our information, especially in the most timely problems of the day which fall mostly in the area of evaluating the potentially hazardous effects of low doses of radiation received at low dose rates. By definition these involve the late somatic effects of radiation which can lead to premature death.

2.2. *Effect of total dose and dose rate.* Although it is tempting to subdivide the important effects of ionizing radiation on animals into immediate and late effects, this subdivision is arbitrary and depends on both the total dose and the dose rate and also on the quality of the radiation.

In model animal systems it has been known for many years that X-rays or gamma rays received at sufficiently high dose rates produce different modes of death as a function of the total dose as shown in Figure 1. Between 300 and 1000 rads, experimental animals die after a latent period of a few days to three weeks predominately from damage to the bone marrow which is reflected in

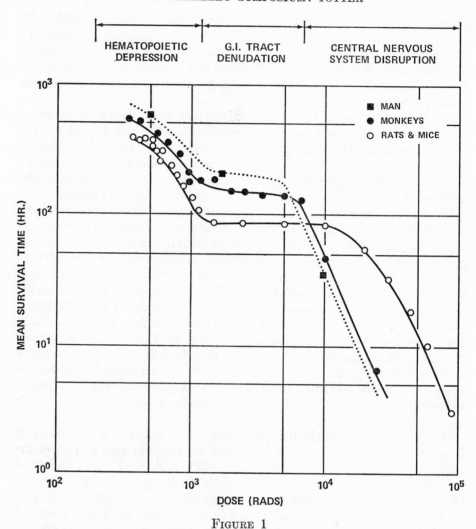

FIGURE 1

Survival time and associated mode of death in relation to dose of acute whole-body irradiation [15].

a marked depression in the production of red and white blood cells. At doses between 1000 and 10,000 rads, animals die after several days due primarily to loss of the epithelium of the gastrointestinal tract. At higher doses of 10,000 to 100,000 animals may die in minutes to a few hours because of damage produced predominately in the central nervous system.

While this figure was not intended to imply that only three organs or tissues are involved in the lethal radiation syndrome, an impressive quantity of data derived from AEC sponsored research does permit some general conclusions to be reached about the early lethal effects of radiation on a number of mammalian systems. Without exception the radiation damage that leads to early death involves systems whose function depends on continuous cellular replacement such as the intestinal epithelium or systems that provide stem cells for a variety of important cell functions such as the bone marrow. The marrow, of course, is the continuous source of cells for such vital functions as respiration, maintenance of continuity of capillary function as well as the blood clotting mechanism, maintenance of body defense against infection by way of removing foreign cells from the circulation or by setting up a permanent immunological defense against infectious organisms.

At doses lower than 100 rads, effects on these same important systems can be detected, but through the process of cell replacement from stem cell populations recovery of the irradiated individual seems complete. The survivors of such populations, however, may express latent damage to a variety of cell systems which can be classified as late somatic effects and will be discussed below.

As a result of a number of studies on mice, some generalizations can be made about changing radiosensitivity with age of the irradiated population. Irradiation during embryonic or fetal development presents at least one unique problem and that is formation of morphological abnormalities in the development of the offspring. Figure 2 summarizes some of the findings of studies on experimental

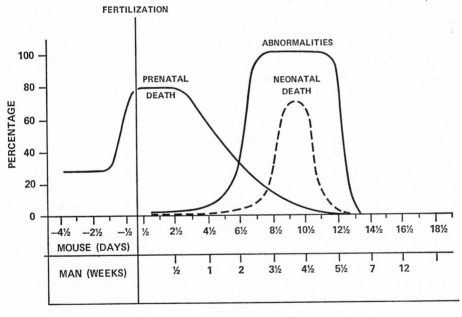

FIGURE 2

Incidence of pre- and neonatal deaths and of abnormal individuals at term after 200 to 400 R X-irradiation of mice at various intervals separated by 24 hours. The corresponding estimated chronology for man is shown [16].

animals and attempts to scale the sensitive periods for man based on the relative time-spans for the two species. It is clear from the data shown that remarkable changes in sensitivity for prenatal and neonatal death occur during the gestation period. Production of a variety of abnormalities also shows a large change per unit dose. As would be expected on the basis of the kinds of abnormalities produced some are found to be lethal after birth. Although these effects on the developing embryo or fetus are reasonably well documented they relate only to total dose.

It is important to assess dose rate dependence for both lethal effects and abnormalities induced in the fetus, and we have initiated such work at ORNL and at the University of Tennessee-AEC Laboratory within the past year. Some effort has been made to establish the minimum dose required for production of specific abnormalities. Some of you, as biometricians, are well aware of the size of the populations required to establish minimum doses for most biological effects. It is of utmost importance to establish the effectiveness of radiation on fetal and juvenile animals for production of late somatic effects. The few data on human *in utero* radiation resulting from retrospective epidemiological studies in several countries are not consistent with each other in concluding an abnormally high radiosensitivity of the fetus for induction of leukemia and other types of cancer. In addition all prospective studies have been negative. We have initiated studies in mice at ORNL and dogs at the University of California at Davis to evaluate experimentally the suggestions derived from the human studies.

2.3. *Quality of radiation and its interrelationship to dose and dose rate.* There are many ways to look at the relative effectiveness of various kinds of radiation. In biophysical terms the most meaningful way is to classify the various ionizing radiations on the basis of the density of ionization or energy loss along the particle tracks as they pass through matter. Often the effectiveness per unit dose is described as a function of the *linear energy transfer* (LET) in tissue. Most of the assessment of relative biological effectiveness (RBE) has involved X-rays or gamma rays as compared with neutrons which produce primarily energetic protons as they are slowed down in tissue-like materials. Most of the data from the fundamental cellular studies as well as whole animal studies indicate that high LET radiations such as fast neutrons are five to ten times more effective per unit dose than low LET gamma and X-radiation. The best explanation to date for this effect is that most cells are incapable of repairing damage produced by the high LET radiations. The best evidence in this regard stems from dose-rate studies which show a dramatic reduction in effectiveness per unit dose with decreasing dose rate for low LET radiations, but little or no dose rate dependence for high LET radiations. In biophysical terms this suggests that several independent physical events are required for production of most somatic effects. This is tantamount to suggesting that a true threshold dose, however small, must exist for low doses of radiation received at very low dose rates. All somatic effects studied to date, whether immediate or delayed,

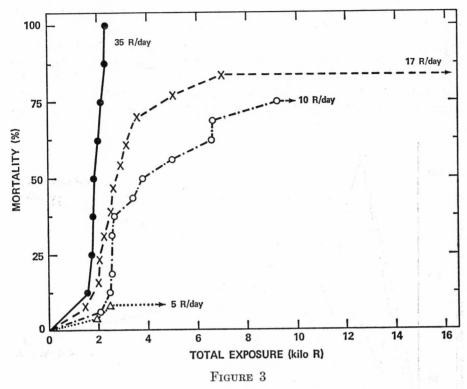

FIGURE 3

Mortality in beagle dogs exposed continuously to either 5, 10, 17, or 35 R/day
in a cobalt-60 gamma ray field. The arrows indicate the total exposure reached
at the time of this report.

show this interesting dose rate dependence for low LET radiations and much
reduced or no dependence on dose rate for high LET radiations.

2.4. *Delayed effects and dose rate dependence.* Most discussions of radiation
effects attempt to separate conceptually early and late somatic effects. A number
of studies performed as part of the DBM program suggest a continuance of
responses as a function of the dose rate at which the radiation is received.
A classic example of this is the rather extensive dose rate study on dogs performed at Argonne National Laboratory (Figure 3). It is clear that survival
time, as a result of continuous exposure until death, shows a continuous increase
as the dose rate is reduced. At least a factor of 20 difference was found between
the LD_{50} at the highest dose rate, 300 R/day, to the lowest so far used, 5 R/day.
Although the survival times are remarkably different, the cause of death reflects
primarily damage to the bone marrow, which ranges from septicemia, to anemia,
to leukemia.

The most massive collection of data on dose rate effects on late somatic effects
on animals come from mouse studies at the ORNL Argonne National Labo-

ratory and Brookhaven National Laboratory. Although data have been collected for a number of late effects including graying or depigmentation of hair, production of lens opacities resembling cataracts, and other disorders such as kidney damage, we will be concerned here with the overall life span shortening and production of cancer. These studies have shown these two effects to be intimately related because a major cause of life span shortening occurs as the result of the overall production of neoplasms. Figure 4 shows the overall life

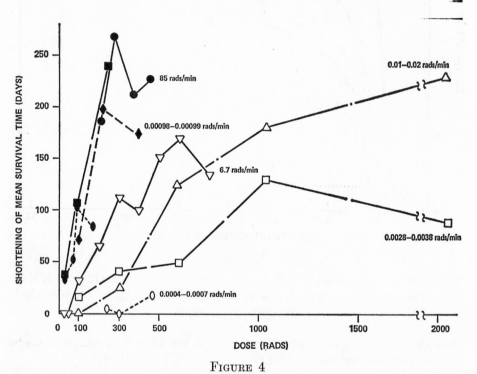

FIGURE 4

Life shortening in females as influenced by dose and dose rate of gamma rays and neutrons. Open symbols represent gamma rays, shaded symbols neutrons.

span shortening per unit dose for gamma rays that have been administered at various dose rates ranging from 85 rads/min to 0.0004 rads/min. It is clear that the effectiveness of gamma rays is systematically reduced as the dose rate is reduced. The shaded symbols show for contrast the results of neutron irradiations, over a similar range of dose rates, indicating more effectiveness per unit dose and independence of dose rate as previously mentioned.

Similarly when radiation induction of specific types of cancer are measured as a function of dose and dose rate, there is a remarkable reduction of the incidence per unit dose with reduction in dose rate, as shown in the next figure (Figure 5). These data are derived from the Oak Ridge study, and a similar

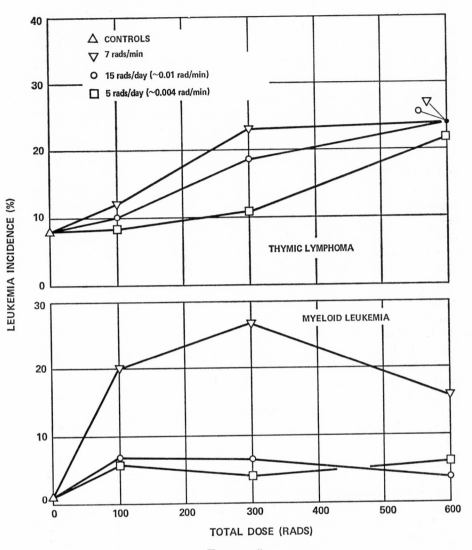

FIGURE 5

Leukemia incidence in relation to dose and dose rate of gamma radiation.

study at Argonne National Laboratory has provided confirmation of the dose rate effect for a number of mouse strains.

Although these studies give evidence for an overall reduction of radiation induced cancer incidence with decreasing dose rate for those radiations of most concern, none of the experiments have attempted to resolve the question of effects of minute doses received at the dose rates most likely to be involved in present or future developments in the nuclear energy program.

However, with the help of our biometrics staff in the National Laboratories mouse experiments are underway to evaluate life span shortening and specific tumor incidence at a total dose, as low as ten rads and at several dose rates for both low and high LET radiations. These studies are designed to detect tumor incidences in irradiated populations as compared with unirradiated populations which are significant at the five per cent level. These experiments may well serve as models for what can be practically done for other potentially hazardous agents being added to man's environment.

3. Human exposures

Studies in experimental animals are of great value in shedding light upon many aspects of radiobiology, but in order to apply the findings to humans there is a great need for direct information on human beings. As I have mentioned this is difficult to obtain and there are always certain drawbacks because of the nature of the sources, but when interpreted in light of results from experimental animals the data can be very useful.

Human experience with radiation exposures is found mainly in the following groups: (1) radium and thorium recipients; (2) uranium miners; (3) Nagasaki and Hiroshima survivors; (4) accidental exposures; (5) therapeutic exposures.

3.1. *Radium and thorium recipients.* Radium is a naturally occurring radioactive alkaline earth element which, when taken into the body, deposits primarily in newly forming bone. During radioactive decay, radium and its decay products emit alpha and beta particles and gamma rays and irradiate the contiguous cells which cover the bone and overlying tissue.

The unfortunate poisoning of human beings provided the opportunity to study the biological effects of radium, including bone cancer development, in a relatively large sample of humans. Radium serves both as a reference point between man and experimental animals for studying relative toxicology and as a reference point for calculating maximum permissible burdens for other radionuclides normally picked up by bone.

For the past twenty years a joint study of more than 400 persons bearing considerable body burdens of radium 226 has been in progress at AEC's Argonne Cancer Research Hospital and the AEC's Argonne National Laboratory. Most of these persons painted radium dials on luminous watches during 1920 to 1930; others received radium chloride by injection or orally as a medicament between 1920 and 1933. From 400 to 500 cases, mainly former dial painters, have been studied extensively at the Massachusetts Institute of Technology to determine the long-term effects of radium and mesothorium in human subjects. The MIT radium studies are being consolidated with those at Argonne Cancer Research Hospital-Argonne National Laboratory as part of a newly established Human Radiobiology project under the direction of Argonne National Laboratory. The names and location of more than 1,000 radium contaminated people are known. Of these, 70 per cent are still living. About 85 per cent of the living and 20 per cent of the deceased subjects have been studied. This long-term study has shown that persons with a considerable body burden of radium have characteristic defects and destructive changes in skeletal structure; skeletal tumors and other rare tumors of the tissues lining the mastoid bone, paranasal sinuses, and oral cavity may also exist.

3.2. *Uranium miners.* The problem of pulmonary malignancy in uranium miners and other hard rock miners has been long recognized. Researchers at St. Mary's Hospital, Grand Junction, Colorado, have developed cytological examinations of miners' sputum to a level which indicates considerable promise as a readily obtainable screening and prognostic procedure. Assisting the St. Mary's group is a group of consultants in pathology who are attempting to establish uniform diagnostic test criteria which hopefully will have usefulness for general medical application.

Three facts make the development of an early test for cancerous lung lesions especially compelling: (a) as chronic irritation affects the lungs, the previously healthy cell lining undergoes degeneration to an identifiable "pre-cancerous" condition; (b) recovery of the abnormal cells has been observed (in sputum samples) where these individuals were removed from further exposure to lung irritants; and (c) with continued exposure to lung irritants these abnormal pre-cancerous cells show changes which are irretrievably headed into malignancy. Thus, timely preventative medicine based on early diagnosis is critical.

At Colorado State University at Fort Collins, human subjects are exposed to mine aerosols both in a controlled test tunnel and in general work areas in an operating uranium mine. Physical characteristics of the exposure aerosols are measured on samples collected currently with the exposures. Deposition patterns of radon decay products in the human respiratory tract are analyzed in a mobile whole body counting laboratory. Depositions in resultant doses are studied with respect to various mine atmospheres and personal control measures. Samples of urine, feces, blood and hair are collected from working miners and analyzed for long lived radon decay products in an attempt to discover bioassay procedures for correlation with measured exposures.

3.3. *Accidental exposures.* Radiation accidents have been very rare, however, for purposes of evaluating and accumulating information on radiation accidents, AEC has had a program for collecting data whenever possible. The sources include accidents and operations for recovery of fissionable materials, radioactive fallout (Marshallese), exposures from industrial radiography sources, exposures from particle accelerator beams, electronic tube exposures, accidental administration of large amounts of radiotherapy to patients through miscalculations or mislabelling, and accidental misplacement of laboratory gamma ray sources. Data have been collected at the Oak Ridge Associated Universities from several accidents. Careful cataloging of the various experiences will allow the AEC in the future to pursue follow-up studies for the remote effects and bring together the combined experience of the physicians who have managed victims with total body irradiation or combinations of total body irradiation and local destructive radiation lesions of the extremities.

On March 1, 1954, a nuclear weapons test detonation at the Pacific Proving

Ground produced fallout which was unexpectedly carried by winds over the inhabited Rongelap Atoll and adjacent islands. During the past five years abnormalities of the thyroid glands have been detected in a number of the Marshallese people of Rongelap Island who were accidentally exposed in 1954. These abnormalities, believed to be late effects of radioiodines deposited in the thyroid at the time of fallout, were detected during routine annual medical surveys of these people by the AEC's Brookhaven National Laboratory and the Trust Territory of the Pacific Islands. The thyroid abnormalities consisted of 18 cases with benign nodules, two others with marked hypothyroidism, and one case of cancer of the gland. Most of the abnormalities have occurred in children exposed to radiation at less than ten years of age. It is estimated that the thyroid glands of the children receive a dose of between 700 to 1400 R of radiation from internally deposited radionuclides plus 175 R of penetrating gamma radiation from outside the body. It is not known whether the one case of thyroid cancer, observed in a 40 year old woman, was related to radiation exposure. The population of the atoll, regularly examined by a medical team from BNL, has a clearly higher than normal incidence of thyroid abnormalities that has been related to the presence of radioactive iodine in fresh fallout.

The early effects of fallout exposure on the Marshallese has been well documented. The 64 people on Rongelap received 175 rads of gamma radiation which proved to be sublethal but caused early nausea and vomiting. Significant depression of formed blood elements lasted over a period of several months. Exposure of the skin resulted in burns and loss of hair. In addition, absorption into the body of radioisotopes of iodine from contaminated food and water with deposition in the thyroid gland proved to be the most serious internal exposure. Recovery of formed blood elements to near normal and healing of skin lesions with regrowth of hair occurred by the end of the first year. No acute or chronic effects of exposure to other internally absorbed radioisotopes has been noted.

3.4. *Therapeutic exposures.* Clinical and laboratory data from human total body irradiation exposures in the United States have been obtained and coded, and are being analyzed by electronic computer methods. The Oak Ridge Associated Universities material on patients obtained from their radiotherapeutic programs since 1957 and from additional human data (including over 3,000 case histories obtained from other hospital centers with similar treatments) have been gathered for a study to determine what dose ranges in man cause loss of appetite, nausea, vomiting, diarrhea, fatigue, weight loss, fever, skin reddening, loss of hair, and blood cell changes in number and type including anemia, chromosome damage, decreased resistance to infection, decreased antibody synthesis, premature aging and development of cancer.

Dose response relationships derived from these data were reported to the Space Radiation Study Panel of the National Research Council of the National Academy of Sciences. These data also appear in a four volume compendium published by the National Aeronautics and Space Administration entitled "Compendium of Human Responses to the Aerospace Environment."

On November 26, 1946, President Truman approved a proposal by the then Secretary of the Navy, Mr. James Forrestal, which instructed the National Academy of Sciences-National Research Council to undertake a long range, continuing study of the biological and medical effects of the atomic bomb on man. It was decided that the AEC should provide the funds; and since that time the AEC has funded for, and has provided, programmatic supervision of this research program which has come to be known as the Atomic Bomb Casualty Commission (ABCC). The first contract with the National Academy of Sciences was executed on April 17, 1948, with renewals at intervals. Since its inception in 1947 through June of 1969 more than 468 scientific papers have been published on various aspects of the ABCC research program and referred by medical journals throughout the world.

The ABCC observations are applicable to the long term effects in man for all types of ionizing radiation. The basic population sample in the ABCC is comprised of approximately 50,000 Japanese selected on a basis of their history and records of exposure and 50,000 equally selected matched control Japanese. Observations on this population which justify definitive statements or conclusions at this time are:

(1) Cataract of a rather specific type involving the posterior subcapsular tissue of the lens developed in about 100 people after exposure. It tends to regress with time, and may be successfully treated by surgery.

(2) The incidence of leukemia in the exposed groups began to rise in 1948 and reached a peak in 1952–53 some 20 to 50 times that in the control population; the incidence subsequently has decreased but remains with variations at a level slightly higher than that expected for the population. In general those persons receiving the higher doses developed the condition sooner, but there was nothing clinically unusual about the disease itself. It is very important to learn whether the incidence of leukemia will increase more rapidly than expected as the population ages, as leukemia is primarily a disease of the aged; or whether it will increase at all.

(3) Cancer of the thyroid gland has shown an increased incidence, particularly in those surviving the higher exposures. Since the natural incidence of this disease is low it is necessary to continue to observe the population in order to accumulate enough cases for statistical validity. Persons receiving lower doses probably will develop the disease sometime in the not too distant future, if they are going to.

(4) A major research effort (1956), followed up by a resurvey (1966), concluded that there were no radiation-induced genetic effects observed in the children born of one or both exposed parents. Since that time more sensitive tests for mutations have been developed which may be employed in the ABCC population if the cooperation of the Japanese can be obtained.

(5) By virtue of a ten-year study at the Oak Ridge National Laboratory the radiation emission fields of both weapons were reconstructed and the combined neutron and gamma-ray exposures were calculated for nearly every survivor

with an accuracy of plus or minus 10 to 15 per cent. Previously the biological effects had to be related to the distance of the person from ground zero. The calculations have been verified independently by Japanese scientists. Reliable dose/effect relations can now be established for the radiation effects being studied. The greater precision of the dose/effect relationships, particularly of the low dose exposures, ought to be useful in examining questions on the linearity or nonlinearity of responses of man to irradiation.

(6) The incidence of leukemia in children born of exposed parents is not different from that of children of unexposed parents. Also, the parents of children who do develop leukemia are equally divided as to having been exposed and not exposed, corrections having been made for the lesser number of parents in the exposed group.

4. Internal emitter program

Studies which deal primarily with experiments related to radiotoxicity of internally deposited radionuclides and their metabolic behavior date back to the Manhattan Project of the early 1940's. Because of its central importance o the development of a national nuclear energy program biological experiments related to the potential toxicity of plutonium can be traced back virtually to its discovery in February 1941. Animal experiments using CF-1 mice to determine the toxicity of injected plutonium were performed at the Metallurgical Laboratory at the University of Chicago in 1944 by Brues, Lisco and Finkel. By 1947 these researchers published several articles in the open literature calling attention to the carcinogenic properties of plutonium when administered in relatively large quantities to experimental animals [10], [11].

I should diverge at this point for a moment to say that the field of radiation protection had been established formally on both the national and international levels in the period 1928 to 1929. Of further interest is the realization that this year we are observing the 76th anniversary of the discovery of X-rays by Roentgen which was followed a year later by Becquerel's discovery of radioactivity.

On the first of December, 1952, the first dog was injected at the University of Utah for the purpose of comparing long term biological effects of a single intravenous injection of radium 226 and plutonium 239 in adult Beagles. Because of prior experience with radium toxicity in man (Dr. Martland reported the first radium recipient fatality in 1925) [12] it was clear that strontium 90, ^{239}Pu and other bone seeking radionuclides could become of interest to the nuclear energy industry. To determine the extent of the potential hazard, several long term investigations were initiated to make use of the human radium data by designing the studies around the following relation:

$$\frac{^{239}\text{Pu Effects on Man}}{^{226}\text{Ra Effects on Man}} \cong \frac{^{239}\text{Pu Effects on Beagles}}{^{226}\text{Ra Effects on Beagles}}$$

Thus the toxicity of ^{239}Pu can be predicted from the information on the ^{226}Ra patients and the animal experiments using ^{239}Pu and ^{226}Ra.

Of course, several factors contribute uncertainty to this design. For example, the surface to volume ratios of bone in the Beagle and Man, the fact that the radium subjects received radium by oral ingestion over a period of months and the uncertainty of the importance of radium 228 and thorium 228 which some radium subjects also received. Because of the latter consideration, ^{228}Ra and ^{228}Th were added to the original experimental design. In 1954, ^{90}Sr was added to the design because of increased interest in radioactive fallout.

The radioactivities administered as a single intravenous injection to groups of 12 or more 17-month old beagles, were:

$$^{90}\text{Sr} - 0.57\text{--}100 \quad \text{microcuries per kilogram}$$
$$^{239}\text{Pu} - 0.0006\text{--}2.8 \quad \text{microcuries per kilogram}$$
$$^{228}\text{Th} - 0.0017\text{--}2.8 \quad \text{microcuries per kilogram}$$
$$^{226}\text{Ra} - 0.0057\text{--}10 \quad \text{microcuries per kilogram}$$
$$^{228}\text{Ra} - 0.19 \text{ to } 10 \quad \text{microcuries per kilogram}$$

Table I shows the tumor response for the ^{226}Ra series. The incidence of osteosarcomas (bone tumors) decreases with injection level-cumulative radiation dose. Also of interest is the increase in time from injection to death (latent period) as a function of decrease in radiation dose. This general pattern has been observed in all the experimental groups to date.

Because of the importance of ^{90}Sr as a potential hazard to man from worldwide nuclear fallout, a major research project was established at the University of California, Davis campus. This effort is complementary to the Utah studies but differs in that ^{226}Ra and ^{90}Sr are given as multiple intakes to simulate continuous intake from nuclear fallout (^{90}Sr) or occupational exposure (^{226}Ra). To achieve uniform labelling of the skeleton with ^{90}Sr, dams were placed on a

TABLE I

OSTEOSARCOMAS IN UTAH BEAGLES (1 MARCH 1969) ^{226}Ra
All injected dogs at levels 5, 4, and 2 have died.

Injection level	Injected μCi/kg	Total	Deaths Osteosarcomas	Osteosarcoma dog averages Years from inj. to death	Rads one year before death
5	10.4	10	9	3.04	10900
4	3.21	13	12	4.36	4530
3	1.07	12	11	6.28	1940
2	0.339	13	5	10.28	837
1.7	0.166	9	1	11.25	458
1	0.0621	12			
0.5	0.0220				
0.2	0.0074				
0	0.0000	12			

diet containing from 0.007 to 3.33 microcuries ^{90}Sr per gram of dietary calcium; their progeny were kept on the same dose level until young adulthood (18 months of age). The dietary calcium was kept at a one per cent level to maintain a constant specific activity.

To date, the lowest level in this experiment which has caused radiation deaths amounts to doses of about 2,000 millirads per day, or about 730,000 millirad per year. For comparison the FRC Radiation Protection Guide assigns a population dose limit of 0.17 rem per year or about one-half millirem per day on the average.

Other Beagles were given eight semi-monthly intravenous injections of ^{226}Ra beginning at 14½ months of age to simulate the brief repeated exposures of the radium dial painters. The amounts injected from 0.024 to 10 microcuries ^{226}Ra per kilogram of body weight, will allow comparison with the Utah series. The injection and feeding regimens were completed in 1970 and late effects will not be known for some years.

Another study on the effects of ^{90}Sr in a large species has been in progress at the former Hanford Laboratories (now the Pacific Northwest Laboratories of the Battelle Memorial Institute) since 1958. About 900 miniature pigs, similar in size to man, with a relatively long life span have been used in this study of the parental and two generations of progeny. Starting at nine months of age, levels of 1, 5, 25, 125, 625, and 3,100 microcuries of ^{90}Sr were fed *each day* to females. The offspring were maintained at levels of 1, 5, 25 and 125 microcuries per day.

Deaths followed in about three months at the highest level (3,100 microcuries per day) and about nine months at the 625 microcuries per day level. Lymphoid metaplasia has been observed in some progeny of the 1, 5, 25 and 125 microcurie per day exposure groups but not in the control groups. Myeloid metaplasia was observed in virtually every parental or offspring exposure group and myeloid neoplasia was found in the 25, 125 and 625 microcurie per day groups. Time to death was inversely related to radiation dose.

Neoplasia of the hematopoietic system has been observed in these and in the Davis studies, both of which involve continuous intake and uniform labelling of the skeletal tissues. In contrast, the major biological effect seen in the Utah studies has been osteosarcoma (bone tumor) formation. This is a good example of the importance of the specific kind of exposure on the kind of biological insult which is produced.

These are only several examples of some of the "older" research programs in the internal emitter field. Space does not permit an exhaustive review.

5. Inhalation carcinogenesis

Airborne radioactivity was implicated as a pulmonary carcinogen early in this century although it had been known for several centuries that pitchblende miners developed fatal pulmonary diseases. The earliest evidence for the pulmonary carcinogenicity of ^{239}Pu in mice was reported by Temple and co-workers in 1959. Because of the importance of the inhalation route of exposure in the nuclear energy industry a sizeable research program has developed.

The AEC research program on inhalation carcinogenesis is mainly related to the following areas:

(1) Development of theoretical and empirical models to describe deposition, retention, translocation and radiation dose rate patterns of inhaled radionuclides within the lung and other tissues of interest.

(2) Investigation of early and late effects of several species of plutonium, other transuranic elements, selected fission products, and other selected materials (for example, rare earth elements) following inhalation of various physical and chemical forms of these nuclides.

(3) Studies of inhaled constituents of uranium mine environments in experimental animals; this program is correlated with the uranium miners by comparative studies of sputum cytology and chemical analyses of lung samples from miners and experimental animals and by comparative measurement of constituents of mine air and animal exposure chambers.

(4) Studies of the cytokinetics of lung cells, macrophage function and transport of radioparticulates and the effects of pharmacologic agents on phagocytosis and clearance mechanisms.

(5) Development of feasible therapeutic means of reducing lung burdens of radioactive materials following deposition.

Over the years our collective experience has indicated that inhalation as a route of exposure to radioactive (and stable) materials in industry must be assigned a high probability. Large projected increases in nuclear power production with attendant increases in nuclear fuel manufacture and processing indicate the need and justification for the program outlined above. Another consideration, despite the very low probabilities of release, is that of high specific activity radioparticulates from nuclear propulsion reactors and thermoelectric generators Systems for Nuclear Auxiliary Power (SNAP) designed for numerous uses. Paralleling these developments will be greatly expanded production of uranium ore, the majority of which will come from mines rather than from open pit sources.

Many reasons exist as to why it is important to conduct this program. Aside from the obvious relation to personnel protection within the nuclear energy industry, much of the information obtained from the program is applicable to the more general problem of environmental pollution from nonradioactive materials. As regards radiation protection, there is a very cogent reason for studying the effect of inhaled radioactive materials. Radiation protection standards and criteria for bone seeking radionuclides, as promulgated by organizations such as the National Council for Radiation Protection and Measurements (NCRP) and the International Commission on Radiation Protection (ICRP), actually are based on a dual system: The observations based upon the human radium recipients and calculated "allowable" exposure estimates, maximum permissible dose (MPD), which can be related to multiples of the radiation dose rate arising from natural background.

Earlier I noted that AEC's central program for the bone-seeking radionuclides is built around a large-scale retrospective epidemiological study of radium toxicity involving several thousand people who were exposed through employment (mostly radium dial painters) or medical treatment. These studies serve as a base line of intercomparison for comparative experiments in which the toxicity and relative hazard of other bone seeking radionuclides are assessed. No such standard based upon human data exists for the lung or for soft tissues in general.

I can only mention several of the many interesting research efforts which are part of this large program in inhalation carcinogenesis. One area of current interest is the third area I previously mentioned, studies of inhaled constituents of uranium mine environments and experimental animals.

The radiobiology of radon and its progeny has been studied in rodents for some years at the University of Rochester Atomic Energy Project. More recently, studies of radon and its progeny have been undertaken in dogs. Four dose levels are being studied in 30 dogs, covering the range of 250 to 3800 working level months (WLM). Average concentrations of radon in its decay products are: radon, 0.6 uCi/liter; radium A, 0.45 uCi/liter; radium B, 0.3 uCi/liter; and radium C-C', 0.18 uCi/liter. Exposures are 20 hours per day for the required number of days, at the rate of five days per week, until the desired exposures are reached. The radon decay products are carried on normal room dust in this experiment. This work is being conducted in conjunction with the National Institute of Environmental and Health Sciences of the U.S. Public Health Service.

Related studies, some also jointly funded with the USPHS, are under way at the Pacific Northwest Laboratory of the Battelle Memorial Institute. This program is also designed to investigate possible causative or contributing factors in the observed increased incidence of lung cancer among uranium miners. One portion of the experiment involves a study of the long term biological separation of long lived alpha emitters in the uranium decay chain. In another study samples of lung tissue from deceased miners are analyzed for numerous nuclides by neutron activation analysis.

The elements scandium and antimony, typically high in concentration among individuals with mining experience, were found to be low in nonminer lung samples. Similar studies are being conducted at the Argonne National Laboratory.

In addition, hamsters have been exposed at the Pacific Northwest Laboratory to various combinations of uranium ore dust together with radon and its progeny or to radon and progeny alone. The observed metaplastic changes were more severe in the hamsters exposed to both agents and no tumors were observed. Supporting experiments with beagles are also beginning at the Pacific Northwest Laboratory of the Battelle Memorial Institute and will include exposure

TABLE II

EXPERIMENTAL DESIGN (PNL-BMI)
Controls were sham-exposed under conditions identical
to those used for the experimental groups.

Exposure Groups	No. Hamsters	No. Dogs
1. Controls	100	9
2. 30 WL Rn daughters	100	
3. 600 WL Rn daughters	100	
4. 600 WL Rn daughters with U-ore dust	100	20
5. Cigarette smoke	—	20
6. Cigarette smoke + 600 WL Rn daughters and U-ore dust	—	20
7. Diesel engine exhaust	100	—
8. Diesel engine exhause + 600 WL Rn daughters with U-ore dust	100	

to radon and its progeny plus combinations of uranium ore, diesel exhaust fumes, and tobacco smoke. The experimental design is shown in Table II.

This program was established to study the effects, in experimental animals under controlled conditions, of four potentially carcinogenic air contaminants to which uranium miners are exposed, that is, radon and its daughters, cigarette smoke, uranium ore dust, and diesel exhaust fumes. It is believed that results from these studies will provide the basis for definitive research to properly assess the interrelationships of the several hazardous materials in mine atmospheres working concomitantly to cause damage to the respiratory epithelium, emphysema, fibrosis, and precancerous or cancerous changes in the lungs.

In another phase of this work simultaneous measurements of human respiratory deposition and the concentration, particle size, and charge of aerosols are being made to define the significant factors relating occupational exposure to hazard. In the relationship of aerosol properties, respiratory deposition and excretion are measured in actual work atmospheres. Field studies have been initiated in a uranium mine and in a uranium fabrication plant.

The dosimetry of selected tissues in radium workers as applied to lung carcinoma in uranium workers is under investigation at the Massachusetts Institute of Technology. The unusually high incidence of carcinomas in the sinus cavities of radium recipients suggest the importance of obtaining the quantitative estimate of the dose rate to overlaying epithelial tissues of the sinus cavities from the presence of radium and its progeny in and around the cavity. It is hoped that results of this dosimetry study will be applicable to the problem of lung carcinomas in uranium miners.

An investigation being conducted at St. Mary's Hospital at Grand Junction, Colorado is designed to study sputum cytology from uranium miners with and without past smoking history. Out-growths of this project will be the development of uranium miner tumor registry and an atlas describing sputum cytology. In an attempt to integrate these observations with experimental data, lung washings from dogs containing plutonium or radon progeny are obtained prior to dog sacrifice at the Pacific Northwest Laboratory of the Battelle Memorial Institute and compared with samples obtained from uranium miners and from confirmed cases of pulmonary neoplasia.

The fate of inhaled lead 210 in human subjects is being investigated at the University of Rochester Atomic Energy Project. This study is related not only to the uranium mining problem as regards long-lived radon progeny such as ^{210}Pb, but also to the general problem of lead contamination in the environment.

6. Radioisotopes movement in the environment

Another aspect of the AEC's concern about environmental health may be illustrated by examples of food chain studies. While the transfer routes of energy and materials through a food chain form a highly complex web, many times it is possible to follow the passage of nutrients using radioactive tags or tracers. We are concerned with five basic questions: (1) What organisms and populations are involved in a particular food chain? (2) What are the pathways of energy and materials in food chains? (3) What are the dynamics of food chains? (4) What concentration of elements (including radioisotopes) occur in food chains? and (5) Can useful predictive mathematical models be made of food chains? Considering these last two items, concentration and models, studies using radioisotopes in food chains may provide suitable models for predicting the fate of pollutants such as pesticides in the environment.

We have sponsored such studies in several areas of the country. For example, researchers at the Ohio State University in 1964 [13] treated four acres of a natural marsh with chlorine 36 ring labeled dichloro-diphenyl-trichloroethane (DDT). The labeled DDT was traced throughout the marsh ecosystem over a four year period. The main concern of the project was with food chain aspects of DDT accumulation. The total quantity of DDT usually was determined from environmental samples of whole organisms though some tissues and organs were assayed. I might mention two findings. The DDT was applied on granules which disintegrated in the water, but little DDT was ever present there at any time because the insecticide was rapidly removed from the water by plankton and other organisms. Secondly, the highest tissue residues of DDT usually was found in the highest tropic levels.

The data from this study has been used by scientists at Battelle's Pacific Northwest Laboratories to construct a computer simulation model to represent the dynamics of DDT in this marsh. I will quote from their report [14]. "An exponentially decreasing input function (representing release of DDT from the granular formulation in which it was applied) determines concentration in water, from which one- or two-compartment models represent the concentration in various forms of flora and fauna. The overall results generally suggest that simple models, of the kind used to study radionuclide transfer through food chains, will provide reasonably good representations of the behavior of DDT in food chains."

The Ohio State Group and a group from SUNY is now studying the dispersion of ^{36}Cl labeled DDT in a meadow area.

Now, I will review briefly some other food chain studies we support as additional examples of our program.

Figures 6 and 7 show models of cycling of cesium 137 in a tulip poplar forest. During a year, the forest produced an average of 1,550 kilocalories of foliage per square meter. Insect leaf feeders in the forest canopy consumed over five per cent of this production, while concentrating radiocesium to levels nearly 50 per cent that of foliage.

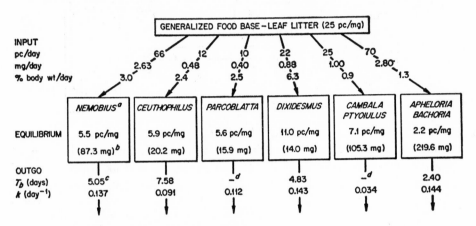

FIGURE 6

Cesium 137 Cycling in Forest Insects. Charted is the radioactive cesium transfer and leaf litter consumption by forest floor insects and millipedes as determined by Oak Ridge National Laboratory studies in food chain dynamics. *Upper arrows* indicate the flow of radiocesium and organic matter from leaf litter to arthropods (crickets *Nemobius* and *Ceuthophilus*, wood roach *Parcoblatta*, and millipedes *Dixidesmus*, *Cambala*, *Ptyoiulus*, *Apheloria*, and *Bachoria*). *Lower arrows* indicate biological turnover of radioactive cesium (*Outgo*) by metabolic processes. From biological half-lives (T_b), the daily loss of radiocesium is calculated. Since the equilibrium concentrations of radiocesium (shown in boxes) remain unchanged, daily *Input* must equal the *Outgo*. *Input* is calculated from elimination coefficient (k) times the *Equilibrium*. Calculated *Input* is shown for radiocesium (picocuries per day) and dry matter (milligrams and per cent body weight per day). Information obtained by this radioactive tracer technique demonstrated a relatively small food requirement for these forest floor animals as well as relatively low cesium 137 body burdens compared to radioactivity levels in their food supply.

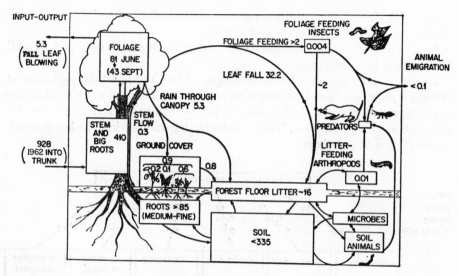

FIGURE 7

Cesium 137 Forest Cycling. Drawing shows the radiocesium cycle in a tagged tulip poplar forest ecosystem (*Liriodendron tulipifera*) at Oak Ridge National Laboratory. This forest (20 × 25 m. or 65 × 82 ft.) was originally tagged several years ago with cesium 137, a long-lived radionuclide. Continuous inventory has followed the seasonal and annual distribution of radiocesium between components of the forest ecosystem. Numbers in boxes are microcuries (1/1,000,000 of a curie) per square meter of ground surface area in summer. Arrows indicate the pathways of radiocesium transfer between compartments. Numbers by the arrows are estimates of total transfer during the growing season.

These examples should not give you the impression that we are concerned only with terrestrial ecosystems. Our support of food chain studies in marine and freshwater systems is equally as large. Let me give you some examples from those areas.

The rates and mechanisms of the chemical and mixing process in the oceans are being studied for us at several sites by University staff and researchers from some of our National Laboratories. At the Pacific Northwest Laboratory these processes are being investigated by use of material chemical elements and radioactive tracers in the ocean. In addition the staff there has used neutron activation to measure the distribution of mercury. Incidentally, they have found significantly higher levels of mercury in Pacific Ocean samples than in those from the Atlantic Ocean. Normal oceanic chemical and biological processes cannot account for the high mercury values. There is some indication that in estuaries, mercury levels are diluted by the entering river.

Several other studies we support deal with marine food chains per se. The immensity of the problem is illustrated by the fact that in the sea about a million species of organisms exist which can be affected by pollutants, including radio-activity. These organisms also cover a wide spectrum of mobility and range of habitat. At the Scripps Institution of Oceanography some progress is being made in amplifying the usually oversimplified idea of the marine food web (Figure 8).

As I indicated earlier, we are interested in the concentration of various ele-ments in food chains. As a final example of our work in this area I will offer a study of cesium 137 concentrations in freshwater, a study conducted at the

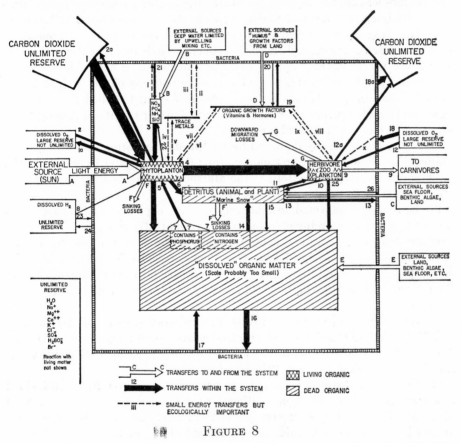

FIGURE 8

Conceptual Relationships in Marine Food Web. Box model illustrating the major components of the sea's biological reservoir, some of the physical and chemical factors which affect them, and the principal routes of material or energy transfer between the components. This Institute of Marine Resources, University of California, diagram is simplified in that it emphasizes the orig-inating part of the marine life cycle and is restricted largely to organisms and processes within the euphotic portion (zone of light penetration) of a water column.

Oak Ridge National Laboratory. Algae were grown in water tagged with cesium 137. Then one group of snails was fed the algae which remained in the tagged water which did not contain algae. The concentration factor for algae was 2,300. For the snails the concentration factor for algae was only 98, and for snails in water alone, the concentration factor was 28. Cesium thus does not invariably concentrate at each successive level of a food chain.

There are many other examples I could give you, and more models and systems analyses in all stages of development. I will not prolong this review. However, before leaving the topic of environmental health I should point out that we are studying the effects of radionuclides in the environment as well as their fate. This program can perhaps be appreciated by the range of environmental studies, involving, for example radiation effects on rodents and rodent relatives, deer, birds, sheep, insects, reptiles, and amphibians, and a host of lower forms, and several plant species. These studies involve not only effects on individuals in nature, but on populations and ecosystems. We have studied the results of exposure to and recovery of ecosystems from ionizing radiation. Among those areas which have been studied are:

(1) Ecosystems directly damaged by nuclear device testing: (a) Trinity Site, Alamagordo, New Mexico, first nuclear test; (b) Binkini and Eniwetok Atolls, (and, indirectly, Rongelap Atoll) in the Marshall Islands; and (c) Nevada Test Site, Nevada, Sedan crater and other areas.

(2) Ecosystems subjected to relatively massive exposures of ionizing radiation from gamma ray sources (cobalt 60 or cesium 137) given in both chronic (continuous or long term) and acute (prompt, or short-term) doses: (a) Brookhaven Ecology Forest, Long Island, Oak-Pine Forest; (b) Yucca Flat, Nevada, desert plant and animal enclosures; (c) Terrestrial Ecology Project, Puerto Rico Nuclear Center, a tropical rain forest; and (d) a short grass prairie near Colorado State University.

(3) Ecosystem subjected to relatively high radiation from radioactive wastes: Oak Ridge, White Oak Lake bed—a meadowland community established after draining of a lake which had been used as a settling pond for radioactive waste from ORNL facilities between 1945 and 1955.

(4) Ecosystem directly damaged by neutron irradiation from unshielded reactors: (a) Marietta, Georgia, southeastern pine forest adjacent to Lockheed Materials Testing Reactory; and (b) ORNL, southeastern pine forest community in vicinity of Oak Ridge Health Physics Research Reactor.

The one result common to the above studies and which may be called a "tentative hypothesis" is that recovery tends to follow "ecologically normal" and predictable successional trends; that is, patterns following damage from radiation are similar to those following fire, logging, or stripping of topsoil and other natural catastrophic events.

Finally, we have expanded our research in another area, research on thermal effluents as potential pollutants. We hope this enlarged program will lead to increasingly definitive predictive models dealing not only with thermal effects, but also thermal in addition to other water quality factors.

While I believe that knowledge now in hand is fully adequate to permit the establishment of practices and procedures in nuclear operations which assure safety of man and his environment, we have a considerable way to go in order to be able to predict quantitatively the effects if any of extremely low levels of exposure produced by the release of small quantities of radioactivity to the environment.

Since we are striving to describe effects produced by exposures which are a small fraction of natural exposures, we are faced with difficult and frequently vexing statistical problems in attempting to isolate effects produced by radiation from changes produced by much larger levels of natural radiation, other environmental stresses, and normal biological processes. Possible synergism between radiation exposure and other environmental agents add to the complexity. I believe that we will make greatest inroads into this problem by achieving, through both basic environmental and molecular and cellular level studies, a better understanding of the nature of radiation induced damage and the repair capability of damaged cell systems, populations and ecosystems.

We face difficult problems also in our efforts to predict the dose to man which results from radioactivity released to the environment. Nuclear reactors release such small quantities of radioactive material that it is difficult to measure them in the environment at any distance from the point of release. We have, however, available and under constant development sophisticated radiation detection methods which permit us to measure extremely low-levels of radioactivity and to construct and improve models for the movement of radionuclides in the environment and to man.

It appears likely that many of the techniques which we have developed and are now using for radiation and radioactivity will be very useful for studying the movements and effects of a variety of pollutants which are becoming an ever increasing source of concern.

It is frequently said that radiation and radioactivity is the most thoroughly studied environmental pollutant. I believe this is so and hope that I have succeeded in giving you some idea of the range of these studies and our approaches to outstanding problems.

REFERENCES

[1] H. J. MULLER, "Artificial transmutation of the gene," *Science*, Vol. 66 (1927), pp. 84–87.

[2] W. P. SPENCER and C. STERN, "Experiments to test the validity of the linear R-dose/mutation frequency relation in *Drosophila* at low dosage," *Genetics*, Vol. 33 (1948), pp. 43–74.

[3] W. L. RUSSELL, "Comparison of X-ray-induced mutation rates in *Drosophila* and mice," *Amer. Natur.*, Vol. 90 (1956), pp. 69–80.

[4] ———, "Studies in mammalian radiation genetics," *Nucleonics*, Vol. 23, New York, McGraw-Hill, 1965.

[5] Report of the United Nations Scientific Committee on the Effects of Atomic Radiation, General Assembly, Official Records: 21st Session, Supplement No. 14 (A/6314), U.N., 1966.

[6] W. L. RUSSELL, "Factors that affect the radiation induction of mutations in the mouse," *Ann. Brasilian Acad. Sci.*, Sup. Vol. 39 (1967), pp. 65–75.

[7] ——, "Effect of the interval between irradiation and conception on mutation frequency in female mice," *Proc. Nat. Acad. Sci.*, U.S.A., Vol. 54 (1965), pp. 1552–1557.

[8] W. J. SCHULL, J. V. NEEL, and A. HASHIZUME, "Some further observations on the sex ratio among infants born to survivors of the atomic bombings of Hiroshima and Nagasaki," *Amer. J. Human Genet.*, Vol. 18 (1966), pp. 328–338.
   H. KATO, W. J. SCHULL, and J. V. NEEL, "A cohort-type study of survival in the children of parents exposed to atomic bombings," *Amer. J. Human Genet.*, Vol. 18 (1966), pp. 339–373.

[9] E. L. GREEN, "Genetic effects of radiation of mammalian populations," *Ann. Rev. Genet.*, Vol. 2 (1968), pp. 87–120.

[10] A. M. BRUES, H. LISCO, and M. FINKEL, "Carcinogenic action of some substances which may be a problem in certain future industries," *Cancer Res.*, Vol. 7 (1946), p. 48.

[11] H. LISCO, M. FINKEL, and A. M. BRUES, "Carcinogenic properties of radioactive fission products and of plutonium," *Radiology*, Vol. 49 (1947), pp. 361–363.

[12] H. S. MARTLAND, P. CONLON, and J. P. KNEF, "Some unrecognized dangers in the use and handling of radioactive substances," *J. Amer. Med. Assoc.*, Vol. 85 (1925), pp. 1769–1776.

[13] R. L. MEEKS, "The accumulation of ^{36}Cl ring-labeled DDT in a freshwater marsh," *J. Wildlife Man.* (1968).

[14] L. L. EBERHARDT, R. L. MEEKS, and T. J. PETERLE, "DDT in freshwater marsh a simulation study," BNWL-1297, UC-48, 1970.

[15] W. H. LANGHAM, P. M. BROOKS, and D. GRAHN, "Biological effects of ionizing radiation," *Aerospace Medicine*, Vol. 36 (1965), pp. 1–55.

[16] A. B. BRILL and E. H. FORGOTSON, "Radiation and congenital malformations," *Am. J. Obstetrics and Gynecology*, Vol. 90 (1964), pp. 1149–1168.

Discussion

Question: J. Neyman, Statistical Laboratory, University of California, Berkeley

Were there any systematic studies, experiments, conducted on the cocarcinogenic effect of radiation?

Reply: J. Totter

There have been some studies involving cocarcinogenesis. In particular, there has been a series of studies conducted at Oak Ridge National Laboratories on the combined effects of partially burned hydrocarbons, viruses, radiation, chromic oxide, and calcium chromate. These studies were supported by the National Cancer Institute. The results showed the very great difficulties involved in conducting meaningful experiments with so many variables when relatively low incidence rates are involved. Part of the data are to be found in the Atomic Energy Commission Symposium Series, Number 18, "Inhalation carcinogenesis," (1970).

Question: Thomas F. Budinger, Donner Laboratory, University of California, Berkeley

In reference to Professor Neyman's question on the availability of data on cocarcinogenesis, the retrospective analysis of Gibson and co-workers (*New England Journal of Medicine*, Vol. 279 (1968), p. 906) is relevant. That study showed an increase incidence in leukemia when *both* virus diseases and irradiation *in utero* occurred. Children who received radiation *in utero* but did not have a childhood viral disease had the same incidence of leukemia as the population at large. Synergism between leukemia virus and extracts of city smog in the transformation of rat or hamster cell cultures was shown recently in Proceedings of the National Academy of Science by Freeman and co-workers (*Proc. Nat. Acad. Sci.*, Vol. 68 (1971), p. 445).

Question: John R. Goldsmith, Environmental Epidemiology, California Department of Public Health

Macht and Lawrence (*Amer. J. Roentgenology*, Vol. 73 (1955), p. 422) studying offspring of radiologists, and comparing these with offspring of other medical specialists have suggestive evidence of greater frequency of congenital defects in offspring of the more heavily exposed and also a greater frequency of male offspring among spontaneous abortuses. Has this data been taken into account in your statements, and are such populations being restudied with care to avoid response bias?

Have populations, such as veterans with intensive and frequent diagnostic radiation for diseases like duodenal ulcer been followed to try to detect late somatic and genetic effects?

Reply: J. Totter

My statements concerning effects in humans are chiefly based on Atomic Bomb Casualty Commission studies which probably provide the most extensive data available.

Question: Burton E. Vaughan, Ecosystems Department, Battelle Memorial Institute, Richland, Wn.

You stated, if I heard clearly, that doses to the Japanese population, in the ABCC study, were calculated to ±15% accuracy. You also stated that it was not necessary to estimate rough distance to ground-zero, for establishing dose. Would you clarify these statements: (1) How are the doses established? (2) Are accurate doses established for all 50,000 people, or some smaller number?

Reply: J. Totter

Estimates of doses to individual Japanese survivors in Hiroshima and Nagasaki have been refined considerably over a simple estimate calculated from their distance from the hypocenter. Survivors have been interviewed exhaustively to determine their precise location at the time of bombing and shielding factors determined as accurately as possible. Furthermore, the yields from the two explosions have been reestimated with much greater reliability than could be ascribed to the early estimates. Doses have been estimated for over 22,000 people who are believed to have received significant exposures. An additional 55,000 doses were estimated for persons in the zero to 9 rad range with a median of zero.

Question: R. J. Hickey, Institute for Environmental Studies, University of Pennsylvania, Philadelphia

Would you please comment on two topics?

(a) There are geographic regions, such as the Malabar Coast, where natural background radiation is very high. Could you provide information pertaining to health effects in such regions?

(b) In a publication by Van Cleave recently, on "Late Somatic Effects of Ionizing Radiation," there is a review discussion on what was called "negative life shortening effects of ionizing radiation." An example involves studies in the fifties by Egon Lorenz. Would you please comment on the concept and evidence regarding "negative life shortening effects?"

Reply: J. Totter

(a) Some studies have been made of the populations exposed to high background radiation in India, in France and in Brazil. In no case were there found to be effects above statistical variation which could be definitely ascribed to radiation. In the Indian studies some indications of effects were found in the group that was exposed to greater than 20 fold normal background, but the sample size is too small to give good statistics. I understand that this material will be presented at the 1971 Geneva Conference on Peaceful Uses of Atomic Energy.

(b) The "negative life shortening" effects of low doses of radiation which have been observed in several instances are thought sometimes to be seen because of great difficulty in providing an entirely suitable control group. Some investigators believe that low doses of radiation especially when continually administered may stimulate repair mechanisms which provide extra protection against "wear and tear" from other causes. So far as I am aware no one has made observations of this sort in humans.

HEALTH INTELLIGENCE FOR ENVIRONMENTAL PROTECTION: A DEMANDING CHALLENGE

J. F. FINKLEA, M. F. CRANMER, D. I. HAMMER,
L. J. McCABE, V. A. NEWILL, and C. M. SHY
ENVIRONMENTAL PROTECTION AGENCY

1. Health research goals

The health research program of the Environmental Protection Agency has three major goals. The first goal is to minimize the adverse effects of the environment upon human health by preventing exposure to harmful new environmental agents, by reducing exposures to existing pollutants and by predicting the adverse effects of proposed environmental control option. The second goal is to quantitate the health benefits of environmental controls and the third is to optimize the environment for man's health and well being. Progress is being made toward each of these goals but the pace must quicken.

A number of Federal programs, many well established and several newly proposed, serve interlocking functions in preventing exposure to harmful new agents; three new approaches are of special interest. The Toxic Substances Act, now being considered by the Congress, would fill an important gap in existing regulatory authority. Essentially, this Act would regulate synthetic organic chemicals, metallic compounds, and intermediary products from industrial processes that do not fall under existing legislation dealing with air pollution, water pollution, pesticides, radiation, noise, food, drugs and cosmetics. One such problem centers upon the polychlorinated biphenyl compounds whose value as heat mediators, plasticizers, and pesticide synergists led to widespread use prior to our understanding that such compounds are persistent environmental pollutants whose ecological and biological effects resemble those of DDT. Another example is the recent concern about the possible adverse health effects of exposure to optical brighteners which are exceedingly useful additives to plastics, textiles, paints and, most importantly, home laundry detergents. The second new approach is the National Center for Toxicological Research which is a joint effort of Food and Drug Administration of the Department of Health, Education and Welfare (DHEW) and the Environmental Protection Agency (EPA). This facility will conduct carefully controlled animal research on the carcinogenicity, mutagenicity and teratogenicity of environmental agents. Research on any one agent will involve large numbers of several species of animals and thus great expense. These research requirements led to the nickname "megamouse facility." Even this substantial effort will screen only a limited number of environmental agents. Clearly, the third approach, development of relatively simple laboratory methods to pre-screen environmental agents prior to animal toxicology studies, is an endeavor of critical importance. Such work is now underway in the National Institutes of Health. Even with progress in each of these areas our posture may still remain essentially reactive as environmental health considerations would have at best a minimal effect upon the early phases of developing technology.

Unfortunately, society lacked the wisdom to prevent exposures to many potentially harmful environmental agents and now such exposures must be reduced. Measures limiting emissions of pollutants from transportation and industrial sources, national ambient air and water quality standards, and cancellation proceedings for several persistent pesticides are current examples. Dose response studies of the health effects of pollutants are vital in executing environmental control because most regulatory legislation specifies that control actions be based upon protection of human health.

Man has seldom predicted the magnitude or subtlety of the adverse health effects of his efforts to control the environment. Failures may continue to occur but systematic prediction efforts are now mandatory. Only the existing pesticide legislation requires an assessment of the impact of proposed control actions. The adverse health consequences of control action may result in more frequent or more severe health impairments than the pollution which a control action is designed to prevent. An obvious example is the restriction of power generation during an acute air pollution episode which might cause voltage reductions and alter the functioning of patient monitoring equipment in intensive care units of hospitals. When such episodes are accompanied by hot weather, as was the case along the eastern seaboard in the summer of 1970, strict emission controls could limit electric power available for air conditioning and thus intensify the mortality peaks known to be associated with heat waves.

Environmental controls based on the need to protect human health should actually protect health and the benefits should be quantitated. Failure to achieve this goal will make it extremely difficult to evaluate the adequacy of environmental controls or the true social cost-benefit relationships. For example, restriction of selected persistent pesticides might be expected to alter tissue pesticide residue levels in biological monitors, including man. Similarly, control of heavy metal emissions into the environment might be accompanied by a reduction in the body burdens. Pollution controls designed to minimize the adverse health effects of the air environment should be accompanied by studies to quantitate improved lung function and decreased respiratory morbidity.

The environment should be optimized in the broadest ecological context for the health and well being of man. This represents an extension of the classic aims of public health that seem very easy to articulate but extremely difficult to pursue because of value conflicts in our society. Specific examples include adding essential trace substances like fluorides to municipal water supplies to help prevent dental caries and adding minerals and vitamins to foods. Sound and light levels, housing, temperature and humidity are other factors in the p hysical environment which can be optimized. An even greater challenge is the social environment. Physical inactivity, faulty diet and damaging behavioral traits including drug abuse and cigarette smoking, have been repeatedly proved important risk factors for disabling and life threatening diseases. Optimumization of the social environment is undoubtedly more difficult but in many respects inseparable from efforts to improve the physical environment.

2. The relationship between health research and environmental controls

The health intelligence needs for environmental controls must be met by the coordinated efforts of different research approaches including toxicology, clinical research and epidemiology. Coordination of Federal environmental health research programs is being fostered by the Office of Science and Technology. Major health research programs are supported by EPA, DHEW, the Atomic Energy Commission (AEC), the National Science Foundation (NSF), and the Department of Defense (DOD). Generally, the research projects of EPA focus upon the effects of pollutant exposure while those of DHEW explore the mechanisms of disease production.

Health effects research must be accompanied by adequate monitoring of environmental pollutants and appropriate covariates. This research should establish dose response relationships between pollutant exposures and adverse effects. Interactions between pollutants must be recognized and the more important ones quantitatively defined. The type and timing of obligated or contemplated environmental control actions are key factors in determining what health research and monitoring is necessary. For example, dose response studies of the acute and chronic effects of acute pollution exposure are necessary in setting standards that trigger emergency environmental control actions. Such studies often require assessment of the effects of dose rate upon some health indicator and thus may demand more intensive sophisticated and expensive environmental monitoring. Establishing a sound basis for long term standards is even more demanding. Different health indicators and a different environmental monitoring regimen are required to quantitate the relationship between long term, continuous or intermittent, low level pollutant exposures and subsequent acute or chronic adverse health effects. In the latter studies demographic and personal covariates are at times overwhelming problems. Despite these difficulties it is axiomatic that rational environmental controls can neither be established or evaluated without adequate health research and environmental monitoring programs.

3. Biological responses to pollutant exposure

Our society has not always considered the full range of biological responses when evaluating environmental pollution. A hierarchical ordering of biological responses may be helpful in relating health intelligence needs and applications.

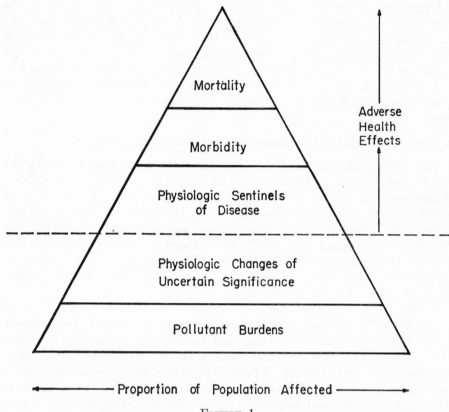

FIGURE 1

A biological response model for pollution exposure.

As shown in Figure 1, five stages of increasing severity may be considered. First, a pollutant burden not associated with any known measurable changes in function. Essentially, all of mankind is so affected. Second, a pollutant burden associated with measurable physiologic changes of uncertain significance. Third, a pollutant burden associated with physiologic changes that are sentinels of disease. Fourth, a pollutant burden associated with morbidity and fifth, one associated with mortality. Adverse health effects might then be defined as preclinical sentinels of disease, illness or death. Generally, health research techniques are adequate to study mortality and pollutant burdens; and better studies of pollutant effects upon morbidity are now underway. However, only limited progress has been made in measuring subtle changes in physiology which follow pollutant exposures and in separating those changes which are sentinels of disease from those of uncertain significance. Such research is extremely important if the present arbitrary rough safety factors are to be replaced by a quantitative appraisal of the risks actually present.

Environmental control actions based on the concept of "no health risk" and health effects defined in terms of "mortality risk" have been proposed. Such a philosophy ignores the spectrum of response and assumes some sort of effects threshold. Equally troublesome is the demand for avoidance of even minor insignificant alterations in physiology. Such "no risk" standards will in fact contain health risks either in terms of morbidity and physiologic changes or in terms of health costs of exceptionally strict environmental controls. An alternate more acceptable approach is to define a socially acceptable risk function in which health costs of pollution and pollution control are but two of many factors.

4. Relationship between selected approaches to health effects research

Laboratory and animal toxicology, clinical research and epidemiology each have unique capabilities for environmental health research; yet there are crosswalks between each approach. A health research program that does not consider each approach and appreciate the necessities of exploiting their biological crosswalks will certainly be wasteful and probably also inadequate. Health research planning has generally been opportunistic and limited to a single approach with only sporadic feeble efforts to exploit biological crosswalks.

The unique capabilities of toxicology include dose response studies where covariates can be carefully controlled, controlled studies of pollutant interactions, lifetime chronic exposure studies, toxicity evaluation prior to human population exposure, studies of sentinel animals, and controlled studies of carcinogenesis, mutagenesis and teratogenesis. Biological crosswalks to other approaches include effects of pollutants on animal models of human disease induced or aggravated by environmental pollution, definition of which physiologic changes are in fact disease sentinels, and the kinetics of absorption, distribution, metabolism and excretion of pollutants.

Clinical research usually involves intensive study of small numbers of normal or diseased human beings in a laboratory setting. Unique capabilities of clinical research include studies of absorption, distribution, metabolism and excretion after pollutant exposures and observation of the acute and chronic effects of accidental exposure to environmental agents. Biological crosswalks included controlled exposure studies limited to part of the dose response range in normal human volunteers, investigation of the benefits of pollutant removal and the effects of pollutant exposures in groups of patients afflicted with specified illnesses, and intensive study of a limited number of naturally exposed studies to identify new biological response indicators. Clinical research studies can verify and link toxicological and epidemiological findings.

Unique capabilities of the epidemiological approach include observation of the health of man in the most relevant, real life setting, studies of the effects of chronic pollution exposure on humans, ability to relate environmental distribution of pollutants and actual population exposures, and assessment of the health benefits of pollution control. Crosswalks between epidemiology and other approaches include appraisal of the disease sentinels observed or predicted from toxicological and clinical studies, assessment of the interactions between pollutant exposures and diseased populations and identification of the acute effects of high level episodic exposure to a complicated mixture of pollutants.

5. Health research program planning for the environmental protection agency

Systematic appraisal of complex environmental control problems demands an overhaul in research program planning. The envisioned approach within EPA is depicted in Figure 2. Parts of this approach are already being imple-

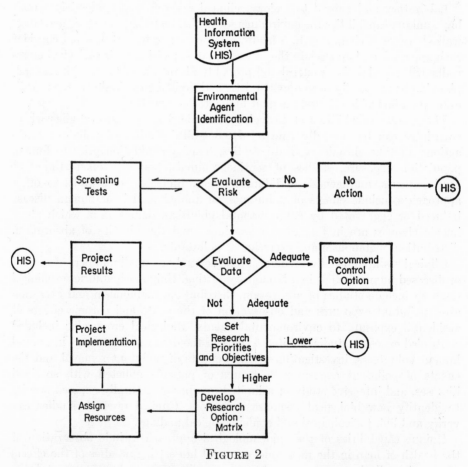

FIGURE 2

Envisioned approach to health research program planning within the Environmental Protection Agency.

mented. An environmental information system, which will include a health information component, must be structured so that potentially harmful environmental agents may be identified at the earliest possible point in time. A task group in the Office of Science and Technology is addressing this environmental information system problem. The potential health risk associated any environmental agent must undergo a preliminary evaluation. Factors involved in this evaluation include the predicted innate toxicity of the agent and assessment of the magnitude and characteristics of the predicted exposed population. Some agents will not be expected to entail any additional risk; thus no control action would be necessary and this decision would be an input for the health information system. Often, there would be so little information available that a series of rapid *in vitro* or *in vivo* screening tests would be necessary before initial evaluation.

The existing effects and exposure data would then be evaluated to ascertain whether an adequate information base existed to recommend an environmental control option. Often the information would be inadequate. In that case, the problem in question would become another member of a set of environmental problems requiring health effects research. Realistically, it is unlikely that all such problems can be a focus for detailed program planning. Decisions regarding research priorities and objects will dictate that no further action be taken on many problems. These decisions would be inputs into the environmental information system. Higher priority problems would be assigned to task force of investigators who would be charged with developing a research option matrix of possible projects. The matrix has dimensions representing research approach, biological response level, exposure level and exposure time. Available resources would be reviewed and assigned to the chosen group of projects. Results of such research would serve as inputs for the health information system and into the data base for decisions on recommended environmental controls.

The research planning mechanism just sketched provides several distinct advantages to the Agency. First, a smooth transition from the reactive posture of today to the needed predictive posture can be accomplished within the framework. Second, the plan could provide a close working relationship among field investigators now geographically dispersed. Third, the needs of the Agency will be real and relevant to the investigator rather than a distant, rather unpleasant imposition.

6. Expenditures for environmental health intelligence and environmental controls

Environmental control expenditures of future years will be based upon the health intelligence gathered by research programs of today. There can be little disagreement that the present health information base is inadequate. Health research expenditures for the units now comprising EPA, excluding environmental monitoring not directly related to health projects, amounted to just

TABLE I

EXPENDITURES FOR ENVIRONMENTAL HEALTH INTELLIGENCE
AND ENVIRONMENTAL CONTROL
Annual average fiscal year 1972–76.

EPA health research expenditures (millions)	Per capita expenditures	
	For health intelligence	For environmental controls
Fiscal year 1971 12.3	6 cents	
Fiscal year 1972 18.0	9 cents	$125

over $12,000,000 in fiscal 1971 and $18,000,000 in fiscal 1972 as seen in Table I. Equivalent per capita expenditures were six cents and nine cents. If additional expenditures by other Federal Agencies were considered, the total in fiscal 1971 might rise to 50 cents. Average yearly costs of scheduled environmental controls have been projected at about $125 per capita per year from fiscal year 1972 through fiscal year 1976. The Environmental Protection Agency and other Federal Agencies intend to work within the framework of national priorities to supplement the environmental health research effort.

7. Summary

Rational environmental control actions must be based upon an adequate pool of health intelligence. Clearly defined health research goals and their relationship to controls have been defined. Appreciation of the spectrum of biological response following pollution exposure and the interlocking nature of different research approaches is mandatory. A systematic approach to setting health research priorities should allow smooth transition from our present "reactive" posture to a "predictive" posture. The challenge demands an accelerated research program within the framework of national priorities.

Discussion

John R. Goldsmith, Environmental Epidemiology, California Department of Public Health

Possibly Professor Neyman's question on experiments concerning interaction of radiation and pollutants can be illuminated by three sets of epidemiological studies. Studies in Great Britain (Reid), and to a lesser extent in the United States, of air pollution effects indicate an interacting role of cigarette smoking and community air pollution on chronic pulmonary disease. Studies of uranium miners have shown a potentiation of pulmonary carcinogenesis of cigarette smoking by radon exposures. They have also demonstrated, incidentally, the prognostic value of sputum cytology. Finally, from Selikoff's work we have evidence of interaction of cigarette smoking and asbestos exposure on development of lung cancer.

R. J. Hickey, Institute for Environmental Studies, University of Pennsylvania, Philadelphia

Could you please clarify a point? You seem to have referred to (a) mutagenesis, (b) carcinogenesis, and (c) teratogenesis as though they were in fact separate and independent processes, though this was possibly not intended. Do you consider these as independent processes, or may they not all be considered properly as variations of mutagenic processes?

Reply: J. F. Finklea

When finally elucidated, the underlying molecular events leading to cancer, congenital abnormalities and mutations may very well be similar processes.

Emanuel Landau, Environmental Protection Agency, Washington, D.C.

The factorial design suggested by Professor Neyman need not be limited to combinations of pollutants. Socioeconomic characteristics may be significant variables inasmuch as there appears to be some evidence from Winkelstein's work that the population at the lowest socioeconomic status is most adversely affected by a given pollutant level. This seems to be observed by British experience, too.

In animal studies, stresses on the animal such as temperature and humidity changes and impairment may be significant in studying effects of specific pollutants or combinations of them.

Harold L. Rosenthal, School of Dentistry, Washington University

Although Professor Neyman's question concerning multivariate experiments is a valid one, it would seem to me that the number of variables is so great that good epidemiological and biostatistical approaches would be more succinct. It seems to me that we have already performed such experiments on this earth with the human and animal populations. This does not mean that animal experiments should not be done or that they can not yield very good information. It only means that many experiments have already been done with human populations—either inadvertently or by design.

T. Sterling, Department of Applied Mathematics and Computer Science, Washington University

Is there a programmatic effort to introduce knowledge of sophisticated statistical designs and data analysis techniques in this model of the Health Information System? (In the sense of a programmatic effort.)

Reply: J. F. Finklea

Yes, in the sense that the proposed health information system would cite the appropriate statistical literature and research involving techniques necessary for environmental health studies.

CHESS, A COMMUNITY HEALTH AND ENVIRONMENTAL SURVEILLANCE SYSTEM

WILSON B. RIGGAN, DOUGLAS I. HAMMER, JOHN F. FINKLEA,
V. HASSELBLAD, CHARLES R. SHARP, ROBERT M. BURTON,
and CARL M. SHY
ENVIRONMENTAL PROTECTION AGENCY

1. Introduction

The Community Health and Environmental Surveillance System (CHESS) relates community health to changing environmental quality. CHESS consists of a series of epidemiologic studies in sets of communities representing consistent exposure gradients to common environmental pollutants. The keystone of the CHESS program is the *coupling* of sensitive health indicators to comprehensive environmental monitoring in sets of communities representing a consistent pollutant exposure gradient, thus allowing temporal and spatial replications of dose response studies.

EPA health research needs are practical and problem oriented. CHESS research is thus pragmatic and our goals are threefold: (1) to evaluate existing environmental standards; (2) to quantitate pollutant burdens in exposed populations; and (3) to quantitate health benefits of pollutant control.

2. Chess historic development and present overview

Obligations to prepare air quality criteria documents and set air quality standards were legislated in the Clean Air Act of 1967. CHESS evolution began in the fiscal year of 1968 (FY 68) with the health appraisal of air quality standards (Figure 1). The CHESS concept developed simultaneously with the growth of a multidisciplinary "critical mass" in FY 1969. Growth for this single medium approach (air) was by initial demonstration of both health indicators and monitoring within established CHESS areas and their subsequent expansion into new areas (FY 1970–71). The recent creation of the Environmental Protection Agency (EPA) signalled a more comprehensive and, now, multimedia approach to environmental hazards. CHESS will be fully operational for air pollution effects by FY 1973 and for multimedia toxic substances by FY 1975. Present CHESS operations consist of three basic, integrated functions, namely, Data Collection, Bioenvironmental Measurements, and Information Synthesis, supported by a fourth function, research and development, and coordinated by a

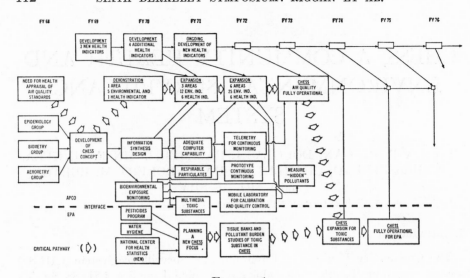

FIGURE 1

Historic development.

fifth function, Program Management (Figure 2). Simultaneous environmental monitoring and measurement of sensitive health indicators in community area sets are the fundamental CHESS components.

3. CHESS area sets

CHESS area sets consist of groups of three or four communities representing an exposure gradient for a pollutant, but similar with respect to climate and socioeconomic traits. Each community within an area set is a defined middle class residential segment of a city containing three or four elementary schools (500 to 1000 children per school) and often a secondary school. CHESS pollutant gradients are as follows:

(1) particulate gradient with low SO_2 (3 Southeast cities),
(2) SO_2 gradient with low particulates (Utah communities),
(3) combined SO_2 and particulate gradient (New York City active, Chicago planned),
(4) photochemical oxidant gradient (Los Angeles Basin),
(5) NO_x gradient (Chattanooga),
(6) trace element and SO_2 gradient (western metal smelter communities).

4. CHESS exposure monitoring

Neighborhood monitoring stations are sited to provide a representative estimate of pollutant exposure for the study population. Supplemental home monitoring of tap water, household dust and soil samples permit even more intimate

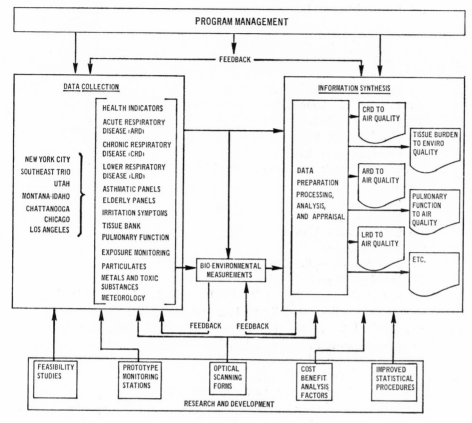

FIGURE 2

CHESS overview.

estimates of environmental trace substance exposure. Study subjects usually live within a 1 to 1.5 mile radius of CHESS stations. Topography, emission sources and local land use are all considered when placing stations. The inlet of the monitoring instruments is usually placed at head level and sheltered from uncommon proximate pollution sources. The CHESS system monitors for the following environmental exposures:

(A) Present CHESS system, all stations
  (1) Total suspended particulate (daily)
    (a) Sulfates (daily)
    (b) Nitrates (daily)
    (c) Organic (monthly)
    (d) Benzapyrene (monthly)
    (e) Trace metals (monthly)
  (2) Respirable particulate (daily)

(3) Dustfall (monthly)
    (a) trace metals (monthly)
(4) Sulfation (Pb O_2 monthly)
(5) Twenty-four hour SO_2 (daily)
(B) Present system, some stations
    (1) Two hour soiling index
    (2) Twenty-four hour NO_2
    (3) Continuous NO_2
    (4) Continuous SO_2
    (5) Continuous oxidants
(C) CHESS-CHAMP: Community Health Ambient Monitoring Program, prototype field testing
    (1) Continuous NO-NO_2
    (2) Continuous SO_2
    (3) Continuous oxidants
    (4) Hydrocarbons
    (5) Mobile unit replication
    (6) Wind speed and direction.

CHESS-CHAMP (The Community Health and Ambient Monitoring Program) is currently collecting daily 24 hour samples and monthly samples for gases and particulates at 30 environmental monitoring stations. Real time pollutant measurements can accurately relate short term environmental variations to acute response health indicators, distinguishing "peak" exposure effects from 24 hour average effects, if any.

Continuous monitors operate in some CHESS-CHAMP stations and a prototype automatic data acquisition continuous monitoring station with magnetic tape storage and "on call" telemetric output is presently being field tested. "On-call" telemetry permits routine instrument performance checks, daily data processing and thus immediate access to data during air pollution episodes. Duplicate sampling of the environment and frequent calibration of all instruments are systematically obtained to ensure accurate and consistent instrument performance in the CHESS-CHAMP system.

5. CHESS health indicators

Relationships between human diseases and pollution exposures are neither simple nor fully understood. However, one may conveniently think of a five stage biologic response spectrum of increasing severity (Figure 1 in Finklea and co-workers): (1) a tissue pollutant burden unassociated with other biological changes; (2) physiologic changes of uncertain significance; (3) physiologic disease sentinels; (4) morbidity; and (5) mortality. CHESS utilizes health indicators which reflect this entire spectrum. The following acute and chronic response indicators are measured in pre-enrolled subject panels as well as community surveys:

(A) *Acute exposure* (less than 24 hours)

    (1) Reversible pulmonary function changes

    (2) Acute irritation symptoms

    (3) Frequency and severity of asthma attacks

    (4) Aggravation of chronic respiratory disease (CRD) symptoms

    (5) Aggravation of cardiac symptoms

    (6) Daily mortality rates

(B) *Chronic exposure* (greater than 24 hours)

    (1) Pollutant burdens (man as an environmental dose integrator)

    (2) Impairment of lung function

    (3) Absenteeism (no longer used)

    (4) Prevalence of chronic respiratory disease (CRD)

    (5) Frequency of lower respiratory disease (LRD)

    (6) Incidence of acute respiratory disease (ARD)

    (7) Mortality studies

Comparison of similar groups is insured by obtaining covariate information such as age, sex, race and smoking status. These study design covariates all relate to morbidity and failure to measure and adjust for them could cause serious confounding effects. They are summarized as follows: (A) demographic—age, sex, ethnic group, socioeconomic, reporting bias; (B) exposure—diet, water, smoking, occupation, migration, indoor-outdoor differences, daily movement; (C) special risk—temporary such as age, pregnancy, illness; permanent such as alpha-1-antitrypsin deficiency or serum IgE levels.

6. CHESS study strategies

Selection of CHESS area sets and pollutant exposure gradients were dictated by the existence of air quality criteria documents for particulate matter, sulfur oxides, nitrogen oxides, photochemical oxidants, hydrocarbons and carbon monoxide published by the National Air Pollution Control Administration. Area sets for individual pollutants are selected from existing exposure monitoring data. A carbon monoxide (CO) CHESS set was not established because short term CO effects are more precisely studied in controlled exposure chambers and long term CO effects are likely to be confounded with the effects of other vehicle emissions products. Nor has an area set with a consistent CO gradient been found.

Middle class neighborhoods are chosen because they are common, have a more homogeneous family and social class distribution, and are migrationally stable, thus providing a higher likelihood of long term participation. Family participants in the surveys for acute upper, acute lower and chronic respiratory diseases and panels for episodes are recruited from elementary school enrollment listings in CHESS neighborhoods. Subjects for the asthma, cardiac and chronic respiratory disease panels are obtained from prevalence survey results and from patient listings of private physicians. As indicated, our broad data acquisition techniques vary in the frequency and in the type of response they measure. The following methods are currently used: (1) exposure monitoring, (2) single time questionnaire, (3) weekly diaries, (4) biweekly telephone contact, (5) spirometry in schools, (6) telephone contact during alerts, (7) tissue collection, (8) vital statistics.

CHESS programs will operate from three to five years in selected areas. Measurement of sensitive health indicators over a period of increased air pollution control is an optimal way to quantitate the health benefits of this control.

CHESS data collection for FY 71 alone will yield a total of 40×10^6 health indicator and 3×10^5 air determination characters for data processing (Table I).

TABLE I

FY 1971 CHESS PROGRAM: DATA COLLECTION SUMMARY

Indicator	Frequency	Population
CRD	biyearly	30,000
LRD	biyearly	30,000
ARD	biweekly	15,000
Pulmonary Function	triyearly	5,000
Asthma	weekly	300
Elderly	weekly	450
Irritation Symptoms	triyearly	12,000
Pollutant Burdens	biyearly	6,000

Rapid reporting is the rule because high priorities are placed on our study results. Recent CHESS findings span the entire biologic response spectrum and are outlined and referenced in Appendices I and II.

Research goals are essential for optimal CHESS functioning and play a critical role in our development. CHESS research and development goals are threefold: (1) to refine exposure monitoring, (2) to improve statistical procedures and (3) to develope and test more sensitive health indicators. Current and future CHESS health-indicator research is outlined in Appendix III.

Estimating environmental exposure doses has always been a problem. In health studies of multimedia toxic substances, this problem increases. Pollutant burden studies of biological accumulators such as pets, plants, and wildlife in addition to humans should be utilized for appropriate metals, pesticides, synthetic organic materials and selected gaseous pollutants. Sample sets of tap water, house dust and soil collected from CHESS panel families provide intimate information about trace metal exposure when coupled to neighborhood environmental monitoring and dietary metal surveys. Personal monitors for all pollutants would permit the best pollutant dose estimates for individual study subjects.

We have addressed our remarks to the central questions of this conference, namely what pollutants to measure, what health indices to measure, available methods of obtaining both types of data, and available study strategies. CHESS permits a systematic, yet flexible, approach to these problems and has already produced answers to some of them.

APPENDIX

A.1. CHESS: recent findings

(A) Pollutant burdens
    (1) Roadside gradients of Cd, Pb and Zn [6]
    (2) As, Cd, Cu, Pb, Ag and Zn in household dust [36]
    (3) Environmental exposure to As, Cd and Pb reflected in hair and consistent over time and within individuals [19], [20]
    (4) Hg in placentas, Cd in cord blood [35]
    (5) PCB highest in urban whites [15]
    (6) Interlaboratory and interwash variation negligible for hair Cd, Pb and Zn determinations [21]

(B) Physiological changes of uncertain significance
    (1) Hair and blood Pb correlate (.40) over an exposure gradient [19]
    (2) Urinary Cd does not increase with age [18]
    (3) Eye irritation highly correlated with oxidant exposure [22]

(C) Physiologic sentinels of disease
    (1) Pulmonary function in children decreased after SO_x, particulate and NO_x exposure [30], [32], [33]
    (2) Pulmonary function in adults decreased after NO_x exposure [35]
    (3) Systolic BP in adults 40 may be increased after Cd exposure, but not diastolic BP or cholesterol [11]

(D) Morbidity
    (1) ARD and LRD in children after exposure in NO_x [25], [32]
    (2) CRD symptoms in young adults but not adolescents more frequent after exposure to SO_x and particulates but not O_x [35]
    (3) Respiratory and eye irritation symptoms induced by acute urban air pollution exposure [4]
    (4) Asthma attacks more frequent after nitrate, SO_x, and particulate exposure [3]
    (5) Cd not increased in toxemia of pregnancy [7]
    (6) No observed effect of chronic oxidant exposure on epidemic influenza in school children [26]
    (7) Epidemiologic evidence linking Cd to hypertension is weak when critically reviewed [18]

(E) Mortality
    (1) No long term effect of As, Pb on survival of "Neal Cohort" [24]
    (2) No long term effect of acute NO_x exposure on survival [17]
    (3) No effect of water hardness and no consistent effect of Cd on cardio-vascular disease mortality [27]
    (4) Possible relationship between chronic urban air pollution exposure and carefully adjusted mortality rates in Chicago and Philadelphia [27]
    (5) Large temperature, influenza and socioeconomic effects on daily mortality [2]
(F) Associations with cigarette smoking
    (1) Increases in ARD, LRD frequency [14]
    (2) Higher influenza attack rates [16]
    (3) Impaired persistence of HI antibody [13]
    (4) Decreased ventilatory function
    (5) CRD symptoms in early adolescence
    (6) Refractiveness to acute air pollution episodes [4]
    (7) No change in ARD, influenza, or antibody persistence among children if parents smoke [10]
(G) Some recent reviews
    (1) General overview of CHESS research [29]
    (2) Overview of human pollutant burden research [12]
    (3) Air pollution episodes—guide for health departments and physicians [5]
    (4) Review of arsenic health effects [1]
    (5) Review of beryllium health effects [28]
    (6) Review of cadmium health effects [18]
    (7) Reviews of environmental lead and human health [8], [31]
    (8) Plasticizers in the environment [23]
    (9) Environmental hazards of optical brighteners [9]

A.2. CHESS: research and development goals

(A) Refined exposure monitoring
    (1) Biological amplifiers (pets, plants, wildlife)
    (2) Personal monitors
    (3) Tap water, housedust, soil
(B) Improved statistical procedures
    (1) Hockey-stick and other dose response functions
    (2) Ridit transformation and linear models for categorical data
    (3) Daily mortality models
    (4) Analysis of truncated and censored data
    (5) Estimating personal exposure

    (6) Multivariate techniques for repeated measurements

    (7) Health information synthesis system

(C) More sensitive health indicators

    (1) Pollutant burdens

        (a) Maternal-fetal tissue sets

        (b) Patients-biopsy, surgery, autopsy

        (c) Special occupations

    (2) Altered physiology of uncertain significance

        (a) Carboxyhemoglobin

        (b) RBC fragility and survival

    (3) Physiologic heralds of disease

        (a) Other PF tests

        (b) Blood lipid patterns

        (c) Blood pressure

        (d) Immune response

        (e) Exfolliative cytology

    (4) Morbidity

        (a) Aggravation of hypertension

        (b) Aggravation of RDS of newborn

    (5) Mortality

        (a) CO and coronary disease

        (b) Area studies linked to SS records

REFERENCES

[1] R. W. Buechley, "The paths of arsenic pollution," CRB In-House Technical Report, September, 1970.

[2] R. W. Buechley, L. E. Truppi, and J. Van Bruggen, "Heat island—death island," CRB In-House Technical Report, August, 1971.

[3] A. A. Cohen, S. M. Bromberg, R. W. Buechley, L. T. Heiderscheit, and C. M. Shy, "Asthma and air pollution from a coal fired power plant," *Amer. J. Public Health*, in press.

[4] A. A. Cohen, C. J. Nelson, S. M. Bromberg, M. Pravda, and E. F. Ferrand, "Symptom reporting during recent publicized and unpublicized air pollution episodes," abstract, 99th Annual APHA Meeting, Minneapolis, October, 1971.

[5] A. A. Cohen, C. M. Shy, F. B. Benson, W. B. Riggan, V. A. Newill, and J. F. Finklea, "Air pollution episodes—a guide for health departments and physicians," *HSMHA Health Reports*, Vol. 86 (1971), pp. 537–550.

[6] J. P. Creason, O. McNulty, L. T. Heiderscheit, D. H. Swanson, and R. W. Buechley, "Roadside gradients in atmospheric concentrations of cadmium, lead and zinc," *Proceedings of the Fifth Annual Conference on Trace Substances in Environmental Health*, in press.

[7] J. P. Creason, J. F. Finklea, and D. I. Hammer, "Relationship of cadmium to toxemia of pregnancy," In-house Technical Report, June, 1970.

[8] R. E. Engel, D. I. Hammer, R. J. M. Horton, N. M. Lane, and L. A. Plumlee, "Environmental lead and public health," EPA, Air Pollution Control Office Publication No. AP-90, March, 1971.

[9] J. F. FINKLEA and K. BRIDBORD, "Environmental hazards of optical brighteners," CRB In-house Technical Report, September, 1971.

[10] J. F. FINKLEA, J. P. CREASON, D. I. HAMMER, S. M. BROMBERG, and W. B. RIGGAN, "Does cigarette smoking by parents alter the ARD immune response of their children," Clinical Research, Vol. 19 (1971), p. 458.

[11] J. F. FINKLEA, J. P. CREASON, S. H. SANDIFER, J. E. KEIL, L. E. PRIESTER, D. I. HAMMER, and W. B. RIGGAN, "Cadmium exposure, blood pressure and cholesterol," abstract, Clinical Res., Vol. 19 (1971), p. 313.

[12] J. F. FINKLEA, D. I. HAMMER, T. A. HINNERS, and C. PINKERTON, "Human pollutant burdens," American Chemical Society Symposium on the Determination of Air Quality, April, 1971, in press.

[13] J. F. FINKLEA, V. HASSELBLAD, W. B. RIGGAN, W. C. NELSON, D. I. HAMMER, and V. A. NEWILL, "Cigarette smoking and HI response to influenza after natural disease and immunization," Amer. Rev. Resp. Dis., in press.

[14] J. F. FINKLEA, V. HASSELBLAD, S. H. SANDIFER, D. I. HAMMER, and G. R. LOWRIMORE, "Cigarette smoking and acute non-influenza respiratory disease in military cadets," Amer. J. Epid., Vol. 93 (1971), pp. 457–462.

[15] J. F. FINKLEA, L. E. PRIESTER, J. P. CREASON, T. HAUSER, and T. HINNERS, "Polychlorinated biphenyl residues in human plasma expose a major urban pollution problem," abstract, 99th Annual APHA Meeting, Minneapolis, October, 1971.

[16] J. F. FINKLEA, S. H. SANDIFER, and D. D. SMITH, "Cigarette smoking and epidemic influenza," Amer. J. Epid., Vol. 90 (1969), pp. 390–399.

[17] K. L. GREGORY, V. F. MALINOSKI, and C. R. SHARP, "Cleveland clinic fire survivorship study, 1929–1965," Arch. Environ. Health, Vol. 18 (1969), pp. 508–515.

[18] D. I. HAMMER, J. F. FINKLEA, J. P. CREASON, S. H. SANDIFER, J. E. KEIL, L. E. PRIESTER, and J. F. STARA, "Cadmium exposure and human health effects: Some epidemiologic considerations," Proceedings of the Fifth Annual Conference on Trace Substances in Environmental Health, in press.

[19] D. I. HAMMER, J. F. FINKLEA, R. M. HENDRICKS, T. A. HINNERS, W. B. RIGGAN, and C. M. SHY, "Trace metals in human hair as a simple epidemiologic monitor of environmental exposure," Proceedings of the Fifth Annual Conference on Trace Substances in Environmental Health, in press.

[20] D. I. HAMMER, J. F. FINKLEA, R. M. HENDRICKS, C. M. SHY, and R. J. M. HORTON, "Hair trace metal levels and environmental exposure," Amer. J. Epid., Vol. 93 (1971), pp. 84–92.

[21] D. I. HAMMER, K. NISHIYAMA, M. PISCATOR, R. G. HENDRICKS, J. P. CREASON, and T. HINNERS, "Cadmium, lead and zinc in hair—effects of environmental exposure, wash techniques and laboratory error," abstract, 99th Annual APHA Meeting, Minneapolis, October, 1971.

[22] D. I. HAMMER, B. PORTNOY, P. F. WEHRLE, V. HASSELBLAD, C. R. SHARP, and R. J. M. HORTON, "A prospective dose-response study of eye discomfort and photochemical oxidants," abstract, 99th Annual APHA Meeting, Minneapolis, October, 1971.

[23] T. R. HAUSER, "Plasticizers in the environment," CRB In-house Technical Report, April, 1971.

[24] W. C. NELSON, M. H. LYKINS, V. A. NEWILL, J. F. FINKLEA, and D. I. HAMMER, "Mortality among orchard workers exposed to lead arsenate spray: A cohort study," DHER In-house Technical Report, 1970, to be published.

[25] M. E. PEARLMAN, J. F. FINKLEA, J. P. CREASON, C. M. SHY, M. M. YOUNG, and R. J. M. HORTON, "Nitrogen dioxide and lower respiratory illness," Pediatrics, Vol. 47 (1971), pp. 391–398.

[26] M. E. PEARLMAN, J. F. FINKLEA, C. M. SHY, J. VAN BRUGGEN, and V. A. NEWILL, "Chronic oxidant exposure and epidemic influenza," Environ. Res., Vol. 4 (1971), pp. 129–140.

[27] C. PINKERTON, J. P. CREASON, C. M. SHY, D. I. HAMMER, R. W. BUECHLEY, and G. K. MURTHY, "Cadmium content of milk and cardiovascular disease mortality," *Proceedings of the 5th Annual Conference on Trace Substances in Environmental Health,* in press.

[28] C. R. SHARP, "Beryllium—A hazardous air pollutant," CRB In-house Technical Report, June, 1971.

[29] C. M. SHY, J. F. FINKLEA, D. C. CALAFIORE, F. B. BENSON, W. C. NELSON, and V. A. NEWILL, "A program of community health and environmental surveillance (CHESS)," American Chemical Society Symposium on the Determination of Air Quality, April, 1971, in press.

[30] C. M. SHY, J. P. CREASON, M. E. PEARLMAN, K. E. MCCLAIN, F. B. BENSON, and M. M. YOUNG, "The Chattanooga school children study 1: Methods, description of pollution exposure and results of ventilatory function testing," *J. Air Poll. Control Assoc.,* Vol. 20 (1970), pp. 539–545.

[31] C. M. SHY, D. I. HAMMER, H. E. GOLDBERG, V. A. NEWILL, and W. C. NELSON, "Health hazards of environmental lead," CRB In-house Technical Report, March, 1971, to be published.

[32] C. M. SHY, V. HASSELBLAD, R. M. BURTON, A. A. COHEN, and M. PRAVDA, "Is air pollution in New York City associated with decreased ventilatory function in children?" abstract, 99th Annual APHA Meeting, Minneapolis, October, 1971.

[33] C. M. SHY, C. J. NELSON, F. B. BENSON, W. B. RIGGAN, and V. A. NEWILL, "The Cincinnati school children study: Effect of atmospheric particulates and sulfur dioxide on ventilatory performance in children," *Amer. J. Epid.,* in press.

[34] W. B. RIGGAN, R. W. BUECHLEY, J. B. VAN BRUGGEN, C. R. SHARP, L. TRUPPI, W. C. NELSON, and V. A. NEWILL, "Daily mortality predictor models: A tool for environmental assessment and pollution control," abstract, 99th Annual APHA Meeting, Minneapolis, October, 1971.

[35] Division of Health Effects Research, unpublished data.

[36] D. I. HAMMER, J. F. FINKLEA, K. BRIDBORD, C. PINKERTON, H. A. HINNERS, and J. P. CREASON, "Household dust as an index of environmental trace substance exposure I," Preliminary Report, presented at the meeting of the Subcommittee on the Toxicology, International Conference of the Permanent Commission and International Association on Occupational Health, Slanchev Bryag, September 20–24, 1971, Bulgaria, to be published.

Discussion

Question: John R. Goldsmith: Environmental Epidemiology, California Department of Public Health

Does the use of the word "after" in statement of "physiological sentinels of disease" of Cadmium on B. P. and in morbidity of asthma and nitrate, SO_x, and particulate exposure, imply a pre-exposure and post-exposure set of observations?

Reply: W. Riggan

No. It generally implies a set of simultaneous observations over an exposure gradient.

Question: B. G. Greenberg, School of Public Health, University of North Carolina, Chapel Hill

I should like to ask the speaker if the surveillance system is looking at early precursors of disease in the social and psychological areas. All of the named measures are physiological in nature, whereas certain other characteristics might be even earlier in occurrence without any demonstrable physical change. Specifically, such measures are irritability, ability to function at optimal levels in remembering digit sequences, adding numbers, and other mentation tasks might be used as response indicators. Have you given thought to including such measures among the response variables?

Reply: W. Riggan

Yes. Both field and laboratory feasibility studies are in progress.

Question: R. J. Hickey: Institute for Environmental Studies, University of Pennsylvania, Philadelphia

I would like to inquire about the "pollutant" referred to as "oxidant." This, as I understand it, is a sort of conglomerate of chemicals which reacts with a KI reagent, and includes ozone, other peroxides, and nitrogen dioxide. There have been reports suggesting that ozone is, or may be, mutagenic. Further, populations are genetically heterogeneous, of course, and in human populations a metabolic defect occurs in some individuals pertaining to the enzyme, catalase. Both acatalasia and hypocatalasia are, I believe, known. Thus there can be, or are, individuals who are particularly sensitive to hydrogen peroxide because of deficiency of catalase. But, so far as I know, NO_2 is extraneous to this system. My question, therefore, has two parts: (a) for a molecular biological approach to statistical studies, can a measure of peroxide be reported free of NO_2, and (b) is there any way that past "oxidant" data may be modified by removal of the NO_2 effect so that only some peroxide measure may be obtained?

Reply: W. Riggan

Your question regarding a biological response to peroxide in oxidant atmospheres is well taken. The measurement of oxidants in the past has been by the use of the KI reagent where a number of compounds may react to both a positive or a negative response. Advanced methodology detects specific ozone concentration, free of the other oxidant compounds. If specific methods were available for monitoring these other compounds routinely during the course of a population health study, it may be possible to deduce the concentration of peroxides. However, these methods are not available at present for routine air monitoring.

It would be very difficult to adjust past oxidant data for NO_2 proportions with any degree of accuracy since NO_2 concentrations vary from both a spatial and temporal standpoint. Any data adjustment would be academic and would not be consistent to the degree of precision and accuracy required for the scientific determination of health effects.

Question: Burton E. Vaughan, Ecosystems Department, Battelle Memorial Institute, Richland, Wn.

In your discussion of a surveillance system, I am troubled by the lack of explicit mention of any data base to be obtained on those organisms needed to assess environmental degradation. Surveillance concerns not just disease. Certain environmental changes have real and expensive consequence to human well being *indirectly* (for example, effects on crops, effects on forests, and so on).

Reply: W. Riggan

Our primary responsibility is health effects research, but we coordinate our efforts with the Division of Ecological Research as well as other agencies whenever feasible.

Question: Colin White, Department of Public Health, Yale School of Medicine

Have you had any problem of over reporting when you used a telephone survey during a pollution alert?

Reply: W. Riggan

We found no over reporting from using a telephone survey during an air pollution alert in New York conurbation. Three areas surveyed were alike in ethnic and socioeconomic characteristics but differed markedly in exposure to air pollution. The first call followed a three day rise in air pollution levels accompanied by the formal issuance of an air pollution forecast. The second call followed a three day period of elevated pollution unaccompanied by publicity. The final call followed three days of low pollution. A series of bias control questions were asked in each call. The response rate of acute symptoms was not affected by publicity, but the response rate was very susceptive to air pollution levels. The response rates to the control health questions were similar among all areas for all three telephone surveys.

STATISTICAL ASPECTS OF A COMMUNITY HEALTH AND ENVIRONMENTAL SURVEILLANCE SYSTEM

WILLIAM C. NELSON, VICTOR HASSELBLAD,
and GENE R. LOWRIMORE
ENVIRONMENTAL PROTECTION AGENCY

1. Introduction

The Community Health and Environmental Surveillance System (CHESS), a program conducted by the Division of Health Effects Research, Office of Research and Monitoring, Environmental Protection Agency, has been described in some detail by Riggan and co-workers [4] and Shy and co-workers [5]. Briefly, CHESS is a continuing series of epidemiologic studies carried out in selected communities representing an exposure gradient for the most common air pollutants. The basic purpose is to relate community health to changing environmental quality. The program involves monitoring of various pollutants and simultaneous surveillance of health indicators known to be sensitive to variations in environmental quality. The CHESS program will be useful in quantitating pollutant burdens, evaluating environmental standards, and documenting the health benefits of pollution control.

Area sets, sensitive health indicators, and environmental monitoring are the three key elements of CHESS. An area set consists of a group of communities selected as representative of a pollution gradient and similar to each other with respect to climate and socioeconomic traits. Each community in an area set is a middle class residential neighborhood.

The health indicators used reflect a broad spectrum of human responses, including no demonstrable effect, increase in body burden, physiologic changes of uncertain significance, physiologic sentinels of disease, acute and chronic morbidity, and death. Indicators used currently include symptoms of chronic respiratory disease in adults, incidence of acute respiratory disease in families, pulmonary function testing and lower respiratory illness of elementary school children, daily symptom reporting of asthmatics and elderly patients with chronic heart or lung disease, and tissue concentrations of selected trace elements.

A pollution monitoring station is established within each study neighborhood. Such factors as topography, emission sources, and land use are considered in monitoring site selection to ensure that the measurements are representative of

125

population exposure. Pollutants are usually monitored on a 24 hour basis, but some continuous instruments are now being used to provide data on short term peaks.

The CHESS program is interdisciplinary and involves clinicians, epidemiologists, statisticians, engineers, chemists, meteorologists, sociologists, and programmers. However, the administration of the studies can be separated into various phases such as field data collection, bioenvironmental laboratory analysis, data preparation, data processing, statistical analysis, and technical report preparation. We have defined the latter four stages as Information Synthesis. Certain phases of Information Synthesis will now be discussed in detail.

2. Data processing aspects

Most health indicator data are collected by questionnaire. Study questions are administered biannually (lower respiratory illness, chronic respiratory illness, pollutant burdens), bimonthly (pulmonary function testing), biweekly (acute respiratory disease), or daily (asthmatic and elderly panels).

The sample sizes for these studies necessarily are large because of the many covariables present, such as age, sex, race, cigarette smoking, and economic status. For the biannual studies, approximately 2,000 families are surveyed in each of the 30 sectors participating in CHESS. About 300 families per sector are enrolled in the biweekly surveys. Each panel study requires approximately 100 individuals per sector. The biannual survey is given first and provides family background information on covariables and chronic disease conditions; this information is utilized in the selection of candidates for the additional studies.

The computer is an essential part of the data processing operation. Not only does it perform the standard large storage and retrieval functions and statistical analyses, but it also aids extensively in the study logistics. The computer selects candidates in priority order for the repetitive studies (acute respiratory disease, asthma and elderly panels). It ensures that study groups are similar as to proximity to environmental monitoring site, socioeconomic status, age, family size, and length of residence.

Much clerical time is saved by using the computer to pre-print necessary identifying information on the questionnaire forms. To lessen an enormous amount of coding and keypunching, optical mark questionnaire forms are now used. The information on these forms can be read directly onto magnetic tape. Even so, a large amount of editing and correcting is still necessary. For each questionnaire, various edit programs are performed to identify errors. The computer also can be used to prepare mailing labels for the questionnaire.

As the program scope expands to include wider use of continuous pollutant monitors and more sensitive health indicators such as electrocardiograms, the importance of the computer system increases still further. Features such as online telemetry and analog-digital conversion will be necessary.

3. Statistical aspects

The appropriate statistical analysis of the CHESS data requires considerable effort. The large populations and the community setting impose severe restrictions on the experimental design. Standard analyses are usually not possible because of repeated measurements, serial correlation, missing observations, or discreteness of the variables.

A pulmonary function study was done on second grade children in eleven selected Cincinnati schools to estimate, separately, socioeconomic and air pollution effects. Cigarette smoking and age variation were designed out of the study. Pollution monitoring stations were set up at each school. Pulmonary function tests, as measured by three-quarter second forced expiratory volume (FEV$_{0.75}$), were performed weekly in November, February, and May. Table I

TABLE I

FEV$_{0.75}$ MEANS AND VARIANCES
FOR 198 SECOND GRADE CHILDREN
IN THE CINCINNATI SCHOOL STUDY

Month	Mean	Variance
Nov.	1.1968	0.04355
Feb.	1.1908	0.04705
May	1.2342	0.05176

shows the mean forced expiratory volume and variance for each month averaged over all schools. Many more children participated than the 198 used for our analyses. However, this group met several selective criteria, including at least three valid readings each month, absence of asthma or acute bronchitis, and socioeconomic status (SES) (determined from personal family interviews) consistent with neighborhood SES.

The simple correlation matrix (Table II) illustrates the consistency of the FEV readings over the three months and the intercorrelation of the covariables

TABLE II

CORRELATION MATRIX OF PULMONARY FUNCTION READINGS
FOR THE CINCINNATI SCHOOL STUDY

	Nov.	Feb.	May	SES	Sex	SO$_x$	Height
Nov. FEV	1.000	0.885	0.857	0.072	0.229	−0.222	0.685
Feb. FEV		1.000	0.900	0.109	0.327	−0.366	0.688
May FEV			1.000	0.192	0.285	−0.338	0.660
SES				1.000	−0.058	−0.168	0.177
Sex					1.000	−0.043	0.219
SO$_x$						1.000	−0.282
Height							1.000

and the particulate sulfate (SO_x) concentrations. Height was used to adjust the FEV for individual body build differences.

A multivariate analysis of variance was used to allow for the repeated measurements and adjust for the covariables. Table III shows the FEV covariances

TABLE III

COVARIANCE AND CORRELATION MATRICES
FOR THE CINCINNATI SCHOOL STUDY

Correlations given below the diagonal.

	Nov.	Feb.	May
Nov.	0.02223	0.01736	0.01818
Feb.	0.7912	0.02165	0.01932
May	0.7488	0.8064	0.02651

(upper triangular), variances (diagonal), and correlations (lower triangular) after adjusting for SES, sex, and height. The variances are approximately half of the unadjusted values. The correlations are reduced only slightly by the adjustment. The p-values associated with the linear hypothesis analysis (Table IV) show each factor to be statistically significant.

TABLE IV

RESULTS FROM LINEAR MODEL ANALYSIS
OF THE CINCINNATI SCHOOL STUDY

Factor	D.F.	U Statistic	p-value
Econ	3,1,193	0.9433	0.0091
Sex	3,1,193	0.9305	0.0025
SO_x	3,1,193	0.8712	<0.0001
Ht	3,1,193	0.5648	<0.0001

The dependent variable in the Cincinnati school study was continuous. In many studies, however, the health indicator is categorical. A useful method for the linear model analysis of discrete data has been developed by Grizzle and co-workers [2]. This method is difficult to apply, however, when the independent variable is continuous, that is, in the mixed model case. One solution is to group the independent variable into categories, which unfortunately, rapidly increases the number of response vectors. For example, a study involving five cities, five cigarette smoking categories, and three age groups produces 75 cells. The analysis requires dealing with several 75 × 75 matrices, which exceeds the memory capacity of many computers. Furthermore, even with a large sample size, some vectors occur with zero variance.

A technique capable of handling these study designs was desired, even at the cost of some sophistication or power. Since the dependent variable is always ordered, the ridit transformation, which involves transforming to the cumulative frequency distribution, was used. While the original use of the ridit by Bross [1] involved a standard or control population, we use the total sample for all areas as our standard. The technique leads to ordinary ANOVA and generalized linear hypothesis analysis. The two-sample ridit tests leading to the standard t test can be shown to be equivalent to the rank sum test.

TABLE V

COMPARISON OF 1000 SIMULATED RIDIT t's WITH STUDENT t's
FOR VARIOUS SAMPLE SIZES

Simulated from multinomial with $p_1 = 0.4$, $p_2 = 0.2$, $p_3 = 0.2$, $p_4 = 0.1$, $p_5 = 0.1$.

Sample sizes		Observed ridit t's (Expected = 50)		Chi-square for goodness of fit (19 degrees of freedom)	p-value
n_1	n_2	$<t_{n_1+n_2-2,\,.05}$	$>t_{n_1+n_2-2,\,.95}$		
5	5	57	53	151.36	.0000
5	10	42	44	15.88	.6653
5	50	38	64	29.88	.0533
10	10	60	55	13.56	.8087
50	50	54	58	18.56	.4854
200	500	45	51	11.16	.9183

Table V shows the result of some empirical sampling. We wanted to see how well the α-levels are preserved if the ridit transformation is used. In particular, we were interested in knowing how the test performs for small sample sizes. A multinomial distribution was simulated with five categories: $p_1 = 0.4$, $p_2 = 0.2$, $p_3 = 0.2$, $p_4 = 0.1$, and $p_5 = 0.1$. Two samples, of size n_1 and n_2 were drawn and tested for differences using the ridit t test. Only the two tails are shown, although 20 intervals, each with expected value of 50, were used. The goodness of fit statistic (19 degrees of freedom) is shown. The results show excellent agreement with the Student t for n_1, $n_2 > 10$. Even for the extreme case of $n_1 = n_2 = 5$, there is no evidence of distortion of the tail probabilities, even though the overall fit is poor.

Further empirical sampling showed ridit procedures using ANOVA and F tests quite comparable to LINCAT programs [2] using chi square analyses. The reason for using ridits is not to replace the rank sum or LINCAT, but there may be times when capacity of programs may be exceeded by those techniques. The ridit is a simple transformation with broad applications. Robustness of the analysis of variance has been reaffirmed even with discrete or dichotomous dependent variables.

Table VI illustrates an ANOVA table developed after a ridit transformation on the dependent variable (chronic respiratory disease prevalence as measured by cough, phlegm, and shortness of breath symptom reporting). The data were collected on fathers of elementary school children in five cities in Montana and Idaho with varying levels of trace metals exposure from mining. The

TABLE VI

ANALYSIS OF SEVERITY OF RESPIRATORY SYMPTOMS IN THE MONTANA-IDAHO STUDY

Factor	D.F.	S.S.	%S.S.	F	p-value
Cities	4	0.3038	0.43	1.64	0.1604
Gradient	1	0.1500	0.21	3.24	0.0719
Age	1	0.1073	0.15	2.32	0.1277
Smoking	1	12.2154	17.15	264.16	$<10^{-6}$
Education	1	0.0060	0.01	0.13	0.7184
Error	1224	56.6004			

symptom responses were placed on a severity scale of one to seven with the most common response, "no symptoms," being one. The purpose of the study was to compare differences in severity of respiratory symptoms with the differences in trace element exposure. In this case, the city differences in severity of disease symptoms were in a direction consistent with the exposure gradient, but were not statistically significant. Cigarette smoking, as expected, was overwhelmingly significant.

In setting pollution control standards, the existence of a threshold concentration for effects is implicitly assumed. The ideal data then are dose response relationships which show no effect until some nonzero threshold concentration is achieved. Most statistical analyses assume a strictly monotonic function as the relationship between two variables, and, therefore, that a significant relationship between health and pollutant exposure occurs at all levels of the pollutant.

As a simple alternative, we hypothesized a segmented line with zero slope up to a point x_0, and with positive slope above x_0. The point x_0 is found by the least squares method. This function has been named a "hockey stick" function. The technique for obtaining the least squares solution to a more general problem is due to Quandt [3].

The hockey stick function was fitted to data on high school cross country runners at a Southern California high school. The per cent of team members with decreased performance (increased time) from their last race is plotted against oxidant value one hour before the race (Figure 1). Fortunately, there was a monitoring station near the school course. Only home meet times were used. For this particular problem, the point estimate of the threshold was 12.2 pphm, with a 95 per cent confidence interval from 6.8 to 16.3 pphm.

We have briefly mentioned the statistical topics of cumulative frequency distribution transformations, mixed model categorical data problems, and fitting the hockey stick function. Another topic that presents problems for us is truncated or censored data. For example, in measuring body burdens for comparison of group means, many persons may have levels below minimum detectable limits. Thus, we may be faced with situations where as many as 50 per cent of the observations are missing. Since we know the number of missing observations, we technically have a censoring problem. We have used the cumulative frequency

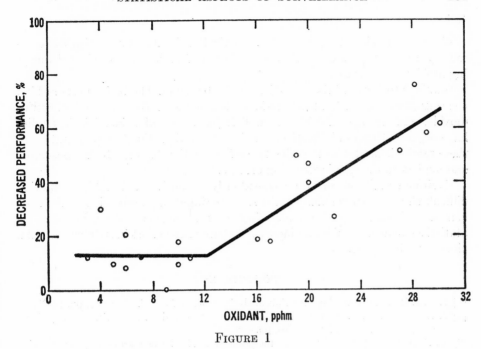

FIGURE 1

Per cent decreased performance *versus* oxidant level, with fitted "hockey stick" function.

transformation to handle this problem, but feel this area needs further development.

Another area requiring more statistical work is the problem of intercorrelated multivariate time series. Our panel studies illustrate the problem. These observations are made on groups of individuals through time. Because both our health indicators and pollutant exposures show seasonal variations, our results can be confounded. We make use of seasonal adjustments or season specific rates. Also we are experimenting with some types of probit analysis, but welcome additional results or suggestions.

Measuring the pollutant exposure of an individual is one of our most difficult problems. At best, we can measure pollution at an elementary school and assume that all children going to that school are exposed to that pollutant level. In some cases, we have to estimate a person's pollutant exposure from his address by using a few pollution values from several miles away. In other cases, we may not know how long he has lived at the address, where he works, whether he smokes, or a number of other important factors. Since humans do not live in controlled chambers, we are forced to estimate pollutant exposure from a few sites. For most pollutants, there are no emission inventories to give additional information on exposure. Statistical models are needed which will estimate exposure from this minimal information.

4. Summary

We have mentioned a few of the statistical topics associated with the community health effects studies. As statisticians, we feel there is no lack of challenging problems in this area.

Solutions to many of these problems must be found. The toxic effects at high concentrations of such atmospheric pollutants as carbon monoxide, sulfur dioxide, and nitrogen dioxide are well known. Several epidemiologic studies have suggested harmful effects of low level pollution. Further quantitation of these results is needed so that the dose effect relationships can be demonstrated and applied to the standard-setting process.

Environmental control is a tremendously expensive undertaking requiring difficult choices between many options. Better inputs are needed before the cost-benefit analysis approach for pollution control management can be effective. Statisticians must shoulder an increasing share of the effort in striving toward a cleaner environment.

REFERENCES

[1] I. D. J. BROSS, "How to use the ridit analysis," *Biometrics*, Vol. 14 (1958), pp. 18–25.
[2] J. E. GRIZZLE, C. F. STARMER, and G. G. KOCH, "Analysis of categorical data by linear models," *Biometrics*, Vol. 25 (1969), pp. 489–502.
[3] R. E. QUANDT, "The estimation of the parameters of a linear regression system obeying two separate regimes," *J. Amer. Statist. Soc.*, Vol. 53 (1958), pp. 873–880.
[4] W. B. RIGGAN, D. I. HAMMER, J. F. FINKLEA, V. HASSELBLAD, C. R. SHARP, R. M. BURTON, and C. M. SHY, "CHESS: a community health and environmental surveillance system," *Proceedings of the Sixth Berkeley Symposium on Mathematical Statistics and Probability*, Berkeley and Los Angeles, University of California Press, 1972, Vol. 6.
[5] C. M. SHY, J. F. FINKLEA, D. C. CALAFIORE, F. BENSON, W. C. NELSON, and V. A. NEWILL, "A program of community health and environmental surveillance studies (CHESS)," *Determination of Air Quality*, New York, Plenum Press, 1972.

Discussion

Question: Burton E. Vaughan, Ecosystems Department, Battelle Memorial Institute, Richland, Wn.

Regarding the "hockey stick" effect shown in your slide. This is basically a correlation, in which other (unmeasured?) stressors presumably also affected running performance.

On the days when oxidant was measured, at say 8, 12, or 20 pphm, do you have evidence to show whether or not temperature, humidity, or some other stressor was also abnormally high? Could one of the latter factors have been the cause of 18% impairment in running performance, for the subthreshold region of your graph?

Reply: W. Nelson

Besides oxidant, hourly monitoring data were obtained for nitrogen oxides, carbon monoxide, particulates, temperature, relative humidity, and wind speed.

The correlations of these environmental factors with decreased running performance were examined. Correlation with oxidant was the most significant, though there was evidence of significant association with some of the other factors. Unfortunately, no extremes of temperature were observed for these races, so this effect could not be accurately assessed.

Question: Alexander Grendon, Donner Laboratory, University of California, Berkeley

The term used to describe the ordinate of your graph on the San Marino athletes was "per cent decrease in performance." It should evidently be, "percent of group who showed decrease in performance." Since the fraction showing a decrease from their last previous level of performance was related to the level of oxidants in air during the latest trial, I ask: was the last previous performance during a period of zero oxidants?

Reply: W. Nelson

There were no periods of zero oxidants observed on the race days. I would agree with you if you are suggesting by your question that, since the performance variable involves change from the last race, a better choice of pollution variable is "change in oxidant from last race." The choice of variables was made in an earlier published report from which we were only estimating a threshold level. However, we also tried using the "change in oxidants" variable and it made no difference for these data.

SURVEY STRATEGIES FOR ESTIMATING RARE HEALTH ATTRIBUTES

MONROE G. SIRKEN

NATIONAL CENTER FOR HEALTH STATISTICS

1. Introduction

Estimation of the incidence and prevalence of rare health attributes in the population is one of the most difficult and persistent methodological problems in the national program for producing health and vital statistics. Speakers at this symposium have called attention to this methodological problem with respect to planning epidemiological studies of pollutant effects. They pointed out that many of the most serious health conditions such as congenital malformations, infant deaths, and numerous severe chronic diseases in which pollutants have been implicated or suspected, affect relatively small numbers of persons.

One objective of this paper is to describe the sample survey methods that have been used by the National Center for Health Statistics (NCHS) to produce national statistics for health conditions with low rates of prevalence and for vital events with low occurrence rates. Since different data systems have evolved in this country for producing vital statistics and for producing morbidity statistics, the methodological problems associated with estimating rare vital events are somewhat different than those associated with rare health conditions and the methods of dealing with these problems have been somewhat different also. Therefore, the matter will be discussed separately for the two data systems.

Another objective of this paper is to describe a new type of estimator that is currently being investigated by NCHS. The estimator is being tested in sample surveys of providers of health services to estimate rare health conditions and in household sample surveys to estimate rare vital events.

2. Rare vital events

National birth and death statistics are predominately by-products of the birth and death registration systems. Vital statistics are derived from the items of information reported on the records of registered births and deaths. Since national vital statistics are based on 100 per cent of the nearly two million deaths registered annually and on a 50 per cent sample of the nearly four million annually registered births, estimating the number of rare vital events in terms of the demographic and medical variables on the records does not present a problem.

135

However, the number of statistical items of information recorded on the vital records is restricted and the items tend to refer to variables that are closely tied to the date that the event occurred. Thus, the death record identifies the decedent's usual place of residence on the date of his death but it does not record his prior places of residence that would be required for ascertaining the decedent's longer term exposure to air pollutants. The number and kinds of items on the vital records are limited because the records are primarily legal rather than statistical records, and they are often used for personal identification purposes. Since 1900 the revisions in the standard certificates of birth and death have been minor [8] despite a dramatic increase in the demand to expand the scope of vital statistics. Consequently, innovative methods are needed to estimate the occurrence of vital events that are not defined in terms of the variables recorded on the vital records themselves.

Two kinds of sample survey techniques have been used to expand the scope of national vital statistics: (1) sample surveys linked to vital records and (2) dual sample surveys. The first is a method to estimate the frequency of rare vital events and the second is a method to estimate the occurrence rates of the rare events.

The design of sample surveys linked to vital records has been described in detail elsewhere [9], [19]. In these surveys, the files of the vital records serve as the sampling frame. They are excellent sampling frames because the files are virtually complete and each record in the files contains information which may be used in the sample selection and estimation process to improve the efficiency of the survey. Furthermore, the vital records provide names and addresses of persons and institutions that serve as the primary sources of information collected in the surveys. For example, the death record identifies the medical certifier of the causes of death, the hospital or institution in which the death occurred, and the "death record informant" who is typically a close relative of the deceased person. The birth record identifies the medical attendant at birth, the hospital in which the birth occurred, and the parents. Effective data collection procedures have been developed [16], [17] which depend primarily on mail surveys with provision as required for personal interview follow-up on subsamples of nonrespondents.

The program of conducting *ad hoc* surveys linked to birth and death records began about 15 years ago [18]. The national statistical program for conducting these surveys on a continuing basis was started about a decade ago [13]. Some examples of national statistics derived from these programs are listed below:

(a) A national survey linked to birth records produced statistics on medical irradiation exposure of the mother during pregnancy [10].

(b) A national survey linked to infant death records produced infant death statistics by fertility history of the mother, and by birth weight of the infant [11].

(c) A national survey linked to records of adult decedents produced mortality statistics by residence history and by smoking history of the decedents [3].

In dual sample surveys, separate surveys are conducted to estimate the numerators and the denominators of vital rates. The numerator of the rate is the number of vital events that occurred during a specified calendar period. The denominator of the rate is the size of the population exposed to the risk of the event during the specified calendar period. The numerator is estimated by conducting a survey linked to vital records. The denominator is estimated by conducting a survey of the exposed to risk population which is usually a household sample survey although it may be a survey linked to vital records. A few examples of the kinds of vital rates that have been derived from dual sample surveys are listed below:

(a) Infant death rates by socioeconomic status. The numerators were based on a survey linked to infant death records and the denominators were based on a survey linked to birth records [11].

(b) Death rates by smoking history and residence history. The numerators were based on a survey linked to death records and the denominators were based on a national household sample survey [3].

3. Rare health conditions

Morbidity statistics are collected in the National Health Survey [12], which comprises a family of sample surveys in which each survey produces distinctive health and related population statistics. For example, the Health Interview Survey (HIS) produces statistics on the social dimensions of illness and the impact of morbidity on the population. On the other hand, the Hospital Discharge Survey (HDS) produces statistics on hospital utilization. Estimates of rare health conditions derived from these surveys are subject to very large sampling errors because the complex designs of the surveys involve relatively small samples. The HIS is based on interviews conducted in a national sample of about 35,000 households annually. The HDS is based on information abstracted from hospital records for a national sample of about 200,000 discharges annually or less than 0.5 per cent of the discharges from short stay hospitals.

The *ad hoc* survey of medical processes is one possible solution to the problem of designing sample surveys to estimate the number and characteristics of persons with specified rare health conditions. In this type of survey, listings of medical sources serve as the sampling frame. The cystic fibrosis survey was a prototype *ad hoc* survey of hospitals and physicians to estimate the incidence, prevalence, and case fatality rates of diagnosed cases of cystic fibrosis, a relatively rare genetic disease affecting roughly 1 in 2500 live births. A sample of medical sources stratified by size of hospital and specialty of physician was selected. In the mail survey, every sample source reported the patients with the disease that it had treated.

The survey presented an interesting estimation problem because cystic fibrosis patients are frequently treated by more than one medical source. Since medical sources were the sampling elements, patients had different probabilities of being reported in the survey, the probabilities being proportional to the number of medical sources in the population who treated them for the disease. Work on this problem led to the development of estimators, which utilize ancillary information on the extent of multi-reporting of the patients by different medical sources.

In the cystic fibrosis survey, the extent of multiple reporting of patients was determined on the basis of the following types of auxiliary data that were collected for each patient reported by a sample source:

(a) The sample source who reported a patient identified other medical sources known to him who also treated his patient.

(b) Nonsample sources who were reported by sample sources as having treated their patients were added to the survey to verify that they had treated these patients and to determine if there were other medical sources who had also treated these patients.

Actually three different unbiased estimators were derived for the cystic fibrosis survey [7] and other estimators have been proposed [2], [5] for dealing with the problem. The cystic fibrosis estimators differed in the way that they utilized the information collected in the survey about multiple reporting of the same patient by different medical sources. The simplest of the three is the multiplicity estimator.

4. Comparison of multiplicity and conventional estimators

There are I_α, $\alpha = 1, \cdots, N$, individuals in the population who have a specified health condition. The problem is to estimate N. There are S_i, $i = 1, \cdots, L$, medical sources from which a sample S_j, $j = 1, \cdots, \ell$, sources is selected without replacement. The estimator of N is

$$(1) \qquad \hat{N} = \frac{L}{\ell} \sum_{j=1}^{\ell} \lambda_j$$

where λ_j represents the information about individuals with the health condition reported in the survey by the jth sample source.

In the conventional survey a counting rule is adoped such that each patient is uniquely reported by a single source. For example, in the cystic fibrosis survey such a rule might state that "each patient is reported in the survey by the one source that has the major responsibility for treating the disease." Under conventional conditions

$$(2) \qquad {}_c\lambda_j = \sum_{\alpha=1}^{N} {}_c\delta_{\alpha,j},$$

$$(3) \qquad {}_c\delta_{\alpha,j} = \begin{cases} 1 & \text{if } S_j, j = 1, \cdots, \ell, \text{ reports } I_\alpha \text{ in the conventional survey} \\ 0 & \text{otherwise.} \end{cases}$$

According to the conventional rule

$$(4) \qquad \sum_{i=1}^{L} {}_c\delta_{\alpha,i} = 1$$

since one and only one source in the population reports I_α. Thus the conventional estimator of N is

$$(5) \qquad \hat{N}_c = \frac{L}{\ell} \sum_{j=1}^{\ell} {}_c\lambda_j = \frac{L}{\ell} \sum_{j=1}^{\ell} \sum_{\alpha=1}^{N} {}_c\delta_{\alpha,j}.$$

In the multiplicity survey (that is, the survey using the multiplicity estimator) a multiplicity rule is adopted such that each patient is reported by at least one source. For example, in the cystic fibrosis survey, the rule stated that "each patient is reported by every medical source that ever treated him for the disease." Under these circumstances

$$(6) \qquad {}_m\lambda_j = \sum_{\alpha=1}^{N} \frac{{}_m\delta_{\alpha,j}}{s_\alpha}$$

where

$$(7) \qquad {}_m\delta_{\alpha,j} = \begin{cases} 1 & \text{if } S_j,\, j = 1, \cdots, \ell, \text{ reports } I_\alpha \text{ in the multiplicity survey} \\ 0 & \text{otherwise} \end{cases}$$

and

$$(8) \qquad \sum_{i=1}^{L} {}_m\delta_{\alpha,i} = s_\alpha = \text{number of sources in the population reporting } I_\alpha.$$

The multiplicity estimator of N is

$$(9) \qquad \hat{N}_m = \frac{L}{\ell} \sum_{j=1}^{\ell} {}_m\lambda_j = \frac{L}{\ell} \sum_{j=1}^{\ell} \sum_{\alpha=1}^{N} \frac{{}_m\delta_{\alpha,j}}{s_\alpha}.$$

Some features of this estimator are particularly noteworthy:

(a) The s_α are needed *only* for I_α that are reported by sample sources.

(b) The survey procedure for determining the s_α does not necessarily require a survey of nonsample sources if this information can be reported by the sample sources.

(c) There is no need to match the patients reported by different sample sources in order to eliminate duplicate reports of the same patient.

(d) The conventional as well as the multiplicity estimates can be derived from the multiplicity survey if the multiplicity rule incorporates the conventional counting rule.

Both $\hat{N}_m$ and $\hat{N}_c$ are unbiased estimators of N provided that every patient is reported by one and only one source in the conventional survey and by at least one source in the multiplicity survey. The multiplicity estimator, however, involves the collection of more data from the sample sources in the survey. Clearly, the number of patients reported per source is greater in the multiplicity than in the conventional survey. Thus,

$$(10) \qquad \frac{1}{L} \sum_{\alpha=1}^{N} \sum_{i=1}^{L} {}_m\delta_{\alpha,i} = \frac{1}{L} \sum_{\alpha=1}^{N} s_\alpha \geq \frac{1}{L} \sum_{\alpha=1}^{N} \sum_{i=1}^{L} {}_c\delta_{\alpha,i} = \frac{N}{L}.$$

In addition, ancillary information is collected in the multiplicity survey in order to determine the s_α of each I_α reported by a sample medical source. This ancillary information is not required in the conventional survey because according to the conventional estimator I_α, $\alpha = 1, \cdots, N$, is linked to a single source.

It has been shown [14] that $\hat{N}_m$ is not necessarily a more efficient estimator than $\hat{N}_c$. Which of the estimators has the smaller sampling variability depends on the particular counting rules adopted in the conventional and multiplicity surveys. The statistician, however, decides which particular counting rule, either conventional or multiplicity, is adopted in each survey. Obviously, the multiplicity estimator should be used selectively in those surveys where it is believed that a multiplicity rule produces a more efficient estimate than a conventional counting rule.

Some guidelines have been developed for selecting efficient multiplicity rules in surveys based on simple random sampling [14] and stratified sampling [15]. Some conditions for making the multiplicity rules more efficient than conventional counting rules are more likely to be met in surveys of rare attributes than in other surveys. For example, if the multiplicity rule satisfies the condition that

$$(11) \qquad \sum_{\alpha=1}^{N} {}_m\delta_{\alpha,i} \leqq 1, \qquad\qquad i = 1, \cdots, L,$$

that is, that none of the sources reports more than one individual, $\hat{N}_m$ is a more efficient estimator than $\hat{N}_c$. The following inequality is derived in the appendix:

$$(12) \qquad R < \frac{1}{N} \sum_{\alpha=1}^{N} \frac{1}{s_\alpha} \leqq 1,$$

where R represents the ratio of the sampling variance of $\hat{N}_m$ to the sampling variance of $\hat{N}_c$ for a simple random sample design.

5. Multiplicity estimators for rare vital events

The prospects of increasing the efficiency of survey estimates of rare health conditions by means of multiplicity estimators, prompted the NCHS to consider applying the multiplicity survey to the problem of estimating rates of rare vital events. In an earlier section of this paper, the dual sample survey method of estimating vital rates was described. According to that method, it will be recalled, estimates of the numerators of vital rates are based on sample surveys linked to vital records and estimates of the population denominators are usually based on household sample surveys. Actually, there is some redundancy in this method because the household survey can produce estimates of the numerators as well as the denominators. However, vital statistics based on household sample surveys are subject to large sampling errors because vital events are relatively rare.

In the single time household survey of population change, vital events that occurred during a preceding reference period are reported by the sample households. In the conventional household survey, counting rules are adopted which assure that each vital event that occurred during the reference period is uniquely linked to one household. For example, the conventional rule for counting deaths links the death that occurred during the reference period to the former dwelling unit of the decedent and the conventional rule for counting births links the surviving baby to its dwelling unit of residence.

Recently, the NCHS began to apply the principles of the design of multiplicity surveys to single time household sample surveys of population change. We have been exploring the feasibility and effectiveness of alternative multiplicity rules for linking vital events to households. One kind of multiplicity rule is based on consanguine relationships and it links the vital event to households containing its relatives. For example, the consanguine rules adopted in the survey might state that "births are reported by parents and grandparents" and "deaths are reported by the spouse, siblings, and children." Accordingly, births would be reported in the survey by households containing either the parents or grandparents and deaths by households containing either the surviving sib, spouse or child. The household reporting a vital event in the survey would also report as ancillary information the number of other households that would be eligible to report the same event in compliance with the multiplicity rule. Thus, the household reporting a death would also report the number of other households containing either the surviving spouse or a sib or a child of the decedent. The ancillary information would be used to determine the multiplicity of the reported event.

An experimental survey was conducted to investigate the effect of alternative consanguine rules on the reliability as well as the validity of birth and death statistics collected in single time household sample surveys of population change. Some preliminary findings of the experiment have been published [20] which compare the completeness of coverage of white deaths associated with different counting rules. The results indicate that coverage of white deaths is more complete in multiplicity surveys based on specified consanguine rules than on conventional counting rules.

6. Summary and conclusions

National health and vital statistics are collected by the NCHS in a family of sample surveys and registration systems. Within the framework of these data collection systems, it is frequently not feasible to produce reliable estimates of rarely occurring vital events and rarely prevailing health conditions. Consequently, special sample survey strategies have been developed for dealing with the problem. These strategies, which are described in this paper, would be applicable to epidemiological studies of pollutant effects.

To some extent these strategies represent the application to health surveys of well known methods for increasing the efficiency of sample surveys to estimate rare items. One method involves assembling sample frames that decrease the rarity of the item and that provide information about the listed units which can be used in the sample selection, estimation and data collection processes to improve the efficiency of the survey. This technique has been applied in surveys to estimate rare events and rare health conditions. For the former, the files of vital records serve as the sampling frame and for the latter, lists of medical sources serve as the sampling frame.

The multiplicity survey described in this report is a relatively new type of sample survey strategy for improving the accuracy of estimates of rare attributes that is being investigated by the NCHS. The multiplicity survey places a premium on counting rules which link several enumeration sources to the same individual with the rare attribute. In contrast, counting rules of the conventional survey prescribe that each individual is uniquely linked to a single source. The estimator of the multiplicity survey has served as an unbiased estimator in surveys with unavoidable duplicate reporting [4] and in surveys where sampling frames contain duplicate listings [6]. Not until recently, however, has the estimator been applied in multiplicity surveys where duplicate reporting is incorporated into the survey as a deliberate strategy to improve the accuracy of the survey estimates.

$$\diamond \qquad \diamond \qquad \diamond \qquad \diamond \qquad \diamond$$

APPENDIX

A simple random sample of ℓ out of L enumeration sources is selected without replacement. Unbiased estimators of N, the number of individuals in the population with a specified attribute, are

$$(A.1) \qquad \hat{N}_\theta = \frac{L}{\ell} \sum_{j=1}^{\ell} {}_\theta\lambda_j, \qquad \theta = c, m$$

where ${}_c\lambda_j$ and ${}_m\lambda_j$ are defined in the text by (2) and (6) respectively. The variance of $\hat{N}_\theta$ is

$$(A.2) \qquad V(\hat{N}_\theta) = \frac{L^2}{\ell} \frac{L-L}{L-1} \left\{ \frac{1}{\ell} \sum_{i=1}^{L} {}_\theta\lambda_i^2 - \left(\frac{N}{L}\right)^2 \right\}.$$

If a conventional rule were selected in the survey

$$(A.3) \qquad \sum_{i=1}^{L} {}_c\lambda_i^2 = \sum_{i=1}^{L} \left(\sum_{\alpha=1}^{N} {}_c\delta_{\alpha,i} \right)^2$$

$$= N + \sum_{i=1}^{L} \sum_{\alpha \neq \beta}^{N} {}_c\delta_{\alpha,i} \, {}_c\delta_{\beta,i}$$

where ${}_c\delta_{\alpha,i}$ is defined in the text by (3). It follows that

$$(A.4) \qquad V(\hat{N}_c) = \frac{L^2}{\ell} \frac{L-\ell}{L-1} \left\{ P(1-P) + \sum_{i=1}^{L} \sum_{\alpha \neq \beta}^{N} {}_c\delta_{\alpha,i} \, {}_c\delta_{\beta,i} \right\}$$

where $P = N/L$ represents the average number of individuals reported per source.

If a multiplicity rule were selected such that no source reported more than one individual with the attribute, the following condition would be satisfied:

(A.5)
$$\sum_{i=1}^{L} \sum_{\alpha \neq \beta}^{N} {}_m\delta_{\alpha,i} \, {}_m\delta_{\beta,i} = 0$$

where ${}_m\delta_{\alpha,i}$ is defined in the text by (7). It follows that

(A.6)
$$\sum_{i=1}^{L} {}_m\lambda_i^2 = \sum_{i=1}^{L} \left(\sum_{\alpha=1}^{N} \frac{{}_m\delta_{\alpha,i}}{s_\alpha} \right)^2 = \sum_{\alpha=1}^{N} \sum_{i=1}^{L} \left(\frac{{}_m\delta_{\alpha,i}}{s_\alpha} \right)^2 = \sum_{\alpha=1}^{N} \frac{1}{s_\alpha}$$

where $s_\alpha \geq 1$ is defined in the text by (8). Consequently, we write

(A.7)
$$V(\hat{N}_m) = \frac{L^2}{\ell} \frac{L - \ell}{L - 1} \left\{ \frac{1}{L} \sum_{\alpha=1}^{N} \frac{1}{s_\alpha} - \left(\frac{N}{L} \right)^2 \right\}$$
$$= \frac{L^2}{\ell} \frac{L - \ell}{L - 1} P \left\{ \frac{1}{N} \sum_{\alpha=1}^{N} \frac{1}{s_\alpha} - P \right\}.$$

Assuming that no source reports more than one individual in the multiplicity survey,

(A.8)
$$R = \frac{V(\hat{N}_m)}{V(\hat{N}_c)} = \frac{P \left\{ \dfrac{1}{N} \sum_{\alpha=1}^{N} \dfrac{1}{s_\alpha} - P \right\}}{P(1 - P) + \sum_{i=1}^{L} \sum_{\alpha \neq \beta}^{N} {}_c\delta_{\alpha,i} \, {}_c\delta_{\beta,i}}$$

$$\leq \frac{\dfrac{1}{N} \sum_{\alpha=1}^{N} \dfrac{1}{s_\alpha} - P}{1 - P} < \frac{1}{N} \sum_{\alpha=1}^{N} \frac{1}{s_\alpha} \leq 1.$$

REFERENCES

[1] W. FELLER, *An Introduction to Probability Theory and Its Applications*, Vol. 2, New York, Wiley, 1966.
[2] L. A. GOODMAN, "Snowball sampling," *Ann. Math. Statist.*, Vol. 32 (1961), pp. 148–170.
[3] W. HAENSZEL, D. B. LOVELAND, and M. G. SIRKEN, "Lung cancer mortality as related to residence and smoking histories, I. white males," *J. Nat. Cancer Inst.*, Vol. 28 (1962), pp. 947–1001.
[4] M. H. HANSEN, W. N. HURWITZ, and W. G. MADOW, *Sample Survey Methods and Theory*, Vol. 1, New York, Wiley, 1953.
[5] NAN-CHANG HSIEH, *Some Estimation Techniques for Utilizing Information from Elements Not in the Sample*, Survey Research Center, UCLA, 1970.
[6] L. KISH, *Survey Sampling*, New York, Wiley, 1965.
[7] NATIONAL CENTER FOR HEALTH STATISTICS, "Design of sample surveys to estimate the prevalence of rare diseases: Three unbiased estimates," *Vital and Health Statistics*, Ser. 2, No. 11 (1968), pp. 1–8.
[8] ———, "The 1968 revision of the standard certificates," *Vital and Health Statistics*, Ser. 4, No. 8 (1968), pp. 1–47.
[9] ———, "Methods and response characteristics, National Natality Survey United States 1963," *Vital and Health Statistics*, Ser. 22, No. 3 (1966), pp. 1–36.
[10] ———, "Medical X-ray visits and examinations during pregnancy, United States 1963," *Vital and Health Statistics*, Ser. 22, No. 5 (1968), pp. 1–41.
[11] ———, "Infant mortality rates by socioeconomic status," *Vital and Health Statistics*, Ser. 22, in print.

[12] ———, "Origin, program, and operation of the U.S. National Health Survey," *Vital and Health Statistics*, Ser. 1, No. 1 (1963), pp. 1-41.

[13] M. G. SIRKEN, "Sampling survey program of the National Vital Statistics Division," *Proceedings of the 9th National Meeting of the Public Health Conference on Records and Statistics*, 1962, pp. 39-41.

[14] ———, "Household surveys with multiplicity," *J. Amer. Statist. Assoc.*, Vol. 65 (1970), pp. 257-266.

[15] ———, "Stratified sample surveys with multiplicity," *J. Amer. Statist. Assoc.*, Vol. 67 (1972), in print.

[16] M. G. SIRKEN, J. W. PIFER, and M. L. BROWN, "Survey procedures for supplementing mortality statistics," *Amer. J. Pub. Health*, Vol. 50 (1960), pp. 1753-1764.

[17] M. G. SIRKEN and M. L. BROWN, "Quality of data elicited by successive mailings in mail surveys," *Proc. Soc. Statist. Sec. Amer. Statist. Assoc.* (1962), pp. 118-125.

[18] M. G. SIRKEN and H. L. DUNN, "Expanding and improving vital statistics," *Pub. Health Rep.* (1958), pp. 537-540.

[19] M. G. SIRKEN, J. W. PIFER, and M. L. BROWN, *Design of Surveys Linked to Death Records*, U.S. Department of Health, Education, and Welfare, Public Health Service, 1962.

[20] M. G. SIRKEN and P. N. ROYSTON, "Reasons deaths are missed in household surveys of population change," *Proc. Soc. Statist. Sec. Amer. Statist. Assoc.* (1970), pp. 361-364.

Discussion

Question: A. C. Hexter, California Department of Public Health

If a source which is supposed to report an event does not (if, for example, physician failed to report one of his patients), would that not lead to bias?

Reply: M. G. Sirken

Yes.

Question: S. Raman, Division of Biostatistics, University of California, Berkeley

Does your scheme of multiple reporting include the case where the same source reports more than once about the same patient?

Reply: M. G. Sirken

Yes.

ENVIRONMENTAL RADIATION
AND HUMAN HEALTH

ERNEST J. STERNGLASS

UNIVERSITY OF PITTSBURGH

1. Introduction

The present paper will address itself to the evidence that low level radiation from nuclear fission products in the environment such as are released by nuclear explosions and power reactors may already have produced serious effects on the health of the world's population far beyond those ever believed possible when our present radiation standards were originally formulated and adopted, especially for the case of the young infant.

Before discussing the latest evidence in some detail, I should like to review very briefly the nature of the early discovery that low level radiation can produce not only genetic but also serious somatic effects in man both at high and low dose rates.

2. Historical background

The earliest indication that low level radiation could produce serious effects in man came from the studies of Alice Stewart at Oxford University in 1958 showing that mothers who had received a series of three to five pelvic X-rays during pregnancy had children who were almost twice as likely to develop leukemia and other cancers before age ten than mothers who had had no pelvic X-ray examinations [1].

This work was independently confirmed in 1962 in a major epidemiological study involving close to 800,000 children born in New York and New England Hospitals by Brian MacMahon of the Harvard School of Public Health [2]. Using these two sets of data, it was possible to show that there appears to exist a direct, straight line relationship between the number of X-ray films given to a pregnant woman and the probability that the child will subsequently develop leukemia, and that there is therefore no evidence for the existence of a safe "threshold level" below which no additional cancers are produced, down to the relatively small dose from a single X-ray. Furthermore, the magnitude of the X-ray dose to the developing fetus *in utero* from one such X-ray was comparable with the dose normally received in the course of two to three years of natural background radiation, or from the fallout produced in the course of the 1961–1963 test series, namely 0.2–0.3 rad [3].

These early findings have since been confirmed by the most recent results of

145

A. Stewart, June, 1970 [4]. This extensive study, based on over 7,000 children born in England and Wales between 1943 and 1965 who developed leukemia or other cancers gave the result that for one rad to a population of one million children exposed shortly before birth, there were an extra 300 to 800 cancer deaths before age ten with a mean number of 572 ± 133 per rad. For a normal rate of incidence of about 700 cases per million children born, this means that only 1.2 rads (1200 mr) are required to double the spontaneous incidence. (See Figure 1.) Furthermore, Dr. Stewart's study showed that when the radiation

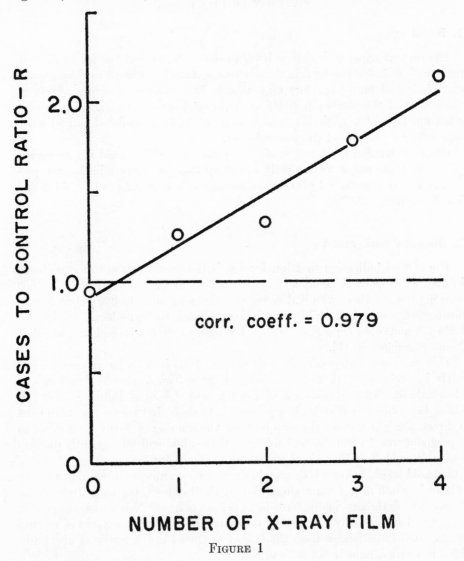

FIGURE 1

Ratio of cancer cases to controls as a function of the number of abdominal X-ray films as reported by Stewart and Kneale (*Lancet*, June 6, 1970).

exposure took place in the first trimester, the excess risk of cancer increased 15 times [4]. This means that a dose of only some 80 mr was found to double the normal cancer risk for the early embryo, much less than the presently permitted 500 mr annual dose to any member of the general population.

3. Fallout and childhood leukemia

It was therefore possible that studies of large populations of children exposed to known incidents where localized fallout occurred in a given area might show detectable increases in leukemia some years later. Such a localized "rain-out" was pointed out by Ralph Lapp [5] as having taken place in Albany-Troy, New York in April of 1953 following the detonation of a 40 kiloton bomb in Nevada. An examination of the data on leukemia incidence published by the New York State Department of Health showed that when plotted by year of death there was a clear increase in the number of cases per year among children under ten years of age at death from about two to three to as many as eight to nine per year some six to eight years after the arrival of the fallout, exactly the same delay in peak incidence as observed in Hiroshima and Nagasaki. Furthermore, the peak contained many children who were not even conceived until a year or more after the arrival of the fallout, suggesting for the first time the existence of an effect prior to conception, see Figure 2 [6].

Due to the relatively small number of cases in Albany-Troy, it was difficult to draw absolutely firm conclusions, and so the situation for New York State as a whole was examined. Again, peaks of leukemia incidence were clearly present some four to six years after known atmospheric tests in Nevada, greatly strengthening the initial observations for Albany-Troy alone [6].

4. Early indications relating fallout and infant mortality

Following the arrival of the fallout in Albany-Troy in 1953, there was also a drastic slowdown in the steady decline of fetal mortality or still births in that area see Figure 3 [6]. Following up this unexpected finding, the fetal and infant mortality statistics for New York State as a whole were examined, followed by those for California and other states. The same slowdown in the decline or even renewed rises in the mortality rates existed to varying degrees depending on the amount of fallout in the milk, beginning in the early 1950's, the declines resuming only two to four years after the end of atmospheric testing [7]. For the U.S. as a whole, the data is shown in Figure 4, where both the infant mortality rates for the total population and the nonwhite population has been plotted together with the data for Sweden. It was then drawn to our attention that I. M. Moriyama of the U.S. National Center for Health Statistics had previously pointed out the levelling trend in the U.S. (beginning in about 1951) as early as 1960 [8], and that he had in fact suggested the possibility that similar upward changes of mortality for all age groups might be connected with the sharp rises in environmental radioactivity from nuclear testing [9].

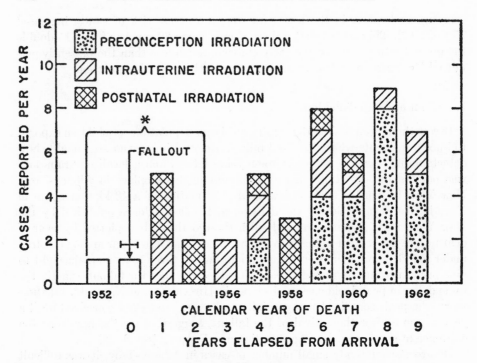

*AVERAGE RATE FOR '52-'55, BEFORE EFFECT OF
FALLOUT COULD APPEAR (2.2 ± 0.8/YR)

FIGURE 2

Number of leukemia cases per year of report for children under ten years of age
in the Albany-Troy, N.Y. area, for which the data is complete, as reported by
Lade (*Science*, Vol. 143 (1964), p. 994). Period from 1952 to 1955 before effect
of fallout could appear (*) gives an average annual number of 2.2 ± 0.8 cases
per year.

Since then, we have extended our studies to other countries in the world, and
especially in northern Europe, which received the fallout from the Nevada tests
in its northeasterly drift across the Atlantic, and the same patterns of slowdown
followed by a renewed decline of infant mortality were found, as shown in
Figure 5. At the same time the levelling trends were much less pronounced in
countries like Canada and France, that were to the north or south of the
path of the Nevada fallout on its northeasterly course, so that they did not
receive as much short lived activity per unit strontium 90 in the milk (see
Figure 6).

We have since established high degrees of correlation between the increases
in infant mortality above the declining base lines, and the measured strontium
90 levels in the milk and therefore in the bone of fetuses, children and young

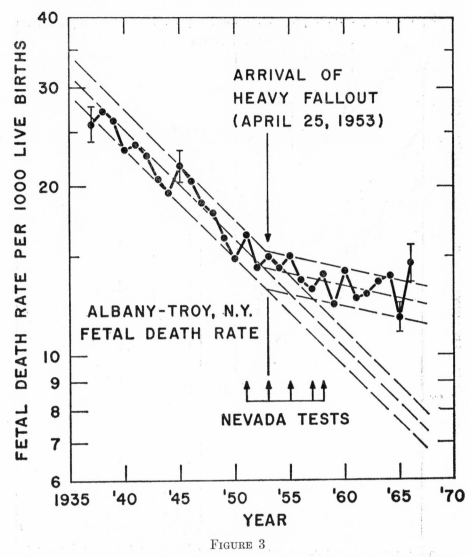

FIGURE 3

Fetal death rate reported per 1000 live births *versus* time before and after the arrival of fallout in the Albany-Troy area.

adults for all the nine regions of the Public Health Service's Raw Milk Network, for which data are available back to 1957–58. (Table I and Figure 7) [10]. These correlations suggest that as many as 400,000 infants up to one year old in the U.S. alone may have died as the result of nuclear testing by 1965.

These results are so startling and so unexpected, that they have naturally encountered considerable skepticism primarily because the technique of trend

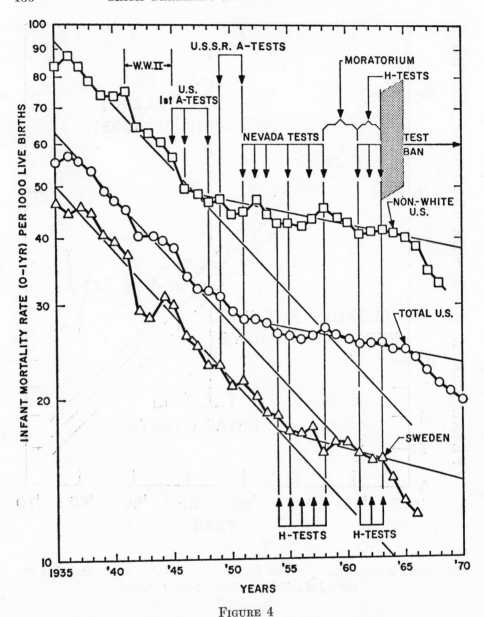

FIGURE 4

Infant mortality rate (0–1 year) per 1000 live births for the U.S. total population, U.S. nonwhite population and Sweden.

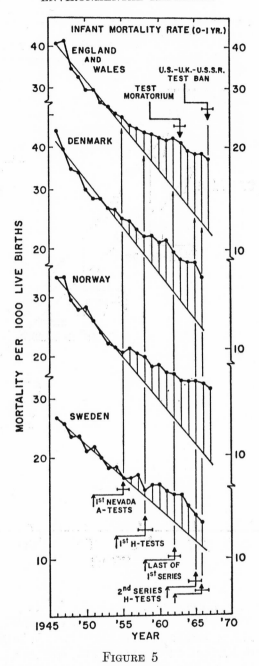

FIGURE 5

Infant mortality trends for northern European countries after World War II.
Note onset of upward deviations peaking some three to five years after major
test series. Least square fits to 1946–1955 trend.

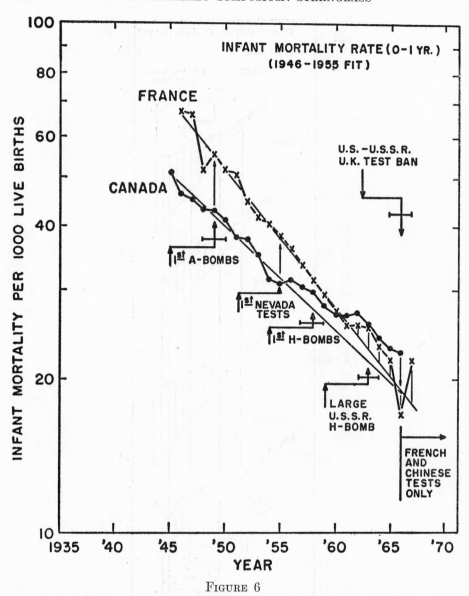

FIGURE 6

Infant mortality trend for France and Canada after World War II. Note smaller upward deviation than for the northern European countries, associated with the fact that the path of the intense, short lived Nevada fallout largely missed Canada to the north and France to the south of the prevailing fallout movement.

TABLE I

Correlation Between Strontium 90 Levels and Excess Infant Mortality, Showing
Effect of Different Levels of Short Lived Isotopes on the Slopes of the Regression
Lines, as Well as the Effect of Unrepresentative Milk Sampling Areas in the Case
of California, Washington State and Ohio

State or country	Correl. Coeff.	Degrees of freedom	t Value	Slope
California	0.964	11	12.00	8.00 ± 0.63
Georgia	0.954	12	10.97	3.08 ± 0.27
Illinois	0.954	12	10.99	3.71 ± 0.32
Ohio	0.976	10	14.13	1.91 ± 0.13
Missouri	0.968	12	13.38	4.00 ± 0.29
New York	0.966	12	12.98	3.51 ± 0.26
Texas	0.967	12	13.19	4.64 ± 0.34
Utah	0.841	12	5.93	3.16 ± 0.56
Washington	0.911	11	7.30	1.09 ± 0.14
U.S. (HASL-214)	0.980	14	18.26	3.15 ± 0.17
England & Wales	0.922	10	7.51	3.74 ± 0.48
New Zealand	0.950	9	9.17	3.83 ± 0.40

analysis as used first by Moriyama to calculate "excess deaths" above normal expectations for all age groups in the U.S. was based on the expectation of a steadily declining infant mortality at least until levels are reached equal to those that had already been attained in other medically advanced nations of the world such as Sweden (see Figure 4). Such an assumption is however justified by the fact that in New Mexico, after the initial test in 1945, there was indeed a return to the same line of steady decline determined by the computer fit to the 1935–50 period, due to the low rainfall and therefore low levels of fallout in the milk after 1950, when nuclear testing was moved north to Nevada (Figure 8). Furthermore, the most recent data on infant mortality show that in a number of rural states such as Maine far from any nuclear facility, infant mortality rates have declined very sharply, reaching the levels predicted on the basis of the 1935 to 1950 rate of decline, as illustrated in Figure 9 for the case of Maine. Nevertheless, such large effects of relatively small amounts of radiation on infant mortality, which is also affected by many other factors, is difficult to accept, and it is therefore important to find other data that is not subject to the same criticism.

5. Fallout and congenital malformations

Such data exist in the case of childhood deaths associated with congenital malformations such as Down's Syndrome, microcephaly and congenital heart defects. For this particular category of infant and childhood deaths, there has

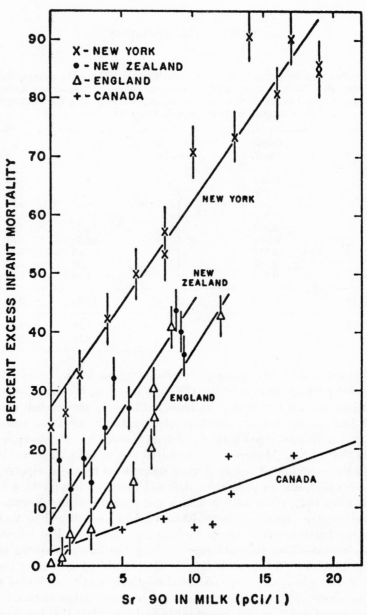

FIGURE 7

Correlation plots for excess infant mortality and strontium 90 in the milk (four year moving average). Note similarity in slope for geographical areas of high rainfall in the path of low altitude tropospheric fallout from tests in Nevada (New York and England) and tests in Australia (New Zealand). In contrast, note the small increase in mortality per unit strontium 90 for Canada, largely missed by the initial pass of the low altitude Nevada and Pacific fallout clouds with their high proportion of short lived isotopes.

154

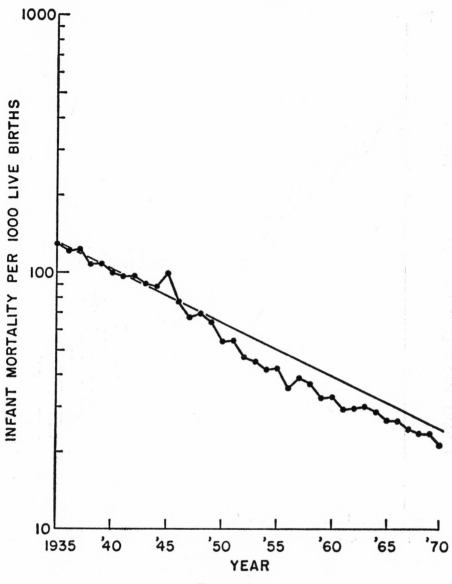

FIGURE 8

Infant mortality trend for New Mexico, 1935–1970. Note the degree to which rates continued to decline parallel to the 1935–50 projection, associated with the very low annual rainfall and geographical location south of the Nevada test site.

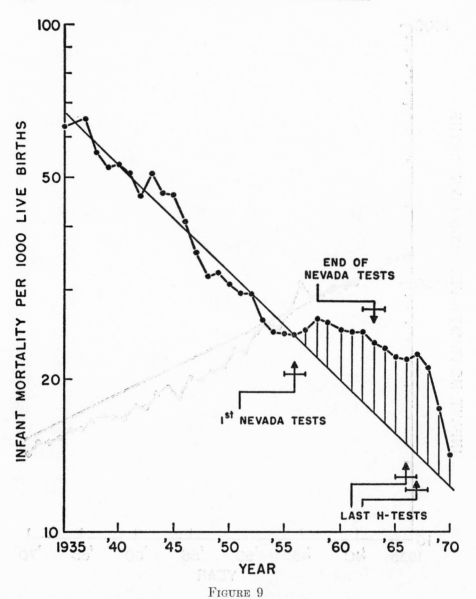

FIGURE 9

Infant mortality for Maine, 1935–1970. Note strong upward deviation beginning
at the time of the Nevada tests (1951), followed by a return to the
projected 1935–50 trend line a few years after the end of
U.S. and U.S.S.R. atmospheric tests.

been only a slight downward trend over the last 20 years, and neither the introduction of new antibiotics, medical care methods nor the gradual improvement in diet and medical care has had significant effects on these mortality rates. As a result, there is here no need to extrapolate a rapid downward trend, and one has for every state and many foreign countries, a well established nearly horizontal base line to the onset of nuclear testing in 1945. Furthermore, it is well known that congenital malformations can be induced by relatively low levels of radiation in animals, and recent studies of such conditions as mental retardation published by the United Nations Scientific Commission on Radiation [11] have established that small amounts of radiation during certain critical periods of embryonic development and organ formation can produce detectable effects in children.

We therefore examined the incidence of deaths among congenitally defective children in relation to children who died of accidents as a control group at various distances from the Nevada test site, where relatively high local fallout was known to have occurred in a number of instances, documented both by the AEC [12] and independent studies by scientists at the University of Utah [13] and the St. Louis Center for Nuclear Information [14].

As an example, Figure 10 shows the annual number of deaths of congenitally defective children up to four years old in Utah directly east of Nevada and therefore generally downwind from the test site as taken from the published figures in the U.S. Vital Statistics, together with the deaths in this age group due to accidents other than those involving automobiles. It is seen that the average number of deaths of congenitally defective children per year in the pretesting period 1937–45 stayed relatively constant at about 75 cases per year. But it rose to a peak of 123 cases per year in 1958, some five years after a particularly large fallout incident in 1953, returning close to the pretesting rate of 80 per year five years after the end of atmospheric tests in Nevada. Such a rise and decline while accidental deaths remained constant is clearly not explainable in terms of a gradual rise in the number of births per year. Altogether, there seem to be some 480 children that are likely to have died of congenital malformations in Utah above expectations, based on a comparison with the number of accidental deaths since the onset of nuclear testing in 1945.

An even more striking peak in deaths of congenitally defective children relative to the number of accidental deaths took place in the five to fourteen year age group shown in Figure 11 for the case of Utah, which includes children who received radiation from the milk and food some time after birth. Again, a four to six year delay is seen to occur between exposure and death, quite similar to the case of Hiroshima and Albany-Troy, New York, corresponding to the fact that children born congenitally defective are much more prone to develop leukemia with its four to six year delay of peak incidence.

The rate of leukemia deaths for all children in the age group five to fourteen, which was shown by Stewart [1] and MacMahon [2] to reflect the effects of perinatal irradiation most strongly is plotted in Figure 12 for the same state.

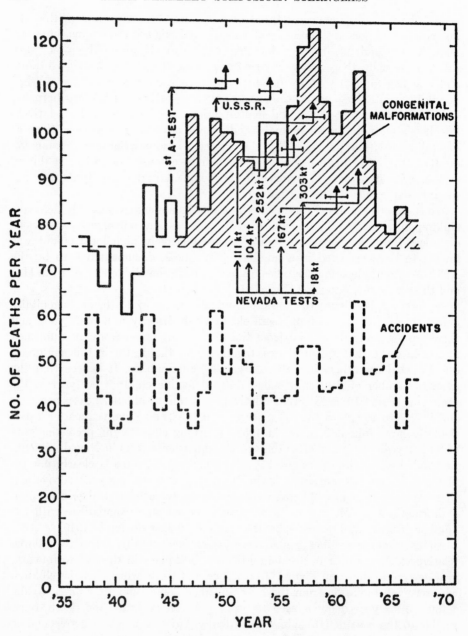

FIGURE 10

Changes in the annual number of deaths of congenitally defective children 0–4
years old in Utah, compared with the number of
non-automobile related accidents.

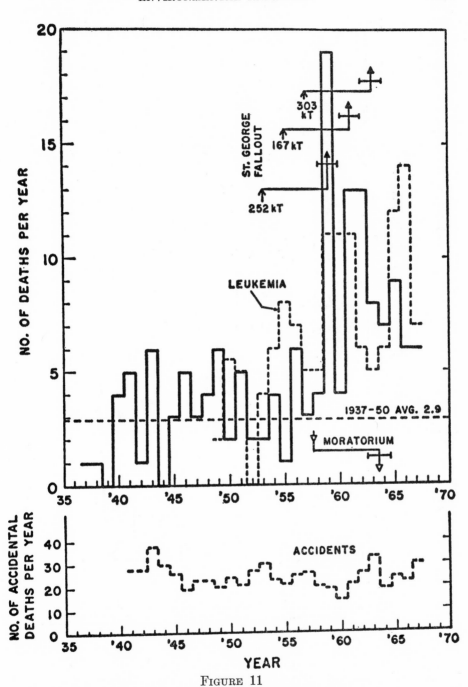

FIGURE 11

Annual number of deaths among children five to fourteen years old
born congenitally defective, compared with the number
of accidental deaths (non-automobile) [39].

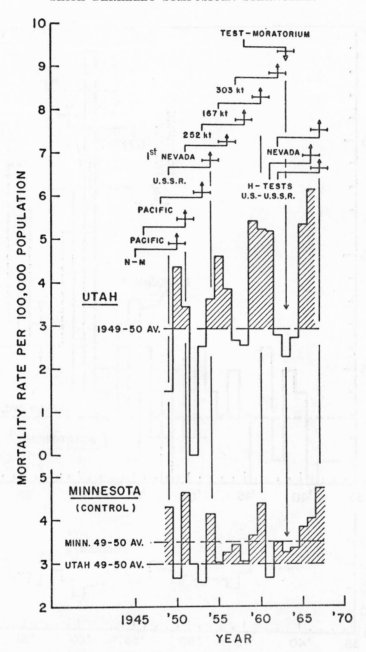

FIGURE 12

Annual rate of leukemia deaths per 100,000 population aged five to fourteen
years for Utah near the Nevada test site compared with Minnesota as control.
Major test series are indicated, together with the four to six year delay expected
for leukemia [39].

It is seen that statistically significant peaks occurred some four to six years after known tests had deposited fallout in Utah, apparently affecting the infants both prior to and after birth. Furthermore, the relative increases were higher than those observed in Minnesota as a control as is to be expected from the great proximity to the test site. Thus, the effects are observed both for annual numbers and rates per 100,000 population, and they confirm the original findings in Albany-Troy.

No other explanation of these striking rises and declines in leukemia and congenital defect mortality rate is known.

As to the reason why such unexpectedly large effects of fallout should be observed when radiation levels were believed to be so low as to be regarded as completely safe, these are evidently connected with the much greater sensitivity of the embryo and infant compared with the adult.

Furthermore, the severity of the effects is also connected with the biological concentration of certain isotopes in the food chain, mainly via the milk, which was not widely recognized at the time when the tests were begun. Another reason is the selective concentration of certain isotopes in various critical organs of the human body, whose biological consequences were not fully appreciated for the sensitive developmental phase of the early embryo and fetus.

Thus, experimental studies on laboratory animals by Walter Müller published in 1967 [15] suggest that strontium 90 and other alkaline earth elements that were long known to seek out bone may also produce biological and possibly genetic effects through their daughter elements such as yttrium 90 into which they decay, and which are known to preferentially concentrate in such vital glands as the pituitary, the liver, the pancreas and the male and female reproductive glands [16], [17].

In any case, we are apparently confronted with still another unanticipated biological concentration effect similar to the surprises we received when we discovered the special hazard of iodine 131 going to the infant thyroid and strontium 90 and 89 going to the bone via the originally unsuspected pathway of milk produced by cows grazing on contaminated pastures.

6. Infant mortality and releases from nuclear reactors—early detection

That similarly unanticipated effects on the developing embryo and infant may have taken place as a result of fission products released from nuclear reactors and fuel processing facilities first became apparent in the course of our state by state study of infant mortality changes following the first nuclear weapons test in New Mexico in 1945.

As shown in Figure 13 each map for the four years following this test showing the per cent changes relative to the trend for the previous five years not only indicated an upward change in infant mortality directly to the east and northeast of New Mexico, but also in the states to the east of the Hanford Plutonium production facility in the state of Washington.

Not only were the Hanford reactors and plutonium production facilities

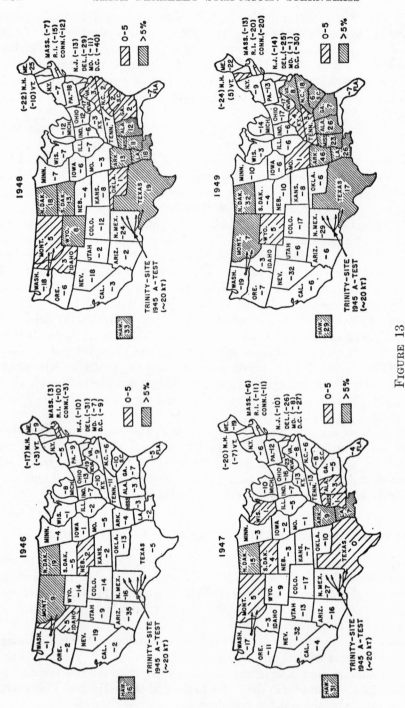

FIGURE 13

Per cent changes in infant mortality for the years 1946–1949 relative to the least square fitted 1940–45 trend just prior to the New Mexico test in 1945 for each state of the U.S. (Based on data from [39].)

operating at very high levels since 1944, releasing into the environment the rare gases that could not be trapped readily, but on a number of occasions, there were serious accidents in the course of extracting the plutonium from the irradiated uranium fuel elements by chemical techniques, when fuel elements burst into flames and discharged large quantities of fission products into the environment [18].

The infant mortality changes were greater in North Dakota than in dry Idaho and Montana, just as they were greater in Arkansas and Louisiana compared with dry Texas closer to the test site in New Mexico. This fits the well known fact that 90 per cent of the fine tropospheric fallout comes down with the rain, since the line of heavy rainfalls passes down through the center of the United States just to the west of the Mississippi from North Dakota in the north to eastern Texas in the south.

This interpretation is further confirmed by a more detailed analysis of infant mortality changes in the counties near the Hanford plant before and after it went into operation between 1943 and 1945. As can be seen from the bar graph in Figure 14, the counties containing the plant as well as those immediately adjacent to the east and south showed sharp rises in infant mortality up to 150 per cent, while the more distant control counties, namely those in which water sampling stations were subsequently established, either rose less than ten per cent or actually declined between 1943 and 1945.

7. Infant mortality near boiling water reactors

A similar pattern of increased infant mortality has now been observed around three commercial nuclear power reactors of the Boiling Water type (BWR), in which the single coolant loop design does not permit as tight a containment of fission products leaking out of corroded fuel elements as in the naval sub-marine type Pressurized Water Reactor (PWR).

As described in recent publications of the Bureau of Radiological Health [19], these reactors have emitted as much as 800,000 curies of fission and neutron activation products in the form of gases per year [20], compared with as little as 0.001 curie per year for the prototype Pressurized Water Reactor at Shippingport, Pennsylvania.

The first of the BWR's studied is the Dresden Reactor located near Morris, Illinois in Grundy County, some 50 miles southwest of Chicago. Since close to two-thirds of the population of Illinois lives within a radius of some 60 miles from this reactor, one might expect to find detectable changes in infant mortality for Illinois as a whole relative to other nearby states that correlate with the rises and declines of emission when fuel elements are changed.

That this appears in fact to have taken place is illustrated by the plot of infant mortality for Illinois compared with Ohio some 200 miles to the east for the period 1959 to 1968 in Figure 15. It is seen that while during the time of Nevada testing, Ohio and Illinois showed the same infant mortality, within

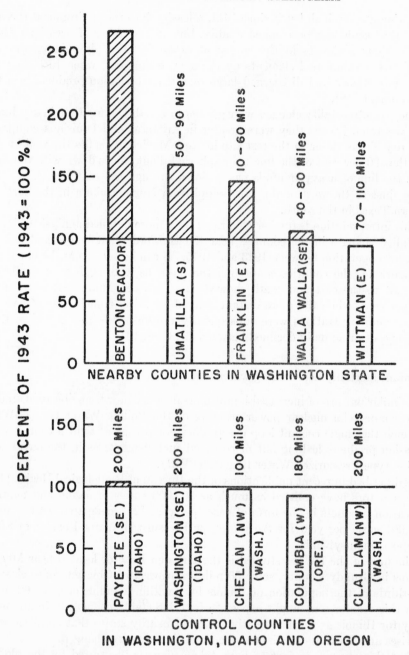

FIGURE 14

Percent change in infant mortality between 1943 and 1945 near the Hanford Reactor in the state of Washington before and after onset of operations in 1944. Control counties are those where water sampling stations were placed. (Based on data from [39].)

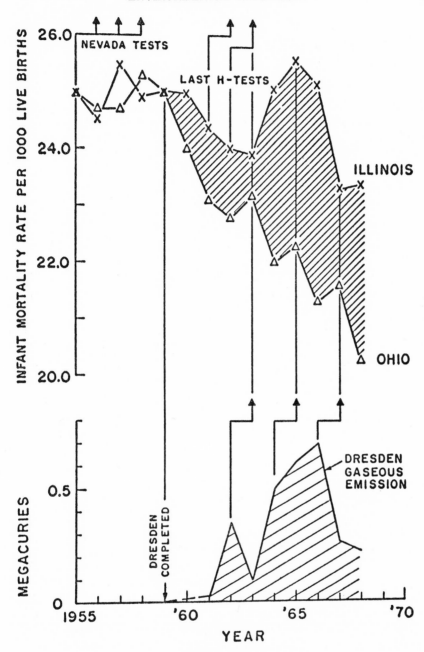

FIGURE 15

Infant mortality in Illinois compared with Ohio for the period 1955 to 1968.
Also shown are the annual releases of gaseous activity from the Dresden Reactor
[39].

a few years after the end of testing, Ohio began a steady decline, whereas Illinois showed a peak highly correlated with the peak of gaseous emissions between 1964 and 1967 (see Table II).

TABLE II

INFANT MORTALITY IN OHIO AND ILLINOIS FOR THE PERIOD 1955–1968 BEFORE AND AFTER ONSET OF GAS EMISSIONS FROM THE DRESDEN REACTOR IN 1961

	Ohio			Illinois			
Year	Live births	Infant deaths	Inf. mort. rate/1000 live births	Live births	Infant deaths	Inf. mort. rate/1000 live births	Curies of gas emissions
1955	222,689	5530	24.8	217,041	5466	24.8	—
1956	234,517	5785	24.7	229,760	5639	24.5	—
1957	243,470	6008	24.7	238,734	6080	25.5	—
1958	234,040	5940	25.4	234,980	5859	24.9	—
1959	232,578	5799	24.9	240,208	6008	25.0	—
1960	230,219	5524	24.0	238,928	5928	25.0	—
1961	229,708	5298	23.1	237,382	5771	24.3	34,800
1962	217,465	4954	22.8	230,878	5538	24.0	284,000
1963	212,583	4938	23.2	225,062	5383	23.9	71,600
1964	209,480	4614	22.0	222,248	5585	25.2	521,000
1965	194,927	4346	22.3	208,188	5340	25.7	610,000
1966	190,444	4066	21.4	201,442	5066	25.4	736,000
1967	185,204	3824	20.6	193,745	4622	23.6	260,000
1968	185,580	3769	20.3	193,520	4536	23.5	240,000

The degree of correlation may be judged from Figure 16 where the difference in infant mortality rates between Illinois and Ohio has been plotted against the annual gaseous discharges. The correlation coefficient is 0.865, and the t test of significance gives $t = 4.565$, which for the seven degrees of freedom gives $P \ll 0.01$.

As in the case of Hanford it is of interest to see whether the effect can also be detected in the nearby states to the east, the direction in which the prevailing winds and weather patterns move. As seen in Figure 17, the infant mortality rate for nearby Indiana does indeed fall exactly between that for Illinois and Ohio on the other side of Indiana after the testing in Nevada ended and the discharges from the Dresden reactor produced significant external doses, comparable with those from distant tests (see Table III).

Likewise in Michigan, just to the north of Indiana, infant mortality began to fall consistently between Illinois and more distant Ohio when the general decline began after the end of nuclear testing in 1963 (see Figure 18 and Table III).

One would also expect on the basis of this hypothesis that a state far to the northwest of Illinois and therefore upwind would show an even more rapid decline after fallout from weapons testing decreased. That this is in fact the

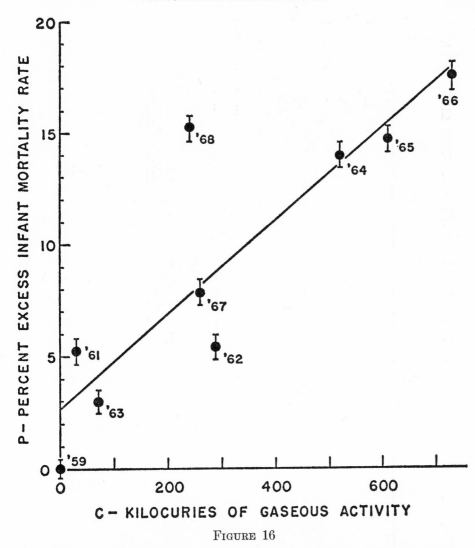

FIGURE 16

Correlation plot for the excess infant mortality in Illinois relative to Ohio vs. the annual average quantities of gaseous activity released from the Dresden Reactor. (1959–1968; no data for 1960.) Least square fitted line shown.

case is seen for the case of North Dakota compared with Illinois in Figure 19 (see Table III).

The rates for Illinois and North Dakota seem to have been identical during the period of heavy Nevada testing and plutonium production at Hanford prior to 1964, despite the great difference in ordinary air pollution and socioeconomic character of the two states. But after the end of nuclear testing by the U.S.

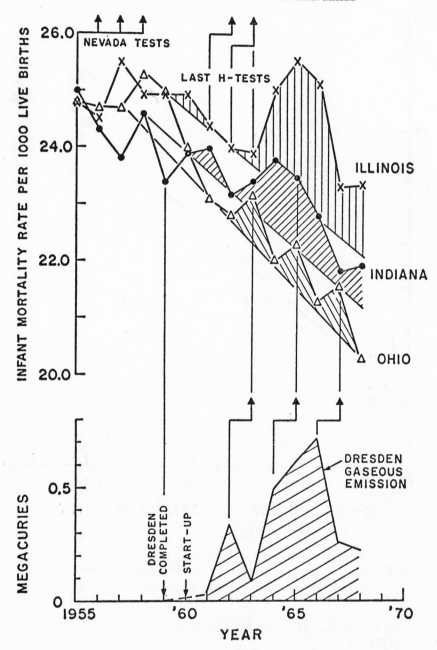

FIGURE 17

Infant mortality for Indiana compared with Illinois and Ohio [39].

TABLE III

INFANT MORTALITY RATES/1000 LIVE BIRTHS

Infant mortality rates for five states in the northern U.S. upwind and downwind from Illinois before and after onset of Dresden emissions in 1961.
Source: [39].

Year	Illinois	Indiana	Michigan	Ohio	North Dakota
1955	24.8	25.0	24.9	24.8	25.1
1956	24.5	24.3	24.5	24.7	24.8
1957	25.5	23.8	24.4	24.7	25.7
1958	24.9	24.6	24.6	25.3	24.9
1959	25.0	23.4	24.4	25.0	23.7
1960	25.0	23.9	24.1	24.0	24.8
1961	24.3	24.0	23.9	23.1	23.2
1962	24.0	23.2	24.0	22.8	22.6
1963	23.9	23.4	23.2	23.2	24.6
1964	25.1	23.8	23.0	22.0	23.1
1965	25.6	23.5	23.6	22.3	21.2
1966	25.1	22.8	22.5	21.3	20.8
1967	23.6	22.3	22.0	20.7	21.0
1968	23.1	22.2	21.8	20.0	17.7

			Population 1960		
	Illinois	Indiana	Michigan	Ohio	North Dakota
	10,081,000	4,662,000	7,823,000	9,706,000	632,000

and U.S.S.R., North Dakota declined rapidly from nearly 25 per 1000 births to under 18 per 1000 by 1968, despite the well known lack of sufficient medical care in rural areas such as North Dakota.

This suggests that although ordinary air pollution is undoubtedly detrimental to health, the radioactivity released by nuclear testing and nuclear plants appears to be significantly more serious in its effects on the early development of the embryo and infant.

In order to further test this hypothesis, the changes in infant mortality in the six counties immediately adjacent to the Dresden plant for the years following the sharpest rise in emission were compared with the changes in six control counties more than 40 miles to the west, Figure 20. They were chosen to be as far away as possible in northern Illinois, not bordering either on the Illinois or Mississippi Rivers that are known to be polluted by radioactive wastes (see Table IV).

The result of this test for 1966 relative to 1964 is shown in Figure 21. Again, the same general pattern is observed as for the Hanford Reactors, the nearby counties showing much greater rises than the more distant control counties of similar rural character and comparable medical care.

In the case of the Dresden reactor, it is possible to carry out a still more crucial test of the biological mechanism that may be involved in bringing

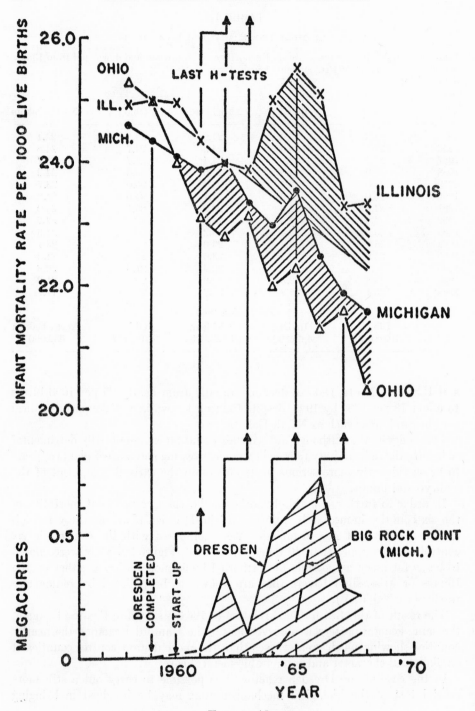

FIGURE 18

Infant mortality for Michigan compared with Illinois and Ohio [39].

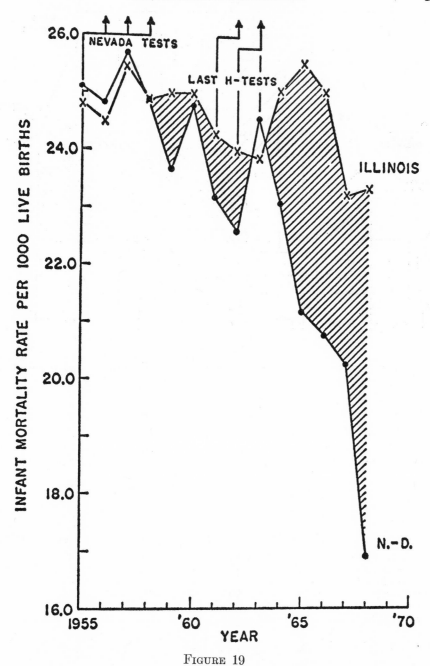

FIGURE 19

Infant mortality in North Dakota compared with Illinois [39].

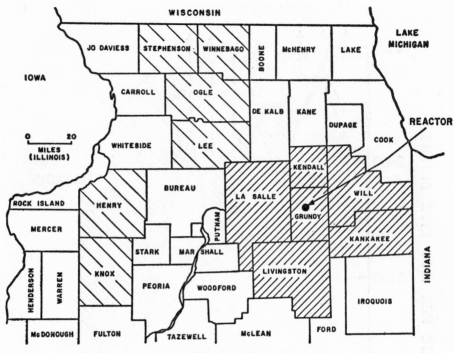

FIGURE 20

Map of northern Illinois showing counties surrounding the Dresden Reactor,
located in Grundy County, and the six control counties to the west.

TABLE IV

DRESDEN REACTOR AREA

Changes in infant mortality in six counties surrounding the Dresden Reactor compared with
six control counties to the west following by one year the period of maximum rise in gaseous
emissions (1963–1965).

		1964 Deaths	births	Rate 1000	1966 Deaths	births	Rate 1000	Per cent change in rates 1964–66	Pop. est. July 1964
Adjacent	Grundy (Reactor)	7	442	15.8	18	474	38.0	+141	23,500
	Livingston (S)	6	728	8.2	12	608	19.7	+140	41,200
	Kankakee (SE)	41	1976	20.7	54	1830	29.5	+ 43	98,500
	Will (NE)	109	4920	22.2	100	4294	23.3	+ 5	214,000
	LaSalle (W)	49	2176	22.5	39	1858	21.0	− 7	112,600
	Kendall (N)	11	460	23.9	7	422	16.6	− 31	20,000
								Avg. + 48	
Control	Ogle (NW)	16	854	18.7	20	808	24.8	+ 33	39,700
	Winnebago (NW)	122	5002	24.4	122	4788	25.5	+ 5	234,000
	Henry (W)	17	930	18.3	16	862	18.6	+ 2	50,000
	Stephenson (NW)	25	978	25.6	20	808	24.8	− 3	47,000
	Knox (SW)	22	1130	19.5	17	946	18.0	− 8	63,700
	Lee (W)	17	658	25.8	9	594	15.2	− 41	39,500
								Avg. − 2	

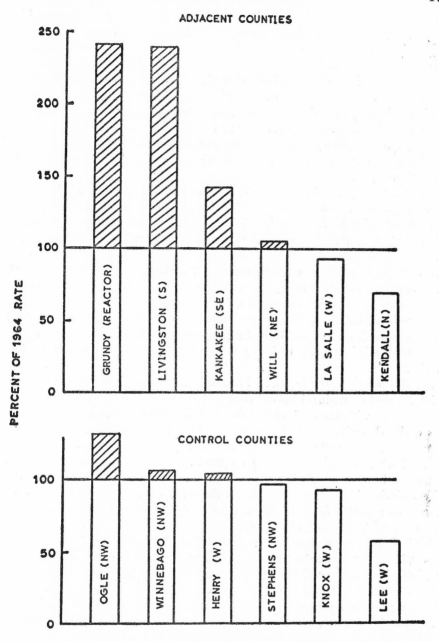

FIGURE 21

Per cent change in infant mortality in the six counties surrounding the Dresden Reactor (<30 miles distant) between 1964 and 1966 compared with the changes in six control counties to the west, following the rise in activity released from 71,600 curies in 1963 to 610,000 curies in 1965 [39].

about such a large effect for relatively small measured external doses, which even in the year of peak release (1966) did not exceed 70–80 mr at the plant boundary when the available measurements of 15–25 mr per year for 1967–68 are used to calculate the dose.

As discussed briefly above and elsewhere [21], the most serious effect is likely to be an indirect one, whereby the radiation acts on the key biochemical processes in such crucial glands controlling metabolism and growth as the pituitary and thyroid glands. Such action could lead to a small decrease in weight at birth, or to a greater frequency of prematurity, such as has in fact been observed in animal experiments and since the early 1950's, among infants born in the United States [22].

Such immaturity at birth results in a reduced ability to fight off infections and to a greater likelihood that a critical chemical or surfactant needed for proper functioning of the lung is missing, leading to respiratory distress and atelectasis [23] so that one would expect a higher mortality in early postnatal life.

To test this hypothesis, one can compare the changes in the fraction of all births that are classified as "premature" or under 2,500 grams for Grundy County as compared with the changes in the control counties to the west. If immature birth is indeed the principal mechanism leading to excessive infant deaths, one would then expect to find a greater rise in the fraction of such births during the period of peak emission in Grundy than in the distant control counties.

That this is indeed the case may be seen in the plot of Figure 22. A peak in the incidence of premature births of close to 140 per cent is seen to have occurred in coincidence with the peak of gaseous emission, declining again as the emissions declined, while the control counties showed no such rise. For Grundy, the increase was from 3.60 to 8.70 per cent of all births (Table V).

Thus, both radioactive releases from nuclear facilities and nuclear detonations seem to produce similar changes in the infant mortality through the indirect biochemical action of fallout on the crucial hormone producing organs of the mother and the fetus, leading to a lowered resistance to the environmental stress most critical shortly after birth.

Identical patterns of rises in infant mortality have now been found for two other Boiling Water Reactors, as shown in Figure 23 for the group of ten small counties 0–40 miles around the Big Rock Point Plant in Michigan and in Figure 24 for the Humboldt Reactor near Eureka in Humboldt County, Northern California [19]. Again there is a sharp halt in the normal decline of infant mortality from its peak during the 1961–62 test series following release of large quantities of gaseous activity comparable to those released at the Dresden Reactor, while more distant areas continue their decline, as shown for the State of Michigan as a whole between 1965 and 67 (see Figure 18 and Table III).

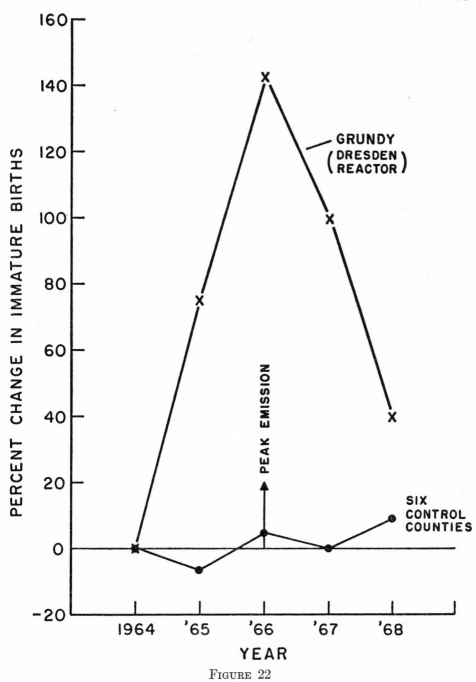

FIGURE 22

Per cent change in the fraction of births under 2500 grams for Grundy County
and the six control counties to the west [40].

TABLE V

Dresden Reactor Area

Changes in the fraction of "premature" or underweight births in the county containing the Dresden Reactor (Grundy) compared with the control counties more than 40 miles to the west. Source: [40].

Counties	Live Births by Years				
	1964	1965	1966	1967	1968
Henry	947	915	862	892	798
Knox	1114	1002	946	895	901
Lee	665	606	594	612	610
Ogle	861	798	808	745	764
Stephenson	974	889	808	793	841
Winnebago	5004	4780	4788	4794	4324
Grundy	445	426	474	457	460
Counties	Premature Births less than 2500 Grams by Years				
	1964	1965	1966	1967	1968
Henry	52	57	49	65	53
Knox	84	68	78	56	66
Lee	46	28	30	41	37
Ogle	51	40	53	46	54
Stephenson	55	70	59	41	59
Winnebago	355	342	353	324	336
Grundy	16	27	42	33	23
Counties	Premature Birth Rate/100 Live Births (under 2500 grams)				
	1964	1965	1966	1967	1968
Henry	5.5	6.2	5.7	7.3	6.6
Knox	7.5	6.8	8.2	6.3	7.3
Lee	6.9	4.6	5.1	6.7	6.1
Ogle	5.9	5.0	6.6	6.2	7.1
Stephenson	5.7	7.9	7.3	5.2	7.0
Winnebago	7.1	7.2	7.4	6.8	7.7
Grundy	3.6	6.3	8.7	7.2	5.0

8. Infant mortality and nuclear fuel processing facilities

As described elsewhere in greater detail [24], the same pattern occurred also for the commercial fuel reprocessing plant operated by the Nuclear Fuel Services Company in West-Valley, N.Y. after it went into operation in April of 1966 [25]. Figure 27 shows that the counties of western New York within a 30–50 mile radius rose sharply in infant mortality the following year, while the more distant counties declined as did New York State as a whole. Like Humboldt County, the nearby areas had shown a peak near the height of weapons testing, then began to decline only to reverse this trend sharply after the onset of large radioactive waste releases.

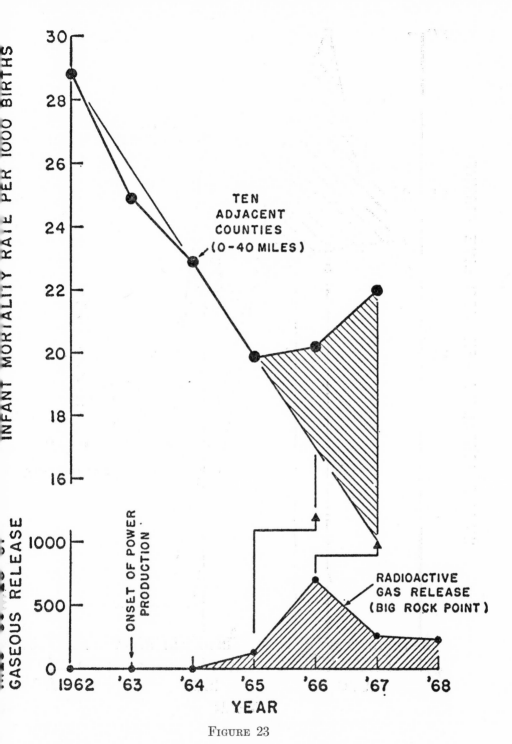

FIGURE 23

Infant mortality rate per 1000 live births for a group of ten counties within a radius of about 40 miles of the Big Rock Point Nuclear Plant in Charlevoix, Michigan, together with the yearly gaseous activity released. The total number of deaths in these counties was 45 in 1966.

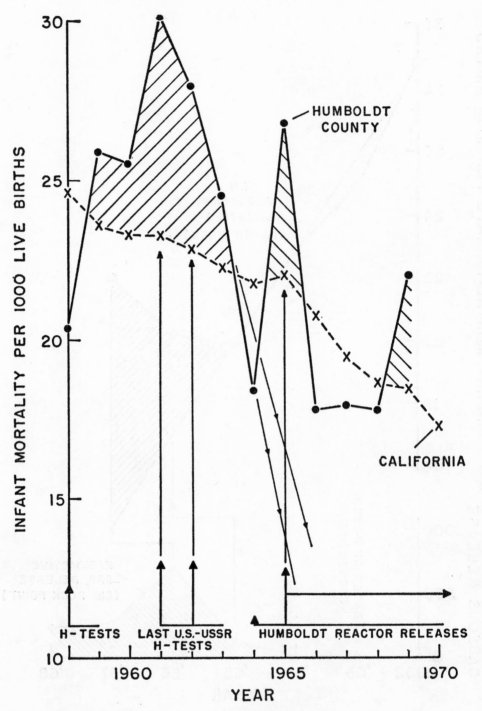

FIGURE 24

Infant mortality rate per 1000 live births in Humboldt County, California, 1958–1969 [39]. Releases from the Humboldt Reactor increased from 5975 curies gaseous waste in 1964 to 197,000 curies in 1965. Further rises took place in 1967–1968. Liquid waste discharges rose steadily to a peak of 3.2 curies in 1968, corresponding to 19.7 per cent of the permissible limit. Note the peaks corresponding to the 1961–62 nuclear tests, and the steady decline of California as a whole after 1961.

178

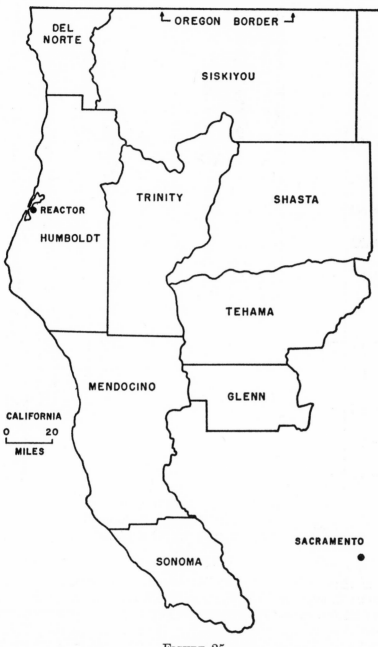

FIGURE 25

Map of northern California showing location of Humboldt Reactor and the counties along the Pacific Coast as well as in the dry area to the east of the coastal mountains.

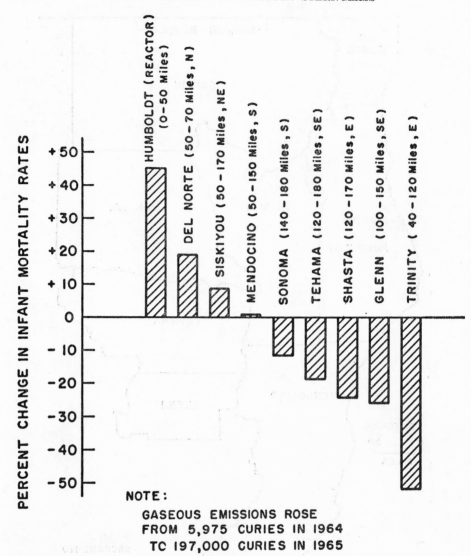

FIGURE 26

Per cent changes in infant mortality rates for the counties surrounding the Humboldt Reactor between 1964 and 1965, when gaseous releases rose from 5975 to 197,000 curies. Only Humboldt and Del Norte County immediately adjacent along the Pacific Coast showed significant rises greater than ten per cent. All other counties either remained constant or declined, especially those separated from Humboldt by the coastal mountain ranges such as Trinity, Shasta, Tehama and Glenn to the east and southeast. (Based on data from [39].)

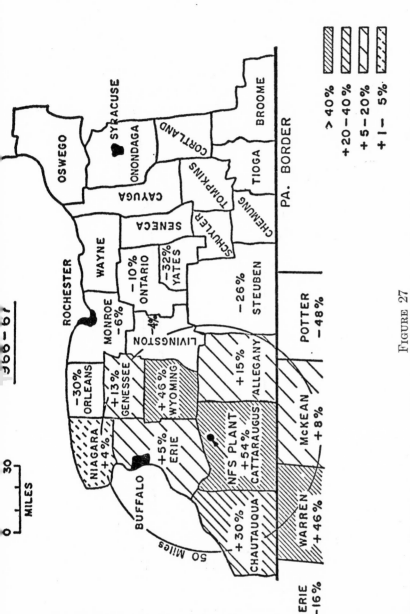

FIGURE 27

Change in infant mortality rates between 1966 and 1967 near the Nuclear Fuel Services Plant in Cattaraugus County, N.Y., after it went into operation in April 1966. Note rises for ring of counties within 40 to 60 miles, and declines at greater distances to the east and northeast in New York State. Counties in Pennsylvania along the Allegheny River flowing south from Cattaraugus County such as Warren and Venango, also showed sharp rises in this period.

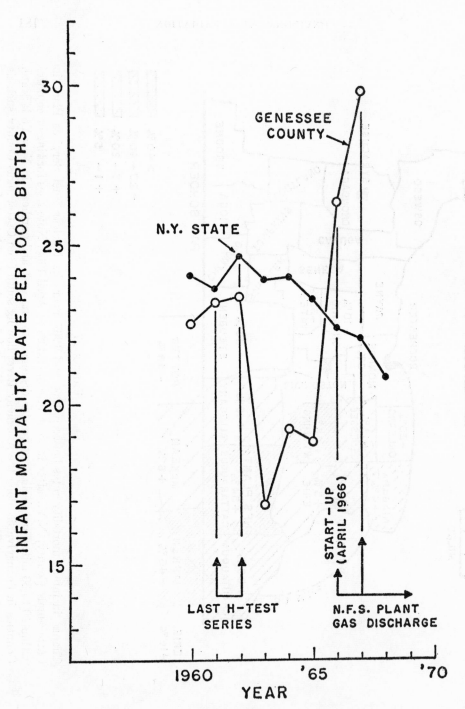

FIGURE 28

Infant mortality rates between 1960 and 1967 for a typical county in western New York State within 40 miles of the Nuclear Fuels Services Plant in Cattaraugus County. Note sharp rise above the rate for New York State as a whole when plant releases started in early 1966.

A typical case is Genessee County, N.Y., shown in Figure 28, where infant mortality rates began to exceed those of the rest of the state only after onset of plant operation. A similar time history was observed for Warren and Venango Counties downstream along the Allegheny River some of whose tributaries come within a few miles of the plant in Cattaraugus County.

9. Infant mortality near Gas Cooled Reactor

That even the relatively smaller radioactive gas releases from a Gas Cooled Nuclear Reactor appear to be capable of producing detectable rises in infant mortality is shown for the case of the Peach Bottom Reactor located on the Susquehanna River in York County, Pennsylvania. Figure 29 again shows the typical drop in infant mortality after cessation of atmospheric tests for the two counties on either side of the plant, namely York and Lancaster and the agriculturally similar control county, Lebanon, 30 to 50 miles to the north. The decline continued until the onset of a large increase in emissions resulting from fuel failure that started in 1968 and reached 109 curies in 1968 [19]. After 1967 York and Lancaster reversed their trend, while the more distant control county merely slowed its rate of decline.

Part of the reason why even the small releases from the Peach Bottom Reactor could have had such a strong effect seems to lie in the fact that the surrounding area is a major dairy farming region, where such biologically important but relatively short lived rare gas daughter products as cesium 138 and strontium 89 known to be produced in large amounts from the escaping xenon 138 and krypton 89 [19] can rapidly enter the body through the locally produced milk and other dairy products. Thus, the number of curies released able to produce serious biological effects can be much smaller than from a fuel processing plant discharging mainly Kr 85 that has no radioactive daughter product.

10. Nuclear air pollution and respiratory disease mortality

But the potential damage is not merely confined to the newborn and young child. There is evidence that suggests that the many radioactive gases presently released from nuclear reactors and nuclear tests may have a serious effect on the incidence of chronic diseases of the respiratory system such as bronchitis and emphysema that equal or even exceed the effects of conventional chemical air pollutants.

This is more strikingly shown in Figure 30, which shows the number of deaths due to respiratory diseases other than influenza and pneumonia per 100,000 population in New Mexico and New York State between 1942 and 1966.

It is clear that between 1945 and 1950, there was a sharp rise of deaths due to noninfectious respiratory diseases such that the incidence of these diseases previously very low in the pollution free air of New Mexico, exceeded the death

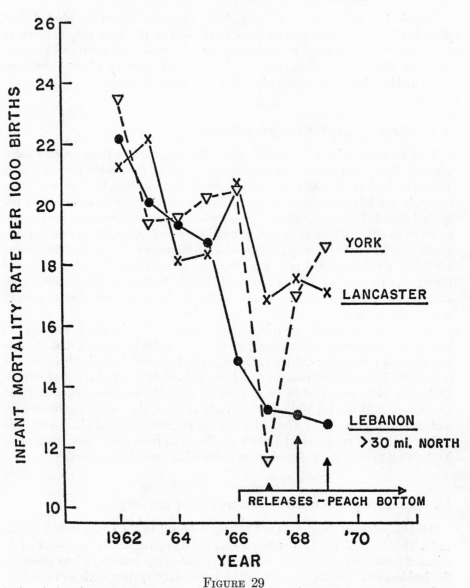

FIGURE 29

Infant mortality rates for the area near the Peach Bottom Reactor, York County, Pennsylvania, before and after onset of significant releases of gaseous activity in 1967–1968, compared with rates in nearby Lancaster, directly adjacent to the east of the reactor, and Lebanon, more than 30 miles to the north of Lancaster. Releases were 0.00126 curies in 1966, 7.76 curies in 1967, 109 curies in 1968 and 100 curies in 1969.

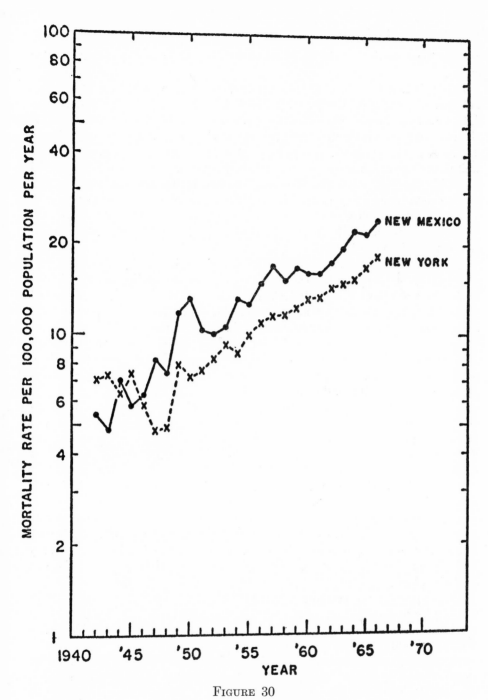

FIGURE 30

Mortality rate due to respiratory diseases other than pneumonia and influenza per 100,000 population for New York and New Mexico between 1941 and 1965. These diseases are principally emphysema, bronchitis and asthma.

185

rate for the same diseases in heavily polluted New York by as much as a factor
of two.

That this is not an isolated case perhaps associated with a sudden influx of
older people into New Mexico after 1950 follows from Figure 31 where similar
data on deaths due to respiratory diseases have been plotted for Wyoming and
Illinois. Again, there is the dramatic rise of chronic obstructive lung disease
deaths in a state of almost no ordinary air pollution such as Wyoming to levels
well above the death rates in heavily industrialized and polluted Illinois. And
a similar situation exists for Wyoming relative to heavily polluted Pennsylvania, where respiratory death rates in 1944 were five times higher than in
Wyoming before nuclear testing began, while in recent years the rate in Wyoming
began to exceed that in Pennsylvania, despite the fact that the chemical pollution is much lower in Wyoming.

Such an apparently strong effect of radioactivity in the dry air of the west
central part of the U.S., fits the observed high beta-radiation activity in the
dusty areas of the western states relative to that in the high rainfall areas east
of the Mississippi, where the activity sinks into the soil to give lower air concentrations but higher strontium 90 levels in the milk [26].

That the operation of Boiling Water Nuclear reactors with their discharge
of large quantities of radioactive gases appears to have had a more serious effect
on the rate of noninfectious respiratory disease than the operation of fossil fuel
plants may also be inferred from Figure 31.

In the decade 1949 to 1959, prior to the start of Dresden releases, the mortality rate for these diseases rose only some 10 per cent despite a 100 per cent
increase in power generated. But in the years following onset of Dresden operations, the rate of rise increased almost ten-fold, exceeding that of either New
York or Pennsylvania. And since the onset of Dresden emissions, respiratory
diseases and bronchitis as a cause of death in infants over 28 days in Illinois
showed the sharpest rise among all causes of death [27].

Laboratory evidence that inhaled fission products such as the rare-earth
isotopes can in fact produce chronic obstructive lung disease in animals such
as fibroadenomas, severe chronic inflammatory changes and added susceptibility
to infectious lung diseases, has recently been reported by H. L. Berke and
D. Deitch [28].

That especially the newborn infant between 0 and 28 days old seems to be
affected by fission products acting on the lung may be seen from a plot of the
rate of respiratory disease deaths among infants in the U.S. other than pneumonia and influenza shown in [8].

This rate rose suddenly by a factor of ten between 1949 and 1957, the
time of onset of heavy atmospheric testing, declining again after the end of
atmospheric testing in Nevada in 1958.

That the rate of increase of respiratory cancer was also affected by the sharp
rise in atmospheric radioactivity from nuclear testing is indicated in Figure 32,
where the relative changes in lung cancer rates per 100,000 population have

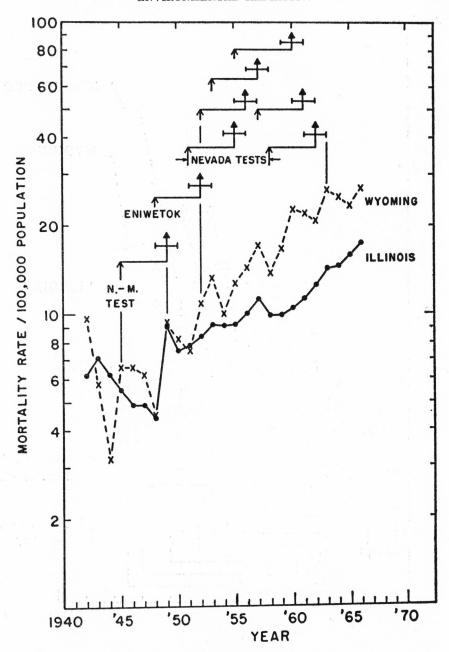

FIGURE 31

Mortality rate per 100,000 population for respiratory diseases other than pneumonia and influenza for Wyoming and Illinois. Note also sharp rise in Illinois after onset of Dresden operation in 1959.

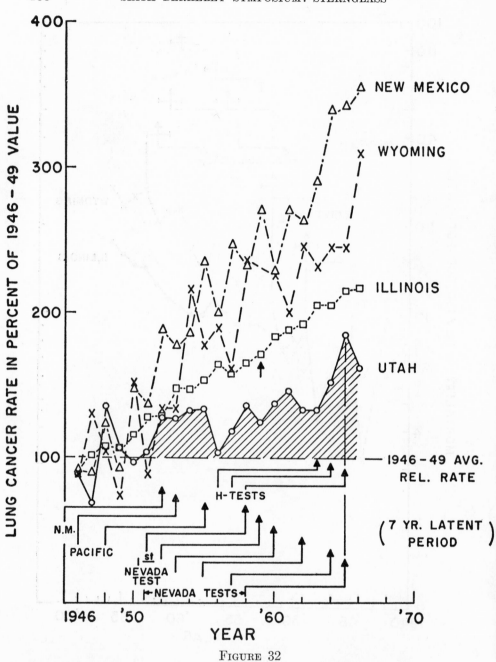

FIGURE 32

Relative changes in the rates of respiratory cancers per 100,000 population in Utah, Illinois, Wyoming and New Mexico, 1946–1966. Also shown are the principal atmospheric weapons test [39]. (1946–49 average rate = 100).

been plotted for Illinois and three dry western mountain states between 1946 and 1966. Using the average rate for 1946 to 1949 as reference rate equal to 100, rises in cancer rates for the lung, trachia and bronchus started to rise sharply some five to nine years after the first nuclear detonation in New Mexico. The greatest relative rise took place in New Mexico, followed by Wyoming, which showed its greatest rise some seven years after onset of Nevada testing.

Illinois, despite its heavy air pollution presumably acting synergistically with the radioactivity in the air rose about half as much as Wyoming and New Mexico by 1966 relative to 1946–49, while Utah, with its lower air pollution, nevertheless approached Illinois in its relative rise some seven to eight years after the peak of atmospheric testing in nearby Nevada. The observed lag of about seven years is consistent with the observed latency period of five to ten years for the uranium miners who developed lung cancer.

11. Infant mortality changes near a Pressurized Water Reactor (PWR)

In view of the proposed large increase in the amount of nuclear generating facilities to be installed near large metropolitan areas such as New York City, it seemed desirable to carry out a study of possible health effects on children in the greater New York Metropolitan Area from the releases of nuclear facilities that have been operating in this region for the past ten to fifteen years.

The most important sources of radioactive effluent close to the New York Metropolitan Area have been the Indian Point Pressurized Water Reactor located in Westchester County along the Hudson River some 20 miles north of New York City, and the Gas Cooled Nuclear Reactor at the Brookhaven National Laboratory near Upton, Suffolk County, Long Island 50 miles east of Manhattan.

The study was based on the available data for infant mortality rates for all the counties of New York State within a radius of 100 miles of New York City as published in the Annual Vital Statistics Reports of the New York State Department of Health [32]. Information on the releases from the Indian Point Unit number 1 were obtained from a report of the U.S. Department of Health, Education and Welfare [19], as well as official AEC summaries of reactor releases [33]. It is important to note that the releases of radioactive gases and liquids, with the exception of tritium, were much higher from the Indian Point Reactor than from Naval-type Pressurized Water Reactors such as Shippingport [19]. Figures on releases of wastes from the Brookhaven National Laboratory as well as on external radiation doses produced by gaseous releases and fallout were obtained from a report by A. P. Hull [34], using the average weekly dose rates at monitoring stations at the northeastern edge of the laboratory grounds and 4.8 miles away to the north.

The basic data on releases taken from these sources is reproduced in Tables VI and VII.

TABLE VI

RADIOACTIVE WASTE DISCHARGES FROM INDIAN POINT UNIT No. 1

Radioactive waste discharges from the Indian Point Pressurized Water Reactor, Unit No. 1. Note the large drop in liquid waste discharges expressed in per cent of permissible levels subsequent to the replacement of the original core in 1966.

Taken from U.S. Public Health Service Report BRH/DER 70-2 (March 1970).
The 1969 data are taken from A.E.C. Report,
testimony of Commissioner J. T. Ramey, Pa. Senate, Oct., 1970.
The last three entries for 1 year average are
based on radionuclide analysis.
N.R. means not reported. Note that
new fuel core installed in 1966.

Year	Gaseous waste noble and act. gases Curies	Tritium in liquid waste Curies	Liquid waste gross β and γ Curies	Liquid waste-gross β and γ as % of permissible limit 1 yr. av.	Liquid waste-gross β and γ as % of permissible limit 2 yr. av.
1963	0.0072	N.R.	0.164	0.26	0.24
1964	13.2	N.R.	13.0	22.0	11.13
1965	33.1	N.R.	26.3	43.0	32.50
1966	36.4	125	43.7	70.1	56.50
1967	23.4	297	28.0	1.55	35.80
1968	59.7	787	34.6	1.65	1.60
1969	600	1100	28.0	1.50	1.58

In order to account for other factors known to affect infant mortality such as socioeconomic, medical care, diet, drugs, pesticides, climate, air pollution, infectious diseases, fallout and various unknown factors that might influence the changes in infant mortality besides low level radiation from plant releases, all mortality changes in the counties near the plant were compared with neighboring counties of similar socioeconomic character having no large sources of radioactive effluent.

Thus, Westchester and Rockland may be compared most closely with Nassau County, Long Island, since it has a similar total population of close to one million, similar suburban character, and closely similar fallout levels as well as similar socioeconomic characteristics.

Furthermore, as shown in the map of lower New York State (Figure 33), it is possible to use progressively more distant counties of New York State stretching in the form of a sector towards the northwest and north as control counties.

In order to correct for the fact that these counties further to the north have a more rural character than Westchester and therefore different socioeconomic situations, medical care and air pollution, one can normalize the infant mortality rates in a suitable fashion and then examine the per cent changes following the onset of emissions. Since a given small dose of radiation is expected to have closely the same relative effect on mortality changes regardless of the

TABLE VII

EXTERNAL BACKGROUND RADIATION DOSE RATES AND WASTE DISCHARGES
AT BROOKHAVEN NATIONAL LABORATORIES (BNL)

Based on data by A. P. Hull [34].
The dose/year of BNL release is the difference between dose
measured at Northeast Perimeter Station and station 4.8 miles north.
The total dose and the fallout dose are measured at 4.8 miles
north of BNL perimeter; the dose measures in 1952 are from
station 3.5 miles south of BNL perimeter.
The year 1951 has lowest background rate at station 4.8 miles
north of BNL perimeter, taken as normal background rate
prior to major weapons testing and releases from BNL.

Year	Total mr/wk	Fallout dose mr/wk.	Fallout dose mr/yr.	Dose/yr. BNL release mr/yr	Liquid waste input to BNL filter bed mCi/yr	Liquid waste released from BNL filter bed mCi/yr.
1949	1.80	0.21	10.9	—	—	—
1950	1.74	0.15	7.8	—	—	—
1951	1.59	0.00	0.0	5.2	160.5	21.5
1952	—	0.03	1.5	3.6	116.6	27.9
1953	1.73	0.14	7.3	3.1	132.9	35.8
1954	1.66	0.07	3.7	5.2	182.1	48.5
1955	1.70	0.11	5.7	13.5	223.8	75.0
1956	1.79	0.20	10.4	7.8	170.0	55.0
1957	1.89	0.30	15.6	10.4	300.8	105.1
1958	2.23	0.64	33.2	20.8	325.1	106.0
1959	2.58	0.99	51.5	6.8	586.6	169.5
1960	1.88	0.29	15.1	3.6	542.9	177.8
1961	1.73	0.14	7.3	7.3	384.4	219.1
1962	2.41	0.82	42.8	5.2	128.9	135.9
1963	3.05	1.46	76.0	29.6	127.5	99.4
1964	2.65	1.06	55.2	28.6	89.0	76.4
1965	2.07	0.48	25.0	15.6	66.8	41.8
1966	1.77	0.18	9.4	12.0	85.1	37.2
1967	1.73	0.14	7.3	4.7	81.2	47.9
1968	1.70	0.11	5.7	2.6	21.5	16.2
1969	1.65	0.06	3.1	0		

absolute rate, this technique allows one to detect changes in time as well as changes with distance from the source despite such differences as medical care and economic level.

The counties with smaller population can then be conveniently grouped into larger units with approximately the same distance from the point of release of the effluent.

The simplest and most direct test is to plot the pattern of mortality among infants born live and 0–1 year at death per 1000 live births for the two counties immediately surrounding the Indian Point Reactor and compare it with the time history in Nassau County 30 to 50 miles away (see Figure 34).

As can be seen from an inspection of Figure [34], for a period of six years

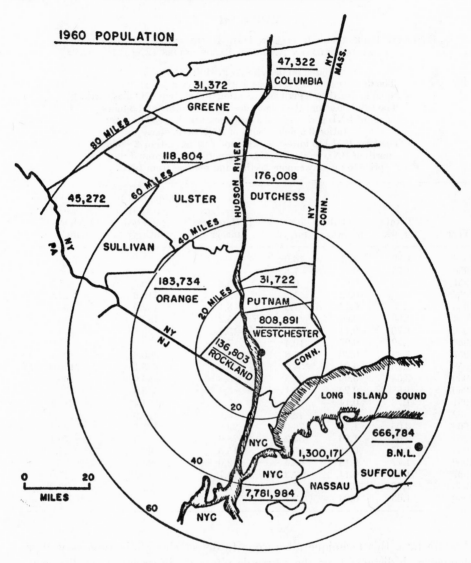

1960 POPULATION

FIGURE 33

Map of lower New York State showing the location of the Indian Point Plant
in Westchester and the Brookhaven National Laboratory in Suffolk.
Population figures are those for 1960.

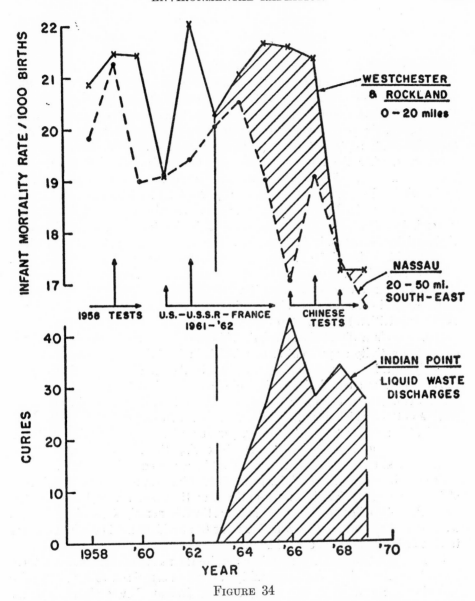

FIGURE 34

Infant mortality rates for Westchester and Rockland Counties compared with the rates for Nassau 1958–1969. Also shown is the liquid radioactive waste other than tritium released from the Indian Point Plant.

prior to the onset of large releases from the Indian Point Plant in 1964, the infant mortality rates for Nassau and Westchester-Rockland were essentially the same within the statistical fluctuation of about five per cent or ± 1.0 per 1000 births that exists for the observed 300 to 500 deaths per year. There were rises apparently associated with the fallout from the large test series in 1958 and 1961–62 prior to the onset of large releases of the Indian Point Plant in 1964 but the two counties showed exactly the same infant mortality rates of 19.1 in 1961, the year of lowest fallout in the air and diet just prior to the resumption of atmospheric testing by the U.S.S.R. in the fall of 1961.

However, after the releases began from the Indian Point Reactor, while Nassau infant mortality moved downward as did most areas of the U.S. following the end of nuclear testing [6] [7], Westchester and Rockland moved upward and remained high for a period of four successive years. Not until after the emissions began to show a tendency to decline following the replacement of the original fuel core in 1966 that had developed serious leaks [19] did Westchester and Rockland infant mortality decline close to where Nassau had moved.

If one now plots the difference in infant mortality between the two counties nearest the reactor and compares it with the annual releases of liquid radioactive waste in the form of mixed fission products (beta and gamma emitters other than tritium) (Figure 35) expressed as per cent excess over the Nassau rate, one finds a direct linear relationship between excess mortality and the amount of activity as per cent of permissible limit of liquid releases.

Applying a least square fitting procedure to the data for the period 1963 to 1969 one obtains a correlation coefficient $C = 0.835$. A still better fit is obtained for the two year average, or $C = 0.974$. The t test of statistical significance gives $t = 9.96$ which for the present case of five degrees of freedom gives $P < 0.01$. Since, as Figure 36 shows, gaseous releases closely followed liquid releases in magnitude, not only areas bordering the Hudson River but also areas exposed to the gaseous releases would be expected to be affected.

As an independent check of this result, it is of interest to compare the changes of infant mortality for the two counties near the reactor with those counties more than 40 miles to the north and northwest, namely Columbia, Greene, Sullivan and Ulster, grouped together so as to provide a total population closer to that of Westchester and Rockland.

In order to allow such a comparison despite the more rural character of these control counties, their infant mortality rate was normalized to equal that for Westchester-Rockland in 1961, the year when Nassau showed the same infant mortality rate as the two counties next to Indian Point. Figure 37 thus shows the per cent changes relative to the year 1961, again both before and after the emissions began.

It is seen that as in the case of the comparison with Nassau County in Figure 34, the control group shows a very similar pattern prior to 1964, but as soon as the releases occurred, a gap between the nearby and the distant counties begins to appear amounting to about four standard deviations by 1966. The control

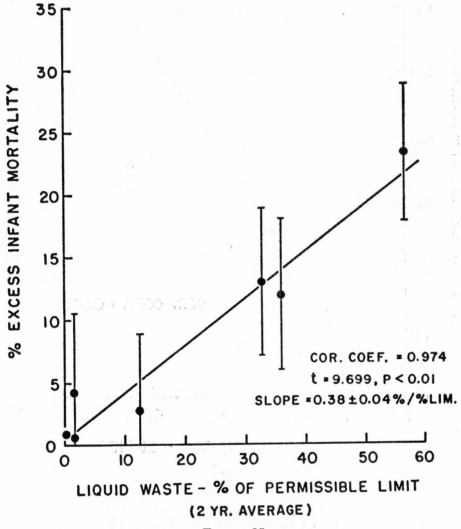

COR. COEF. = 0.974

t = 9.699, P < 0.01

SLOPE = 0.38 ± 0.04%/%LIM.

LIQUID WASTE - % OF PERMISSIBLE LIMIT
(2 YR. AVERAGE)

FIGURE 35

Per cent excess infant mortality for Westchester and Rockland Counties relative to Nassau *versus* the annual amounts of liquid waste discharged from Indian Point, expressed in per cent of permissible limit.

counties show a rapid decline in infant mortality while the nearby counties show a rise followed by years of failure to decline.

Once again, one can examine the correlation between the excess in the infant mortality of the exposed counties as compared to the more distant control counties, as shown in Figure 38. As in the case of the use of Nassau as a control,

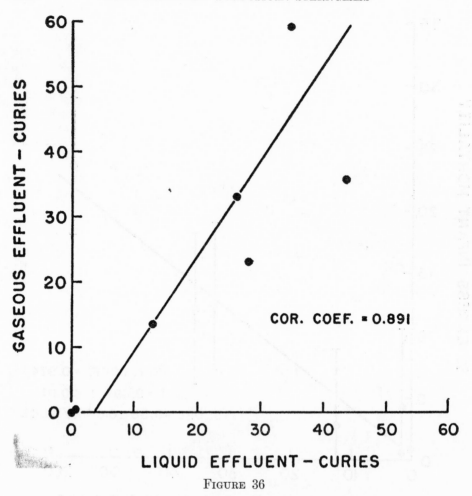

FIGURE 36

Correlation between liquid and gaseous effluent in the form of noble and activation gases from the Indian Point Plant 1963–1968
as reported in the P.H.S. Publication BRH-DER-70–2.

there is a strong, positive correlation between excess mortality and the quantity
of radioactive wastes discharged in per cent of permissible limit. The correlation
coefficient is found to be 0.957 and $t = 7.37$, which for the five degrees of freedom
leads again to a small probability $P < 0.01$ that this association is a pure chance
occurrence. Furthermore, the amount of change per unit radioactive discharge
is found to be closely the same using this group of controls as when Nassau
County was used, within the accuracy of the data.

Using the same normalization procedure for the group of intermediate counties
to the north of Westchester and Rockland, namely Dutchess, Orange and

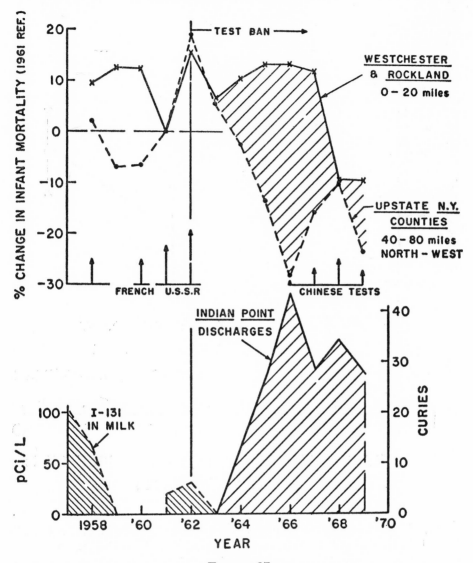

FIGURE 37

Changes in infant mortality relative to 1961 for Westchester and Rockland compared with four upstate control counties 40 to 80 miles north. Also shown are Indian Point liquid releases and iodine 131 in New York City milk in average monthly concentrations (pCi/liter).

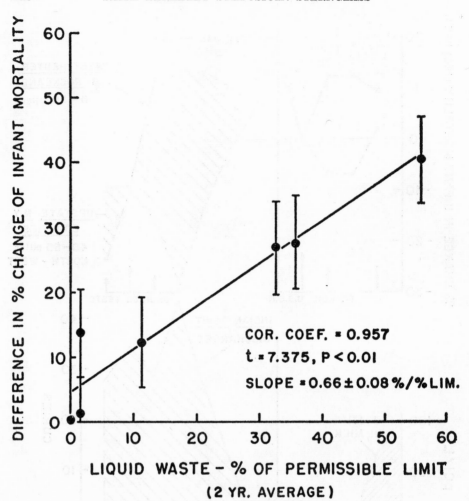

FIGURE 38

Correlation between per cent excess infant mortality for Westchester and Rockland relative to upstate control counties and liquid waste
discharges from the Indian Point Reactor.

Putnam, it is now possible to test whether they show a pattern intermediate
between the nearby and more distant counties during the period of peak emissions from the Indian Point Plant.

The result for the year of peak emission (1966) is shown in Figure 39, where
the three groups of counties have been plotted according to their average distances from the Indian Point Plant in Westchester County. Not only do the
intermediate counties show the required intermediate position in the change

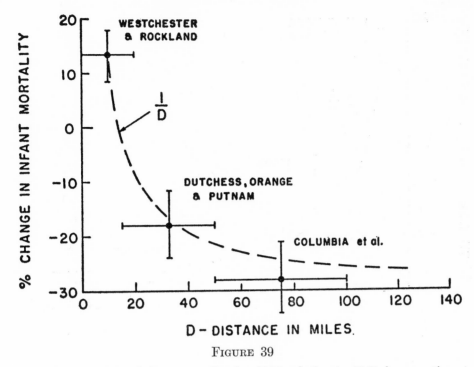

FIGURE 39

Per cent changes in infant mortality by 1966 relative to 1961 for counties at increasing distances from the Indian Point Plant moving north.

of infant mortality, but the three groups show a dependence on distance consistent with an inverse first power law expected for long lived gases diffusing from a stack [35].

As a further test of the hypothesis that the infant mortality changes are associated with releases from the Indian Point Plant, one can make the same plot for Nassau and Suffolk counties to the southeast as shown in Figure 40, and again the pattern of declining mortality fits the hypothesis.

It is of interest to see whether despite its much poorer socioeconomic pattern, air pollution problems and medical care, New York City shows a decline in infant mortality during the time that Westchester and Rockland showed a rise above the 1961 level. Using the same normalization procedure, the infant mortality for New York City shown in Figure 41 is in fact found to decline after 1964, though not as rapidly as the more remote counties to the north and east. Thus, the pattern of infant mortality changes following the onset of radioactivity releases from the Indian Point Plant as shown in the bar-graph of Figure 42 is consistent with a causal effect of the releases on infant mortality, similar to the effects already noted for seven other nuclear reactors and fuel processing facilities.

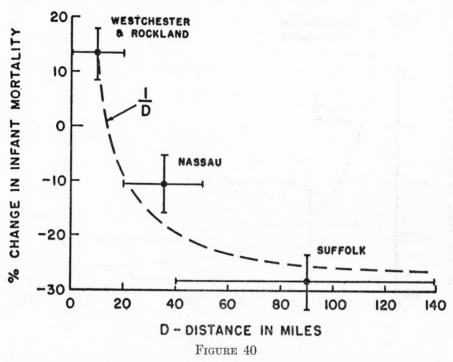

FIGURE 40

Per cent changes in infant mortality by 1966 relative to 1961 for counties at increasing distances from the Indian Point Plant moving southeast.

Taking either the control counties to the north or to the east as a reference, the excess infant mortality associated with a release of 43.7 curies per year of mixed fission products in liquid waste and 36.4 curies of noble and activation gases is 41 per cent. For the year 1966, this represents an excess mortality of approximately 100 infants 0–1 year old in Westchester and Rockland Counties combined out of a total of 367 infants that died in their first year of life during 1966.

For New York City, assuming that the relative changes shown for 1966 in Figure 42 can be attributed to the plant releases, the excess mortality would be approximately 26 per cent. This would mean that out of the total of 3,686 infant deaths in 1966 some 750 probably died as a result of the operation of the Indian Point Plant. Thus, although New York City is more distant than Westchester and Rockland, due to its large population, the total number of additional deaths is some seven times larger than for the nearby counties.

12. Effects of low level fallout from nuclear testing in Long Island, N.Y.

These results are so serious that it is essential to apply still further tests in an effort to see whether the observed association is likely to be of a causal nature.

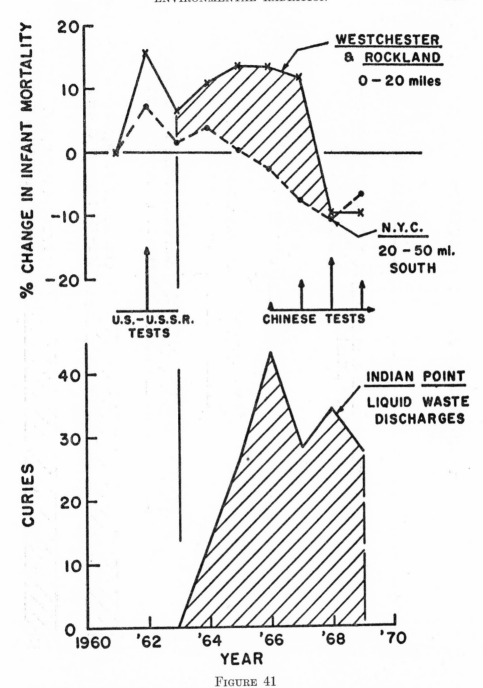

FIGURE 41

Changes in infant mortality for Westchester and Rockland compared
with New York City relative to the 1961 rates.

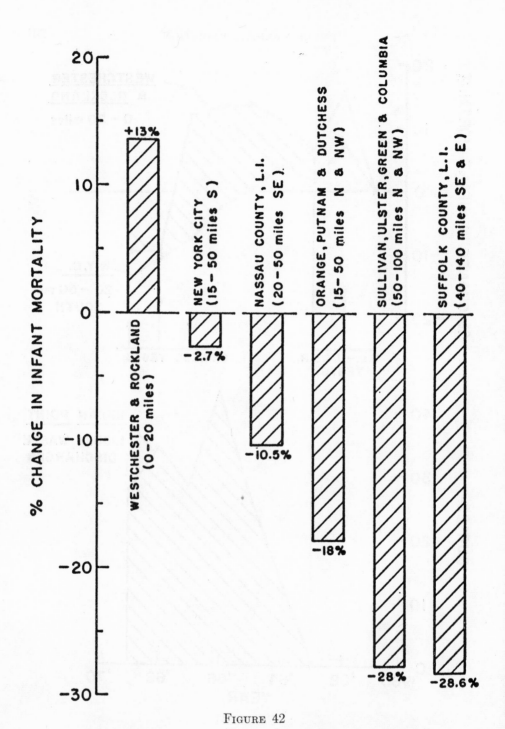

FIGURE 42

Per cent changes of infant mortality for the year of peak releases from the
Indian Point Plant by 1966 relative to 1961 for all New York
counties within a radius of 100 miles.

202

Thus, if low levels of radiation near a nuclear plant, typically well below the 500 mr per year allowable to any individual or of the order of a few millirads per year, can indeed produce such serious effects on the early embryo, then effects should be seen for the low level fallout radiation measured at Brookhaven over a period of many years [32].

Assuming that Nassau County on Long Island just west of Suffolk County received essentially the same fallout levels as Brookhaven, it is possible to see whether the changing levels of annual fallout dose were in fact accompanied by corresponding changes in infant mortality in Nassau.

The data on infant mortality rates for Nassau are shown in Figures 43 and 44 for the period following the first large H-bomb tests in the Pacific in 1954, together with the annual external gamma radiation dose as measured at Brookhaven [34] (see Table VII).

It is seen that as the radiation dose rose from about 6 mr/year in 1955 to 51.5 mr/year in 1959, infant mortality rose 17 per cent from 18.1 to 21.2 per thousand live births. This first rise was followed by a second peak associated with the 1961–62 test series, again followed within a year by a renewed peak in infant mortality.

Using the line connecting the points for 1955 before the rise and 1966 after the end of large-scale testing as a reference, it is possible to arrive at estimates for the yearly excess infant mortality and compare them with the measured external gamma dose.

The result of this comparison is shown in Figure 44. It is seen that the excess infant mortality in Nassau is indeed highly correlated with the changing levels of fallout radiation varying up and down as fallout levels rose and declined repeatedly. The correlation coefficient is found to be 0.797, with a t value of 4.172, corresponding to $P < 0.01$, making it a highly significant association.

The slope of the line is found to be 0.22 ± 0.05 per cent per mr/year. Thus, this data suggests that a dose of as little as 1 millirad of fallout per year radiation from the ground, or only about one per cent of natural background radiation, leads to almost a $\frac{1}{4}$ per cent increase in infant mortality.

13. Leukemia in Nassau County, Long Island, associated with fallout

As still another test of the hypothesis that such small levels of radiation can in fact lead to detectable rises in leukemia even when given over a period of months instead of in a few seconds as occurs for diagnostic X-rays, one can examine the changes in leukemia in Nassau County and compare them with the changes in external gamma radiation from fallout.

Since the typical latency period for leukemia is some four to six years for the infant irradiated *in utero* or early postnatal life [1], [2], [3], [4], the comparison must be carried out with the radiation level existing five years earlier.

The leukemia data for Nassau County are shown in Figure 46, together with the measured external radiation dose five years prior to the reported leukemia mortality.

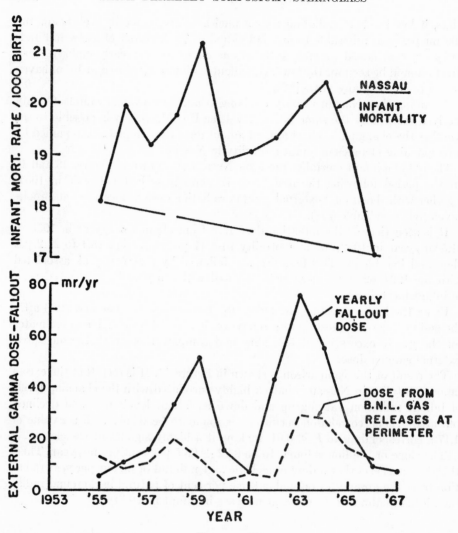

FIGURE 43

Infant mortality in Nassau County during period of peak nuclear testing in the atmosphere compared with external radiation levels measured at Brookhaven. Also shown are annual doses from gaseous releases measured at the northeast perimeter of the Brookhaven Laboratory.

Inspection of Figure 45 shows a striking parallel behavior for the two quantities. This is confirmed by the calculated correlation, which is strong and positive with a correlation coefficient of 0.819, $t = 3.503$ corresponding to $P < 0.02$. (Figure 46).

The slope obtained by the least square fit is 0.49 ± 0.13 per cent/mr/year,

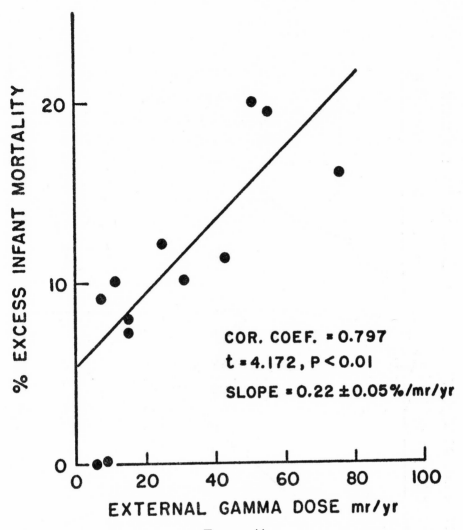

FIGURE 44

Excess infant mortality in Nassau County relative to the 1955–66 baseline *versus* the external gamma radiation dose measured at Brookhaven National Laboratory. The slope of the least square fitted line corresponds to a 22 per cent increase for a dose of only 100 mr per year.

comparable with the slope relating the per cent increase of infant mortality and fallout radiation.

From this result, one can calculate the doubling dose, or the dose for a 100 per cent increase, of 204 ± 54 mr per year, equal to 51 ± 13 mr in any three month period. Considering that this represents only external dose, a total

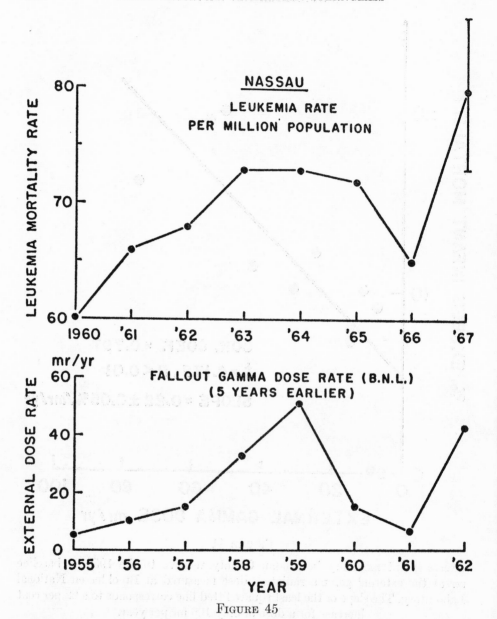

FIGURE 45

Leukemia rate per million population for Nassau County compared with the measured external gamma radiation rate from fallout five years earlier.

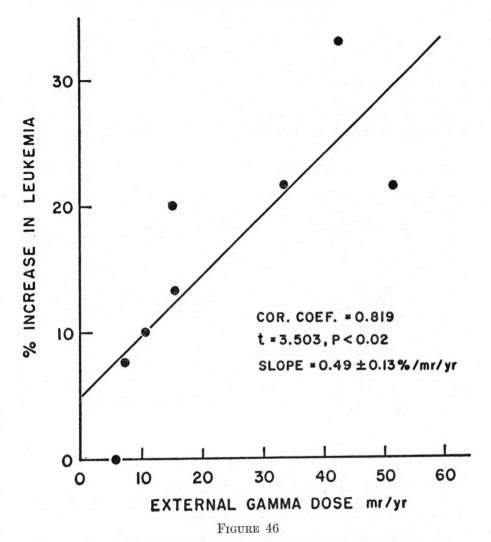

FIGURE 46

Correlation between the per cent increase in leukemia rates in Nassau County
and the annual dose from external fallout radiation. The least square fitted line
corresponds to an increase of 49 per cent for a dose of 100 mr per year.

doubling dose of 80 mr to the early embryo in the first trimester as obtained
from the study of diagnostic X-ray effects is therefore not unreasonable for
fallout radiation as well. Thus there is no evidence for any decreased effectiveness of protracted as compared with high dose rate radiation from diagnostic
X-rays in the early embryo and fetus.

One should therefore not be surprised to find similar changes in infant mortality that involve subtle genetic defects leading to slight immaturity at birth,

which by itself tends to increase greatly the chance of death from respiratory
or infectious diseases [22]. Such changes in immaturity or lowered weight at
birth have in fact been observed in the county in which the Dresden Reactor
is located and among children born in the U.S. since the early 1950's [22], the
time when large scale nuclear testing began, a trend that has only recently begun
to reverse itself.

In fact, mortality for all age groups showed sharp upward changes beginning
in the early 1950's as first pointed out by I. M. Moriyama [9].

14. Infant mortality near Brookhaven National Laboratory, Long Island, N.Y.

These considerations therefore lead one to expect that the gaseous and liquid
effluent from the Brookhaven Gas Cooled Reactor may also have led to detectable changes in infant mortality in Suffolk County.

That this appears in fact to have been the case is shown in Figure 47, where
the infant mortality in Suffolk County is plotted together with the reported
radioactive effluent produced and discharged at Brookhaven. The anomalous
rise of infant mortality in Suffolk between 1953 and 1960 relative to Nassau is
strongly associated with the reported activity produced at Brookhaven and
the fraction released into the streams [34] which in turn reaches the wells serving
for irrigation and drinking water supplies in Suffolk County. Both before and
after this period, Suffolk and Nassau showed the same infant mortality rates.
And with the drastic reduction in releases that took place since the peak of
activity in 1959, infant mortality in Suffolk County dropped from a high of 24.1
in 1960 to 19.3 in 1963, an unprecedented drop of 25 per cent in only three years.

15. Infant mortality and underground tests

On a number of occasions, significant quantities of radioactivity have escaped
into the atmosphere from underground nuclear detonations. The most serious
of these occurred at the Nevada Test Site on December 18, 1970, in the course
of a test of a so-called "tactical" weapon announced to be in the 20 kiloton
range [36].

A radioactive cloud was reported to have risen to an altitude of some 8,000
feet, which drifted off towards the north and northwest [36] or towards Idaho
and Montana.

As tabulated in Radiological Health Data and Reports [37], the levels of
gross beta activity in surface air, the total deposition of beta activity on the
ground and the levels of cesium 137 in the milk increased during the month of
December 1970 over large areas of the northwestern and central United States
as well as southern Canada shown for the case of milk in Figure 48.

Upward changes in the levels of cesium 137 in the milk were recorded for
many states relative to the previous 12 months average. The highest levels were
recorded in Montana, the rise being 13 pC/l, corresponding to an increase of

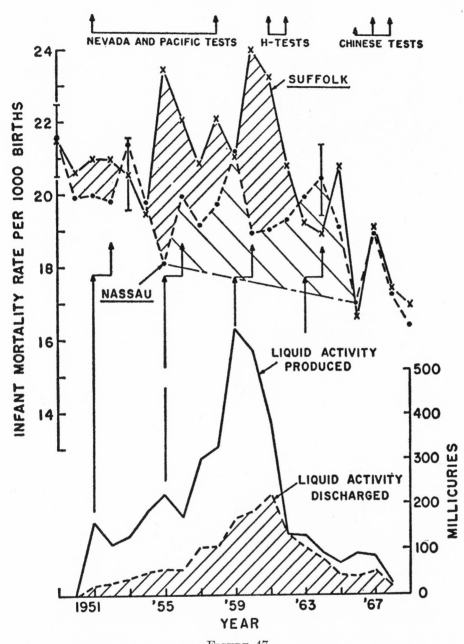

FIGURE 47

Infant mortality rates for Suffolk and Nassau Counties, 1949–1969, compared
with the releases of liquid radioactive waste from the
Brookhaven National Laboratory in Suffolk County.

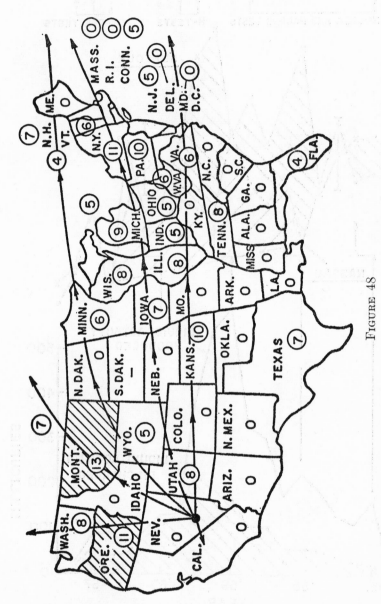

FIGURE 48

Rises in cesium 137 in the milk for the U.S. and Canada, in December 1970 relative to the previous 12 month average. Source: Radiation Health Data and Reports (April 1971). States showing no rises or declines are marked 0. Note largest increase in Montana.

186 per cent. No rises or actual declines took place throughout the entire southern United States, with the exception of Texas, Tennessee and Florida. Rises in the cesium content of milk took place also across a number of states in the northern United States and southern Canada, together with rises in ground deposition for areas more than 3,000 miles from the test site as shown in Figure 49 for Canada, although maximum air concentrations or ground deposition did not always result in maximum milk concentrations due to differences in precipitation and cattle feeding practices.

The Monthly Vital Statistics Reports published by the National Center for Health Statistics of the U.S. Department of Health, Education and Welfare for the first three months of 1971 were examined for changes in infant mortality rates in each state. Figure 50 shows that for the months of January and February 1971 relative to the average for January and February 1969 and 1970, infant mortality rose most sharply in the states immediately to the north and northeast of Nevada, while it declined for most of the more distant states of the South and East, with the exception of Maine, Connecticut and Kentucky where localized precipitation presumably led to higher contamination levels.

A similar pattern was found to hold for the months of January, February and March combined relative to the corresponding periods in 1969 and 1970, with a general decline in infant mortality excesses relative to the first two months following the detonation.

Thus, the most recent accidental release of radioactivity into the environment for which much more complete documentation of radiation levels exists strongly supports the hypothesis of a direct causal connection with infant mortality originally observed for the states downwind from the very first atomic test in New Mexico in July of 1945.

16. Summary and conclusion

The evidence of rises in infant mortality, congenital defects and childhood cancers associated with nuclear testing has recently been corroborated by similar rises in infant mortality in the vicinity of four different types of nuclear facilities known to release quantities of radioactive gases into the environment that led to nearby environmental activity levels comparable to those measured during nuclear weapons tests.

In both types of low level exposure, infant mortality was associated with increased frequency of underweight birth or immaturity. Thus it appears that low level radiation acting on the early embryo, fetus, and young infant can not only lead to significant rises in diseases previously known to be produced by radiation such as congenital defects and cancer, but it also appears to act indirectly so as to produce small decreases in maturity at birth that in turn can increase the chance of early death from various causes such as respiratory distress and infectious diseases.

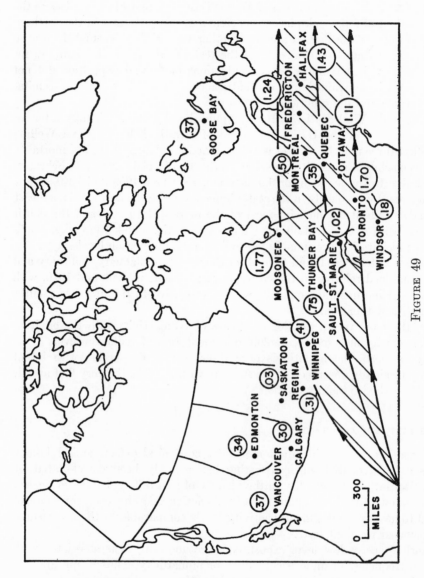

FIGURE 49

Total ground deposition of beta activity for December 1970 in Canada. Deposition is measured in units of nanocuries per square meter.

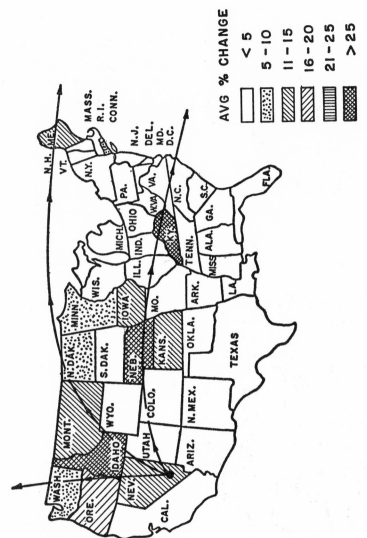

FIGURE 50

Pattern of infant mortality increases for January and February 1971 relative to the same months in 1969 and 1970. Note especially the rises along the northernmost path observed by ground deposition in Canada.

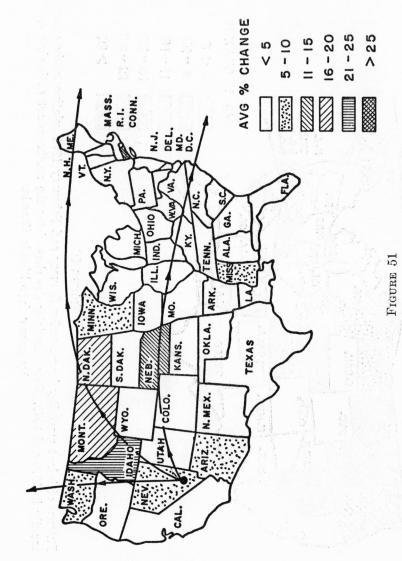

FIGURE 51

Pattern of infant mortality increases for period January–March 1971 relative to same period in 1969 and 1970. Note persistence of earlier pattern at lower intensity north and northeast of the Baneberry test site in Nevada.

In view of the present findings, it appears likely that both infant mortality and chronic diseases for all ages having genetic components and involving subtle disturbances of the cell chemistry may have been more seriously affected by low level environmental radiation than had been expected on the basis of high level radiation studies on laboratory animals carried out mainly with external X-rays and gamma rays.

REFERENCES

[1] A. STEWART, J. WEBB, and D. HEWITT, "A survey of childhood malignancies," *Brit. Med. J.*, Vol. 1 (1958), pp. 1495–1508.

[2] B. MacMAHON, "Pre-natal X-ray exposure and childhood cancers," *J. Nat. Cancer Inst.*, Vol. 28 (1962), pp. 1173–1191.

[3] E. J. STERNGLASS, "Cancer: relation of prenatal radiation to development of the disease in childhood," *Science*, Vol. 140 (1963), pp. 1102–1104.

[4] A. STEWART and G. W. KNEALE, "Radiation dose effects in relation to obstetric X-rays and childhood cancers," *Lancet*, Vol. 1, pp. 1185–1188.

[5] R. LAPP, "Nevada test fallout and radioiodine in milk," *Science*, Vol. 137 (1962), pp. 756–758.

[6] E. J. STERNGLASS, "Evidence for low-level radiation effects on the human embryo and fetus," *Radiation Biology of the Fetal and Juvenile Mammal*, AEC Symposium Series, Vol. 17, 1969, pp. 693–717 (Proceedings of 9th Hanford Biology Symposium).

[7] ———, "Infant mortality and nuclear tests," *Bull. Atomic Sci.*, Vol. 25 (1969), pp. 18–20.

[8] I. M. MORIYAMA, "Recent changes in infant mortality trend," *Pub. Health Rep.*, Vol. 75 (1960), pp. 391–406.

[9] ———, "The change in mortality trend in the United States," National Center for Health Statistics, Ser. 3, No. 1 (1964), pp. 1–45.

[10] E. J. STERNGLASS, "Infant mortality and nuclear testing; a reply," *Quart. Bull.*, *Amer. Assoc. Phys. Med.*, Vol. 4 (1970), pp. 115–119.

[11] United Nations Scientific Committee on the Effects of Radiation, 24th Session, Supplement No. 13 (A/7613), 1969.

[12] "Major Activities in the Atomic Energy Programs," Semi-Annual Reports, January 1952 and later years.

[13] C. W. MAYS, "Iodine-131 in Utah during July and August 1962," *Hearings on Fallout, Radiation Standards and Counter-Measures, Joint Committee on Atomic Energy*, Part 2 (1963), pp. 536–563; also, R. C. PENDLETON, R. D. LLOYD, and C. W. MAYS, *Science*, Vol. 141 (1963), pp. 640–642.

[14] E. REISS, in *Hearings on Fallout Radiation Standards and Counter-Measures, Joint Committee on Atomic Energy*, Part 2 (1963), pp. 601–672.

[15] W. A. MULLER, "Gonad dose in male mice after incorporation of strontium-90," *Nature*, Vol. 214 (1967), pp. 931–933.

[16] E. SPODE, "Über die Verteilung von Radioyttrium und radioaktiven seltener Erden im Saugerorganismus," *Z. Naturforschung*, Vol. 13b (1958), pp. 286–291.

[17] E. H. GRAUL and H. HUNDESHAGEN, "Studies of the organ distribution of yttrium 90," *Strahlentherapie*, Vol. 106 (1958), pp. 405–457.

[18] D. H. SLADE (editor), *Meteorology and Atomic Energy*, USAEC, Div. Tech. Inf. (T1D-24190), 1968. (See Chapter 1, Section 2.2, p. 5ff.)

[19] "Radioactive waste discharges to the environment from nuclear power facilities," U.S. Dept. of H. E. W., P. H. S., Bur. Rad. Health (BRH-DER 70-2), Rockville, Md., March 1970.

[20] "Radiological surveillance studies at a boiling water nuclear power station," U.S. Dept. H. E. W., P. H. S., Bur. Rad. Health (BRH-DER 70-1), Rockville, Md., March 1970.

[21] E. J. STERNGLASS, "A reply," *Bull. Atomic Sci.*, Vol. 26 (1970), pp. 41–42; 47.

[22] H. C. CHASE and M. E. BYRNES, "Trends in prematurity in the United States," *Amer. J. Public Health*, Vol. 60 (1970), pp. 1967–1983.

[23] J. H. KNELSON, "Environmental influence on intrauterine lung development," *Arch. Internal Med.*, Vol. 127 (1971), pp. 421–425.

[24] E. J. STERNGLASS, "Infant mortality changes near a nuclear fuel reprocessing facility," University of Pittsburgh, Nov. 1970, presented at the Liscensing Hearings, Davis-Besse Nuclear Plant, January 7, 1971, AEC Docket No. 50-346.

[25] B. SHLEIEN, "An estimate of radiation doses received in vicinity of a nuclear fuel reprocessing plant," U.S. Dept. H. E. W., Bur. Rad. Health (BRH-NERHL 70-1), Rockville, Md., May 1970; also, BRH-NERHL 70-3, July 1970.

[26] *Radiological Health Data and Reports*, published monthly by the Bureau of Rad. Health, Rockville, Md.

[27] "Ten leading causes of infant death," *Illinois Vital Statistics*, 1963–1968, Table D, Illinois Department of Health, Springfield, Ill.

[28] H. L. BERKE and D. DEITCH, "Pathological effects in the rat after repetitive exposure to europium 152-154," *Inhalation Carcinogenesis*, AEC Symposium Series, Vol. 18, 1970. pp. 429–431.

[29] J. I. MOSKALEV, L. A. BULDAKOV, A. M. LYAGINSKAYA, E. P. OVCHARENKO, and T. M. EGOROVA, "Experimental study of radionuclide transfer through the placenta and their biological action on the fetus," *Radiation Biology of the Fetal and Juvenile Mammal*, AEC Symposium Series, Vol. 17, 1969, pp. 153–166.

[30] T. M. FLIEDNER, R. J. HAAS, F. BOHNE, and E. B. HARRISS, "Radiation effects produced in pregnant rats and their offspring by continuous infusion of tritiated thymidime," *Radiation Biology of the Fetal and Juvenile Mammal*, AEC Symposium Series, Vol. 17, 1969, pp. 263–282.

[31] D. F. CAHILL, and C. L. YUILE, "Some effects of tritiated water on mammalian fetal development," *Radiation Biology of the Fetal and Juvenile Mammal*, AEC Symposium Series, Vol. 17, 1969, pp. 283–287.

[32] *Annual Statistical Reports*, New York State Department of Health, H. S. Ingraham, Commissioner, Albany, N.Y., available through 1967.

[33] Testimony of Commissioner J. T. Ramey, Hearings before the Pennsylvania Senate Select Committee on Reactor Siting, October 1970.

[34] A. P. HULL, "Background radiation levels at Brookhaven National Laboratory," report submitted May 15, 1970, at the Licensing Hearings, Shoreham Nuclear Plant (SEC Docket No. 50-322).

[35] M. J. MAY and I. F. STUART, "Comparison of calculated and measured long term gamma doses from a stack effluent of radioactive gases," *Environmental Surveillance in the Vicinity of Nuclear Facilities* (edited by W. C. Reinig), Springfield, Illinois, C. C. Thomas Co., 1970, p. 234.

[36] ASSOCIATED PRESS, Mercury, Nevada, December 18, 1970.

[37] *Radiation Health Data and Reports*, Vol. 12, No. 4 (1971), pp. 171–234, published by the Environmental Protection Agency, Rockville, Md.

[38] N. C. DYER and A. B. BRILL, "Fetal radiation dose from maternally administered ^{59}Fe and ^{131}I," *Radiation Biology of the Fetal and Juvenile Mammal*, AEC Symposium Series, Vol. 17, 1969, pp. 73–88.

[39] DEPARTMENT OF HEALTH, EDUCATION, and WELFARE, BUREAU OF STANDARDS, *U.S. Vital Statistics*, Washington, D.C., Government Printing Office.

[40] ILLINOIS DEPARTMENT OF HEALTH, *Illinois Vital Statistics, 1963–1968*, Springfield, Illinois.

Discussion

Question: Alexander Grendon, Donner Laboratory, University of California, Berkeley

The brevity of the discussion period did not allow me to make any comment. Part of this comment was made after the close of the morning session; the rest is what I would have said, given time.

(1) You chose as your base the period 1935–1950. Sulfa drugs were introduced about 1935; penicillin about 1945. These treatments caused a sharp decline in the death rate. Why would you expect that sharp decline to continue rather than expecting a return to the gradual decline seen pre- 1935 and again post- 1950?

(2) In one of your curves, included in the paper you prepared for the Health Physics Society meeting (Figure 13), you show a high correlation between radio-active gas releases from Dresden and per cent excess infant mortality rates in the same years. Since radiation does not kill *promptly*, except at rates $>10^6$ times those possible here, why relate mortality to emissions in the *same* year rather than, say, to emissions in the preceding year or earlier? I displaced your data by one year and, by eye, observe approximately zero correlation.

(3) In citing infant mortality around the Humboldt Reactor and relating it to gaseous emissions, you chose the years 1964–1965 to prove your point. The following data were supplied by the California Department of Public Health (slide shown):

Year	Humboldt infant mortality per 1000	Gaseous releases Ci($\times$100)	Del Norte infant mortality	Humboldt fetal mortality
1963	24.5	7	10/398	29
1964	20.0	60	7/369	27
1965	27.0	1,970	8/347	26
1966	18.4	2,820	8/281	21
1967	17.2	8,960s	4/285	19

You chose the only pair of years in which the data seem to support your hypothesis. How do you justify that choice? And did you know that there was a *rubella* epidemic in Humboldt County in 1965, cases having risen from 49 in all of 1964 to 626 in the first nine months of 1965?

Reply: E. J. Sternglass

(1) Indeed, in those areas of the United States where, due to low rainfall, there was very little fallout, such as New Mexico after 1945, infant mortality did continue to decline without any levelling off whatsoever. After 1963 in many states having no nuclear facilities (such as Maine), infant mortality resumed its decline, recovering or actually exceeding the rapid rate of decline characteristic of the period 1935–1950. Thus, it is clear that the halt in decline was a temporary phenomenon highly associated with fallout.

(2) For the case of fetal and infant mortality, deaths occur primarily in the

period up to a few days to a few weeks after birth. Therefore the delay period between exposure to radiation and the observed deaths can be as short as a few months. The deaths are of an indirect type such as hyaline membrane disease, not of the true radiation type observed at very high doses.

(3) With respect to *rubella* epidemics, such outbreaks of infectious diseases are precisely what are found whenever radiation exposure is high. It is found that the periods of very high fallout are followed by a weakening of the newborn expressed in a subsequent abrupt rise in all infectious diseases. Such a rise in *rubella* incidence and in the incidents of other diseases occurred around the Dresden Reactor following the period of its emission peak. Therefore, finding the same correlation in the case of the Humboldt Reactor actually confirms the hypothesis.

Furthermore, the evidence for increased premature births during the active period of the Dresden Reactor, and its decline during the inactive period contributes additional support to the hypothesis of an indirect effect.

The choice of the period 1964–1965 as the period of investigation of the Humboldt Reactor area was not made by me. The mortality rose for the first time according to the Bureau of Radiation and Health's publication, "Radioactive waste discharges to the environment from nuclear power facilities" (Figure 4-2, [19]) when the emission level rose from as low as a few hundred curies per month in 1964 to as high as 100,000 curies per month in August 1965. This, in fact, is why a comparison was made between the years 1964 and 1965; there had been a thousand-fold increase in the rate of emission. Cumulatively, this meant 197,000 curies per year, resulting in the forced shutdown of the reactor from September 1965 to December 1965. Subsequently, the reactor was not permitted to be operated at such high emission levels again until the end of 1966. Therefore, 1966 showed a sharp drop in infant mortality, fully confirming the hypothesis.

Toward the end of 1966, emission levels once more increased (see Figure 4-2 [19]), and so for 1966 as a whole, the total emission was approximately one third higher than in 1965. But the children who were irradiated did not begin to show the effects until 1967, when (instead of declining to 10–15 deaths/1000/year as expected from the 1961 to 1964 rate of decline), infant mortality remained high in 1967–1968, while emission levels were allowed to reach 100,000 curies/month once again. And so, by 1969 there was another sharp rise in infant mortality. The rate for Humboldt County was 22 infant deaths/1000, exceeding the rate for the rest of California which was 18 infant deaths/1000. Thus, the data for Humboldt County fully confirm the hypothesis that the large releases from boiling-water reactors result in anomalously high infant mortality in the adjacent areas.

Question: J. Neyman, Statistical Laboratory, University of California, Berkeley

While the graphs exhibited by Dr. Sternglass are impressive, the question in my mind is whether the indicated increases in infant mortality are really caused by radiation. It is well known that, as time goes on, the environment is increasingly affected by all kinds of pollution. Therefore, it is possible that the increases in mortality indicated by Dr. Sternglass represent a sum total of all the pollutants and, possibly, are not due to radiation.

In particular, with reference to the first slide of Dr. Sternglass illustrating an increase of deaths in parallel with the increase of X-ray films taken during pregnancy, the question arises whether the extra X-ray pictures were taken *because* of some difficulties in pregnancy which later manifested themselves in deaths of the newborn.

In general, causal relations can be established only through controlled experiments. With observational studies, a comprehensive statistical analysis concerned with many suspected pollutants and many localities may represent an approximation to an experiment.

Reply: E. J. Sternglass

With regard to X-ray examinations possibly being performed on disease-prone individuals, note that extensive examination of the question has been accomplished (see Brian MacMahon). In a study of 800,000 cases X-rayed in New England hospitals, it was concluded that no such association existed. Furthermore, the fact that there exists a direct relationship between the number of X-rays and the increase in risk could not be explained by a suggestion that the X-rayed individuals were cancer-prone. Since the number of X-rays taken was the same for all types of patients, and related to how many views the physician requested, and how many repeats were necessary due to improper exposures, there can be no association between the number of X-rays and any inherent characteristics of the mother irradiated.

With respect to causal association of low level radiation and leukemia and cancer incidence, there is no question that this has been established in laboratory experiments on all types of animals, and for the case of man one has the bombing effects in Hiroshima and the fallout incidents in Utah and Albany, N.Y. where everyone was irradiated regardless of prior medical history or tendencies. The direct dose response in Hiroshima makes it highly unlikely that the association of radiation and cancer, leukemia, and similarly genetically caused problems could be other than causal. Furthermore, the relationship between low level radiation and reduction in birth weight has been demonstrated in the laboratory. There is no question that low levels retard growth in animal litters, and, by inference from studies of children exposed in Hiroshima and accidentally to therapeutic radiation, in man. Furthermore, it has been observed in a series of about 1000 pregnant mothers irradiated in the course of regular pelvic X-rays (which had been prescribed for all pregnant women studied by Dr. M. L. Griem at the University of Chicago in the mid 1940's), that their offspring showed a significant increase in all types of illnesses and defects. Note that no selection of patients had been made in this study; every pregnant woman who requested normal prenatal care was irradiated. Similarly, no selection based on difficulties before or during pregnancy was involved in the case of some 800 pregnant women given tracer-doses of iron-59 in a study of nutritional requirements by

P. F. Hahn and co-workers at Vanderbilt University in the late 1940's. Here again a significant excess of leukemia and other cancer cases were observed among the 679 offspring (P = 0.03), despite the extremely small fetal doses that ranged from as little as 35.9 millirads to 1,780 millirads, comparable to doses received in many fallout situations and diagnostic X-rays [38]. Thus, it seems extremely unlikely that any other conclusion can be drawn than that low level radiation produces cancer, leukemia, and retarded growth leading to increased sensitivity to respiratory and other infectious diseases.

Question: T. Sterling, Department of Applied Mathematics and Computer Sciences, Washington University

It is true that Dr. Sternglass has used some poor data and some good data. It is also true that he has deferred to the judgment of recognized experts. For example, Dr. Stewart's data is based on a self selected sample and possibly biased. The thymus data (that is, children radiated for "thymus" disease when infants, and followed in a prospective study) has many of the same flaws pointed out ad nauseum by Saenger, Silverman and myself (E. L. Saenger, F. N. Silverman, T. D. Sterling, and M. E. Turner, "Neoplasia following therapeutic irradiation for benign conditions in childhood," *Radiology*, Vol. 74 (1960), pp. 889–904). *But*, the confrontations of Doctors Sternglass and Totter are part of a pattern. In the last 15 years we have been in controversy—to mention just a few: the birth control pill, low dose radiation, smoking, and now, pesticides and others. These controversies are based on a solid base of opinions and very little "hard" data. Constantly we review reviews. What are needed are (1) mechanisms to review data, and (2) authoritative statements of the kinds of data needed for inferences in the areas of controversy in question.

The American Statistical Association has had a splendid record for performing such public services through special commissions. Another source for authoritative criteria would be the newly formed section on Biostatistics of the National Academy of Science—National Research Council. They ought to be asked to consider such questions and act on them.

Reply: E. J. Sternglass

I am in full agreement that this information should be examined by the NAS, and it is my understanding that they are in the process of doing so. As to the data by Stewart and MacMahon, it is unlikely that some unrecognized factor crept into their selection of patients. That the very factors which epidemiologists are trained to recognize should have escaped the scrutiny of such noted epidemiologists is a remote possibility, which in the case of as serious a question as this seems out of place.

Question: J. R. Totter, Division of Biology and Medicine, Atomic Energy Commission

I have never before commented on Dr. Sternglass' presentation because I felt that he had so obviously and flagrantly misused data that his work should not be dignified by serious scientific comment. As statisticians or experimental biologists, you know that even in very well controlled experiments the data never fit the hypothesis being tested so beautifully as the data Dr. Sternglass has presented. I believe, therefore, that he has convicted himself of selection of data to fit his hypothesis. I shall mention only three examples.

He presented Stewart and Kneale's dose response curve for childhood cancers but did not mention the data from ABCC which appears to be totally at variance with it. Dr. Stewart's data was based on about 1800 person rads while the ABCC data was from about 35,000 person rads and, furthermore, was a relatively random sample.

Secondly, he used as a base for infant mortality the years 1935–1950. The slope of the line before 1935 was approximately the same as during the "fallout" years.

Thirdly, he did not mention the rubella epidemic during 1964–1965 in connection with the Dresden data.

Reply: E. J. Sternglass

With regard to the data of Stewart and Kneale, it should be pointed out that rad doses under conditions of medical examinations with well defined X-ray factors are far more precisely determined than could be done retrospectively for the cases from Hiroshima and Nagasaki. Furthermore, Stewart and Kneale's data was based on 15×10^6 children born in England and Wales of which ten per cent were irradiated, yielding a study population of 1.9×10^6 irradiated infants, of which 13,000 developed cancer. Some 7000 of the latter were followed up. By way of contrast, in the case of the Hiroshima-Nagasaki survivors, a mere handful of leukemia cases were born to known survivors of radiation exposure the dose of which could be accurately estimated. Thus, the ABCC study is vastly less reliable than fetal X-ray studies. Finally, the ABCC studies suffer from the fact that fallout doses which descended on suburbs and the surrounding area were considered to be zero, while E. T. Arakawa (see *New England Journal of Medicine*, Vol. 263 (1960), pp. 488–493) estimated that doses of 100 rads were experienced in the suburbs of Nagasaki from fallout alone. Thus, there exists a great uncertainty in both the exposure of the proximate Japanese population, and even of the so-called control population which supposedly (according to the ABCC) was not irradiated. From all these considerations, we must conclude that the ABCC data are far inferior to that obtained in fetal X-ray studies.

As to the choice of the baseline, 1935–1950, the choice was not mine. It was selected by Moriyama of the USPH Service, the National Center for Health Statistics, in his analysis of changes in trends in infant mortality which he first identified in the 1960's. Furthermore, this slope has now resumed its early value since the end of testing, and the totally unexpected renewed decline of infant mortality has followed the gradual decrease in radioactivity in the environment.

With regard to the question of the occurrence of *rubella* epidemics as an alternative explanation of the rises in infant mortality, note that such epidemics of childhood infectious diseases are in fact expected on the basis that the fetus and embryo are weakened by exposure to radiation so that the newborn has reduced ability to fight off infections of all types.

STATISTICAL STUDIES OF THE EFFECT OF LOW LEVEL RADIATION FROM NUCLEAR REACTORS ON HUMAN HEALTH

MORRIS H. DeGROOT

CARNEGIE-MELLON UNIVERSITY

1. Possible effects of nuclear reactors

Government policy with regard to the construction and operation of nuclear power plants is of great public concern, not only because of the possibility of a serious accident at one of these plants, but also because of the possibility that radioactive discharges from these plants during their routine operation may affect the health of nearby populations. In particular, because of the vulnerability of the human fetus, it is possible that exposure of a population to these discharges may be reflected in the infant mortality rate, the fetal death rate, the prematurity rate, and similar health indices of the population.

Since several nuclear reactors have been in operation in the United States for at least five years, and some for more than ten years, the relevant data for a statistical study of this problem are largely available in published records. A study of this type would necessarily be retrospective in nature and confined to short term effects of low level radiation. If these effects are discernible, then they should be reflected in certain relationships between the health indices mentioned above for a given population and various measures of radioactivity in the environment.

2. Populations to be considered

Annual infant and fetal mortality rates, as well as prematurity rates, are typically available on a county by county basis in the published vital statistics of each state. It is suggested for simplicity, therefore, that counties form the basic units of population to be considered. Thus, for a given reactor, annual health indices for the county containing the reactor and for nearby counties would be investigated over a period both before and after the reactor became critical for possible relations with measures of the total annual radioactive

This research was supported in part by the National Science Foundation under grant GP-23708.

discharges from the reactor. Obviously, the counties that might be affected by a given reactor can lie in more than one state.

Furthermore, the health indices for a given county containing or near a reactor can be compared with the corresponding indices in certain "control" counties which are located far from the reactor but which are similar to the given county with regard to other characteristics.

3. Variables to be considered

The basic purpose of the type of study being discussed here is to relate health indices such as the annual infant mortality, fetal mortality, prematurity, and fertility rates for a given population to measures of the annual amounts and compositions of radioactive gaseous and liquid discharges from a given nuclear reactor. It is clear, however, that many other variables besides the radioactive discharges from the reactor can affect these health indices.

Some of the variables which ideally should be included in the study are the distribution of the population by age, sex, and race; meteorological data pertinent to the times of discharge of gaseous effluents and to the geographic distribution of the population; socioeconomic indices such as income, housing, education, and the quality of medical care; the sources of food and water; natural background radiation levels; radioactive fallout from bomb tests; levels of air pollution, both SO_2 and particulate matter; and personal characteristics, such as smoking and dietary habits. Obviously, the list could be extended almost indefinitely and, equally obviously, it will be very difficult to obtain the relevant data for many of them.

In addition, besides simply looking at the overall infant mortality rates for a given population and its various stratifications, it would be valuable to look at these rates for various specific causes of death. Although certain causes of death can more easily be associated with radiation effects than others, this analysis may not be as straightforward as it might at first appear. For example, it is possible that an infectious disease could cause the death of an infant whose susceptibility has been increased by exposure to low level radiation, but not cause the death of an infant who has not been so exposed. Even accidental deaths must be considered. One great hazard of being rushed to the hospital with an acute respiratory ailment is the high speed and often reckless ambulance or automobile ride that one must undergo.

Clearly, when analyzing data of the type being discussed here, one must always interpret changing rates with caution. Although there has been a general decrease in the infant mortality rate in the United States during the period from 1952 to 1967, there has also been a general increase in the prematurity rate over that period. This increase might reflect the increasingly deleterious effects of radiation, air pollution, or other environmental agents, or of changing practices of prenatal care. On the other hand, it might merely reflect changing practices in the reporting of birth weights, or even the beneficial effects of changing medical practices which convert potential fetal deaths into premature live births and consequently also bring about a decrease in the fetal death rate.

4. Results of preliminary regression analyses

In order to get a feeling for the possible magnitudes of the effects of radio-active wastes from nuclear reactors on infant mortality, and for the relative difficulty or ease with which these effects can be identified, some preliminary multiple regression analyses were carried out for the following four reactors: (1) the Dresden reactor in Grundy County, Illinois; (2) the Shippingport reactor in Beaver County, Pennsylvania; (3) the Indian Point reactor in Westchester County, New York; and (4) an experimental reactor at Brookhaven National Laboratory in Suffolk County, New York. In these regression models, the infant mortality rate in a given county containing or near a nuclear reactor, or the logarithm of this rate, was regressed on the amounts of radioactive gaseous and liquid wastes from the reactor and on either the infant mortality rate in a specified reference population or simply on a general linear trend in time. It must be emphasized that none of the multitude of other environmental agents and relevant variables listed earlier in this paper were specifically included in the models.

The general outcome of these preliminary studies is the only one that could have been anticipated, in view of the smallness of the effects and the simplicity of the model. Namely, the studies are inconclusive. They neither establish nor disprove the existence of an effect. They do, however, lead to the inescapable recommendation that more comprehensive and detailed studies of these questions are urgently needed.

The regressions that were carried out will briefly be summarized here. Time series of annual infant mortality rates for the United States as a whole, Illinois, Pennsylvania, New York, and the counties containing or near the four reactors studied are readily available from the published vital statistics of the federal government and the individual states, and will not be reproduced here. Time series of annual gaseous and liquid discharges from Dresden, Shippingport and Indian Point were obtained from the report "Radioactive Waste Discharges to the Environment from Nuclear Power Facilities" by Joe E. Logsdon and Robert I. Chissler, March, 1970, Bureau of Radiological Health, Environmental Health Service, Public Health Service, U.S. Department of Health, Education, and Welfare, Rockville, Maryland 20852. The time series of annual sand filter bed discharges and background radiation levels for the Brookhaven reactor were obtained from a report entitled "Background Radiation Levels in Brookhaven National Laboratory" by Andrew P. Hull, which was presented in March, 1970, at licensing hearings for the Shoreham nuclear power station on Long Island. None of these data are included in this paper because they are available in the sources cited and because the primary purpose of my reporting on these regression studies here is not to convince the reader of the validity and strength of particular conclusions that are reached. Rather, the purpose is to indicate that it is not possible to derive strong conclusions about either the existence or the nonexistence of an effect from the simple regression models used here, and to urge that a full scale statistical study of these problems be carried out.

5. Dresden

The Dresden reactor is located in Grundy County, Illinois, and began emitting radioactive discharges in 1960. Infant mortality rates in Grundy County were studied from 1950 to 1967, the most recent year for which these rates have been published, in order to include a relatively modern time period of reasonable length in which the reactor was inoperative as well as a time period of reasonable length in which it was active.

A relation of the following form was studied:

$$(1) \qquad M_t = \beta_0 + \beta_1 t + \beta_2 X_{2t} + \epsilon_t.$$

In this relation, the index t represents the particular year being studied in the period from 1950 to 1967. For simplicity, only the final two digits of the year were used for identification, so that the year 1958, say, would be represented by the value $t = 58$. The interpretation of the other variables is as follows: M_t is the infant mortality rate in Grundy County in the year t (that is, the number of infant deaths per 1000 live births in that year), and X_{2t} is a two year moving average (for the years t and $t - 1$) of the liquid discharge (less tritium) from Dresden measured in curies. It was originally intended also to include the yearly gaseous discharges from Dresden in equation (1), but the gaseous and liquid discharges were highly correlated over the entire period. Hence, only the liquid discharges were used in this model.

The least squares estimates of β_0, β_1, and β_2 turn out to be $\hat{\beta}_0 = 55.4$, $\hat{\beta}_1 = -0.606$, and $\hat{\beta}_2 = 1.59$. Their estimated standard deviations are 30.6, 0.546, and 0.943, respectively, which yield the following t-statistics: 1.81, -1.11, and 1.68. Each of these t-statistics has 15 degrees of freedom. Thus, one might find in these values mild evidence of a positive relationship between liquid discharges and infant mortality superimposed on a general downward trend. The peak liquid discharge from Dresden was more than ten curies in 1966 (two year moving average), and it is seen by using the least squares estimate $\hat{\beta}_2$, that this value corresponds to an infant mortality rate of 15.9 deaths per 1000 live births above the overall linear trend.

It must be emphasized that none of these estimates are very reliable. Grundy County has a population of only 22,000 and the average number of infant deaths per year during the period being studied was only 11.4. Furthermore, it must be kept in mind that even if a definite relationship between infant mortality and radioactive discharges was established by these techniques, one would still be unable to conclude definitely that by actually reducing the discharges in future years, the infant mortality rate would be reduced. In fact, the discharges may simply be surrogates for some other variables which are the actual causative agents. However, it is fair to state that each scientist would regard such statistical evidence as at least favoring to a certain extent the hypothesis that the discharges are affecting the infant mortality rate.

When a similar analysis is carried out for LaSalle County, which is directly to the west of Grundy and has a population of 110,000, no evidence of a relationship is found. The least squares estimates are $\hat{\beta}_0 = 24.3$, $\hat{\beta}_1 = -0.029$, and $\hat{\beta}_2 = -0.222$. The corresponding t-statistics are 1.95, -0.13, and -0.58, respectively.

The next step was to replace M_t in (1) by log M_t, since it is generally believed to be more appropriate to try to fit a linear trend to log M_t rather than to M_t itself. The results obtained were little changed from before. The t-statistics corresponding to $\hat{\beta}_2$ for Grundy and LaSalle Counties became 1.42 and -0.63, respectively. These values are not much different from their previous values.

Here, the fitted value of M_t is equal to the product of a factor exp $\{\hat{\beta}_0 + \hat{\beta}_1 t\}$ representing the general trend in time, and a factor exp $\{\hat{\beta}_2 X_{2t}\}$.

For Grundy County, we now have $\hat{\beta}_0 = 4.26$ and $\hat{\beta}_1 = 0.022$, and the least squares estimate of the general trend factor for 1966 is therefore 16.6. Also, we now have $\hat{\beta}_2 = 0.061$. Thus, the effect of the factor due to the liquid discharge of 10 curies in 1966 (two year moving average) is to multiply the estimated infant mortality rate for that year by exp $\{0.61\} = 1.84$. This factor therefore corresponds to an infant mortality rate of 13.9 deaths per 1000 live births above the general trend for that year. This estimated increase is not much different from the estimated increase of 15.9 found from the linear model without logarithms.

One important consideration that makes an analysis of this type somewhat questionable, is that although the radioactive liquid discharge from Dresden was 0 for each year in the 1950's and only began to be positive in the 1960's, the populations of Grundy and LaSalle Counties may have been exposed to relatively large levels of radiation during the 1950's from bomb tests that were not present in the 1960's. Thus, in fact, the exposure of the population to radiation may actually have decreased when the bomb tests of the 1950's ceased and the Dresden reactor became active in the 1960's, rather than having increased, as is implicitly assumed in the models being used here.

In order to overcome this difficulty, the linear trend $\beta_0 + \beta_1 t$ in (1) was replaced by $\beta_0 + \beta_1 X_{1t}$, where X_{1t} is the infant mortality rate in the entire United States for the year t. In other words, it was felt that the ups and downs in the infant mortality rate in the United States through the years would reflect the general exposure of the population to radioactive fallout from bomb tests as well as other pollutants and other transient and sporadic effects. Thus, rather than assuming a linear trend, we assume that the expected infant mortality rate in Grundy County is a linear function of the infant mortality rate in the United States plus a multiple of the discharges from the Dresden reactor.

The model used is therefore

$$(2) \qquad M_t = \beta_0 + \beta_1 X_{1t} + \beta_2 X_{2t} + \epsilon_t.$$

Regressions were also carried out in which X_{1t} was (i) the infant mortality rate in Illinois, rather than in the United States as a whole, (ii) the infant mortality rate in Illinois minus Cook County, since Chicago forms most of Cook County and was thought to have its own special characteristics; and (iii) the total infant mortality rate in Boone, DeWitt, Logan, McDonough, and Warren Counties in Illinois, which were chosen because they matched Grundy County to some extent with regard to their rural nature and their size, and were not near the reactor.

When X_{1t} is the United States infant mortality rate, the least squares estimates of β_0, β_1, and β_2 in (2) are $\hat{\beta}_0 = -28.4$, $\hat{\beta}_1 = 1.85$, and $\hat{\beta}_2 = 1.57$. Their estimated standard deviations are 45.8, 1.70, and 0.938, respectively. The t-statistic calculated from $\hat{\beta}_2$ is therefore 1.67. It is curious to note that this value is almost identical to the value found from equation (1).

When X_{1t} is the Illinois infant mortality rate, we find $\hat{\beta}_0 = -87.1$, $\hat{\beta}_1 = 4.36$, and $\hat{\beta}_2 = 0.937$, with estimated standard deviations 71.1, 2.85, and 0.552, respectively. The t-statistic calculated from $\hat{\beta}_2$ is therefore 1.70, again almost identical to the previous values.

When X_{1t} is the infant mortality rate in Illinois minus Cook County, we find $\hat{\beta}_0 = -51.6$, $\hat{\beta}_1 = 3.04$, and $\hat{\beta}_2 = 1.46$, with estimated standard deviations 39.9, 1.65, and 0.651, respectively. The t-statistic for $\hat{\beta}_2$ is therefore now 2.24.

When X_{1t} is the infant mortality rate in the group of matched counties, we find $\hat{\beta}_0 = 20.6$, $\hat{\beta}_1 = 0.0393$, and $\hat{\beta}_2 = 0.758$, with estimated standard deviations 16.5, 0.644, and 0.627, respectively. The t-statistic for $\hat{\beta}_2$ is now 1.21. It is interesting to note that the infant mortality rate in Grundy is almost totally unrelated to the infant mortality rate in the matching counties.

Finally, when M_t in (2) is replaced by the infant mortality rate in LaSalle County, we find that $\hat{\beta}_2$ is again negative in each of these regressions.

In summary, regardless of which regression model was used to study the infant mortality rate in small Grundy County, where the Dresden reactor is located, the coefficient $\hat{\beta}_2$ of the amount of radioactive liquid discharge was always found to be positive although the corresponding t-statistics were of modest magnitude. In neighboring LaSalle County, $\hat{\beta}_2$ was always found to be negative.

6. Shippingport

The Shippingport reactor, which is located in Beaver County, Pennsylvania, started up and began discharging tritium in 1958. It began emitting measurable radioactive gaseous and other liquid discharges the following year. The infant mortality rate in Beaver County, which has a population of 206,000 was also studied for the period from 1950 to 1967. The following regression model was used:

(3) $$M_t = \beta_0 + \beta_1 t + \beta_2 X_{2t} + \beta_3 X_{3t} + \beta_4 X_{4t} + \epsilon_t,$$

where X_{2t} is a two year moving average of the gaseous discharges from Shippingport, X_{3t} is a two year moving average of the liquid discharges (less tritium), and X_{4t} is a two year moving average of the tritium discharges. Both X_{2t} and X_{3t} are measured in millicuries and X_{4t} is measured in curies.

No evidence of a positive relationship between the discharges and the infant mortality rate was found, and some of the estimated coefficients are negative. In particular, the least squares estimates are $\hat{\beta}_0 = 55.0$, $\hat{\beta}_1 = -0.569$, $\hat{\beta}_2 = -0.0093$, $\hat{\beta}_3 = -0.0023$, and $\hat{\beta}_4 = 0.032$. The corresponding t-statistics, each with 13 degrees of freedom, are 5.28, -3.02, -0.57, -0.24, and 1.13, respectively.

When equation (3) is applied to Allegheny County, Pennsylvania, which is directly to the southeast of Beaver and has a population of 1,628,000, including Pittsburgh, the estimates are $\hat{\beta}_0 = 38.1$, $\hat{\beta}_1 = -0.239$, $\hat{\beta}_2 = -0.0060$, $\hat{\beta}_3 = 0.0050$, and $\hat{\beta}_4 = -0.021$. The corresponding t-statistics are 12.6, -4.37, -1.26, 1.79, and -2.60. The stability of the infant mortality rates in Allegheny County, because of its large population, is reflected here in the relatively large magnitudes of the t-statistics. Among the coefficients, $\hat{\beta}_2$, $\hat{\beta}_3$, and $\hat{\beta}_4$ of the discharges, however, the t-statistic with the largest magnitude corresponds to the negative coefficient $\hat{\beta}_4$ of the tritium discharge. Thus, evidence of a positive relationship is again lacking.

Similar results were obtained when the linear term in time in (3) was replaced by the infant mortality rate in either the United States or Pennsylvania.

7. Indian Point

The Indian Point reactor, which is located in Westchester County, New York, started up and began emitting radioactive liquid discharges in 1962. The infant mortality rate in Westchester County, which has a population of 853,000, was studied for the period from 1950 to 1967 and regression analyses were carried out similar to those described for the Dresden reactor. Equation (1) was studied first with M_t denoting the infant mortality rate in Westchester and X_{2t} denoting the two year moving average of liquid discharges (less tritium) from Indian Point measured in curies. The gaseous and liquid discharges were again highly correlated (both were 0 over much of the period and then they rose together), so only liquid discharges are included in the regression equation. The least squares estimates of β_0, β_1, and β_2 are $\hat{\beta}_0 = 31.7$, $\hat{\beta}_1 = -0.178$, and $\hat{\beta}_2 = 0.059$, with estimated standard deviations 3.51, 0.062, and 0.028, respectively. Thus, the value of the t-statistic corresponding to $\hat{\beta}_2$, with 15 degrees of freedom, is 2.11.

The liquid discharges from Indian Point were more than 35 curies in both 1966 and 1967, and it is seen by using the least squares estimate $\hat{\beta}_2$ that this value corresponds to an infant mortality rate of 2.03 deaths per 1000 live births above the overall linear trend.

When a similar analysis is carried out for smaller Rockland County, which is across the Hudson River from Westchester and has a population of 192,000, the results are $\hat{\beta}_0 = 25.9$, $\hat{\beta}_1 = -0.064$, and $\hat{\beta}_2 = -0.037$, and the corresponding t-statistics are 1.75, -0.25, and -0.32, respectively.

When M_t is replaced by $\log M_t$ in (1), the results are little changed. The t-statistics corresponding to $\hat{\beta}_2$ for Westchester and Rockland Counties are 2.14 and -0.25, respectively. These values are not much different from their previous values. For Westchester County we now have $\hat{\beta}_0 = 3.54$, $\hat{\beta}_1 = -0.0081$, and $\hat{\beta}_2 = 0.0028$. For 1966, the factor of the fitted infant mortality corresponding to the general trend is therefore $e^{3.00} = 20.1$. The factor corresponding to the liquid discharge of 35 curies in that year (two year moving average) is $\exp \{0.098\} = 1.10$. Thus, the estimated increase in the infant mortality rate corresponding to the liquid discharge for that year, above the general trend, is 2.01 deaths per 1000 live births. Again, this value is in close agreement with the value found from the linear model without logarithms.

Next, equation (2) was studied for models in which M_t is the infant mortality rate in Westchester County and X_{1t} is the infant mortality rate in the United States. The least squares estimates are $\hat{\beta}_0 = 6.395$, $\hat{\beta}_1 = 0.571$, and $\hat{\beta}_2 = 0.065$, and the values of the corresponding t-statistics, again with 15 degrees of freedom, are 1.12, 2.68, and 2.11, respectively.

When X_{1t} is the infant mortality rate in New York State, the estimates are $\hat{\beta}_0 = 2.231$, $\hat{\beta}_1 = 0.802$, and $\hat{\beta}_2 = 0.045$, and the corresponding t-statistics are 0.13, 1.14, and 1.05, respectively.

When M_t is taken to be the infant mortality rate in Rockland County in these two models based on equation (2), the estimates of β_2 are $\hat{\beta}_2 = -0.046$ and $\hat{\beta}_2 = 0.055$. The corresponding t-statistics are -0.37 and 0.37.

Although these data can perhaps be interpreted as evidence at least mildly favoring the existence of a positive relationship between radioactive liquid discharges from Indian Point and the infant mortality rate in Westchester County, it must be emphasized that these discharges were 0 until 1962 and then increased monotonely from 1962 to 1967. Clearly then, this simple pattern might be present in the corresponding time series of many other environmental agents, one or more of which might actually be affecting infant mortality in the magnitude being attributed here to the Indian Point reactor. The fact however that these effects seem to be slightly more established in Westchester County than in Rockland County does provide some evidence, albeit weak, against this possibility.

8. Brookhaven

The final reactor to be studied was an experimental reactor at Brookhaven National Laboratory in Suffolk County, New York, for the period from 1951, the year that radioactivity was first used at the Laboratory, to 1968. The population of Suffolk County is 666,000. The following model was used:

(4) $$M_t = \beta_0 + \beta_1 t + \beta_2 X_{2t} + \beta_3 X_{3t} + \epsilon_t,$$

where M_t was the infant mortality rate in Suffolk County in year t, X_{2t} is the two year moving average of the concentration of the sand filter bed discharge in picocuries per liter, and X_{3t} is an average of two offsite background radiation measurements made in year t and the two measurements made in year $t - 1$, measured in milliroentgens per week.

The least squares estimates turn out to be $\hat{\beta}_0 = 27.6$, $\hat{\beta}_1 = -0.142$, $\hat{\beta}_2 = 0.015$, and $\hat{\beta}_3 = -0.265$. The corresponding t-statistics, with 14 degrees of freedom, are 7.89, -2.19, 4.17, and -0.30. The striking aspect of this relation is the large t-statistic corresponding to the coefficient $\hat{\beta}_2$ of the concentration of the liquid discharge from the sand filter bed. From this relation it is found that an increase in the gross beta concentrations of the liquid releases of 300 pCi/liter, the observed value for 1961 (two year moving average) corresponds to an increase in the infant mortality rate of 4.5 deaths per 1000 live births.

These figures must again be interpreted with the greatest caution since the total amount of radioactivity in the liquid releases from the Brookhaven reactor is small, the maximum value being 219 millicuries in 1961. One interesting possibility suggested by this observation is that the actual composition of these releases may be as important as their total amount in affecting health.

It should also be noted that the background radiation levels bear essentially no relation to the infant mortality rates in Suffolk County.

When M_t is the infant mortality rate in Nassau County, which is to the west of Suffolk County on Long Island and has a population of 1,300,000, the estimates of the regression coefficients are $\hat{\beta}_0 = 24.9$, $\hat{\beta}_1 = -0.148$, $\hat{\beta}_2 = 0$ (to six decimal places), and $\hat{\beta}_3 = 1.66$. Only the years from 1951 to 1967 were included in this analysis, since the infant mortality rate in Nassau County in 1968 was not immediately available. The values of the t-statistics, with 13 degrees of freedom, for these four coefficients are 9.34, -2.79, -0.13, and 2.46, respectively. Thus, there is no evidence whatsoever of a relation between the filter bed discharge and the infant mortality rate in Nassau County, but there is now a relation in the observed data between off site background radiation levels and the infant mortality rate.

As before, when M_t is replaced by $\log M_t$ and the above analyses are carried out, the results are little changed.

9. Summary

It should be emphasized again that the results of these preliminary regression studies are inconclusive. They do not present strong evidence that there is a relationship between the exposure of a population to low level radiation from nuclear reactor discharges and the infant mortality rate in the population, and they do not present strong evidence that there is no such relation. The four reactors studied have different designs, and the inconclusive nature of these studies perhaps suggests that the actual composition of the discharges might be important, as well as whether and how these discharges enter the food chain. Some of the many other variables mentioned earlier in this paper, but not included in the regression models, are likely to be very influential.

The simple studies carried out here and their inconclusive results do lead, therefore, to a very strong and important recommendation. A large scale statistical study is urgently needed to aid in resolving this vital issue. Of course, statistical analysis can neither strictly prove nor disprove the hypothesis that exposure of a population to low level radiation increases the infant mortality rate. However, these analyses can substantially raise or lower the probability that the hypothesis is correct. Indeed, a large scale statistical study, such as the study of the effect of smoking on human health, could go far toward bringing the scientific community into agreement on this question.

In my classes, I usually define a scientist to be a person who can keep clearly in mind the distinction between the subjective utility that he assigns to any specific hypothesis and the subjective probability that he assigns to that hypothesis. In other words, a scientist must never let his hope or desire that there is no relation between low level radiation and infant mortality affect his professional evaluation of the probability that such a relation might exist. Statistical studies performed by interdisciplinary teams of scientists, in this strict sense, could provide information that will be of great help in reaching decisions regarding nuclear reactors that might critically affect large segments of the world's population.

I am indebted to Dr. Ernest J. Sternglass, who initially stimulated my interest in this topic, for many helpful conversations. I am also indebted to Dr. Lincoln J. Gerende for his kind permission to use freely material he had prepared for a research proposal submitted jointly by him, Dr. Kenneth D. Rogers, and myself to the office of the Attorney General of Pennsylvania. I am further indebted to Dr. Gerende and Dr. Floyd H. Taylor for several valuable discussions of this project. Finally, I am indebted to William J. Franks, Jr., who did most of the groundwork and all of the computations for this paper, and whose assistance has been of great value.

Discussion

Question: P. Armitage, London School of Hygiene and Tropical Medicine, London

Isn't the analysis very sensitive to the true nature of the time trend? If the trend is really quadratic (as might be expected) with the curvature, may not the X factor be taking the place of the quadratic term?

Reply: M. DeGroot

The possible effects of the curvature of the trend were investigated by fitting a linear model to the logarithm of the infant mortality rate, a model for which there is some theoretical justification, as well as to the infant mortality rate itself. As I describe in this paper, the results for the two models were in close agreement and the magnitude of the effect of the radioactive discharge was almost the same for both models. The effects of trend curvature are also greatly reduced in those models where the rate in a given county is regarded as a linear function of the rate in some control population such as the state.

Question: V. L. Sailor, Brookhaven National Laboratory

It should be pointed out that the data used by Dr. DeGroot in his analysis of the Brookhaven Laboratory situation (liquid waste) does not have a plausible connection with infant mortality. The liquid waste flows into a stream which flows to the east through a completely uninhabited area to Peconic Bay away from the high density of population. The magnitude of the emissions are so small that they can no longer be detected a few miles off site, nor do the biota show activity. The total amount released over a period of twenty years was about 1¾ curies. During the same period Suffolk County had more than 100,000 times as much radioactivity deposited on it from weapons tests fallout. Gaseous radioactive release from Brookhaven was far greater (millions of curies per year), but these releases do not correlate with infant mortality since when the gaseous releases were high, the mortality rate was dropping. When gaseous releases were reduced, the mortality rate increased.

Reply: M. DeGroot

It is true that the total amount of liquid waste from Brookhaven National Laboratory was small compared to other contaminants. It is possible, therefore, that this discharge, which was zero until 1951, built up to its peak in 1961, and then steadily diminished, is simply acting as a surrogate for some other factor which seriously affects infant mortality but which was not explicitly identified in the analysis. On the other hand, it may well be true that the important consequences of radioactive discharges are derived not simply from the total amount, but rather from the actual composition of the effluent and the way in which various elements enter the food chain or otherwise reach the embryo.

Furthermore, the effect of radioactive releases on infant mortality cannot be measured simply by noting whether infant mortality went up or down in a given year, since there are obviously many other factors affecting infant mortality. The relevant measure of the effect of radioactive releases must be given in terms of whether or not infant mortality was higher in the given year than *it would have been* if these releases were not present but all the other factors were. It is this type of measure that the statistical methods described in this paper attempt to evaluate.

Question: J. Neyman, Statistical Laboratory, University of California, Berkeley

I am curious about the possible change in the socioeconomic composition of the population in a given county that might have occurred after a nuclear facility went into operation.

Also, how variable were the year to year numbers of live births in a given county. Did these numbers exhibit some temporal trend, and could there be any danger of some spurious correlations?

Reply: M. DeGroot

Dr. Neyman has raised two very interesting questions about my paper. First, he is quite correct that the construction of a nuclear reactor at a given site might well lead to changes in the socioeconomic composition of the population near that site which in turn lead to changes in the infant mortality rates. It is difficult to check this possibility because the relevant census data are published only every ten years. My own guess is that although there might be such changes in the immediate vicinity of the reactor (say within a few blocks), it is less likely that the composition of the county as a whole will shift because of the reactor. Of course, it may shift for other reasons in accordance with certain population trends or patterns, which is equally damaging to the analysis. However, I should think that the particular counties considered in my paper, rural Grundy as well as relatively populated Beaver, Westchester, and Suffolk, retained their same general character over the entire period studied. This question clearly requires further and more careful investigation.

Second, Dr. Neyman is again completely correct that a regression analysis based on rates is a tricky business when both the numerators and denominators are random variables, especially if the distribution of the number of live births in the denominator may be changing with time. Here, however, the yearly time series of the number of births and deaths in the various counties do not reveal any "substantial" changes over the period studied. Perhaps more reassuring, a glance at the graph of the time series of the infant mortality rate for each county seems to indicate that the variability of the annual rate remains roughly the same over the entire period.

EPIDEMIOLOGIC STUDIES OF CARCINOGENESIS BY IONIZING RADIATION

JOHN W. GOFMAN and ARTHUR R. TAMPLIN
LAWRENCE LIVERMORE LABORATORY and
UNIVERSITY OF CALIFORNIA, BERKELEY

1. Do we really need human epidemiologic data for pollutants?

In general, we should like to express our lack of sympathy for the expressed purpose of this Symposium, which is the planning of epidemiological studies for the evaluation of effects of major pollutants on humans. Carcinogenesis and leukemogenesis are two particularly worrisome long term effects which deserve consideration with respect to any pollutant. From our experience with ionizing radiation as a pollutant we have derived some lessons that we believe are extremely important to understand if society is to avoid paying a very high, probably unacceptable, price for the introduction of environmental pollutants. One such lesson centers around the prevalent notion that human epidemiological evidence concerning carcinogenesis should be required *before* technological promoters are willing to admit the serious potential hazards of a pollutant. Ionizing radiation is a classic example of this fallacious notion.

In our opinion it is *neither* appropriate *nor* good public health practice to demand human epidemiologic evidence to evaluate carcinogenic or leukemogenic hazard of a pollutant. First, in a civilized society, there *should never* exist an ideal set of human epidemiologic data. What epidemiologic data do become available are always subject to serious reservations with respect to equivalence of controls and exposed groups upon variables other than the specific pollutant variable under study. The net result is that controversy persists interminably. Peculiarly, but not unexpectedly in the face of promotional bias, the presumption is all too commonly made that, where uncertainty exists about the magnitude of effect, it is appropriate to continue the exposure of humans to the potential pollutant. It would indeed be sad if this Symposium helped contribute to this pernicious philosophy, which can only be described as that characteristic of a society bent upon ecocide in the name of ostensible technological progress.

In the case of radiation as a pollutant, we may consider some of the major epidemiological samples that have become available for study and relate the reservations that have been raised concerning acceptance of the results derived

This work was supported in part by the U.S. Atomic Energy Commission.

from the study of these samples. Approximately 100,000 survivors of Hiroshima and Nagasaki atomic bombing have been under followup study with respect to cancer and leukemia. Dosimetry reconstruction is difficult, at best, considering the nature of the event during which the radiation exposure occurred. Further, the associated possible injurious factors other than radiation were expected, in general, to be highly correlated with radiation exposure. Another large sample available for epidemiological study is the series of some 11,000 cases of ankylosing spondylitis in Great Britain, treated with X-irradiation. No satisfactory control series of spondylitics, untreated by X-rays, but otherwise equivalent, is available. Hence, questions can properly be raised about using the population at large as a reference sample. And the use of drugs for pain relief in addition to radiation therapy leads to the question of effects due to the drugs alone or to synergistic effects between drugs and radiation (see Collen and Friedman's contribution to this Symposium).

It can be pointed out that a vast experience with experimental animals of several species has proved cause and effect relationship between radiation exposure and carcinogenesis and leukemogenesis. Therefore, the real significance of the human studies is to ascertain comparability of dose response relationships for humans versus other species, rather than establishment of whether the observed association of radiation and cancer in these human population samples is *causal*.

We believe the appropriate approach to the study of leukemogenic or carcinogenic potential of pollutants is the study of dose response relationships in several mammalian species. And until or unless scaling laws are established among species, including humans, it should be *assumed*, for public health purposes, that the human is *at least* as sensitive as the *most* sensitive experimental species studied. In the ionizing radiation case, abundant experimental animal data have accumulated over the past quarter century demonstrating that radiation can provoke cancers of essentially all organs, provided the radiation is delivered to susceptible cells. Moreover, reasonable dose response data were available through such studies (Gofman and Tamplin [11]). Had these experimental animal data been utilized properly, the recent surprise concerning the higher than anticipated cancer hazard of ionizing radiation need not have occurred.

Having expressed our serious disapproval of the concept that human epidemiological studies should represent an approach to the study of pollutant effects, we should like to review here the treachery inherent in such studies, how they led to an earlier underestimate of the carcinogenic effect of radiation, and the residual uncertainties which still exist in assessment of the magnitude of the carcinogenic response to ionizing radiation in humans.

1.1. *Carcinogenesis and leukemogenesis in humans exposed to ionizing radiation.* Direct evidence that virtually all forms of human cancer can be induced by ionizing radiation has accumulated over several decades, often, however, with poor assessment of dose response relationships. By now, acute and chronic myelogenous leukemia, other acute leukemias, multiple myeloma, bone sarcoma, skin carcinoma, lung cancer (bronchiogenic and other varieties), thyroid cancer, breast cancer, stomach cancer, pancreas cancer, malignant lymphoma, colon cancer, cerebral tumors, neuroblastoma, Wilms tumor, maxillary and other sinus carcinomas, and pharynx cancer have all been shown to be inducible in humans by ionizing radiation. (Gofman and Tamplin, [7]). One disease (presumed malignant), chronic lymphatic leukemia, does not, thus far, appear to be radiation-induced (Lewis [21]). The implications of this finding remain unclear. For those remaining varieties of human cancer, other than the ones just listed, no evidence indicates they are not radiation-inducible. Within the evidence available, fortunately limited, there are simply no adequate data concerning radiation-induction.

Recently we presented three generalizations concerning induction of human cancer and leukemia by ionizing radiation. (Gofman and Tamplin, [7], [9]). These generalizations follow:

GENERALIZATION 1. All forms of cancer, in all probability, can be increased by ionizing radiation, and the correct way to describe the phenomenon is either in terms of the dose required to double the spontaneous mortality rate for each cancer or, alternatively, of the increase in mortality rate of such cancers per rad of exposure.

GENERALIZATION 2. All forms of cancer show closely similar doubling doses and closely similar percentage increases in cancer mortality rate per rad.

GENERALIZATION 3. Youthful subjects require less radiation to increase the mortality rate by a specified fraction than do adults.

Others (Stewart and Kneale, [31]) had clearly stated the outlines of these generalizations based upon the irradiation of infants *in utero*. Court-Brown and Doll [3] had done so based upon irradiation of adults. Additional study (Gofman, Gofman, Tamplin, Kovich, [13]) provides no reason to suggest a change in any of these generalizations; rather, it provides supplementary support for the generalizations.

The second of these generalizations led us to predict that for every leukemia induced by ionizing radiation, the *sum* of the number of cancers induced would stand to leukemia *as does the sum of spontaneous cancer mortalities to leukemia mortality*. Since the sum of spontaneous cancer mortalities is some twenty times that of leukemia mortality (Table III) over a fair share of the human adult life span, we predicted the sum of cancer mortalities per unit of radiation would be twentyfold that of leukemia. This caused a furor in the "radiation community," since the International Commission on Radiological Protection (ICRP) [17] had predicted in 1966 only *one* cancer mortality per leukemia mortality from radiation (exclusive of thyroid carcinoma which shows a low mortality rate in the cases which do occur). The error in the ICRP estimate represents a classic illustration of the pitfalls in the epidemiologic approach that had been used. Leukemia happens to occur earlier, post-irradiation, than do other cancers. Thus, since the ICRP was studying population samples in the relatively early years post-irradiation, the cancer mortality was seriously underestimated.

Data are available for adults from the study of the irradiated ankylosing spondylitis cases in Great Britain [3]. These subjects were irradiated primarily in early adulthood and then followed for periods up to 27 years. This study provides a good basis for testing the prediction that the sum of cancer mortalities is some 20 times that of leukemia mortality following irradiation. It is obvious that such a comparison test requires that radiation dosages be equivalent for all sites compared, or that appropriate dosage corrections be made before comparison of cancer mortalities with leukemia mortality. The Court-Brown and Doll data are presented in Table II, including partial followup through 27 years.

TABLE I

INCREASE IN CHILDHOOD CANCER AND LEUKEMIA FROM *In Utero* RADIATION
Radiation delivered in the form of X-rays during diagnostic pelvimetry.
Estimated dose <2 rads.

Type of cancer	Radiation induced increase
Stewart-Kneale data (1968)	
Leukemia	50% increase over spontaneous mortality rate
Lymphosarcoma	50%
Cerebral tumors	50%
Neuroblastoma	50%
Wilms' tumor	60%
Other cancers	50%
MacMahon data (1962)	
Leukemia	50%
Central nervous system tumors	60%
Other cancers	40%

TABLE II

CHANGE IN RATE OF INDUCED MALIGNANT DISEASE WITH DURATION
OF TIME SINCE EXPOSURE IN IRRADIATED ANKYLOSING SPONDYLITICS
(From data in Table VI of Court-Brown and Doll, 1965.)

	Cases per 10,000 man-years at risk	
Years after irradiation	Leukemia and aplastic anemia	Cancers at heavily irradiated sites
0–2	2.5	3.0
3–5	6.0	0.7
6–8	5.2	3.6
9–11	3.6	13.0
12–14	4.0	17.0
15–27	0.4	20.0
Total of expected cases in 10,000 persons in 27 years calculated from the rates given	67	369

TABLE III

RATIO OF SPONTANEOUS CANCER MORTALITY RATES
TO LEUKEMIA MORTALITY RATES
(Derived from *U.S. Vital Statistics* for 1966.)

Males Age group (years)	Ratio, $\left(\dfrac{\sum \text{Spontaneous cancer mortality rates}}{\sum \text{Leukemia mortality rates}}\right)$
40–44	15.9
45–49	22.9
50–54	28.5
55–59	28.7
60–64	29.2
65–69	29.1
70–74	23.5

In these studies 40 per cent of the total bone marrow (the expected site of origin of the leukemias) is estimated to have received irradiation. The spondylitis treatment is directed to the spine, not to other bone sites containing marrow. The mean bone marrow dose is 880 rads (for spinal marrow).

The "heavily irradiated" sites in those studies represent the sites receiving spray irradiation incident to the planned spinal irradiation. Dolphin and Eve [4] estimated that these "heavily irradiated" sites received approximately seven per cent of the mean spinal marrow dose.

From Table II, the observed ($\sum$ Cancer Mortalities/Leukemia Mortality) = (369/67).

The $\sum$ Cancer Mortalities must be multiplied by (100/7), to correct dosage for "heavily irradiated" sites to be equivalent to that for the spinal marrow.

The Leukemia Mortality must be multiplied by 2.5 to correct for the fact that only approximately 40 per cent of the total bone marrow received irradiation.

Therefore, for true total body irradiation the *Corrected Ratio* for radiation-induced malignant diseases, ($\sum$ Cancer Mortalities/Leukemia Mortality) = (369/67)(14/2.5) $\cong$ 31.

Since the spondylitis patients were irradiated in early adulthood, the period of followup is approximately in the 40 to 70 year age region. From *U.S. Vital Statistics*, 1966, we can derive the ratio, ($\sum$ Spontaneous Cancer Mortality Rates/$\sum$ Leukemia Mortality Rate) for this age range. These values are presented in Table III.

In the spondylitis patients, the sites designated as "heavily irradiated" include lung, stomach, colon, pharynx, esophagus, pancreas, lymphatic tissue. The major contributing sources to cancer mortality are, therefore, included. Possibly the ratio ($\sum$ Radiation-Induced Cancer Mortalities/$\sum$ Leukemia Mortality), determined here to be approximately 31 might be increased some if remaining tissue sites had been irradiated. The ratio ($\sum$ Spontaneous Cancer Mortality Rates/$\sum$ Leukemia Mortality Rate) is in the neighborhood of 20 to

30, for the relevant age range. Within the errors of such data as those for the spondylitis cases, the similarity of ratios for the spontaneous and the radiation-induced cases can be taken as strong support for Generalization 2 presented above, and as *grossly* at variance with the earlier ICRP prediction.

By now, however, this whole controversy has all but subsided. An ICRP Task Force (1969) has presented the Court-Brown and Doll data, together with the dose correction shown above (application of the Dolphin-Eve correction). Hamilton [15] stated that his own estimate of the ratio, ($\sum$ Radiation-Induced Cancer Mortalities/Radiation-Induced Leukemias), is within a factor of five of that of the authors, but he failed to take into account the dosage corrections which are, of course, absolutely essential in the treatment of the ankylosing spondylitis data. When the Hamilton estimate is appropriately corrected for the dose difference between bone marrow and the "heavily irradiated" sites (where cancers arise), his revised estimate would be *entirely* in accord with our own estimate. Mole [24] has recently published an estimate that the sum of radiation-induced cancer mortalities is "an order of magnitude" greater than radiation-induced leukemias. In a personal communication in 1970, Mole indicated to us that he had *not* applied the full Dolphin-Eve dosage correction, and this almost certainly explains the residual factor of two differences between his estimates and our own.

Thus, the so-called "radiation controversy," at least with respect to the ratio ($\sum$ Cancer Mortalities/$\sum$ Leukemia Mortality) for total body radiation, is essentially over. The controversy did pinpoint a valuable epidemiological pitfall, namely, the serious underestimate of cancer hazard from ionizing radiation resulting from the use, by standard setting bodies, of epidemiologic data for a time interval *before* the serious carcinogenic effects had developed. And the long observation periods required should alert us to the futility of hopes of learning of carcinogenic effects of new pollutants through human epidemiologic studies on a time scale that can be practically useful.

1.2. *Dose response relationships: ionizing radiation-induction of cancer and leukemia.* The ultimate objective, for a pollutant such as ionizing radiation, is an estimate of the human cost in premature death through cancer and leukemia, resulting from fairly chronic low or moderate dose irradiation. It is self evident that dose response relationships are required for such an estimate. Less immediately evident are some of the more subtle characteristics of the dose response relationships, characteristics which are crucially determinative of the magnitude of expected human cost.

One such characteristic is the time of onset of the carcinogenic response following exposure. Closely related is the duration of the response period in an exposed population. A second characteristic is the nature of the dose response curve over a wide range of doses. This becomes especially important because much of the available epidemiologic data covers a dose region higher than that anticipated for population exposure. Dose rate is an ancillary feature deserving consideration. A third characteristic is the variation in dose response relationship as a function of *age at exposure*.

1.2.1. *Time of onset of carcinogenic response and its duration.* A valid parameter commonly employed to assess carcinogenic response to ionizing radiation is the *radiation-induced age specific mortality rate* from any particular malignancy or group of malignancies. It would be ideal if this parameter were readily available both from the experimental animal and human data, but this is not always the case. Following radiation exposure (of humans and experimental animals) there is a period of time which elapses before any provably induced mortality from cancer or leukemia is observed. In short lived mammals, like the laboratory rat, this period is on the order of magnitude of months; in the human, of years. Most workers have referred to this apparently silent period as a *latent period*. It is not at all certain that such a latent period is truly as long as has generally been suspected. What is more likely is that the dose response curve shows at first a gentle slope upward with time, followed by a more steep slope, and then followed by what may be called a "plateau" region (Figure 1a). In studies involving relatively few subjects, the low incidence in the gentle slope region can appear to be a period free of effects, and this may well be why the impression has arisen of a long latent period. In most of the data available for analysis, the quantitative features of this segment of the response versus time curve are poorly defined.

Of additional great importance would be knowledge concerning *duration* of the "plateau" region of the response versus time relationship. Unfortunately, the available data simply do not allow, for any particular malignancy, satisfactory construction of this curve to ascertain how long the "plateau" region persists. For chronic myelogenous leukemia [36] the data suggest that once the excess mortality rate from radiation is perceived, it persists year after year for some 10 to 15 years, whereupon the excess mortality rate drops toward a lower value. In that same study the radiation-induced excess acute leukemia mortality rate showed no significant decline from the peak (or "plateau") value even after 20 years beyond irradiation. In the study on patients with spondylitis treated by X-rays [3], the 15 to 27 year period post-irradiation showed a *higher* excess mortality rate than any earlier periods of observation. There is no evidence within that study, of a return toward spontaneous mortality rates from malignant disease for the irradiated subjects.

Both the Japanese studies and the spondylitis studies should, in the next ten years, provide very valuable clues concerning the *duration* of the plateau region of response. For the present, however, no valid data are available to determine plateau duration. Indeed, and regrettably, the data for experimental animals, with respect to this issue, are no better than the sparse human data. As will be noted in the subsequent discussion of estimating long term population effects of low or moderate dose radiation, the duration of the plateau region is an extremely crucial parameter in determining the human cost expected. Furthermore, the

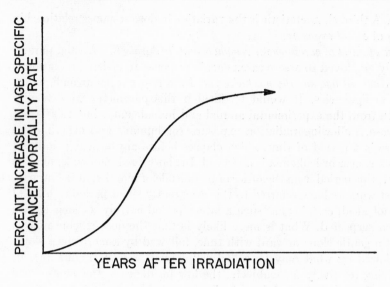

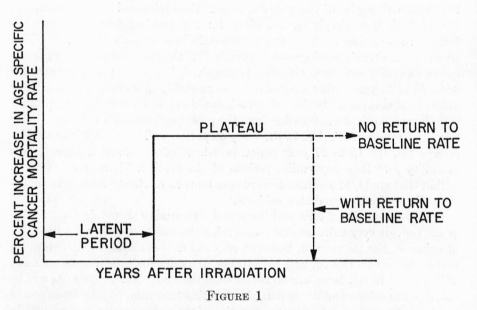

FIGURE 1

Dose response *versus* time curve; actual shown in upper panel, ideal shown in lower panel.

shape of the early part of the dose response curve (the so-called latent period region) is also an important parameter in determining the total magnitude of expected population cost.

In the absence of definitive data on these two issues, we shall idealize such dose response curves using simplifying assumptions which are in reasonable accord with what experience is available. Figure 1b presents such an idealized diagram describing the main features of the dose response relationship. The gently sloping part of the response curve is there replaced by an idealized "zero" response; followed by an abrupt rise to a flat plateau region. The duration of the flat plateau region is then available as a parameter for study, which is all that can be done at this time in the absence of definitive data.

In order to explore the consequences of variation in major parameters (length of "latent period" and duration of "plateau"), the following assumptions will be used:

ASSUMPTION 1.   *A single latent period of five years for in utero irradiation is assumed to agree with the estimates of Stewart and Kneale* [30], [31].

ASSUMPTION 2.   *A single latent period of 15 years is assumed for all forms of cancer for all irradiation beyond birth (except in the extreme case).*

Three general case types will be discussed, first, that with *no* return toward the spontaneous mortality rate (plateau, extending throughout the remaining life span for the population at risk); and second, that with an idealized abrupt return to spontaneous mortality rates after a 30 year plateau region. And third, an extreme case with a latency period of ten years (instead of 15 years) for all post-natal radiation and a plateau duration of 20 years. Both these changes have the effect of reducing the expected consequences of irradiation. We refer to such cases as "extreme" because it appears doubtful that the gently sloping part of the dose response curve is any shorter than ten years for the majority of radiation-induced malignancies (aside from leukemia, which *appears* shorter than all others), and second, because what evidence is available suggests that the plateau region is most likely to be *greater* than 20 years in duration.

It is essential to consider the manner of description of the radiation-induced excess age specific mortality rates. Commonly, results are presented either as excess cases per 1000 population at risk, or as the percentage increment in cancer mortality over the spontaneous age specific mortality rates. In some cases data are available for individual malignancies; in others, all cancers are presented as a sum. There is no theoretical reason for preference of absolute or percentage increments in age specific mortality rates. Both expressions suffer the defect that data derived from one population sample (for example, Japanese subjects) may not be directly applicable to another population sample (for example, United States subjects). We are far, far from having sufficient epidemiologic data to address such questions.

We have mentioned earlier the desirability of having age specific mortality rates for all ages of interest. We are far from that goal. Instead, available to us are radiation-excess mortality rates over a span of years of observation of exposed population samples. Therefore, expressed either as absolute or per cent increment in mortalities, the data allow only an average value for this span of years. In the absence of further evidence, we are here treating the plateau region as a *fixed* percentage increment in cancer mortality per unit of radiation over the *entire plateau region*. Only extensive further data can definitively test the validity of this particular approach. In its favor is the conservative nature of this treatment for public health purposes. Let us consider the implications of this treatment. Since *spontaneous* age specific cancer mortality rates change with age (rising steeply with age beyond 20 years), the assumption of a *fixed* percentage increment for radiation-induced excess over the whole plateau implies that the *absolute* increase in age specific mortality rate induced by radiation also changes with age. Thus, if the plateau region represents a 50 per cent increase in mortality rate, there will be 1000 extra deaths per 10^6 persons per year where the spontaneous mortality rate is 2000 deaths per 10^6 persons per year. At a later age, with a spontaneous mortality rate of 4000 deaths per 10^6 persons per year, the *absolute* increment due to radiation would be 2000 deaths per 10^6 persons per year. Thus, a constant percentage increment in the plateau response region *implies* that absolute radiation-induced age-specific mortality rate increments will increase over a span of ages.

"Spontaneous" cancer mortality rates include all known and unknown causes of cancer. Therefore, in an epidemiologic study, radiation-induced cases resulting from natural radiation background plus medical radiation exposures are included in the "spontaneous" cancer mortality rates for the population sample under study. Thus, if calculations are presented concerning the percentage increase in cancer mortality rate per rad of *additional* exposure to such a population, the true "spontaneous" base rate must be lower than that which includes the radiation effect from such sources as medical or natural radiation. Therefore, the true radiation percentage increment per rad is actually larger than that presented. For calculational purposes this does not introduce any significant complications. However, if the effect per rad is high, then the observed per cent increment per rad is *stated* to be lower than it truly is, simply because the spontaneous rate already is inflated by that mortality due to natural plus medical (and other) radiation for the population sample under study.

1.2.2. *Dose response relationships over a range of doses.*   One cannot be certain that the time aspects of the dose response relationship are identical over all dose ranges to be considered. Earlier impressions have been that the "latent period" (the gently sloping region described in Section 1.2.1.) might be longer at lower radiation doses. This speculation was weakly supported, if at all. The kind of study which led to this impression of a longer latent period at lower doses generally included small population samples at the lower doses, such that the expectancy of cancer at the low doses was often measured as a small fraction of one case [37]. Under these circumstances the probability of observing zero cases, in a small population sample, is very high. The *observation* of zero cases led to the false impression of a long latent period. In this manner the myth arose, concerning "practical thresholds" at low doses, that low doses of radiation might not be carcinogenic simply because the latent period could exceed the life span of the exposed population. Finkel and co-workers [6] recently demolished this myth very effectively, based upon a study of some 3200 mice exposed to radium 226. They saw no evidence of a variation in latent period with dose and indicated that they believed no other investigators would see variation either if they had an *adequate* population sample in the low dose region.

In the absence of any evidence to the contrary we shall assume latent periods and duration of plateau to be *independent* of the dose range under consideration. We shall, further, consider the effects calculated in the plateau region of the idealized diagram of Figure 1b. The epidemiologic data available cover a wide range of doses of radiation, with much of the human data, at least up to recently, having been obtained at moderate or high doses. Our interest, for purposes of evaluation of radiation is, in general, for doses in the low to moderate range. It is, therefore, essential to know the nature of the dose response curve over a wide range of doses if the epidemiologic data are to be utilized for predictive purposes in the case of population exposures.

A priori, in such problems, there is no way to predict the nature of the dose response curve. In principle, three generalized dose response patterns, connecting observations at high doses with those to be anticipated at low doses are conceivable (see Figure 2).

Curve A may be taken as one representative curve of a family of curves that are convex upward. Clearly, curves of this family express pessimism in that they predict a higher response at low doses than would be anticipated from a linear dose response relationship, such as curve B. Curve C, on the other hand, is a representative of a family of curves concave upward. This curve may be considered the "optimistic" curve from the viewpoint of a radiation-associated technology. The optimism arises because there can be a low dose region where the excess mortality due to radiation may be extremely low.

Early in the history of study of radiation carcinogenesis data were available, for humans and experimental animals, only for the fairly high dose region, and the shape of the entire curve down to very low doses was unknown. During that period, most responsible scientists and radiation study groups such as the International Commission on Radiological Protection made the prudent assumption of a linear relationship of radiation dose versus excess cancer mortality rate (curve B). While this did represent a conservative approach consistent with sound public health principles, it must be emphasized that this was by no means the *most* conservative position. Any of the family of curves, represented by curve A (Figure 2) represents a *more* conservative relationship for connecting available high dose points with the low dose region. But all these considerations describe an era that is now past. Abundant new data, in humans and experimental animals, have now become available, permitting description of the dose response relationship over a wide range of doses. These new data all point unmistakably to the correctness of curve B, the linear relationship between

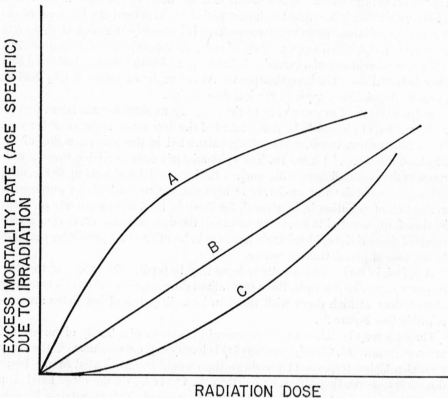

FIGURE 2

Three generalized dose response patterns: (A) higher response at low doses; (B) linear dose response relationship; and (C) low response at low doses.

excess cancer mortality and radiation dose, over a very wide range of doses for a variety of cancers and benign tumors. While one can understand the disappointment of radiation-industry promoters over the disappearance of the fondly regarded curve C, it is not possible to condone their lack of appreciation of the existence of all this new evidence.

Let us consider the specific new evidence that has appeared in recent years. (1) Shellabarger, Bond, Cronkite and Aponte [28] have demonstrated linearity both for breast adenocarcinoma and breast fibroadenoma development in rats exposed to X-rays or gamma rays down to total doses of 15 rads. (2) Upton and co-workers [34] have demonstrated linearity for mouse mortality from thymic lymphoma down to total doses of ten rads. Studies at lower doses are in progress. (3) Finkel, Biskis, and Jinkins [6] have demonstrated linearity for osteosarcoma development in the mouse with radium 226 injection over a wide range of doses. This is a landmark study, since it is refreshingly characterized by the experimental design of providing an adequate number of experimental animals in the low dose region. The authors [7], [8] have pointed out the fallacious conclusions derived from the study of inadequate numbers of humans, exposed to radium 226, who developed osteosarcoma. (4) Hempelmann [16] has indicated linearity in the production of human thyroid adenomas by X-rays, including data points down to 20 rads total dose to the thyroid gland. (5) Beebe, Kato, and Land [1] have extended the leukemia studies in survivors of the Hiroshima-Nagasaki bombings. They have demonstrated linearity in the production of human leukemia with radiation dose, down to total doses of 20 rads. (6) Stewart and Kneale [31] have demonstrated linearity between cancer and leukemia induction in children during the first ten years of life and irradiation by X-rays *in utero* in the process of diagnostic obstetric radiography. Their observations covered the range of approximately 2.0 rads, thus providing direct human evidence in the extremely low dose region. (7) Mays and Spiess [29] have demonstrated linearity in the production of osteosarcoma both in human adults and children as a result of radium 224 injection. Their experimental data extend down to 90 rads estimated dose. These studies are grossly at variance with the claims of Evans [36] of a "threshold" for osteosarcoma in humans by alpha emitters at a dose of 1000 rads.

Taken overall, these recent and diverse publications leave very little reason to doubt a linear dose response relationship for cancer and leukemia induction by radiation. It has been an interesting phenomenon, indeed, to observe the antics of the promoters of radiation-associated technologies during the evolution of all these data. Starting with their hope that linearity would fail below 100 rads, they have been forced to retreat steadily to 50 rads, then 25 rads, and now they find themselves faced with linearity down to the region of a fraction of one rad. Hope springs eternal.

To be sure, for any particular set of data, one could always argue that *perhaps* there is a deviation from linearity somewhere below the dosage represented by the lowest experimental point. There exists, however, no rational support for such an assumption, since it would require a fundamental change in the mechanism of radiation carcinogenesis in the region below the linearity region. Further, such an assumption, in the *absence* of evidence supporting it, represents an unsound approach to the protection of the public health. The *in utero* data [31] extending down to approximately 0.3 rads, militate strongly against further serious consideration of nonlinearity in the very low dose region.

From the point of view of mechanism, linearity between radiation dose and carcinogenic response suggests that a single event phenomenon is involved in the production of the critical change which results in the development of cancer. If a single event produces the carcinogenic change over a wide range of doses, for a variety of cancers, for several mammalian species, there appears little reason to expect a fundamental change in such mechanism at still lower doses.

Since linearity appears well established for a variety of cancers, we shall here consider the dose response relationship, in the plateau region, as being linear for *every* type of cancer and leukemia (Figure 3) for prediction purposes. The excess age specific mortality rate, for any cancer, can be expressed, for a linear dose response relationship, as a percentage increase per rad over the spontaneous mortality rate for that particular cancer. Such percentage increment is simply the slope of the linear plot of Figure 3. For illustrative purposes, assume the slope, for a particular cancer, were determined to be one per cent per rad. It follows then, for a linear relationship, that 100 rads will produce 100 × 1, or a 100 per cent increase in cancer mortality above the spontaneous cancer mortality rate. That dose which increases the spontaneous cancer mortality rate by 100 per cent is commonly *defined* as one doubling dose of radiation for production of that particular cancer. Thus, if a is the slope of the line in Figure 3, then the doubling dose is defined as $100/a$ (for this particular cancer). The doubling dose notation does *not* in any way imply a geometric progression in excess cancer mortality rate with increasing radiation dose. Rather, one doubling dose adds 100 per cent to the spontaneous age specific mortality rate, two doubling doses add 200 per cent, three doubling doses add 300 per cent, and so forth. It is simply a matter of convenience as to whether radiation carcinogenesis, for any particular cancer, is described as the per cent increment in cancer mortality rate per rad or as the

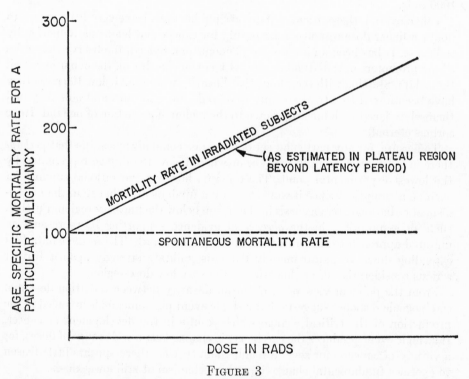

FIGURE 3

Linear dose response relationship in the plateau region where the age specific mortality rate is a percentage increase per rad.

dose in rads, the doubling dose, required to add an *excess* mortality rate equal to the spontaneous age specific rate. Nothing about the doubling dose notation infers or suggests that the doubling dose is the same for all forms of cancer. This is a matter for experimental evidence to decide. Before considering the question of variation of doubling dose with form of cancer induced, it is necessary to turn our attention to the variation in sensitivity to radiation carcinogenesis with age at irradiation.

1.2.3. *Variation in carcinogenic dose response relationship with age at radiation exposure.* Our considerations thus far have led us to description of radiation carcinogenesis as follows: (1) *dose response relationship, at a specified age, is linear* (Figure 3), *characterized by a particular percentage increment in cancer mortality per rad for a particular form of cancer;* and (2) *dose response relationships will be treated for the idealized "plateau" region of the curve of the response* versus *time after radiation exposure.*

For any particular cancer, occurring *at a specified age,* how does the slope of the dose response line vary with *age at irradiation?* It is clear that any refined effort to assess population response to continuous or intermittent radiation exposure required consideration of this particular question. To the best of our knowledge there exists no theory that provides the answer to this question. We must, therefore, have recourse to empirical data.

First, we have the data for *in utero* radiation provided by Stewart and Kneale [30], and MacMahon [22]. These data, presented in Table I, describe the increase in cancer and leukemia mortalities during the first ten years of life following irradiation *in utero.* Inspection of the data leads to a best estimate of a 50 per cent increase in mortality rates for a variety of cancers and for leukemia for radiation associated with diagnostic pelvimetry. The similarity in percentage increase in cancer mortality for diverse cancers and for leukemia, for such radiation, is striking. Stewart and Kneale [31] indicate that approximately four X-ray films lead to 100 per cent increase in such childhood cancers and leukemia mortality rates. During the period of accumulation of their evidence, each X-ray film represented less than 0.5 rad delivered to the infant *in utero.* Conservatively, therefore, one estimates that two films, or 1.0 rads, are required for an approximate 50 per cent increase either in cancer or leukemia mortality rates. (The true value may be somewhat higher than the conservative 50 per cent increase per rad.)

None of the Stewart studies address the issue of effects of *in utero* radiation upon the development of cancer or leukemia beyond the first ten years of life. Both from the Hempelmann studies [16] and the Hiroshima-Nagasaki studies (Jablon and Belsky, [19], involving the irradiation in early infancy, we have conclusive evidence that carcinogenesis extends far beyond the first ten years of life. It would be surprising, therefore, if such were not the case for *in utero* irradiation as well. In any event, our treatment of the data for estimating population exposure specifically explores the effect of various durations of the plateau response region. Utilizing the Stewart-Kneale and MacMahon data, we shall use a 50 per cent increase in age specific cancer mortality rates per rad for irradiation received *in utero* and shall assume this value holds for all cancers and leukemias.

For infancy and childhood irradiation, there are two major sources of information: (a) data for thyroid cancer induction in U.S. children irradiated in early infancy; and (b) data for various cancers in Japanese subjects in Hiroshima and Nagasaki, who were between 0 and 9 years of age at the time of bombing.

For the thyroid cancers occurring in irradiated children, Pochin [26] provided an estimate that the absolute increment is one case per 10^6 persons per year per rad of exposure of the thyroid gland. Carroll, Hadden, Handy and Weeben [2] reported the spontaneous thyroid cancer rate as approximately five to ten cases per 10^6 persons per year in the age range 10–20 years. Combining these data, we have previously estimated 10 to 20 per cent increase in thyroid cancer per year per rad for irradiation in infancy. (Gofman and Tamplin, [9]).

Jablon and Belsky [19] have recently provided data for cancers (other than leukemia) in persons exposed to atom bombing at 0–9 years of age. For those receiving 100 rads or more, the cancer mortality rates (during the period 10 to 24 years beyond exposure) was 8.4 times that observed for persons receiving less than 10 rads. The mean dose for the (100 rad or more) group was not given, but it must lie between 100 and 200 rads. So, 100 to 200 rads represent 7.4 doubling doses ($8.4 - 1.0 = 7.4$). Therefore, the doubling dose for cancer production in these 0–9-year old children (at exposure) lies between 14 and 28 rads. This corresponds to a 3.5 to 7.0 per cent increase in cancer mortality rate per rad. The per cent increment in leukemia mortality rate per rad was even higher, as observed in a group of children 0 to 14 years of age at the time of bombing [19]. A variety of cancers were represented in the Jablon and Belsky data, but the limitations of numbers did not allow for treatment of individual types of cancers, (see also [20]).

From several sources, data are available concerning the percentage increase in specific site cancer mortality rates per rad for persons irradiated in early adulthood [32]. These include data for subjects receiving radiation under widely different conditions. Included are: (1) breast cancer (Nova Scotia women [23] receiving fluoroscopic radiation and Japanese survivors of atomic bombing); (2) thyroid cancer (Japanese survivors of atomic bombing); (3) lung cancer (spondylitis cases and Japanese survivors of atomic bombing); (4) leukemia (spondylitis cases and Japanese survivors of atomic bombing); (5) stomach cancer (spondylitis cases); (6) colon cancer (spondylitis cases); (7) pancreas cancer (spondylitis cases); (8) bone cancer (spondylitis cases); (9) lymphatic and other hematopoeitic organ cancer (10) miscellaneous cancers (spondylitis cases); and (11) pharynx cancer (spondylitis cases).

The range of values determined for percentage increase in cancer mortality rate per rad of exposure was between one and five per cent with an estimated best value of approximately two per cent per rad. Ideally, one would want to have these values determined for groups irradiated at a specified age, and one would wish to be certain that the observations were strictly referable to the plateau region of the response *versus* time curve, rather than possibly including some data referable to the latent period. But such ideal data are unavailable. Hence, we shall use a two per cent increase in cancer mortality rate per rad as a "best" value, we shall consider that it applied to all cancers (the major ones are all represented in the data), and we shall relate this value to irradiation at approximately 20 to 30 years of age. As will be noted below, the overall data indicate the sensitivity to cancer induction when expressed as the per cent increase over spontaneous cancer mortality rates per rad of exposure, is a steeply declining function of age at irradiation. Therefore, it is entirely possible that the range of one to five per cent increase in cancer mortality rate per rad might be narrowed appreciably but for differences in age at irradiation for the young adult groups tabulated. Inaccuracies in dosimetry may also account for part of the range of values observed. In any event, the average value of two per cent per rad for irradiation in the age range of 20–30 years will be seen below to be consistent with trends noted over a very broad span of ages at irradiation.

Beebe, Kato and Land [1] have recently presented data for cancer mortalities during the 1962–1966 period for Hiroshima-Nagasaki survivors who were between 25 and 55 years of age at the time of bombing (1945). It appears quite clear, from their studies, that there is a *markedly* lower sensitivity for cancer induction per rad compared with that for younger subjects. These workers estimate a 20 per cent increase in cancer mortality risk per 100 rads, or 0.2 per cent per rad for this older group of subjects.

Summarizing all the evidence just described, we have the following estimates of sensitivity to radiation-induction of cancer and leukemia as a function of *age at irradiation:*

in utero	~50%	increase in mortality rate per rad
0–9 years of age	3.5–20%	
20–30 years of age	~2%	
~50 years of age	~0.2%	

There can be no doubt that risk of induction of excess cancer mortality rates per rad, described as per cent increase over spontaneous mortality rate, declines steeply with increasing age *at* irradiation. Within the totality of available epidemiologic evidence now available the estimates just listed provide about as much description of this declining function as is now possible. For purposes of estimation of the consequences of population exposure, these estimates can be reasonably approximated by the step function presented in Table IV. It can be shown that the precise values in the step function are *not* the dominant parameters that determine the consequences of population exposure. Of far greater importance is the *duration* of the plateau region of the response *versus* time curve.

TABLE IV

VARIATION IN CANCER INDUCTION PER RAD WITH AGE

These estimates represent a step function approximation
in reasonable accord with the data points
available in the text.

Age at irradiation (years)	Increase in cancer mortality rate *per rad* (in Plateau Region) (per cent)
In utero	50
0–5	10
6–10	8
11–15	6
16–20	4
21–30	2
31–40	1
41–50	0.5
51–60	0.25
61 and beyond	Assumed negligible

2. The carcinogenic consequences of population exposure to environmental ionizing radiation

The major parameters required to evaluate the consequences of population exposures to ionizing radiation have been identified in the foregoing discussion. That the epidemiologic data are far less than ideal for quantitative evaluation is undeniable. A humane society should consider itself fortunate that better data are *not* available.

The various sources of potential ionizing radiation exposure include natural radiation, radiation from weapons testing fallout, radiation from a variety of peaceful atomic energy programs, and radiation from diagnostic medical and dental exposure. Since the signing of the atmospheric test ban treaty, weapons testing fallout has become a small source, and should decline further, unless nonsignatories to that treaty increase weapons testing appreciably.

Peaceful atomic energy programs are currently *allowed* to deliver an average dose of 0.17 rads per year to the U.S. population. At present, so far as measurements allow dose estimates, it appears that such programs deliver only a small fraction of this "allowable" average dose. Nevertheless, with the burgeoning growth of the nuclear electric power industry plus numerous proposals for utilization of "peaceful" nuclear explosives (Project Plowshare) plus growing radioisotope utilization, the exposure to the population from the "peaceful" atom will undoubtedly grow. So long as 0.17 rads per year remains permissible by Federal Regulations, there is good reason to believe the full exposure may ultimately be reached. It is, therefore, of special importance to calculate the cancer and leukemia expectation for such an average exposure to the U.S. population.

Medical and dental exposures to X-rays have resulted in a steadily increasing average population dose of ionizing radiation. Medical diagnostic X-ray exposure has recently been estimated to provide approximately 0.10 rads as an average population somatic tissue dose (Morgan [25]). We are in full accord with Morgan that advantage should be taken of modern technology to reduce such exposure drastically, especially since Morgan has estimated that a ten-fold reduction in average exposure could be accomplished without any loss in diagnostic X-ray information.

Natural radiation provides an average population exposure in the neighborhood of 0.125 rads per year. Such features as radioactivity content of building materials, radioactivity in rocks of the earth, and elevation above sea level account for variation in such natural doses among population subsamples. Through a strange system of logic, or better, illogic, it is commonplace for promoters of radiation-associated technologies to arrive at the wholly absurd conclusion that doses comparable to natural radiation cannot be carcinogenic because natural radiation "has always been with us."

The above sources of ionizing radiation represent primarily low Linear Energy Transfer (LET) radiation. Primarily the radiations are X-rays, gamma rays, and beta rays. Carcinogenic effect per rad will be essentially identical for all these radiation sources. One could estimate population consequences per millirad per year, for natural radiation exposures, for medical exposures, or for the 0.17 rads per year permitted as an average population exposure for peaceful atomic energy activities. Since the concern of this Symposium is with matters related to environmental pollution, it is particularly appropriate to estimate the consequences of the 0.17 rad per year average allowable population exposure. The U.S. Government [5] has decreed this much population pollution to be permissible (Federal Radiation Council, 1960). The scientific and lay communities should be especially interested in the carcinogenic consequences of this permissible pollution by ionizing radiation. It should be evident that the consequences of natural, medical, or weapons fallout exposures can be derived from the consequences of 0.17 rads per year by direct application of the linearity of dose *versus* response.

We have previously estimated the cancer plus leukemia consequences of exposure to 0.17 rads per year to be approximately 32,000 extra cancer plus leukemia deaths per year, at equilibrium, for the U.S. population at its current size of 2×10^8 persons, [10], [11]. That estimate was based upon the average two per cent increase in cancer mortality rate per year per rad of exposure observed for young adults, coupled with a 30 year duration of the plateau region. With the more extensive data available in the past year concerning sensitivity variation with age, a more refined estimate is now possible. Moreover, it is important to explore the implications of both a longer and shorter duration of the plateau region, as well as the implications of variation in "latent" period. As we shall see, the estimate of 32,000 extra deaths per year is by no means overly conservative, since this number can rise several fold if it turns out that the plateau region extends throughout the life span of exposed populations.

2.1. *Cancer hazard for average population exposure (total body irradiation) at 0.17 rads per year.* Three general cases will be considered here, the case where the plateau persists indefinitely after latent period, where the plateau region persists 30 years with subsequent return to spontaneous cancer mortality rates, and the extreme case where the plateau region persists 20 years with a latent period of 10 years for post natal radiation (in contrast with 15 years for the first two cases).

CASE 1. *Plateau persists indefinitely after latent period.*

The calculation is based upon the consideration of the total per cent increment in radiation-induced cancer mortality rate at a particular specified age as made up of the sum of contributions from radiation received at ages less than the specified age. The procedure will be illustrated below.

For *in utero* irradiation we have stated above that a five year latent period will be assumed.

In Case 1 calculations, a 15 year latent period is assumed for all post natal irradiation.

Radiation received in any particular year of life begins to contribute to cancer mortality rate only after the latent period is over. Thus, radiation in the first year of life *starts* contributing to cancer mortality in the 16th year of life. Radiation in the 10th year of life starts contributing to cancer mortality in the 25th year of life.

For *in utero* irradiation at 0.17 rads per year, approximately 0.12 rads would be received in the course of a pregnancy. At 50 per cent increase in cancer mortality rate per rad, we calculate 50 × 0.12, or a six per cent increase in cancer mortality rate for the *in utero* radiation exposure. Now, since we have assumed a five year latent period for *in utero* radiation, there is obviously *zero* cancer mortality increment during the first four years of life. For the fifth year of life and beyond, however, the six per cent increment in cancer mortality rate would apply for each year that the plateau region persists. In Case 1, under consideration here, this would be for the remainder of the life span of the exposed population.

For irradiation in the first year of life (0.17 rads), the sensitivity factor to be taken from Table IV is ten per cent per rad. Thus, 10 × 0.17 = 1.7 per cent increase in cancer mortality rate. However, since we are taking the latent period for post natal irradiation to be 15 years, it follows that irradiation in the first year of life does not begin to add its increment in cancer mortality rate until the 16th year of life. For Case 1, this increment would be effective for all subsequent years for the exposed population, since indefinite persistence of the plateau is assumed.

Therefore, for the 16th year of life, there is six per cent from the *in utero* irradiation plus 1.7 per cent from irradiation in the first year of life, for a total increment of 7.7 per cent in radiation-induced cancer mortality rate. For the 17th year of life, we have 6 per cent from *in utero* radiation, 1.7 per cent from 1st year irradiation, plus 1.7 per cent from the 2nd year irradiation, for a total of 9.4 per cent increment in cancer mortality rate from the irradiation received *in utero* plus the first two years of post natal life.

The increment in cancer mortality for irradiation in each subsequent year of life is calculated in the same manner as the product of the sensitivity factor from Table IV (for that year of life) by the 0.17 rads. The *total* increment in cancer mortality rate for any particular year of life is the *sum* of all contributions *to that year* from irradiation at earlier years, taking into account that no increment is derived until the latent period is over for that particular year's irradiation. In this manner, a value for total per cent increment in cancer mortality rate becomes available for every year of life, taking into account, appropriately, irradiation received at all earlier periods of life. For ease of comparison with *U.S. Vital Statistics*, these annual values are averaged for five year age intervals.

In assessing impact of irradiation upon the population, we can consider just the per cent increase in age specific cancer mortality rate. The values just calculated provide this result. Or, alternatively, and of possibly greater interest, is the absolute increase in number of cancer deaths per year at each age for the population at risk. We are now immediately in a position to make this estimate.

From *U.S. Vital Statistics*, the absolute number of spontaneous cancer deaths per year for each age interval is provided (1966 data used here). Now, let us suppose for a particular age that the combined increment due to all prior radiation is a 15 per cent increment in cancer mortality rate over the spontaneous cancer mortality rate. And let us suppose, further, that for this particular age, the spontaneous cancer mortality rate is 1000 cases per year. The radiation-induced increment is then $(15/100) \times (1000)$, or 150 radiation-induced cancer deaths for the population at this particular age.

In a similar manner, a tabulation of absolute numbers of radiation-induced cancers by age interval can be built up, separately for males and females. Finally, the total annual number of radiation-induced cancer fatalities can be calculated by summation over all age intervals for males plus females. This tabulation, for Case 1 calculations, is provided in Table V. The result, a prediction of some 104,000 annual additional cancer fatalities is more than three times worse than our earlier estimate. We are, of course, not at all surprised at this result, for we had indicated earlier that taking sensitivity as a function of age into account could make for a much more serious prediction. Additionally, Case 1 calculations consider the plateau region to extend indefinitely, whereas our earlier calculations were based upon a 30 year duration of plateau.

It can be further noted that if the real effect is as large as shown in Table V (and no reason exists to reject the Case 1 analysis), then the contribution of natural plus medical radiation must constitute a quite appreciable segment of the so-called "spontaneous" cancer mortality rates. One could consider a second iteration on the total calculation, correcting the "spontaneous" mortality rates

TABLE V

RADIATION-INDUCED CANCER MORTALITY BY AGE AND SEX

5 year latency for *in utero* radiation.
15 year latency for all other radiation.
Plateau constant after latency period.
Exposure: 0.17 rads/year.

Total spontaneous cancer mortality per year = 303,691 cases.
Total radiation-induced cancer mortality per year = 104,259 cases.
Per cent increase in cancer which would occur with 0.17 rads
average annual exposure = 34.3 per cent.

Age interval	Per cent increase in cancer mortality rate	Annual spontaneous cancers (male)	Annual radiation induced cancers (male)	Annual spontaneous cancers (female)	Annual radiation induced cancers (female)
(Years)					
0–4	0	827	0	720	0
5–9	6	826	50	606	36
10–14	6	673	40	482	29
15–19	9.4	820	77	546	51
20–24	17.2	754	130	508	88
25–29	23.3	796	186	733	171
30–34	27.8	1,145	318	1,418	394
35–39	30.5	2,104	641	2,890	881
40–44	32.2	4,163	1,340	5,565	1,791
45–49	33.4	7,109	2,372	8,732	2,914
50–54	34.2	12,363	4,231	11,950	4,089
55–59	34.8	17,594	6,123	14,359	4,997
60–64	35.2	22,469	7,909	15,780	5,555
65–69	35.5	25,275	8,968	17,921	6,358
70–74	35.7	25,698	9,169	18,746	6,689
75–79	35.8	21,221	7,589	16,650	5,954
80–84	35.8	13,318	4,763	12,141	4,342
85 and beyond	35.8	7,793	2,787	8,996	3,217
Total		164,948	56,703	138,743	47,556

downward (by subtracting the contribution from natural plus medical radiation) and correcting the per cent increment per rad upward as a result of the lower true "spontaneous" mortality. These two effects would tend to balance out, so that the final calculations of population risk would not be seriously altered. It would, however, point up the major contribution of natural plus medical radiation to the existing cancer mortality rate, wholly aside from increments due to peaceful atomic energy programs.

CASE 2. *Plateau region persists 30 years, with subsequent return to spontaneous cancer mortality rates.*

It is possible that once the increased cancer risk due to irradiation is fully developed (the plateau region), such risk may not persist indefinitely. It is difficult to know, within presently available epidemiological data, how many years the plateau lasts, if it does indeed only last a limited period for cancer. A calculation based upon a 30 year plateau period is provided here. In this calculation, the contribution of radiation received in any particular year of life is credited for 30 successive years, following the latent period. After this, the contribution of that particular radiation is cut off. Thus, for example, the per cent increment in cancer mortality rate from radiation received during the 1st year of life begins to be credited starting in the 16th year of life, and is credited for each subsequent year of life out to the 45th year of life. Beyond the 45th year of life, no crediting toward radiation-induced cancer mortality is given for irradiation in the first year of life. Similar calculations are made for irradiation in each subsequent year of life. Otherwise, procedures of calculation are similar to those for Case 1, Table V (five year latent period for *in utero* radiation; 15 year latent period for all post natal irradiation). The calculations for Case 2 are presented in Table VI.

TABLE VI

CASE 2: RADIATION-INDUCED CANCER MORTALITY BY AGE AND SEX

5-year latency period for *in utero* irradiation.
15-year latency period for all other irradiation.
Plateau: 30 years beyond latency period.
Exposure: 0.17 rads/year.

Total spontaneous cancer mortality per year = 303,691 cases.
Total radiation-induced cancer mortality per year = 74,013 cases.
Per cent increase in cancer which would occur with 0.17 rads average annual exposure = 24.4%.

Age interval (years)	Per cent increase in cancer mortality rate	Annual spontaneous cancers (male)	Annual radiation-induced cancers (male)	Annual spontaneous cancers (female)	Annual radiation-induced cancers (female)
0–4	0	827	0	720	0
5–9	6	826	50	606	36
10–14	6	673	40	482	29
15–19	9.4	820	77	546	51
20–24	17.2	754	130	508	87
25–29	23.3	796	185	733	171
30–34	27.8	1,145	318	1,418	394
35–39	24.5	2,104	515	2,890	708
40–44	26.2	4,163	1,091	5,565	1,458
45–49	26.0	7,109	1,863	8,732	2,288
50–54	25.4	12,363	3,140	11,950	3,035
55–59	24.9	17,594	4,381	14,359	3,575
60–64	24.6	22,469	5,527	15,780	3,882
65–69	24.4	25,275	6,167	17,921	4,373
70–74	24.6	25,698	6,322	18,746	4,612
75–79	24.4	21,221	5,178	16,650	4,063
80–84	24.5	13,318	3,263	12,141	2,975
85 and beyond	24.0	7,793	1,870	8,996	2,159
Total		164,948	40,117	138,743	33,896

It is evident, on comparison of Table V with Table VI, that reduction of the plateau duration provokes a marked drop in the expected mortalities (104,000 down to 74,000). However, both values are extremely high and should raise grave concern about the nature of the societal benefits that might be worth permitting population exposures as high as 0.17 rads per year as the average exposure. No comfort whatever is to be drawn from repeated assurances that abound from nuclear promoters to the effect that "we'll never give you the full allowable exposure" while at the same time they staunchly defend retaining such an allowable exposure. Good intentions are materially aided by codification into Federal Regulations.

The calculations should be especially illuminating to the sponsors of this Symposium addressing the issue of designing epidemiologic studies for the evaluation of societal impact of environmental pollutants. A quarter of a century into the atomic era, the epidemiologic data indicate that our permissible doses could lead to a public health calamity—a 25 to 35 per cent increase in annual cancer mortality rate. No evidence at this time militates against the most pessimistic calculation (Case 1). We have commented elsewhere that this late realization based upon epidemiologic data could all have been averted by judicious use of experimental animal data decades ago (Gofman and Tamplin [11]).

It is of interest to speculate upon possibilities that might have resulted in the Case 1 or Case 2 calculations leading to a serious overestimate of the cancer hazard. For example, one might consider the possibility that dosimetric or other errors had led to an overestimate of the percentage increment in cancer mortality rates per rad at all of the ages listed in Table IV. We believe it is unlikely that such an overestimate could be as much as two-fold. Moreover, one might also, under such circumstances, consider that the seriousness of the results is underestimated as a result of dosimetric errors.

CASE 3.   *The extreme case: plateau region persists 20 years, latent period of 10 years; post natal irradiation.*

It is important to ascertain what the prospects for "optimism" may be with regard to carcinogenic consequences of population exposure to radiation. Therefore, we may consider the possibility that the duration of the plateau region of the response *versus* time relationship is materially shorter than 30 years. From the epidemiologic evidence available, admittedly still scanty, we would estimate that it is highly unlikely for plateau duration to be less than 20 years. (Radiation-induced cancers have been described occurring 30 to 40 years after exposure.) But since this should lessen greatly the expected consequences, we shall test here a 20 year duration for the plateau region. It is also evident that if the latent period were shorter than 15 years, the net carcinogenic effect would be reduced further, because the large per cent increments in cancer mortality rate for irradiation early in life would not be carried as far forward into the later age spans where the spontaneous cancer mortality rates are high and, hence, the products of per cent increment by spontaneous mortality rates are also high. The procedure of calculation is precisely the same as that employed for Case 1 and Case 2 except for the alterations in the two parameters, plateau duration and latent period for postnatal irradiation. The results are presented in Table VII. The final estimate for population exposure at an average of 0.17 rads per

TABLE VII

Case 3: Radiation-Induced Cancer Mortality by Age and Sex

5 year latency period for *in utero* radiation.
10 year latency period for all other radiation.
Plateau: 20 years beyond latency period.
Exposure: 0.17 rads/year.

Total spontaneous cancer mortality per year = 303,691.
Total radiation induced cancer mortality per year = 9,428.
Per cent increase in cancer which would occur with 0.17 rads average annual exposure = 3.1%.

Age interval	Per cent increase in cancer mortality rate	Annual spontaneous cancers (male)	Annual radiation induced cancers (male)	Annual spontaneous cancers (female)	Annual radiation induced cancers (female)
(Years)					
0–4	0	827	0	720	0
5–9	6	826	50	606	36
10–14	9.4	673	63	482	45
15–19	17.2	820	141	546	94
20–24	23.3	754	176	508	118
25–29	21.8	796	173	733	160
30–34	21.1	1,145	241	1,418	299
35–39	15.0	2,104	315	2,890	434
40–44	10.2	4,163	425	5,565	568
45–49	6.6	7,109	471	8,732	576
50–54	4.6	12,363	566	11,950	550
55–59	3.3	17,594	577	14,359	474
60–64	2.2	22,469	503	15,780	347
65–69	1.6	25,275	402	17,921	287
70–74	1.2	25,698	311	18,746	225
75–79	1.0	21,221	212	16,650	167
80–84	1.0	13,318	133	12,141	121
85 and beyond	1.0	7,793	78	8,996	90
Total		164,948	4,837	138,743	4,591

year is 9428 extra cancer deaths per year. While this is a marked reduction compared with the estimates for Case 1 and Case 2, the seriousness of such radiation exposure levels is self-evident. We would doubt that a more "optimistic" set of parameters than those for the Case 3 calculation is likely to be justified.

3. Life shortening by radiation-induced cancer

A variety of pronouncements have greeted estimates of the serious carcinogenic hazard of population exposures to doses in the neighborhood of 0.17 rads per year. One such we have dealt with above, namely, the statement that, after all, this dose is comparable in magnitude with natural radiation, which humans have endured on earth for the entire history of the species. No further comment is required. A second is that even though the calculated cancer deaths may indeed occur, they will occur so late in life as to be inconsequential. Grendon has championed this approach, readily provable to be false, [14]. A variant of this approach is that of Sagan [27] who has pointed out that, even if the calculated cancers did occur, the average life shortening for the exposed population would be very small. In fact, it has become fashionable of late to estimate the deleterious effect of environmental hazards in terms of average life shortening for the exposed population. We hear "Wouldn't people be willing to give up a few minutes, hours, or days of life span so we can all enjoy 'clean, cheap, and safe' nuclear electricity?" This approach to evaluation of life shortening is exceeded in its scientific fallacy only by its immorality in public deception.

If those who die prematurely of cancer due to irradiation are averaged in with those who do not, the *apparent* loss of life expectancy appears quite small. What *really* matters is the average loss of life expectancy for those who *do* develop radiation-induced cancer. Their loss of *decades* of life expectancy is not easily recompensed by a "loan" from those who do not become victims. The losses in life expectancy for the victims are readily estimated. If the victims of radiation-induced cancer had *not* been irradiated, there is *a priori* every reason to assume they would have experienced the usual life expectancy associated with their age group at victimization. Thus, from 1971 estimates, a man at age 25 years has a life expectancy of 45.5 years. If he dies at 25 years of age of radiation-induced cancer, he has lost 45.5 years of life expectancy. In Table VIII are presented the calculated losses of life expectancy by age group for the persons developing radiation-induced cancers, as well as the average loss of life expectancy for *all* the cases of radiation-induced cancers as a group. For males developing radiation-induced cancers, the average loss of life expectancy is 13.1 years. For females, the loss is 13.7 years. Such average losses hardly are in accord with Grendon's assertion that the radiation-induced cancers occur so late in life as to be inconsequential. For men in the age group of 65 to 69 years, the life expectancy (as of 1971) is 11.5 years. If these men lose their life through radiation-induced cancer at 67 years, they have lost 11.5 years. One wonders whether Grendon has checked with such members of the population to ascertain that these "old" people need not care about losing 11.5 years of life.

Let us return to the Sagan view of only a minor loss of life expectancy (hours or days). If the man-years of life expectancy are distributed into the *entire* U.S. male population of 95,919,000 men instead of into the 56,703 victims of radiation-induced cancer, the average loss of life expectancy is computed to be 2.8 days. This practice of hiding the serious loss in life expectancy for the victims of an environmental poison by averaging the loss over the larger group of nonvictims deserves strong condemnation. The sole effect of the practice is to obscure the real hazard of an environmental poison from the public, carried through on

TABLE VIII

LOSS OF LIFE EXPECTANCY FROM RADIATION-INDUCED CANCER
(Data from Table V)

Life expectancies are somewhat higher for females than males, so the use
of male life expectancies here leads to a slight *underestimate* of the loss of life
expectancy for females with radiation-induced cancers.

Note: The use of data from Table V (the Case 1 estimate) leads to the *lowest* estimate of loss
of life expectancy. For Case 2 (Table VI) and Case 3 (Table VII), the radiation-induced excess
cancer mortalities are more prominent at earlier ages. Hence, for either of these the life expectancy loss would be appreciably *higher* than the 13 year estimate for Case 1.

Age group (in years)	① Number of radiation-induced cancers	② Average Loss of life expectancy (years)	① × ② (Man-years of loss of expectancy)	③ Number of radiation-induced cancers	③ × ② Woman-years of loss of expectancy
0–4	0	66.1	0	0	0
5–9	50	62.0	3,100.0	36	2,232.0
10–14	40	57.2	2,288.0	29	1,658.8
15–19	77	52.5	4,042.5	51	2,677.5
20–24	130	47.8	6,214.0	88	4,206.4
25–29	186	43.2	8,035.2	171	7,387.2
30–34	318	38.6	12,274.8	394	15,208.4
35–39	641	34.0	21,794.0	881	29,954.0
40–44	1,340	29.5	39,530.0	1,791	52,834.5
45–49	2,372	25.3	60,011.6	2,914	73,724.2
50–54	4,231	21.3	90,120.3	4,089	87,095.7
55–59	6,123	17.7	108,377.1	4,997	88,446.9
60–64	7,909	14.4	113,889.6	5,555	79,992.0
65–69	8,968	11.5	103,132.0	6,358	73,117.0
70–74	9,169	9.1	83,437.9	6,689	60,869.9
75–79	7,589	6.9	52,364.1	5,954	41,082.6
80–84	4,763	5.1	24,291.3	4,342	22,144.2
85+	2,787	~3.0	8,361.0	3,217	9,651.0
Total	56,703		741,263.4	47,556	652,282.3

Average loss in life expectancy (males) $= \dfrac{741,263.4}{56,703}$, or 13.1 years.

Average loss in life expectancy (females) $= \dfrac{652,282.3}{47,556}$, or 13.7 years.

behalf of the promoters of the technology responsible for the distribution of the
poison.

The ridiculous nature of this approach to calculation of loss of life expectancy
would be obvious to everyone if we considered an issue like the death of young
Americans in Vietnam. After all, when those Americans who are at home are
averaged in with those who are killed in Vietnam, the *average* loss of life expectancy is small, the deaths are not tragic, for, on the average, everyone is just losing
days from their life. The public would not stand for such nonsense. Why they

are so readily brainwashed by pseudoscientific evaluation of loss of life expectancy for environmental poisons escapes understanding.

4. Are there possible mitigating factors which could reduce the estimated hazard of population exposure?

We have considered above the crucial parameters, such as latent period and duration of carcinogenic response plateau, which can determine in a major way the magnitude of expected population cost. We must address a few other concepts, since the uninitiated may hear that such concepts provide a reasonable basis for expecting a lesser hazard. As will become evident, there is essentially no reason to expect any lessening of hazard. Among these concepts are: (a) a possible threshold, (b) a possible "practical" threshold, (c) protraction of radiation, and (d) repair of radiation injury.

4.1. *Thresholds: absolute and "practical."*   In the discussions above it was demonstrated that abundant new data concerning the low dose region of radiation exposure indicate linearity of dose *versus* carcinogenic response over a wide range of doses. There really never has existed *any* acceptable evidence for an absolute threshold of exposure below which radiation carcinogenesis will not occur. It is to the credit of all radiation study groups that they have consistently rejected supposed evidence for radiation thresholds with respect to carcinogenesis. The linearity of dose *versus* response, now demonstrated down to very low doses, indicates there is no reason to expect any evidence for an absolute threshold ever to develop.

One total *non sequitur* has often been introduced into discussions concerning a possible threshold. That concerns the development of signs and symptoms of acute radiation sickness following radiation exposure. Everyone cognizant with this field has known for decades that acute radiation sickness is *not* linearly related to radiation dose, whereas carcinogenesis now appears definitely so related. The underlying mechanism in acute radiation sickness relates to whether or not cell *replacement* can operate rapidly enough to prevent such phenomena as mucosal ulceration or leukopenia. At radiation doses where cellular replacement *is* rapid enough, radiation sickness just does not occur. For carcinogenesis, not a shred of evidence has ever been adduced that cellular replacement can avert cancerous change.

The modification of the threshold concept to the "practical" threshold we have dealt with above. There is no basis for expecting any help from this concept.

4.2. *Protraction of radiation.*   It is very commonly stated, with appallingly little evidence, if any, that if radiation is delivered slowly, the carcinogenic effect is lessened. A little later this was modified to the statement that protraction protects against carcinogenesis from low LET radiation (such as beta rays, X-rays, or gamma rays), but not high LET radiation (such as neutrons or alpha particles). A variety of experiments have been cited as direct demonstrations that protraction of radiation affords protection against carcinogenesis, [34].

Almost invariably such experiments contrast acute delivery of radiation *early* in life with protracted radiation extending from early in life through a significant part of the life span of the experimental animal. In some of the specific cases reported, the author has himself demonstrated a marked diminution in carcinogenicity of radiation with increasing age at irradiation [33]. In other studies, this point is entirely neglected. In the material presented throughout this communication the steep decline in carcinogenicity per rad with age in *humans* has been documented. Thus, the most probable interpretation of experiments contrasting acute versus protracted irradiation is simply that protraction provides part of the irradiation *at older* ages and, hence, cancer induction is lessened. All that this re-emphasizes is the extreme seriousness of radiation as a carcinogen early in life. Whether there truly exists *any* residual mitigation from radiation protraction is uncertain within present evidence. Certainly such bodies as the International Commission on Radiological Protection have acted with wisdom, from the public health viewpoint, in refusing to count upon protraction of radiation to lessen carcinogenic hazard.

We feel strongly that it would be appropriate to go further, for any environmental pollutant, and state the following principle: "If under any dosage rate schedule a pollutant shows a certain magnitude of toxic effect, that toxic effect should be assumed to be *at least as high* for any other dosage rate schedule, until and unless definitively proven otherwise."

Adherence to such a public health principle might reduce the danger from those individuals all too ready to spew forth clichés, such as, "Maybe the poison won't be so bad if we give it slowly."

In the carcinogenesis field there is one special circumstance that deserves special consideration here. This is the case, either in humans or experimental animals, of a cancer whose incidence does *not* increase spontaneously in a monotonic fashion with increasing age. While most of the familiar cancers of adult life do show monotonically increasing incidence rates with increasing age, this is not true for several human cancers that occur in childhood (for example, neuroblastoma, Wilms' tumor). Some of these childhood cancers show a peak incidence in the first decade of life and a declining incidence thereafter. There is every reason to suspect that certain cancers of experimental animals may have a similar age related incidence pattern.

Earlier in this communication we presented a generalization (Generalization 1) which stated "the correct way to describe the phenomenon (cancer induction by ionizing radiation) is either in terms of the dose required to double the spontaneous mortality rate for each cancer, or, alternatively, of the increase in mortality rate of such cancers per rad of exposure." Let us consider what might occur if one happened to do dose protraction *versus* acute radiation studies on a cancer having a peak incidence at one age period. If Generalization 1 is correct, then the results obtained by dose protraction could appear to be a lesser incidence of the cancer simply *because* of its spontaneous age incidence pattern, and be wholly unrelated to any "protection" resulting from slow delivery of the radiation. We suspect that in time such an experiment will be done, and the results misinterpreted, to society's detriment.

4.3. *Repair of radiation injury.* Lastly, we must consider the phenomenon known as "repair." We hear commonly stated that DNA repair mechanisms exist and, hence, low dose radiation may not be as harmful as a carcinogen as had been suspected. No serious student of biology doubts the existence of DNA excision-repair or of such phenomena as light-stimulated thymine dimer repair. However, the *existence* of such phenomena by no means argues in any way for mitigation of radiation carcinogenesis. There is no evidence whatever that has been adduced relating such repair to ionizing radiation carcinogenesis.

When we observe the induction of cancer by ionizing radiation, we are, as yet, totally in the dark concerning the mechanism operative in production of the cancer. Whatever such mechanism may be it is entirely conceivable that a large part of the carcinogenic damage of radiation may get repaired. What we are *observing* is the net, unrepaired carcinogenic damage. The only conceivable way that any such hypothetical *carcinogenic* repair could help at low dose would be for *more efficient* repair to exist at low doses or slow delivery of dose than for high doses or rapid delivery of dose. If the *fraction* of unrepaired carcinogenic damage by radiation were independent of total dose and/or dose rate, then the very existence of any such repair mechanism would be wholly irrelevant as a possible mitigating factor for population consequences of low dose rate exposure. And since (a) we know of no such carcinogenic repair mechanism, and (b) nothing whatever is known about variation in efficiency of an unknown repair mechanism as a function of dose and dose rate, it should be clear that all this represents the sheerest of speculative fancy. The linearity of dose response in carcinogenesis by radiation argues strongly *against* repair of carcinogenic damage at low doses with decreasing repair at successively higher doses.

Injection of speculative fancy into a serious matter of public health protection is irresponsible. Relating DNA repair phenomena to mitigation of carcinogenic injury by radiation, in the absence of any demonstration that these phenomena are in any way related to each other, seems equally irresponsible.

5. A re-look at the purposes of this symposium after consideration of the potential population consequences of low dose radiation exposure

Do we really want to design epidemiologic studies to evaluate the population effects of pollutants, or potential pollutants, past, present, or future? Radiation, to paraphrase many nuclear enthusiasts, is one of the most intensively studied environmental poisons. Yet, for those who have had the patience to read through this communication, certain points, we hope, will stand out. Twenty-five years into the atomic era, and 75 years after Roentgen's discovery of the X-ray, we realize that, while the risk of cancer is high, certain parameters, still not possible to evaluate within present epidemiologic data, may make the cancer risk *more than three times* higher than our pessimistic estimates of 1969. Are there rational humans who will be able to understand setting an allowable radiation guide for population exposure which may provoke a public health hazard one-third the magnitude of the entire cancer problem? We can only hope that the lessons of the radiation story will lead to a radical change in human approach to the questions of environmental pollutants.

Statisticians and epidemiologists, of course, are inclined to look forward to doing what statisticians and epidemiologists are professionally prepared to do. Unfortunately, this is true also about physicists, chemists, and engineers.

The purpose of this Symposium *implies* that, for the host of potential pollutants now being introduced into our environment, enough epidemiologic evidence will, in the course of time, accumulate so that the statisticians and epidemiologists can do their thing. This means that the statisticians and epidemiologists have capitulated *in toto* to the dictum that progress means we must expose humans to by product poisons of industry in the future as we have in the past. And *then* the effects will be studied. If our radiation experience is any guide at all concerning the time scale over which we will learn the effect of our folly, and there is every reason to believe for carcinogenesis or genetic injury that the time scales will be similar, then the chances for humans surviving this approach are slim indeed.

We think it might have been more important if this gathering of statisticians and epidemiologists had met instead to lend their talents and wisdom to a concerted human effort to work toward to total recycling economy, in which essentially zero pollution is the objective instead of the building up of a reservoir of epidemiologic evidence of the effects of pollutants on humans. Indeed, such a thrust might even lead to the revolutionary idea of "Why do some of these nonsensical activities labelled 'Progress' at all?"

6. Summary

Ionizing radiation is a potent leukemogen and carcinogen. The demand for epidemiologic evidence of human injury has resulted in a belated appreciation of the true magnitude of the serious carcinogenic hazard of population exposure to radiation. Even now, a quarter of century into the evaluation of the epidemiologic evidence, certain parameters of crucial character remain indeterminate. Should these parameters turn out to have unfavorable values, the seriousness of the hazard may truly be even larger than recent pessimistic estimates. We question, therefore, the wisdom of epidemiologic studies of *human* exposure for new potential carcinogens being introduced into our environment.

Refined estimates presented here suggest that our earlier estimate of 32,000 extra cancer deaths per year for exposure to the still permissible 0.17 rads per year (average for U.S. population from the "peaceful" atom) are not at all conservative. The true cancer risk may be closer to 100,000 extra deaths per year, representing a 30 per cent increase over the current spontaneous cancer mortality. Fortunately, atomic energy programs have not *yet* progressed to a point where such allowable exposure are being experienced.

The National Council on Radiation Protection has recently stated that the current standards for radiation exposure are satisfactory (1971). We would not for one moment challenge the fact that the exposure standards are satisfactory to the membership of the National Council on Radiation Protection any more than we would challenge the concept that possession of 10,000 nuclear missiles is satisfactory for the Department of Defense. What escapes our understanding, however, is how one might go about evaluating the quantitative nature of the nebulous relationship between the interests of the membership of the NCRP and the public's interest in good health.

Medical uses of X-rays presently are a major source of population exposure and are undoubtedly responsible for a significant part of our currently experienced cancer mortality rate. Morgan's suggestions for feasible reduction in medical X-ray exposure, without loss of medical diagnostic information, deserve immediate action [25].

Natural radiation, while in large part not directly within our control, is comparable in responsibility to medical X-rays in the quantitative fraction of cancer mortality rate currently being experienced. No rational basis exists for the frequently heard suggestion that natural radiation can be used as a benchmark for estimation of "safe" exposures. Natural radiation must be estimated as possibly responsible for taking a toll of several tens of thousands of lives annually by premature cancer and leukemia in the USA alone. Here again we must agree with Morgan, that man may decide to look carefully at the radioactivity of certain "natural" building materials before using them for home construction.

Life expectancy loss experienced by those who will become the victims of *allowable* population radiation exposure will average more than 13 years. The assertions of "only a few days of loss of life" are arrived at by the absurd and dangerous practice of distribution of the man-years lost in life expectancy into the larger group of nonvictims of radiation carcinogenesis.

Epidemiologic investigations are extremely interesting and carry, for the investigators, the thrill experienced in solving murder mysteries and other challenging problems. We have extreme doubt that the planning of appropriate epidemiologic investigations for future environmental pollutants is likely to be any real contribution to the public health. There has to be a more rational approach to the question of potential environmental carcinogens—like not introducing them into the environment at all.

REFERENCES

[1] G. W. Beebe, H. Kato, and C. E. Land, "Mortality and radiation dose, atomic bomb survivors, 1950–1966," presentation at the IVth International Congress of Radiation Research, Evian, France, June 29–July 4, 1970.

[2] R. E. CARROLL, W. HADDON, JR., V. H. HANDY, and E. E. WEEBEN, SR., "Thyroid cancer: cohort analysis of increasing incidence in New York State, 1941–1962," *J. Nat. Cancer Inst.*, Vol. 33 (1964), pp. 277–283.

[3] W. M. COURT-BROWN and R. DOLL, "Mortality from cancer and other causes after radiotherapy for ankylosing spondylitis," *Brit. Med. J.*, Vol. 2 (1965), pp. 1327–1332.

[4] G. W. DOLPHIN and I. S. EVE, "Some aspects of the radiological protection and dosimetry of the gastrointestinal tract," *Gastrointestinal Radiation Injury* (edited by M. F. Sullivan), Amsterdam, Excerpta Medica Foundation, 1968, pp. 465–474.

[5] FEDERAL RADIATION COUNCIL, "Staff report no. 1. Background material for the development of radiation protection standards," Washington, D.C., 1960, Part V, pp. 26–30.

[6] M. P. FINKEL, B. O. BISKIS, and P. B. JINKINS, "Toxicity of radium-226 in mice," *Radiation-Induced Cancer* (Proceedings of a Symposium, Athens, Greece, 28 April–2 May, 1969, Organized by International Atomic Energy Agency in Collaboration with the World Health Organization), Vienna, Austria: *also* International Atomic Energy Agency, 1969, pp. 369–391.

[7] J. W. GOFMAN and A. R. TAMPLIN, "Low dose radiation and cancer," *IEEE Trans. Nuc. Sci.*, Vol. NS-17 (1970), pp. 1–9.

[8] J. W. GOFMAN and A. R. TAMPLIN, "Studies of radium exposed humans II: further refutation of the R. D. Evans' claim that the linear, non threshold model of human radiation carcinogenesis is incorrect," testimony (on Bill S3042) presented before the Subcommittee on Air and Water Pollution, U.S. Senate, 91st Congress, 1970, pp. 326–350.

[9] J. W. GOFMAN and A. R. TAMPLIN, "Federal radiation council guidelines for radiation exposure of the population-at-large—protection or disaster?," *Underground Uses of Nuclear Energy*, Part 1 (Hearings before the Subcommittee on Air and Water Pollution, U.S. Senate, 91st Congress, November 18, 1969), Washington, D.C., U.S. Government Printing Office, 1970, pp. 58–73.

[10] J. W. GOFMAN and A. R. TAMPLIN, "A proposal for at least a ten-fold reduction in FRC guidelines for radiation exposure to the population-at-large: supportive evidence," *ibid.*, 1970, pp. 319–325.

[11] J. W. GOFMAN and A. R. TAMPLIN, "Nuclear energy and the public health," *Nevada Engin.*, Vol. 6 (1970), pp. 1–16.

[12] J. W. GOFMAN and A. R. TAMPLIN, "The question of safe radiation thresholds for alpha emitting bone seekers in man," *Health Phys.*, Vol. 21 (1971), p. 47.

[13] J. W. GOFMAN, J. D. GOFMAN, A. R. TAMPLIN, and E. KOVICH, "Radiation as an environmental hazard," presentation at the 1971 Symposium on Fundamental Cancer Research, The University of Texas, M. D. Anderson Hospital and Tumor Institute, Houston, Texas, March 3, 1971, in press.

[14] A. GRENDON, "Radiation protection standards," in *Environmental Effects of Producing Electric Power*, hearings before the Joint Committee on Atomic Energy, 91st Congress, 2nd Session, January 27–February 26, 1970, Part 2, Vol. II, p. 2371.

[15] L. D. HAMILTON, "Biological significance of environmental radiation: calculation of the risk," presentation at the 1971 Spring Meeting of the American Physical Society, Washington, D.C., April 29, 1971.

[16] L. H. HEMPELMANN, "Risk of thyroid neoplasms after irradiation in childhood," *Science*, Vol. 160 (1968), pp. 159–163.

[17] INTERNATIONAL COMMISSION ON RADIOLOGICAL PROTECTION, Publication No. 8, *Radiation Protection: The Evaluation of Risks from Radiation*, Oxford, Pergamon Press, 1966, Table 15, p. 56.

[18] INTERNATIONAL COMMISSION ON RADIOLOGICAL PROTECTION, Publication No. 14, *Radiosensitivity and Spatial Distribution of Dose*, Oxford, Pergamon Press, 1969, Appendix III, pp. 56–106.

[19] S. JABLON and J. L. BELSKY, "Radiation-induced cancer in atomic bomb survivors," presentation at the Xth International Cancer Congress, Houston, Texas, May 1970.

[20] S. JABLON, J. L. BEISKY, K. TACHIKAWA, and A. STEER, "Cancer in Japanese exposed as children to atomic bombs," *Lancet*, No. 7706 (1971), pp. 927–931.

[21] E. B. LEWIS, "Ionizing radiation and tumor," *Genetic Concepts and Neoplasia* (23rd Annual Symposium on Fundamental Cancer Research, 1969, University of Texas, M. D. Anderson Hospital and Tumor Institute, Houston, Texas), Baltimore, Williams and Wilkins Co., 1970, pp. 57–73.

[22] B. MACMAHON, "Pre-natal x-ray exposure and childhood cancer," *J. Nat. Cancer Inst.*, Vol. 28 (1962), pp. 1173–1191.

[23] I. MACKENZIE, "Breast cancer following multiple fluoroscopies," *Brit. J. Cancer*, Vol. 19 (1965), pp. 1–8.

[24] R. H. MOLE, "Radiation effects in man: current views and prospects," *Health Phys.*, Vol. 20 (1971), pp. 485–490.

[25] K. Z. MORGAN, "Never do harm," *Environment*, Vol. 13 (1971), pp. 28–38; *also* NCRP Report 39, "Basic radiation protection criteria," published by National Council on Radiation Protection and Measurements, Washington, D.C., 1971, p. 97.

[26] E. E. POCHIN, "Somatic risks—thyroid carcinoma," *Internat. Comm. Rad. Protec.*, Publication 8, Oxford, Pergamon Press, 1966, p. 9.

[27] L. SAGAN, "A positive word for nuclear power," *This World*, section of the *San Francisco Examiner and Chronicle*, January 10, 1971.

[28] C. J. SHELLABARGER, V. P. BOND, E. P. CRONKITE, and G. E. APONTE, "Relationship of dose of total-body ^{60}Co radiation to incidence of mammary neoplasia in female rats," *Radiation-Induced Cancer* (Proceedings of a Symposium, Athens, Greece, 28 April–2 May, 1969. Organized by International Atomic Energy Agency in Collaboration with the World Health Organization), Vienna, International Atomic Energy Agency, 1969, pp. 161–172.

[29] H. SPIESS and C. W. MAYS, "Bone cancers induced by ^{224}Ra (ThX) in children and adults," *Health Phys.*, Vol. 19 (1970), pp. 713–729.

[30] A. STEWART and G. W. KNEALE, "Changes in the cancer risk associated with obstetric radiography," *Lancet*, No. 7532 (1968), pp. 104–107.

[31] A. STEWART and G. W. KNEALE, "Radiation dose effects in relation to obstetric x-rays and childhood cancers," *Lancet*, No. 7658 (1970), pp. 1185–1188.

[32] A. R. TAMPLIN and J. W. GOFMAN, "Biological effects of radiation," *'Population Control' Through Nuclear Pollution*, Chicago, Nelson-Hall Co., 1970, pp. 7–27.

[33] A. C. UPTON, T. T. ODELL, JR., and E. P. SNIFFEN, "Influence of age at time of irradiation on induction of leukemia and ovarian tumors in RF mice," *Proc. Soc. Exper. Biol. Med.*, Vol. 104 (1960), pp. 769–772.

[34] A. C. UPTON, "Comparative observations on radiation carcinogenesis in man and animal," *Carcinogenesis, a Broad Critique* (20th Annual Symposium on Fundamental Cancer Research, 1966, University of Texas, M. D. Anderson Hospital and Tumor Institute, Houston, Texas), Baltimore, Williams and Wilkins Co., 1967, pp. 631–675.

[35] A. C. UPTON, R. C. ALLEN, R. C. BROWN, N. K. CLAPP, J. W. CONKLIN, G. E. COSGROVE, E. B. DARDEN, JR., M. A. KASTENBAUM, T. T. ODELL, JR., L. J. SERRANO, R. L. TYNDALL, and H. E. WALBURG, JR., "Quantitative experimental study of low-level radiation carcinogenesis," *Radiation-Induced Cancer*, Vienna, International Atomic Energy Agency, 1969, pp. 425–438.

[36] O. J. BIZZOZERO, JR., K. G. JOHNSON, and A. CIOCCO, "Radiation-related leukemia in Hiroshima and Nagasaki, 1946–64. I," *New England Journal of Medicine*, Vol. 274 (1966), pp. 1095–1101.

[37] R. D. EVANS, A. T. KEANE, R. J. KOLENKOW, W. R. NEAL, and M. M. SHANAHAN, "Radiogenic tumors in radium and mesothorium cases studied at M.I.T.," *Delayed Effects of Bone-Seeking Radionuclides* (edited by C. W. Mays), Salt Lake City, University of Utah Press, 1969, pp. 157–194.

Discussion

Question: David L. Levin, National Cancer Institute

In regard to the cancer/leukemia ratio and the extrapolation of risk down to low levels of irradiation, we have also looked at the Court-Brown and Doll data on spondylitics and the Hiroshima-Nagasaki data. We have found that the cancer/leukemia ratio, when adjusted for difference in dose to the target organs is indeed higher than often quoted. Also, we have made calculations on the lifetime risk of developing cancer from irradiation levels of 170 mrads based upon extrapolations from high dose situations (X-ray therapy and atom bomb) to low dose situations. Using different methodology than that described today, we computed risks to be of the same general level as those shown by Dr. Gofman. We are now continuing our calculations trying to improve our estimate based upon demographic studies.

Reply: J. Gofman

We are, of course, pleased to hear that Dr. Levin and his colleagues arrive at the same general conclusions as we do. This confirms our statement that the so called "radiation controversy" concerning cancer risk is over. Dr. Levin is certainly correct that there are other calculation methodologies. We have tried and presented several approaches. It turns out that, since we all have the same epidemiologic data at our disposal, the rules of arithmetic and algebra require that the same general conclusions will be reached.

Often the issue is confused by those who don't really disagree with the calculation of the cancer hazard. It is just that some people don't worry about adding 100,000 extra deaths per year from cancer. This is more a problem of central nervous system physiology and pathology than it is one of epidemiology.

Question: E. B. Hook, Birth Defects Institute, Albany Medical College

How do the calculations of a presumed increase in tumor incidence following an increase of 0.17 rads per year compare with the known "background" dose of radiation and known tumor incidence?

Reply: J. Gofman

As discussed in the presentation, we would calculate the tumor incidence due to natural background radiation would stand to that for 0.17 rads/year as does 0.125 rads per year to 0.17 rads/year. No one has experimentally or by an epidemiological study segregated the cancer cases due to natural background radiation from so-called "spontaneous" cancer cases. There is every reason to believe that background radiation accounts for its expected share of the total observed cancer mortality rate.

Question: E. Tompkins, Human Studies Branch, Environmental Protection Agency

Would you please explain your statement that there is no reason to believe that the number of films (as reported by Stewart and Kneale) is associated with disease of the mother.

How do you explain the difference in distribution of number of films by trimester?

Reply: J. Gofman

Let us consider first those women who received X-rays specifically for diagnostic obstetric radiography. This group constituted the bulk of the Stewart-Kneal sample. Within this group of women selected to receive diagnostic obstetric radiography, I know of no reason to believe the radiographer would take a number of films, to achieve pelvic dimension measurements, that would be correlated with pre-existing disease in the mother.

With respect to the difference in distribution of film number by trimester, this is not at all surprising. Films taken before the third trimester almost certainly were taken for an indication other than diagnostic pelvimetry. It is commonplace to find that the number of films required for one particular diagnostic purpose may differ from that required for another.

Question: Alexander Grendon, Donner Laboratory, University of California, Berkeley

You answered a previous question by saying that the component of the natural cancer incidence due to background radiation is also multiplied by added radiation. Doesn't this constitute a square law relation rather than the linear relation postulated by you?

Reply: J. Gofman

I have said nothing at all that suggests a square law relationship. I believe the difficulty, Mr. Grendon, resides in your interpretation of what I said. I said that when the cancer mortality increment per rad is expressed as a percentage of the "spontaneous" cancer mortality rate, the "spontaneous" rate incorporates the contribution of natural plus medical radiation. Such expression of the increment is simply a statement of *observation* in a particular population sample. For purposes of determining the real per cent increment per rad over nonradiation "spontaneous" cancer mortality rate, the appropriate procedure would be to subtract out the medical plus natural radiation contribution to the observed "spontaneous" rate. I believe this is adequately clarified in the text. I had no intention of suggesting that additional radiation multiplies the effect of background radiation.

Question: Unnamed discussant

Could you comment on the quality of the Hiroshima data on leukemia as compared to that of Dr. Stewart's data and that of Dr. MacMahon? Is it not true that the effects of trauma, dietary and medical care effects on the population exposed to the bomb radiation as compared to the controls and the subsequent heavy mortality by abortion and congenital defects among the exposed fetuses and infants make these results far more questionable than the studies of infants exposed in peace time to diagnostic X-rays?

Reply: J. Gofman

Yes, I prefer the Stewart and MacMahon data to the Hiroshima-Nagasaki data on leukemia in childhood following *in utero* radiation both for the reasons you cited and for other reasons.

Perhaps we should clarify for the audience what the issue is here. Recently Jablon [48] published data on subjects irradiated *in utero* in the Japanese atom bombings. Jablon indicated that for the exposure received by this population sample, the observed occurrence of childhood cancer and leukemia was far less than that to be expected from the Stewart or MacMahon data. I have several very serious reservations about the Japanese data.

(1) The bulk of the man-rads of exposure in the Japanese sample arises from those cases with very high exposures (100–200 rads). We are all familiar with the observations in experimental studies that, while the carcinogenic dose response relationship is linear over a large range, one does arrive at high doses where the carcinogenic response levels out and then drops drastically. Many refer to this as "the other side of the dose response curve." From the Stewart evidence, it appears that the doubling dose for *in utero* induction of leukemia and cancer is of the order of one rad. This would mean that the bulk of the man-rad exposure of the Japanese sample occurred in infants receiving 50 to 200 doubling doses. There is every reason to suspect that this would place them well over on the "other side of the dose response curve." Hence this factor alone should lead to a lesser carcinogenic/leukogenic response in the Japanese sample than anticipated based upon the Stewart evidence.

(2) The Japanese sample of infants exposed *in utero* was characterized by an enormous mortality during the first year of life. No data are provided concerning the nature of these mortalities. If the risk of subsequent mortality from cancer or leukemia is correlated with risk of mortality in the first year of life (and we know nothing of the existence or nature of such relationships), it is conceivable that the Japanese sample was depleted, by enormous first year mortality, of the most likely candidates for subsequent cancer or leukemia. The discussant, for example, pointed out the issues of dietary and medical care effects in the Japanese sample. This is certainly appropriate, and it is entirely possible that those with enough radiation injury to develop leukemia later may be especially susceptible to earlier death from malnutrition, for example. This effect would not have occurred in the Stewart or MacMahon population samples.

(3) The Jablon data on children exposed between 0 and 9 years of age at the time of bombing show a marked increase in carcinogenic and leukemogenic risk. It would be very surprising that the carcinogenic/leukemogenic risk in the Japanese *in utero* cases would be absent.

(4) I would take serious issue with Dr. John Totter's statement (in discussing Dr. Sternglass' paper this morning) that the Japanese data represent an unbiassed sample compared with the Stewart data. For the obvious reasons listed above, I would draw the opposite conclusion to that of Dr. Totter. In so doing, I agree with the discussant who raised this question.

Question: Prem S. Puri, Department of Statistics, Purdue University

You presented a table showing the mortality figures by age for individuals who were exposed to radiation at Hiroshima, Japan. Do you have some information on the possible radiation effects on the mortality of the descendants in the next generation of these individuals?

Reply: J. Gofman

I believe it is far too early to be able to comment on mortality effects among the descendants of the survivors of the atom bombing in Japan.

Question: J. Martin Brown, Stanford Medical Center

Is it not true that the data of Stewart, Kneale and of MacMahon show that the radiation induced excess over the spontaneous rate has disappeared eight to 10 years after the pelvic X-rays of the fetus? Won't this make a big difference to your calculation of the radiation induced incidence of cancer in view of its marked dependence on the length of the plateau period?

Reply: J. Gofman

I do not believe that either the Stewart-Kneale or the MacMahon data really allow for one to draw the conclusion that the radiation induced excess disappears at eight to ten years or at any other time, for that matter. I do understand how that impression can have arisen, however. The overall mortality curves (spontaneous) for cancer show a peak at about the middle of the first decade of life and then a decline to a *relatively* low level until the early 20's of age when the upturn begins, due to the appearance of the various malignancies of adult life. In the Stewart-Kneale and MacMahon children the radiation induces the *same* kinds of cancers that occur spontaneously in the children. Since there is about a 50 per cent increase due to obstetric radiography in such cancers and leukemias over their spontaneous occurrence, it is not surprising that the radiation induced cases would show a peaking just as do the spontaneous cases. The decline from the peak among the radiation induced cases gives the *impression* that the radiation induced excess is disappearing. This, I would consider, is totally illusory.

If one could study this phenonmenon into later years (adulthood), radiation induced cases might rise in incidence as the spontaneous incidence rises, and, for all we know, this effect might persist throughout the lifetime of the exposed population. A *very* different study from those of Stewart-Kneale or MacMahon would be required to address this issue. Incidentally, for children irradiated between 0 and 9 years of age, the radiation induced cancers are *rising* in incidence 20 years post irradiation in the Japanese survivors, as would be expected from the rising spontaneous incidence with age increase.

But let us presume the *in utero* effect did decrease after 8 or 10 years. From Table V, the average increase in cancer mortality rate is calculated to be approximately 34 per cent of the spontaneous mortality rate. If the 6 per cent due to *in utero* irradiation is subtracted, the radiation induced rate would be approximately 28 per cent of the spontaneous mortality rate.

Question: B. G. Greenberg, School of Public Health, University of North Carolina, Chapel Hill

I wonder, Dr. Gofman, if the ordinate in your graph, mortality from, say, leukemia, has been adjusted for other deaths on a competitive risk basis. If not, the graph seems to imply that the best protection for a person who has been irradiated is to expose him to an additional overwhelming dose in order to bring his mortality risk down.

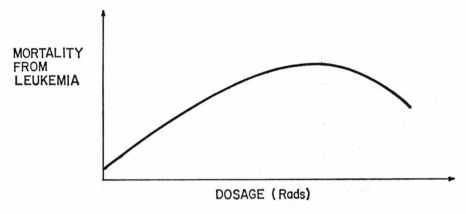

MORTALITY FROM LEUKEMIA

DOSAGE (Rads)

Reply: J. Gofman

Of course there would be an increasing mortality with increasing radiation dose for a whole variety of radiation induced causes of death. Among the survivors of the increased dose, the leukemia risk would still be a lower fraction of the total irradiated population than at lower doses. The competitive risk of other mortalities must be considered. If we wish to focus on keeping just leukemia mortality down, I would suggest that execution of the entire irradiated population sample by a firing squad would be even more effective than the supplemental radiation suggested by Dr. Greenberg. (See Figure 2, Curve A.)

Question: Thomas F. Budinger, Donner Laboratory, University of California, Berkeley

Dose rate effect has been shown for *carcinogenesis* [45], *survival* ([42], [41]), and *genetic mutations* [44]. Exposure in the range of 170 mrem distributed through one year is more than 10^6 times less dose rate than the dose rate of exposures from which Dr. Gofman draws his conclusions. Therefore, from a biophysical standpoint it is difficult to accept his figures as applicable to any anticipated population exposure. These studies as well as curvilinear dose response relationships in radiation carcinogenesis (for example, see [46],) suggest some repair mechanism is present. A reasonable assumption is that DNA and chromosomes are involved in somatic and genetic mutation. We now have ample biophysical data to show DNA breaks are repaired very efficiently by at least two cellular mechanisms (for example, see [43]). In fact, the history of radiation damage and successful or unsuccessful repair is readily seen by following chromosome aberrations [40], [38]. Radiation insult repair is dependent on linear energy transfer [39] and for high LET exposures we would expect, and indeed find, linear dose effect response and little or no dose rate effect on DNA repair, chromosome aberrations, or carcinogenesis. Thus, only for high LET irradiation is a linear extrapolation valid. If we all were being exposed to neutrons or penetrating high Z charged particles, I would agree with Dr. Gofman's figures. Anticipated radiation exposures are low LET at low dose rates; thus, I do not see a public health threat as great as the threat of trace elements and other man-made contaminants.

REFERENCES

[38] W. C. Dewey and R. M. Humphrey, "Restitution of radiation-induced chromosomal damage in Chinese hamster cells related to the cell's life cycle, *Exper. Cell Res.*, Vol. 35 (1964), p. 262.

[39] M. M. Elkind, *Curr. Top. Rad. Res. Quart.*, Vol. 7 (1970), p. 1.

[40] H. J. Evans, "Repair and recovery from chromosome damage after fractionated X-ray dosage," *Genetic Aspects of Radiosensitivity: Mechanisms of Repair*, IAEA, 1966, pp. 31–48.

[41] D. Grahn, "Biological effects of protracted low dose radiation exposure of man and animals," *Late Effects of Radiation*, London, Taylor & Francis, LTD., 1970, p. 101.

[42] E. J. Hall, R. J. Berry, J. S. Bedford, in *Dose Rate in Mammalian Biology*, AEC Conference Report Conf.-680410 (1968), pp. 15.1–15.20.

[43] R. B. Painter, *Curr. Top. Rad. Res. Quart.*, Vol. 7 (1970), p. 45.

[44] W. L. Russell, *Nucleonics*, Vol. 23, p. 53.

[45] A. C. Upton, *Radiation Injury*, Chicago, University of Chicago Press, 1969, p. 79.

[46] ———, "The dose-reponse relation in radiation-induced cancer," *Cancer Res.*, Vol. 21 (1961), p. 717.

Reply: J. Gofman

I believe I have disposed of Dr. Budinger's major concerns in the body of this communication. A few of the specifics raised by Dr. Budinger are either *non sequiturs* or reflect such unsatisfactory public health principles that they deserve comment here.

(a) Dr. Budinger states that dose rate effect for carcinogenesis has been shown by Upton, [45]. This is simply not true. No studies of Upton are free of the criticism amply discussed in the text that the chronic exposures extend into later life of the experimental animal where sensitivity to carcinogenesis drops. Indeed, the studies of Upton, which Dr. Budinger quotes, are provably suspect on these grounds based upon Upton's own data. (This issue is thoroughly discussed in one of our earlier papers, Gofman and Tamplin [47].)

(b) Dr. Budinger comments that "exposure in the range of 170 mrem distributed throughout one year is more than 10^6 times less dose rate than the dose rate of exposures from which Dr. Gofman draws his conclusions. Therefore, from a biophysical standpoint it is difficult to accept his figures as applicable to any anticipated population exposure."

There *may* be sound grounds upon which our figures are unacceptable, but I truly am surprised that Dr. Budinger states "from a biophysical standpoint" it is difficult to accept them. It becomes necessary to point out a few biophysical principles to Dr. Budinger.

First, the existence of a linear relationship between carcinogenic response and dose (see references in text) implies, *biophysically*, a single event phenomenon for radiation induction of cancer. This *biophysical* point should remove most of Dr. Budinger's concern.

Second, let us consider Dr. Budinger's factor of 10^6 in dose rate. Especially, let us consider the biophysics involved. Dr. Budinger would appear to be considering that 170 mrem delivered over the course of a year by environmental pollutants delivering low LET radiation to be oozed into tissue at a slow, smooth rate. Unfortunately, the biophysical reality is considerably different from the picture of Dr. Budinger.

Let us consider, biophysically, the delivery of 170 mrem to cells over one year versus delivery in a fraction of a second, as from an X-ray machine. We shall find that Dr. Budinger's factor of 10^6 melts away with great speed.

At 1 MEV or less, X-rays and gamma rays deliver energy to tissues through their photoelectric or Compton conversion to electrons. Therefore, the entire group of low LET radiations (X-rays, gamma rays, β particles) can be covered by consideration of β particle (electron) interaction with cells.

$$1 \text{ rad} = 100 \text{ ergs/gram} = 6.25 \times 10^7 \text{ MEV/gram.}$$

Let us consider 1 MEV β particles as representative.

This means 1 rad represents 6.25×10^7 β particles delivering their energy per gram of tissue.

The range in tissue for 1 MEV β particles is approximately 4000 microns.

For a cell of 20 micron diameter, a 1 MEV β particle traverses 200 cells, on the average. Therefore, 6.25×10^7 β particles traverse 1.25×10^{10} cells.

For cells of approximately 20 micron diameter, volume is approximately $4 \times 10^3 \mu^3$, and 1 gram of tissue represents approximately $10^{12} \mu^3$.

So there are $10^{12}/(4 \times 10^3) \cong 2.5 \times 10^8$ cells per gram of tissue cells.

If one rad represents traversal of 1.25×10^{10} cells, then each cell is traversed $(1.25 \times 10^{10})/(2.5 \times 10^8) = 50$ times.

Therefore for 170 mrads, each cell is traversed $(0.17)(50) = 8.5$ times on the average.

Each traversal of a cell by a single beta particle occurs in a time frame of a fraction of a second. Now, if we are to compare *equivalent doses*, 170 mrads, delivered instantaneously versus over the span of one year, for the same kind of radiation, for example, 1 MEV β particle (the same would be true for X-rays or gamma rays), it follows, obviously that the exact same number of β particles, on the average, must traverse the cells whether instantaneously or over the span of a year. For the instantaneous delivery, let us assume all the β particles (between 8 and 9 of them, average 8.5) all are delivered together exciting effects over a short interval, t (time frame, seconds or less). For the one year delivery, there will still be 8.5 β particles delivered per cell, and *each* β particle will exert effects in precisely the same short time interval, t, except that the individual events will be separated from each other by a little over a month, on the average. Therefore, the "slow" delivery, so far as cellular events are concerned occupies (8.5 t) instead of t. Therefore, the maximum *real* difference in rate of delivery of energy to the cell is 8.5 fold. Thus, Dr. Budinger's "factor of 10^6," drops to a factor of 8.5, a drop of more than 100,000 fold. The interjection of the issue of about a month between events is a red herring. I do not believe that, for carcinogenic events occurring 5 to 25 years later, Dr. Budinger would argue that irradiation in September of a particular year would be much different from irradiation in May.

In all likelihood, even the factor of 8.5 fold is illusory. For a single cell, not all of the 8 or 9 events occur in the same region of the cell's volume. For a *particular* region within a cell receiving irradiation, 170 mrads may, therefore, represent the effect of only *one* β particle traversal, *not* the traversal of 8 or 9. So the *actual* time frame of events for the cellular level may be precisely the same at a particular sensitive site whether all the 170 mrads is delivered instantaneously or spread over one year. This would melt Dr. Budinger's factor of 10^6 down to a factor of *unity*, for the two regimes of irradiation. But we needn't quibble as to whether Dr. Budinger's concern is irrelevant by a factor of 100,000 or a factor of 1,000,000.

There are additional data, indeed available from Upton's work (A. C. Upton and G. E. Cosgrove, Jr. "Radiation-induced leukemia," *Experimental Leukemia*, (edited by M. A. Rich) New York Appleton-Century Crofts, 1968, pp. 131–158) which show conclusively, at least for thymic lymphoma in the mouse, that even at doses like 20 rads, there is no difference between "instantaneous" and "slow" delivery of radiation with respect to thymic lymphoma development. One of the good features of this experiment of Upton's is that the two regimes of irradiation, instantaneous and slow, were delivered at comparable age periods in the life span of the mouse, a feature *not* controlled in the vast majority of acute versus chronic irradiations. What Upton and Cosgrove showed was that there was *no* difference for thymic lymphoma incidence whether 200 rads total was delivered as 10 exposures, each of 20 rads (7 to 25 rads/min), spaced 30 days apart or whether delivered at 0.5 millirads/minute, which consumed the same approximate overall time period (approximately 300 days).

Using the same type of calculations as above, where 20 millirads represents one ionizing event (50 events per rad), the Upton data would indicate that one event per 40 minutes gave no different result, for thymic lymphoma induction, from 1000 events in *one* minute (20 rads in one minute $\cong$ 1000 events).

(c) Dr. Budinger brings up DNA repair and chromosome aberrations. These are very interesting phenomena that every knowledgeable scientist realizes do exist. But neither Dr. Budinger nor anyone else has suggested any relevance of these phenomena for the question of radiation dose rate and carcinogenesis. If Dr. Budinger knows the events in radiation carcinogenesis well enough to assert that DNA repair has anything to do with dose rate and cancer production, I urge him strongly to publish these findings. At present, without such evidence, the mere mention of phrases like "DNA repair," or like "DNA and chromosomes are involved in somatic mutation" is about as useful in assessing the question of dose rate and carcinogenesis as are yesterday's stock market quotations. We wouldn't deny that the stock market quotations are interesting.

(d) Lastly, we must respond to Dr. Budinger's statement, "Anticipated radiation exposures are low LET at low dose rates; thus, I do not see a public health threat as great as the threat of trace elements and other man-made contaminants."

We have seen above that Dr. Budinger's factor of 10^6 in dose rate for low LET radiations melts away at least by 100,000 fold, when the biophysics is considered. But, whatever the magnitude of carcinogenic effect of ionizing radiation may be, it is difficult to understand the philosophy implied in Dr. Budinger's statement that other poisons may be worse. They may very well be. If one unnecessary poison kills 10,000 people per year while another kills 30,000 people per year, shall we exonerate the first unnecessary poison because it unhappily didn't reach the *top* of the best seller list?

REFERENCES

[47] J. W. GOFMAN and A. R. TAMPLIN, "The mechanism of radiation carcinogenesis (a) An explanation for the illusory effect of protraction of radiation in reducing carcinogenesis for low LET radiation," *Environmental Effects of Producing Electric Power,* hearings of the Joint Committee on Atomic Energy, 91st Congress, 2nd Session 1970, Part 2, Vol. II, 1970, pp. 2068–2098.
[48] S. JABLON and H. KATO, "Childhood cancer in relation to prenatal exposure to A-bomb radiation," *Lancet,* Vol. 2 (1970), pp. 1000–1003.

INFANT MORTALITY AROUND THREE NUCLEAR POWER REACTORS

EDYTHALENA A. TOMPKINS, PEGGY M. HAMILTON,
and DANIEL A. HOFFMAN
ENVIRONMENTAL PROTECTION AGENCY

A part of the continuing epidemiologic research program on the effects of radiation exposure to humans is the occasional use of published vital statistics data to test the validity of the magnitude of a specific risk which has been proposed. The difficulties inherent in using vital statistics data to test a hypothesis of an association of a specific disease entity with radiation exposure are well recognized by our staff. None of the disease conditions which have been identified as being possible long term effects of exposure to radiation are unique, and therefore, no inferences about changes in rates of a specific disease following exposure can be made without looking at all of the factors which are known to influence the occurrence of that disease. However, occasionally, the magnitude of risk which has been proposed to be associated with radiation exposure is so large that it is possible to do a qualitative study. It can be hypothesized that if the risk of a specific effect associated with exposure is large enough, a change in the rate in the vital statistics data should be detectable despite all the other factors which might be influencing that effect.

In the papers presented by Dr. Ernest J. Sternglass and Dr. Morris H. DeGroot at this symposium, it has been suggested that the developing fetus is so uniquely sensitive to radiation damage that relatively very low levels of exposure to radiation from the operation of nuclear power plants are reflected in fairly large increases in infant mortality. If this is true, one can hypothesize that a comparison of infant mortality rates around a nuclear power plant before and after the beginning of operation should show differences in rates graded according to the distance from the plant. Increases in rates can be expected close to the facility with no change or decreases in rates at remote distances.

Three nuclear power facilities were selected for study: the Humboldt Bay Plant Unit 3 at Eureka, California; the Dresden Nuclear Power Station Unit 1 at Morris, Illinois; and the Big Rock Point Nuclear Power Station at Charlevoix, Michigan. These reactors were selected because they were among the first plants to be constructed and therefore provide data for several years of infant mortality experience since they began operation. In addition, they are all boiling water reactors with subsequent higher rates of radioactive gas discharge to the environment as compared to discharges from pressurized water reactors. These plants, therefore, have a history of the highest potential exposure to the population in the area of the reactor of any of the operating facilities, and thus provide the best opportunity for detecting an increase in infant mortality associated with this exposure, if it exists.

The method of study was identical for each reactor. Four concentric geographical bands were defined around each reactor site: one from 0 to 25 miles, one 25 to 50 miles, one 50 to 100 miles, and one from 100 to 200 miles. The wider bands were used more distant from the reactor to provide larger populations at risk to offset the smaller increased risk which would be hypothesized at the low exposure levels calculated for these distances. The concentric bands were then further divided into compass sectors, northeast, southeast, northwest, and southwest to permit evaluation of the effect of prevailing winds on the distribution of the gaseous wastes.

The appropriate population of live births and infant deaths within each band and sector was then estimated for each reactor. When a total county was included in one band sector, the published vital statistics for the county were used. When the vital statistics for a county needed to be divided between band sectors, it was done by crude population weighting. Using the Bureau of the Census data for 1960, the population of each minor civil subdivision within a county was determined and assigned to the relevant band sector. The percentage of the total population of a county which resided in each band sector in 1960 was then calculated, and the same percentage was used to allocate the live births and infant deaths for the county to the appropriate band sector. When the vital statistics for a municipality within a county were reported separately, these data were handled similarly to the county data.

Having obtained the number of live births and infant deaths for each band, and each band sector, four infant mortality rates were calculated: the rate of deaths under one year of age per 1000 live births (infant mortality) for the total experience during the five years before the reactor started operation and a similar rate for the five years after operation began. The same two rates were calculated for deaths under 28 days of age per 1000 live births (neonatal mortality.) More than half of infant mortality deaths occur within the first 28 days, and most of these deaths are due to conditions associated with prematurity [5]. The five year period was used to minimize the effect of small number variability and real annual fluctuations caused by such factors as local epidemics.

The infant mortality rates around the Big Rock Point plant are shown in Table I. The live births, deaths under one year of age, and the mortality rates per thousand live births are shown for each concentric band, going out from the reactor site, and for the compass sectors within each band for the five years before the reactor went operational, 1957 through 1961, and the five years after operation began, 1963 through 1967. The differences and directions of change in rates in the two time periods are also shown. It should be noted that three band sectors are contributing practically no data to these rates. The Big Rock Point plant is located on the shores of Lake Michigan so the northwest sectors

TABLE I

Infant Mortality—Big Rock Point Nuclear Power Station

Distance & direction from reactor	1957–1961			1963–1967			Difference in rates
	Live births	Deaths <1 yr.	Rate/1000 live births	Live births	Deaths <1 yr.	Rate/1000 live births	
<25 miles							
NE	1,619	44	27.1	1,246	25	20.1	−7.0
SE	1,414	40	28.3	1,270	25	19.7	−8.6
SW	802	20	24.9	702	13	18.5	−6.4
NW	5	0	—	5	0	—	—
Total	3,840	104	27.1	3,223	63	19.5	−7.6
25–50 miles							
NE	2,751	71	25.8	1,825	44	24.1	−1.7
SE	1,893	50	26.4	1,646	35	21.3	−5.1
SW	4,138	94	22.7	3,539	79	22.3	−0.4
NW	20	0	—	23	0	—	—
Total	8,802	215	24.4	7,033	158	22.5	−1.9
50–100 miles							
NE	6,425	168	26.1	5,573	129	23.1	−3.0
SE	11,347	283	24.9	8,979	182	20.3	−4.6
SW	4,354	85	19.5	3,117	72	23.1	+3.6
NW	5,645	137	24.3	3,906	84	21.5	−2.8
Total	27,771	673	24.2	21,575	467	21.6	−2.6
100–200 miles							
NE		Canada			Canada		
SE	196,068	4,711	24.0	170,541	3,700	21.7	−2.3
SW	198,151	4,349	21.9	166,254	3,624	21.8	−0.1
NW	20,506	432	21.1	17,309	392	22.6	+1.5
Total	414,725	9,492	22.9	354,104	7,716	21.8	−1.1

under 50 miles are principally lake, and the northeast sector in the 100–200 mile band is completely in Canada and was excluded.

The figures in this table reflect certain patterns of infant mortality which are common throughout the country for these time periods. First, that the number of births is lower in the second time period than in the first which is reflecting the decreasing birth rate in this country. Secondly, most of the mortality rates in the second time period are lower than in the first time period. Again, this is consistent with the continuing declining infant mortality rates in the United States. And finally, the areas with the highest starting infant mortality rates in general show the largest absolute decrease in rates. This latter phenomenon is believed to reflect better application of developed medical practice. As the preventable or treatable conditions are no longer leading to death, the rates are asymptotically approaching the so-called hard core causes of infant mortality—congenital malformations and the diseases of early infancy such as birth injuries, postnatal asphyxia, and the premature delivery of infants. A significant medical breakthrough in dealing with these conditions will have to be made before further sharp decreases in infant mortality can be expected [5].

There is no evidence in the pattern of the rates of infant mortality after the plant began operation, nor in the differences in the rates before and after operation to suggest an increase in infant mortality associated with the operation of the Big Rock Point reactor. The rates for neonatal mortality, which most strongly reflect the effects of prematurity and congenital malformations are shown in Table II and similarly show no trend or differences in rates which can be associated with the operation of the reactor.

TABLE II

NEONATAL MORTALITY—BIG ROCK POINT NUCLEAR POWER STATION

Distance & direction from reactor	1957–1961			1963–1967			Dif- fer- ence in rates
	Live births	Deaths <28 days	Rate/1000 live births	Live births	Deaths <28 days	Rate/1000 live births	
<25 miles							
NE	1,619	35	21.6	1,246	20	16.1	−5.5
SE	1,414	29	20.5	1,270	16	12.6	−7.9
SW	802	14	17.4	702	8	11.4	−6.0
NW	5	0	—	5	0	—	—
Total	3,840	78	20.3	3,223	44	13.6	−6.7
25–50 miles							
NE	2,751	53	19.3	1,825	33	18.1	−1.2
SE	1,893	35	18.5	1,646	21	12.8	−5.7
SW	4,138	78	18.8	3,539	58	16.4	−2.4
NW	20	0	—	23	0	—	—
Total	8,802	166	18.8	7,033	112	15.9	−2.9
50–100 miles							
NE	6,425	118	18.4	5,573	104	18.7	+0.3
SE	11,347	199	17.5	8,979	128	14.2	−3.3
SW	4,354	75	17.2	3,117	47	15.1	−2.1
NW	5,645	107	19.0	3,906	63	16.1	−2.9
Total	27,771	499	18.0	21,575	342	15.9	−2.1
100–200 miles							
NE		Canada			Canada		
SE	196,068	3,528	18.0	170,541	2,869	16.8	−1.2
SW	198,151	3,252	16.4	166,254	2,705	16.3	−0.1
NW	20,506	336	16.4	17,309	280	16.2	−0.2
Total	414,725	7,116	17.2	354,104	5,854	16.5	−0.7

Similar data for the Humboldt Bay plant at Eureka, California are shown in Tables III and IV. This reactor went operational in 1963, and the two time periods used are 1958 through 1962, and 1964 through 1967. Only four years of experience are included in the later time period because 1968 mortality data

TABLE III

INFANT MORTALITY—HUMBOLDT BAY PLANT UNIT 3

Distance & direction from reactor	1958–1962			1964–1967			Dif-ference in rates
	Live births	Deaths <1 yr.	Rate/1000 live births	Live births	Deaths <1 yr.	Rate/1000 live births	
<25 miles							
NE	6,411	191	29.8	3,646	73	20.0	−9.8
SE	3,262	72	22.1	1,849	37	20.0	−2.0
Total	9,673	263	27.2	5,495	110	20.0	−7.2
25–50 miles							
NE	1,555	34	21.9	875	18	20.6	−1.3
SE	1,561	35	22.4	873	18	20.6	−1.8
Total	3,116	69	22.1	1,748	36	20.6	−1.5
50–100 miles							
NE	4,696	138	29.4	2,798	59	21.1	−8.3
SE	5,757	164	28.5	4,203	88	20.9	−7.6
Total	10,453	302	28.9	7,001	147	21.0	−7.9
100–200 miles							
NE	32,523	760	23.4	22,441	471	21.0	−2.4
SE	64,857	1,544	23.8	49,239	1,073	21.8	−2.0
Total	97,380	2,304	23.7	71,680	1,544	21.5	−2.2

TABLE IV

NEONATAL MORTALITY—HUMBOLDT BAY PLANT UNIT 3

Distance & direction from reactor	1958–1962			1964–1967			Dif-ference in rates
	Live births	Deaths < 28 days	Rate/1000 live births	Live births	Deaths <28 days	Rate/1000 live births	
<25 miles							
NE	6,411	127	19.8	3,646	45	12.3	−7.5
SE	3,262	43	13.2	1,849	22	11.9	−1.3
Total	9,673	170	17.6	5,495	67	12.2	−5.4
25–50 miles							
NE	1,555	20	12.9	875	10	11.4	−1.5
SE	1,561	22	14.1	873	10	11.4	−2.7
Total	3,116	42	13.5	1,748	20	11.4	−2.1
50–100 miles							
NE	4,696	89	19.0	2,798	37	13.2	−5.8
SE	5,757	115	20.0	4,203	60	14.3	−5.7
Total	10,453	204	19.5	7,001	97	13.8	−5.7
100–200 miles							
NE	32,523	531	16.3	22,441	332	14.8	−1.5
SE	64,857	1,109	17.1	49,239	751	15.2	−1.9
Total	97,380	1,640	16.8	71,680	1,083	15.1	−1.7

are not yet published. Again the data show decreasing birth rates, decreasing infant mortality rates, and the largest decrease in rates occurring in the area with the highest starting rates. But neither the pattern of rates after the reactor began operation, nor the differences in the rates before and after operation suggest any change in infant or neonatal mortality associated with the operation of the reactor.

Despite the sizes of the geographical bands and the accumulation of five years experience, some of these rates are based on relatively few deaths, particularly in the 25–50 mile band. This band is primarily the coastal mountain region of northern California and southern Oregon which is sparsely populated. Also, the reactor is located on Humboldt Bay and as the two western sectors are in the Pacific Ocean, only two compass sectors in each band are contributing data.

Tables V and VI show the data around the Dresden Nuclear Power Station. This plant began operation in 1960 so the two time periods used are 1955 through 1959, and 1961 through 1965. The Dresden reactor is located in Morris, Illinois,

TABLE V

INFANT MORTALITY—DRESDEN NUCLEAR POWER STATION UNIT 1

Distance & direction from reactor	1955–1959			1961–1965			Difference in rates
	Live births	Deaths <1 yr.	Rate/1000 live births	Live births	Deaths <1 yr.	Rate/1000 live births	
<25 miles							
NE	29,430	775	26.3	31,045	794	25.6	−0.7
SE	1,429	33	23.1	1,639	36	22.0	−1.1
SW	3,358	72	21.4	2,950	56	19.0	−2.4
NW	2,782	58	20.8	3,159	66	20.9	+0.1
Total	36,999	938	25.4	38,793	952	24.5	−0.9
25–50 miles							
NE	643,007	16,711	26.0	621,494	16,150	26.0	±0.0
SE	12,143	296	24.4	12,381	291	23.5	−0.9
SW	14,576	329	22.6	12,979	285	22.0	−0.6
NW	8,607	175	20.3	8,220	174	21.2	+0.9
Total	678,333	17,511	25.8	655,074	16,900	25.8	±0.0
50–100 miles							
NE	126,947	3,250	25.6	131,063	3,182	24.3	−1.3
SE	37,768	921	24.4	34,155	836	24.5	+0.1
SW	50,807	1,126	22.2	45,820	999	21.8	−0.4
NW	35,732	855	23.9	35,081	786	22.4	−1.5
Total	251,254	6,152	24.5	246,119	5,803	23.6	−0.9
100–200 miles							
NE	123,956	2,792	22.5	125,849	2,723	21.6	−0.9
SE	48,893	1,069	21.9	45,725	988	21.6	−0.3
SW	68,780	1,587	23.1	60,848	1,379	22.7	−0.4
NW	63,884	1,368	21.4	61,095	1,314	21.5	+0.1
Total	305,513	6,816	22.3	293,517	6,404	21.8	−0.5

TABLE VI

NEONATAL MORTALITY—DRESDEN NUCLEAR POWER STATION UNIT 1

Distance & direction from reactor	1955–1959			1961–1965			Dif-fer-ence in rates
	Live births	Deaths <28 days	Rate/1000 live births	Live births	Deaths <28 days	Rate/1000 live births	
<25 miles							
NE	29,430	566	19.2	31,045	598	19.3	+0.1
SE	1,429	21	14.7	1,639	25	15.2	+0.5
SW	3,358	61	18.2	2,950	41	13.9	−4.3
NW	2,782	50	18.0	3,159	47	14.9	−3.1
Total	36,999	698	18.9	38,793	711	18.3	−0.6
25–50 miles							
NE	643,007	11,945	18.6	621,494	11,680	18.8	+0.2
SE	12,143	203	16.7	12,381	208	16.8	+0.1
SW	14,576	245	16.8	12,979	215	16.6	−0.2
NW	8,607	126	14.6	8,220	136	16.5	+1.9
Total	678,333	12,519	18.4	655,074	12,239	18.7	+0.3
50–100 miles							
NE	126,947	2,373	18.7	131,063	2,333	17.8	−0.9
SE	37,768	677	17.9	34,155	639	18.7	+0.8
SW	50,807	853	16.8	45,820	759	16.6	−0.2
NW	35,732	626	17.5	35,081	583	16.6	−0.9
Total	251,254	4,529	18.0	246,119	4,314	17.5	−0.5
100–200 miles							
NE	123,956	2,202	17.8	125,849	2,066	16.4	−1.4
SE	48,893	762	15.6	45,725	724	15.8	+0.2
SW	68,780	1,172	17.0	60,848	1,052	17.3	+0.3
NW	63,884	1,136	17.8	61,095	1,030	16.9	−0.9
Total	305,513	5,272	17.2	293,517	4,872	16.6	−0.6

about 30 miles southwest of Chicago, and the patterns of live births, as well as infant and neonatal mortality rates, are quite different in this strongly urbanized area from those in the relatively sparsely populated areas of the upper Michigan peninsula and northern coastal California. Although the three distant bands reflect the decreasing birth rate of the country, the under 25 mile band has an increase in number of live births and is apparently reflecting a growth in population during the time period under study. The three compass sectors in this band as well as the southeast sector in the 25–50 mile band and the northeast sector in the 50–100 mile band which have had an increase in number of births all include suburban counties around Chicago. These data are consistent with the known population growth surrounding large metropolitan areas.

It should also be noted that none of the bands have had any significant change in mortality rates between the two time periods studied. As most of these rates are dominated by the Chicago metropolitan area, the data may simply be reflecting the higher rate of hard core causes of infant mortality which has been associated with urban places. It might also be reflecting a change in the ratio of nonwhite births in the population with their associated increasing rate of prematurity [2]. Whatever the real explanation is for the differences in the pattern of rates around the Dresden nuclear power facility as compared to that around Big Rock Point and Humboldt Bay, there is no evidence for an increase in infant mortality rates associated with the operation of this nuclear plant.

Although the data around each of these reactors do not support a hypothesis of an increase in infant mortality in these areas, it was felt that two possible artifacts in the data should be investigated. The first was the remote possibility that the lack of association was a chance occurrence resulting from the choice of the arbitrary geographical bands which were used. Bands at different distances could have been defined and the births and deaths reapportioned. However, analysis of the data within compass sectors around each reactor should accomplish the same result. Using the mean distance from the reactor for each band as the x-value and the difference in mortality rates as the y-value, a regression line was fitted, the slope of the line estimated and then tested to see if it was significantly different from zero. This was done for the infant mortality rates and the neonatal mortality rates for each sector around each reactor. Only one slope was significantly different from zero at the .05 level—the southeast sector of the Big Rock Point Power Station. Examination of the data revealed that all of the differences in this sector were negative and the magnitude of the difference in rates was decreasing with distance from the reactor. This was not consistent with the hypothesis being tested.

Surveillance measurements around operating boiling water reactors have indicated that most of the radioactivity dispersed in the environment comes from the stack effluents [4]. Therefore, the second possible artifact was that effects on the rates in the downwind direction could be washed out by the lack of effects in other directions. The annual records of the direction of the wind at the plant were obtained for each of the reactors under study and are shown in Table VII. The slopes of the regression lines for the prevailing downwind sector and the sector which was downwind the minimum period of time for each reactor were tested for equality. No difference in the slopes was found.

TABLE VII

WIND DIRECTION FREQUENCY (PERCENTAGES)

Southwest includes winds from S, SSW, SW, WSW; Northwest includes W, WNW, NW, NNW; Northeast includes N, NNE, NE, ENE, and Southeast includes E, ESE, SE, SSE.

	Big Rock Point	Humboldt Bay	Dresden
Southwest	24.3%	20.6%	33.5%
Northwest	23.8	22.9	31.7
Northeast	17.7	39.8	15.0
Southeast	21.2	12.1	18.1
No Wind	13.0	4.6	1.7

Although no evidence for an increase in infant mortality associated with the operation of these three nuclear facilities could be detected, one must seriously question whether one would expect to, based on our knowledge of radiobiology and the dose involved. All of the experimental evidence on radiation effects indicate that the risk of effects is a function of dose, the lower the dose, the lower the risk of an effect. This relationship holds because injury to tissue which may lead to an observable effect is a direct result of the absorption of energy in the tissue. As injury results only from this absorption of energy, independent of the source of the energy, one cannot look for effects from exposure to one source of radiation and ignore other sources of exposure. Medical uses of radiation and natural background radiation are the two primary contributors to radiation exposure in this country [1]. Many surveys have shown that medical exposure of fetuses and infants is relatively rare, so the contribution of exposure to this population of infant from medical uses is probably very small and can be ignored. However, the contribution from natural background radiation must not be ignored.

Carefully conducted studies around the Dresden reactor have provided us with data which can be used to evaluate the relative contribution to the population exposure of radiation from the gaseous discharges from the plant and from natural background [3]. Calculated exposures to people due to naturally occurring radioactive materials in the ground and from cosmic rays around Dresden ranged from 46 to 110 millirads per year, with an average exposure of 80 millirads per year. The maximum calculated exposure resulting from gaseous discharges from the reactor was only 10 millirads per year at a distance of $1\frac{1}{2}$ to 3 miles from the plant. Therefore, the maximum calculated exposure from the reactor effluents represents only a 12.5 per cent increase over the average exposure from background and is considerably less than the variability within the background exposure.

Radiation from the gas plume decreases continuously with distance from the reactor due to horizontal and vertical dispersion of the gas and radioactive decay of the short lived radionuclides. At a distance of 16 miles from the reactor, the additional exposure to the population from air effluents was calculated to be 1 millirad per year, and at 29 miles it had fallen to 0.4 millirad per year. Beyond this distance no activity above background could be detected even in the direct line of the plume, although one could calculate probable exposures beyond this point which would be diminishingly small. Thus, one is in the position of trying to detect an effect attributed to exposure due to radioactivity from the nuclear reactor which, for the mass of the population studied, is only a very small fraction of the exposure everyone in the population received from natural sources.

One must also question the rationale of using infant mortality as a health indicator of low dose, low dose rate radiation exposure. Studies of children exposed *in utero* to radiation for medical reasons have provided no evidence of increased infant mortality resulting from these exposures, although the dose is considerably higher than even the maximum potential exposure from the operation of a nuclear power plant. Studies in animals at comparable doses have also been negative.

When considering infant mortality as a possible effect of postnatal irradiation of the young child, the very definition of infant mortality, that is that death must occur within the first year of life, implies that one is expecting acute lethal effects. Laboratory studies of various animal species have demonstrated that the LD_{50}—the dose at which 50 per cent of the exposed population will die in a given time period—is lower for the young animal than the adult. However, the LD_{50} for man is accepted to be between 300,000 and 400,000 millirads. Even a greatly increased sensitivity of the human infant to radiation as compared to other animal species would not lead one to expect to observe this acute effect at doses of less than ten millirads.

Although infant mortality does not appear to be a profitable health indicator for the study of radiation effects, the need for study of low level effects is paramount. There is good evidence that radiation does increase the risk of certain malignancies, and in particular leukemia, at relatively high doses. There is also good evidence in animals that relatively high doses of radiation early in the gestation period can increase the incidence of congenital malformations. Before it will be possible to quantitate the risks of these and other effects associated with low doses, new methods of study must be devised. The effects to be investigated occur relatively rarely, and when investigating low dose effects, the size of population required for definitive study, using established statistical and epidemiologic techniques, is so large that studies are simply not feasible. It is hoped that as a result of this conference, ideas for the development of new approaches to the study of the occurrence of rare diseases will be generated.

In conclusion, the patterns of infant mortality around three boiling water nuclear power plants do not support a hypothesis of an increase in infant mortality associated with the operation of these three reactors. However, in the spirit of this conference, three factors should be mentioned which must be considered in planning a study of the health effects of pollutants including radiation.

(1) Pressurized water reactors and newer reactors with hold-up tanks for gaseous discharges have much lower discharge rates than the Dresden Nuclear Power Station Unit I. Therefore the potential exposure to a population is considerably lower than the maximum of ten millirads calculated for Dresden.

(2) The malignancies which have been associated with exposure to radiation have long latent periods. In the case of leukemia, the peak occurrence in the Japanese was at 6 to 12 years [6] after exposure and there is a suggestion that lower exposures may have longer latent periods. Other malignancies may just now be showing up at increased frequencies 26 years after exposure. It must be remembered that this evidence is coming from acute exposures in the range of somewhere below 100 up to around 300 rads—or 100,000 to 300,000 millirads.

(3) Background radiation is highly variable. One must consider the gross average differences of exposures to populations to be studied resulting from

differences in altitude and basic geological formations. But differences in exposure of individuals within the populations due to the variability of the concentration of natural radionuclides in the soil as well as in the building materials of structures in which they live and work must also be considered.

◇   ◇   ◇   ◇   ◇

The authors wish to thank Mary McMahon, Edith Levinson, and Theda Bell for their help in obtaining the data and James Godbold for analysis.

REFERENCES

[1] *Atomic Energy Clearing House*, Vol. 17, No. 26 (1971), p. 35.
[2] H. C. CHASE and M. E. BYRNES, "Trends in "prematurity": United States, 1950–1967," *Amer. J. Pub. Health*, Vol. 60 (1970), pp. 1967–1983.
[3] B. KAHN, *et al.*, "Radiological surveillance studies at a boiling water nuclear power reactor," *BRH/DER* 70-1, U.S. Dept. of HEW, Public Health Service, March, 1970, Rockville, Md.
[4] J. E. LOGSDON and R. I. CHISSLER, "Radioactive waste discharges to the environment from nuclear power facilities," *BRH/DER* 70-2, U.S. Dept. of HEW, Public Health Service, March, 1970, Rockville, Md.
[5] I. M. MORIYAMA, "Recent change in infant mortality trend," *Pub. Health Rep.*, Vol. 75 (1960), pp. 391–406.
[6] Report of the United Nations Scientific Committee on the Effects of Atomic Radiation, General Assembly Official Records: Nineteenth Session Supplement No. 14 (A/5814), United Nations (1964), New York, p. 83.

Discussion

John W. Gofman, Division of Medical Physics, University of California

Mrs. Tompkins, you indicate that the new reactor will have even lower releases than current reactors. We could even assume that new reactors will release no radioactivity. This does *not* address the problem of nuclear power, since all the steps in the cycle of nuclear power create the real problem.

A 1000 megawatt reactor generates 22 megatons of long lived fission products *per year* of operation. If the AEC optimistic projection of 1000 reactors by the year 2000 occurs, we will generate 22,000 megatons (TNT equivalent) of long lived fission products.

Assuming the engineering is 99.99 per cent perfect at every step along the way, this means 0.01 per cent release of radioactivity. Therefore 0.01 per cent of 22,000 megatons is 2.2 megatons of long lived fission products *per year* distributed to the U.S. biosphere. This is clearly not acceptable, even if we are so optimistic as to accept 99.99 per cent containment.

J. Neyman, Statistical Laboratory, University of California, Berkeley

Data produced by Mrs. Tompkins suggest conclusions contrary to those indicated by the data of Dr. Sternglass. This is a conflict of facts which is impossible to solve at a meeting like the present, but possible to clarify at a leisurely conference of only a few interested people.

This incident illustrates the reason why I suggest that, if a broad comprehensive statistical study is ever attempted, it should be conducted not just by one but by several statistical groups, working independently on the same data.

E. J. Sternglass, School of Medicine, University of Pittsburgh

There would appear to be a number of reasons why the methodology chosen in Mrs. Tompkins' paper tends to show no effects of plant emissions from boiling water type of reactors, while the studies of Dr. DeGroot as well as our own do show an effect on infant mortality.

The first and major reason would appear to be the choice of concentric circles in the search for a spatial gradient, since this choice implies that the entire area studied is assumed to be homogeneous with regard to the major parameters known to influence, or assumed to affect, the rate of infant mortality. These include the following:

(1) Plant emissions are not geographically uniform into all directions of the compass because

(a) Prevailing winds tend to favor certain directions.

(b) Release practices by the operator are often designed to minimize the exposure of the larger population centers lying in certain wind directions.

(c) Since fallout particles are mainly brought down by rainfall, widely different patterns of annual rainfalls, as for instance along a narrow coastal strip in the case of the Humboldt plant in California, will result in very different exposure patterns for the population at risk, both from the plant emissions and test fallout.

(d) Major geographical features, such as mountain regions along the coast, will have major effects on the radiation exposure from heavy gases, and particularly in regions only a few miles apart.

(e) The presence of large bodies of water or rivers into which liquid discharges take place will greatly influence the radiation exposure in a way that will often tend to dominate. Thus, again as in the case of Humboldt, heavy exposures along a coastal strip will tend to be masked by a pattern of circular regions that include areas far from the coast.

(2) Socioeconomic and medical factors are not homogeneous with directions around a nuclear plant, particularly when such a plant is located to one side of a major metropolitan area as in the case of the Dresden plant. The inclusion of upwind rural areas to the west with polluted and socioeconomically much poorer areas to the east of Dresden in southern parts of Chicago generally downwind will tend to mask out any decrease with increasing distance, as we have found when one takes these factors into consideration. Another problem in the methodology is the use of five year time periods before and after onset of operations. since peak emissions often occur only for single months or years, and the five year period before releases began was a period of heavy nuclear weapons testing when infant mortality would be expected to be high.

POPULATION EXPOSURE TO RADIATION: NATURAL AND MAN-MADE

V. L. SAILOR

BROOKHAVEN NATIONAL LABORATORY, UPTON

1. Radiation in the environment

Environmental radiation can be detected with great sensitivity. With modern instrumentation and calibration techniques the exposure level can be quantitatively measured to a precision of better than five per cent, even at the extremely low levels of natural background. By proper use of spectroscopic techniques, it is also practical to distinguish between particular natural and man-made sources [27].

Only during the past two decades has the radiation environment of mankind been surveyed extensively [1], [60], [61], primarily to monitor fallout from weapons tests. Considerable data were available more than 40 years ago, but those studies were directed toward an understanding of cosmic rays, rather than environmental exposures [36]. Summaries and bibliography can be found in the annexes of [11], [22].

Numerous studies have been made of the biological effects of radiation. The scope can be appreciated by examining the UNSCEAR reports (for example, see bibliography in [60] pp. 67–83, 108–117, and 183–206). Although laboratory experiments have been limited to plants and animals, several groups of humans, inadvertently exposed, have also been studied. With rare exception, the observations have been based on exposures which were extremely large and at high rates by comparison with environmental levels. No data exist which give dose response curves at such low doses. To be conservative, all standards setting bodies *assume* that the high level, high rate, dose response curves extrapolate linearly to zero dose, that is, that no threshold exists below which radiation is harmless. However, it must be emphasized that the nonthreshold, linear response is an *assumption* and not a scientific fact. A major objective of a statistical study would be to obtain better information on the shape of the low dose response curve.

Despite the sensitivity and precision for measurement of radiation, and despite the extensive knowledge of biological effects, the health hazards associated with environmental radiation are difficult, if not impossible, to evaluate.

Research supported by the U.S. Atomic Energy Commission.

291

There are several reasons for these difficulties. At such low exposures radiation-induced "afflictions" are rare, and must somehow be discerned from among large populations. This is not straightforward since radiation-induced "afflictions" are not caused uniquely by radiation, but can also be induced by other agents in the environment. Furthermore, since radiation is present everywhere on earth it is impossible to observe a control population with zero exposure.

If we regard natural radiation as *noise*, and man-made radiation "pollution" as the *signal*, then, in normal circumstances, we are working with a very unfavorable signal to noise ratio. Only when the man-made exposure is massive does the signal to noise ratio favor detection of a cause-effect relationship. Incidentally, it should be noted that man-made radiation pollution differs from many other pollutants. Radiation has always been present in the environment and all living things have always been subject to exposure. On the other hand, pollutants such as DDT, Pb, and Hg have been previously either nonexistent or very rare in the biosphere.

2. Population dose distributions from natural radiation

2.1. *Sources of radiation.* Ever since man first walked the surface of the earth, he has been subject to varying amounts of exposure. Ever since he started collecting attractive rocks, selecting building material for dwellings, mining ores from the ground, he has unwittingly modified his exposure. There are two primary sources of natural radiation: cosmic rays and radioactive isotopes. Both fluctuate with time and geographical location. Typical doses from natural sources are shown in Table I.

TABLE I

TYPICAL WHOLE BODY DOSES TO STANDARD MAN FROM NATURAL SOURCES

Values listed for cosmic rays omit neutron component which lies
in range of approximately 0.7 to 7 mrem at sea level.

Source	Dose (mrem/year)
Internal	
Potassium 40 in human body	20
Other radionuclides in body	
(carbon 14, radon 222, radium 222, 228 and so on)	3
External out of doors	
γ-rays from soil and rocks	50
Cosmic rays (sea level, 50° geomagnetic lat.)	28
Cosmic rays (Denver)	67

The "standard man" (the characteristics of "Standard Man" can be found in Appendix III, p. 408 and following, of H. Cember [9]) contains 140 grams of potassium in equilibrium which has a specific activity of approximately 32 disintegrations per second per gram due to the presence of the radioisotope potassium 40 (0.0118 per cent isotopic abundance). The annual whole body dose from potassium contained in the body is about 20 mrem. Other natural radioactive materials in the body contribute three mrem/year. (The unit of radiation dosage used here is the millirem (mrem) which is 1/1000 of one rem. For most purposes one rem (roentgen equivalent man) is equal to one rad which is defined as the dose which deposits 100 ergs of energy in 1 gram of tissue. The dose in rems equals the dose in rads multiplied by the relative biological effectiveness (RBE); that is, rems = rads $\times$ RBE where the RBE for electrons is taken as unity. Other forms of radiations can have larger RBE depending on the details of their interaction with tissue.)

Rocks and soil contain varying amounts of natural radioactivity (potassium 40, members of the thorium 232, uranium 235 and uranium 238 decay series, products of cosmic ray interactions, and miscellaneous long lived natural radioisotopes). A typical annual dose from rocks and soil is 50 mrem. This is due mostly to the γ-ray components of the decays, since β- and α-particles are absorbed by relatively small thicknesses of material and have only short range in air.

Cosmic rays originate from outer space and the sun. The primary rays consist of extremely energetic charged particles, mostly protons and α-particles. As the primaries enter the atmosphere they interact with nuclei to produce a variety of secondary radiation consisting of muons, electrons, γ-rays, protons and neutrons. The intensity of the primaries and secondaries are attenuated as they pass downward through the atmosphere, the energy being gradually dissipated by particle production, nuclear reactions, and ionizing events. Thus the composition and intensity is a strong function of altitude. The magnetic field which surrounds the earth, deflects and traps some of the less energetic components producing a variation with geomagnetic latitude.

At sea level, in middle latitudes the dose due to the ionizing component is about 28 mrem/year. The neutron component contributes an additional 0.7 to 7 mrem/year, the uncertainty being associated in assessing the RBE of this component (see [61] pp. 14–18). As the altitude is increased there is less protection from the atmosphere, hence the dose increases, for example, at the elevation of Denver being more than double that of sea level.

2.2. *Geographical and temporal variations.* Except for the internal dose from the body burden of potassium, the natural radiation dose varies widely with location and time.

Many regions of the earth are rich in radioactive ores and in such regions the rocks and soil contribute higher than average external doses. In addition, the radioactive content of vegetation and animal life depend on the soil and water, consequently the human body burden depends on the source of foodstuffs and drinking water. For example, approximately 1,000,000 people in Illinois and Iowa consume drinking water containing unusually high content of lead 210 (RaD), polonium 210 (RaF), radium 226, and so forth [18], [51]. The bones of these people contain as much as four times the normal amounts of radium and daughter products. Other areas in the United States having relatively high radioactive content in rock are regions of Vermont and New Hampshire [29], parts of Connecticut [35], Manhattan Island [28], areas in the Rocky Mountain states [47], and a section of North Carolina [27]. A few "hot spots" have even been located along coastal beaches [31]. Other inhabited areas of the world have even higher natural radiation levels, for example, the Black Forest of Germany, the states of Espirito Santo and Rio de Janeiro, Brazil and Kerala, India. (These anomalies are discussed in [1], [60], [61].)

Several radioisotopes are present in the atmosphere as gases or attached to dust particles. Among these are the isotopes produced by cosmic rays, and the noble gases radon and thoron which are formed as daughter products in the thorium and uranium decay series. Radon and thoron diffuse from the soil at rates which depend on such factors as snow cover, moisture content of the soil and vegetation cover. The radioactive content of a local volume of air varies with meteorological conditions, being enhanced during periods of inversion (particularly in valleys), and at a minimum after heavy rainfall.

Cosmic ray dose rates at sea level fluctuate with time by as much as ten per cent during the course of the solar cycle [33]. This variation increases at higher altitudes, becoming approximately 20 per cent at 20 km. As mentioned earlier, the cosmic ray dose increases with elevation. Near sea level the dose approximately doubles for each 1400 meters increase in elevation. Table II compares the cosmic ray dose for several locations in the U.S.

TABLE II

APPROXIMATE COSMIC RAY DOSES FOR SEVERAL AMERICAN CITIES
The doses listed were obtained from graphs in [43].

Location	Nominal elevation (feet)	Dose rate (mrem/year)	Difference from sea level (mrem/year)
Albuquerque	4958	60	32
Atlanta	1050	33	5
Baltimore	20	28	0
Chicago	579	31	3
Denver	5280	67	39
Kansas City	750	32	4
Oklahoma City	1207	34	6
Phoenix	1090	33	5
Salt Lake City	4260	54	26
San Francisco	65	28.5	0.5

A broad maximum in the dose rate occurs at 20–25 km altitude as a consequence of the various mechanisms for converting the primaries [43]. At flight altitudes for commercial jet aircraft the dose rate is in the range 0.3–0.7 mrem per hour. During a transcontinental round trip flight the average passenger receives an approximate five mrem dose.

2.3. *The effect of shelter.* The average American spends much of his time indoors where the walls and the roof of the building reduce exposure to external radiation. Often this reduction is partially offset or overwhelmed by exposure from the natural radioactivity in the structural materials. A typical wooden house will reduce exposure to about 70 per cent of the outdoor level [29]. Brick, concrete and stone buildings typically contribute exposures 50 to 100 mrem/year greater than outdoor values at the same site; however, much larger indoor exposures have been observed (see Table III), and might be more common than generally suspected.

TABLE III

Gamma Exposure Rates Inside Various Buildings

Cosmic ray component included.

Location	Structure	Exposure rate Typical (mrem/year)	Extreme	Ref.
East Germany		106	1200	[44]
New York City	brick	79–118		[54]
Grand Central Station	Millstone Point granite	25– 75	525	[34], [39], [50]
United States	wood	60		[42]
" "	concrete	130		[42]
Aberdeen, Scotland		81	110	[55]
Cornwall, U.K.	granite	145		[63]
Sweden	wood	48– 57		[19]
"	brick	99–112		[19]
"	concrete	158–202		[19]

Within a building exposure rates can vary dramatically, for example, in a multistory building the lower floors are better shielded against cosmic γ-rays. In rooms with poor ventilation radon and thoron content of the air can build up to very high levels. Finishing details of walls such as tiles can produce significant changes—increases or decreases.

At the present time, there is no systematic effort to select building material on the basis of low radioactivity.

2.4. *Population distribution of dose (natural sources).* As the average person moves about during the course of his activities, he is subject to a wide range of exposure from natural sources. The doses to average man are not monitored. Only a few studies have been made in which individuals were monitored over a relatively long period of time [29], [49], [52].

There are enough data to obtain a rough estimate of the population dose distribution. Figure 1 is a histogram constructed from the product of the population and the mean outdoor dose for various localities. (Most of the data for constructing the histogram were measured by the USAEC Health and Safety Laboratory (HASL). The author is indebted to H. L. Beck and J. E. McLaughlin

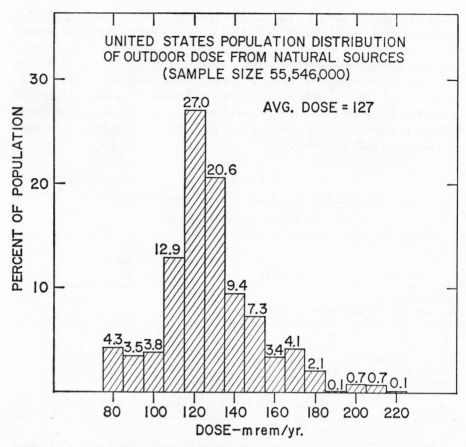

FIGURE 1

Population dose distribution. The histogram was prepared by multiplying the observed dose at each locality by the population of that same locality. The data include 23 mrem from natural radioactivity in the body. Effects of fallout from weapons testing are excluded.

for making such data available.) No corrections were made to account for time spent indoors. The measured exposure rates include natural radiation from the earth and cosmic rays, and internal radiation from potassium, and so forth, of "standard man," but discriminate against radiation from weapons fallout. The weighted average is 127 mrem/year, the full width at half maximum is about 15 mrem/year. The minimum exposure is of the order of 75 mrem/year for beach locations and exceeds 215 mrem/year in the Rocky Mountains.

Since Americans tend to be mobile, and have varying tastes in housing, an estimate of the annual dose to a given individual will have considerable uncertainty, but the value will probably lie inside the distribution curve shown in Figure 1. Any statistical study which attempts to relate radiation dose with biological effect must recognize the "fuzziness" of the base line.

3. Exposures to man-made radiation

3.1. *Medical X-rays.* By far the largest population exposure to man-made radiation comes from medical X-rays. Techniques and condition of equipment show wide variation, consequently the doses received from a given procedure spread over a wide range. The dose delivered to a patient is usually not calculated, measured, nor recorded. In general, no systematic records are kept on the lifetime accumulated doses to individuals.

Because of the lack of systematic records, the task of compiling statistical data on population doses is very complex. Many laborious surveys have been made [2], [5], [6], [13], [17], [21], [45], [60], [64] which provide estimates of average population doses, as well as age and sex distributions for various procedures. I have not been able to locate data which can be used to construct dose distribution among individuals. A Public Health Service analysis [5] concluded that the genetically significant U.S. population dose for the year 1964 was 55 mrad per person per year. There is evidence that this has increased to approximately 95 mrad/person/year as of 1970 [62]. These doses should be about doubled to be comparable to the *whole body doses* cited in Tables I to III.

Compared with other advanced countries of the world, the average U.S. X-ray dose from diagnostic procedures is quite large as shown in Table IV which was compiled by Dr. Karl Z. Morgan of ORNL [38].

TABLE IV

GENETICALLY SIGNIFICANT DOSE (mrem/year)
FROM MEDICAL DIAGNOSIS
IN VARIOUS ADVANCED COUNTRIES

Table borrowed from [38].

United States	95
Japan	39
Sweden	38
Switzerland	22
United Kingdom	14
New Zealand	12
Norway	10

It is obvious that the history of medical X-ray exposures differs widely among individuals. Many persons in the United States live their entire life without any exposures while others receive massive doses. Since individual records are not generally maintained (indeed the ICRP and ICRU recommended against the maintenance of such records [21], we must regard this contribution of human exposure as additional "noise" which any statistical study of radiation pollution must duly take into account.

3.2. *Fallout from weapons testing.* For more than a decade the United Nations Scientific Committee on the Effects of Atomic Radiation (UNSCEAR) has complied and evaluated data on world wide fallout of radioactivity from weapons testing [60], [61]. The latest evaluation of the population dose commitment up to the year 2000 from all tests to date gives a value of approximately 200 mrem or about five mrem/year [16].

Several cases of localized fallout have been recorded which have been subjected to intensive special studies [24]. I will not attempt to discuss this complex subject.

3.3. *Commercial nuclear power.* At the present time more than 20 large commercial nuclear power stations and one fuel reprocessing plant are in operation in the United States. These emit radioactive isotopes into the atmosphere and waterways. The operation of each of these installations has been subjected to close surveillance by a variety of agencies including the U.S. Public Health Service, the AEC Division of Compliance, state health departments and private contractors. Off site doses are generally too small to measure in comparison to natural background, and hence, must be calculated from known emission rates, meteorological conditions, population distributions, and so on. The surveillance studies include sampling of agricultural products, foliage, wild life, water supplies, waterways and marine biota to detect any possible buildup of activity in foodchains (for example, see [11] and [22]).

The calculated off site radiation doses have been extremely small, for example, at the Dresden I Nuclear Power Station, Illinois, during 1968 they were less than 14 mrem/year ([22], p. 53). In a later report, the same authors state: "The radiation exposure from discharged radionuclides was computed to be 1 per cent of the annual average concentration limit for air at the site boundary; and 0.1 per cent of the annual average concentration limit in Illinois River water at the point of discharge" [23]. Probably the largest off site doses which have been detected were at the Humboldt Bay Power Plant (Unit No. 3) near Eureka, California, where the maximum observed values were approximately 50 mrem in 1965 and approximately 35 mrem in 1966 [3].

The newer Boiling Water Reactors (BWR), descendants of the Humboldt Bay and Dresden I reactors, have improved facilities for handling radioactive wastes. In particular, provisions are made for longer delays in discharging gases which further reduces off site doses because a larger fraction of the radioisotopes decay before release. Emissions from Pressurized Water Reactors (PWR) and the only High Temperature Gas Cooled Reactor (HTGR) (Peach Bottom I) have generally lower than BWR [32] since even longer delays in release of gases is practical. This is a consequence of the fact that the BWR must handle far greater quantities of gas because the pressure of the primary steam loop drops below atmospheric at the condenser encouraging air leakage into the condensate via pump seals, and so on. This air mixes with the radioactive gases and thus greatly increases the volume of gas to be processed. Systems are being designed which will deal with this situation, introducing delays of several days by various absorber beds.

Recently the Atomic Energy Commission has proposed new regulations which will limit off site doses to five mrem per year [12]. This limit is the sum of the contributions of all light water reactors located at a given site if more than one is in operation. If these new rules are adopted, existing plants which fail to meet the new limits must be modified to comply within three years. In view of the actual operating record and the new rules, it appears reasonable to assume that nuclear power plants will be limited to off site doses which do not exceed five mrem/year. Let us consider what we can expect the *real* dose to the population to be. The five mrem/year limit applies to *any location* off site. Under normal meteorological and geographical circumstances the maximum off site dose occurs at some point on the site boundary, and is due primarily to noble gases of long half life. Beyond the site boundary, the dose will decrease as a result of dilution of the effluents. J. B. Knox has considered the dose distribution for a variety of situations [25]. It is possible, of course, to assess the geographical distribution of the off site dose in great detail by taking into account the observed meteorological conditions and recorded plant emissions.

In order to gain some understanding of the dose distribution let us consider an extremely simplified model, but retain the notation used by Knox. We shall assume that the population density is constant and that the dose $D(r, \theta)$ is a function of direction θ, and distance r from point of emission. If $D_0(\theta_{max})$ is the limiting dose at the site boundary in the direction θ_{max} which gives the maximum dose average over the year, then the boundary dose in every other direction will be $D_0(\theta) < D_0(\theta_{max}) < 5$ mrem. Assuming that the wind direction is uniform in the area under consideration, we can write $D(r, \theta) = D_0(\theta)R(r)$, where $R(r)$ will depend on the amount of vertical diffusion. It is plausible that $R(r)$ will have a dependence between the extremes of (r_0/r) and $(r_0/r)^2$ where r_0 is the distance to the site boundary. The mean dose within any annulus of width Δr_i is given by

$$(3.1) \qquad \bar{D}_i = \int_{r_i}^{r_i+\Delta r_i} D_0 R(r) pr\, dr \Big/ \int_{r_i}^{r_i+\Delta r_i} pr\, dr$$

where p is the population density. As a practical example take $r_0 = \frac{1}{4}$ mile. When the integrals are evaluated out to a 50 mile distance for the two extremes of $R(r)$, the histograms shown in Figures 2 and 3 are obtained. For $R \propto (r_0/r)$ the mean dose to the population within the 50 mile radius is 0.05 mrem/year (see Figure 2) with two-thirds of the people receiving less than 0.05 mrem/year and less than five per cent receive doses greater than 0.1 mrem. For the inverse square case (Figure 3), the mean dose is 1.3×10^{-3} mrem/year and three-quarters of the population receives less than 5×10^{-4} mrem/year.

Knox gives a more realistic function $R(r)$ which lies between the two extremes of Figures 2 and 3. The three cases are compared in Figure 4.

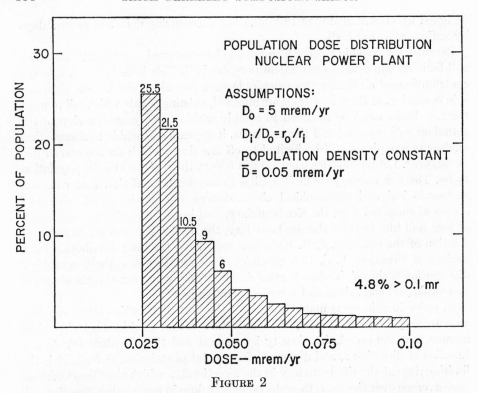

FIGURE 2

Population dose distribution in 50 mile radius about a Nuclear Power Plant. Model assumes a uniform population distribution and a 5 mrem/year maximum fence post dose. This plot assumes that the dose decreases as the inverse of the distance.

Knox considers the effect of multiple facilities surrounding a city such as might be expected in future years. For a case involving 16 plants equally spaced on a 50 *km* radius, his calculations give a maximum annual dose of approximately 0.27 mrem provided the 5 mrem fence post limit is observed [25].

It appears that commercial nuclear power, even taking into account a large anticipated growth, will not contribute a significant increment to the radiation environment of the population.

4. Special biological hazards

The biological effect of radiation depends on the nature of the radiation and location of the source (internal or external). The manifestations from radioisotopes incorporated in the body depend on the biochemistry of the element and on its chemical compound when ingested. Some elements concentrate in particular organs. An obvious example is the case of radioiodine which con-

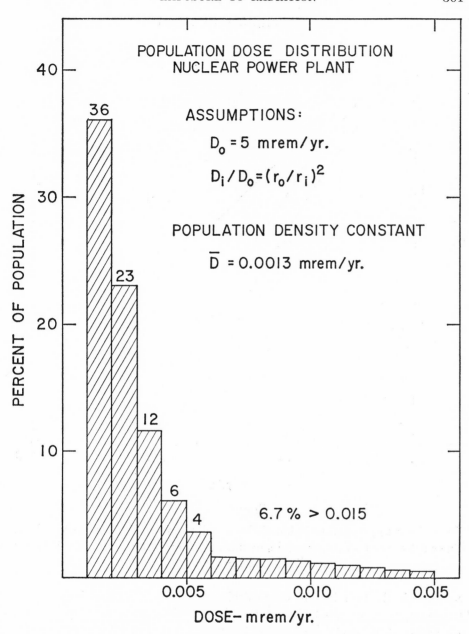

FIGURE 3

Population dose distribution in 50 mile radius about a Nuclear Power Plant. The same assumptions apply as for Figure 2, except this plot assumes that the dose decreases as the inverse square of the distance.

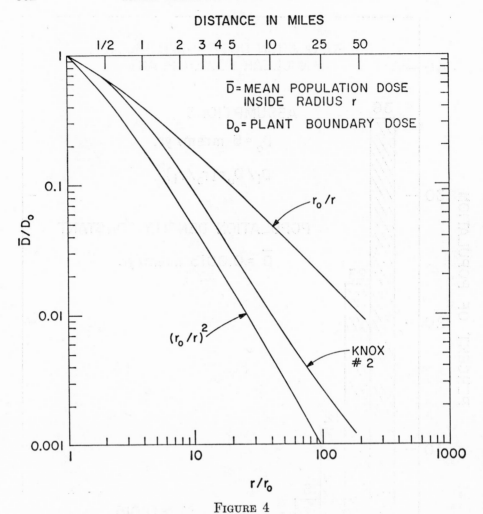

FIGURE 4

Mean doses to population inside circle of radius r from Nuclear Power Plant. The models used to construct Figures 2 and 3 are compared with the more realistic model of Knox [25].

centrates in the thyroid. The dose calculations take such factors into account (for example, see [9], pp. 204–208). In order to make realistic dose calculations, the effects of individual isotopes have been extensively studied and evaluated by ICRP and NCRP [41]. The process of evaluating the special hazards of individual isotopes is a continuing one and revisions to the "standards" are issued as necessary (for example, Subcommittee #24 of the National Council on Radiation Protection and Measurements is currently reappraising the hazards of radioactive nucleic acid precursors). After the maximum body burden has been established, practical working limits are set by studying the pathways to man [7], [8], [10], [48].

In recent times popular articles on environmental topics frequently give the impression that our understanding of biological effects and pathways is very limited (for example, see [46]). Obviously, such complex evaluations cannot be perfect in every respect, and it is prudent to continue to improve our understanding. Cember expressed the situation quite eloquently ([9], p. 177), "Radiation ranks among the most thoroughly investigated etiologic agents associated with disease. Although much still remains to be learned about the interaction between ionizing radiation and living matter, more is known about the mechanism of radiation damage on molecular, cellular, and organ system levels than is known for most other environmental stressing agents. Indeed, it is precisely this vast accumulation of quantitative dose-response data that is available to the health physicist that enables him to specify environmental radiation levels for occupational exposure, thus permitting the continuing industrial, scientific, and medical exploitation of nuclear energy in safety."

Tritium is often cited as an example of an unusual hazard [20], [30], one that reactor operators are unaware of or ignoring [46]. Lyon implies that tritium undergoes concentration in biological chains [30] (this would require *isotopic* separation in the biological processes with an incredibly high separation factor). C. W. Huver has discussed "the abundant evidence illustrating the serious biological effects of tritium" [20] and advocates that routine release of tritium effluents be prohibited. These views are not shared by most authorities. Particularly it is important to distinguish between the hazards associated with tritium in various chemical compounds. The usual environmental effluent is in the form of tritiated water, which behaves quite differently in biological systems than for example would tritiated precursors of nucleic acids (bicarbonate, formate, glycine, hypoxanthine zanthine, orotic acid, and so on). V. P. Bond has recently reviewed the hazards associated with tritiated water effluents [4] and concludes that, "A given dose (for example, an MPD) of radiation from the beta rays of tritium (from either inhalation or ingestion) has the same radiobiological and radiation protection meaning as the given dose from X-rays or gamma rays (same dose rate pattern), and no added significance or potential hazard is to be attached by virtue of the fact that the dose may have been derived from tritium."

E. J. Sternglass has postulated a special toxicity associated with radiostrontium for which the existing ICRP-NCRP recommendations do not provide. According to the Sternglass hypothesis when the skeletal burden of Sr 90 decays the daughter product, yttrium 90, "concentrates in such vital glands as the pituitary, the liver, the pancreas and the male and female reproduction glands" [57]. This contributes to increased infant mortality and congenital defects, and "it also appears to act indirectly so as to produce small decreases in maturity, at birth that in turn can increase the chance of early death from various causes such as respiratory and infectious diseases" [58]. Sternglass elaborated on this postulated sequence as follows: "But what happens apparently is now this: A young woman who drinks the milk over a period of years accumulates strontium 90 in her body, which is known. It is shown in the UN report that it builds up for a period of some years. It reaches a peak and then at the same time the yttrium which is created all the time by the decay of the strontium will now be circulating in her blood. Now the key element that seems to be happening is that the mother in effect is a source of yttrium herself because she has stored up strontium 90 which unfortunately stays bound in the bone for a long time. Now she is pregnant. What will happen is that in the early critical phases of organ development and maturation, the process of maturation of the embryo could be ever so slightly arrested or slowed down with the result that one has babies which are quite normal in every apparent respect but slightly underweight. And such underweight babies have a much greater chance of dying of all the normal causes like respiratory disease, hyaline membrane disease, all the types of infections that affect young babies" [58].

In support of this thesis, he cites the data of Muller [40], Spode [56], and Graul and Hundeshagen [15]. He also cites his own analysis which purports to show correlations of infant mortality with Sr 90 from fallout, and with the emission of short lived noble gas daughter products Sr 89 and Cs 138 from the Peach Bottom Reactor in Pennsylvania.

The Sternglass hypothesis has some flaws. The experiments of Muller, which showed slower decrease of Y 90 in the testes than expected through normal secretion, were done by intraperitoneally injections of Sr 90 and hence the body distribution of Sr 90 and its daughter product Y 90 were not the usual equilibrium distribution that would occur from ingestion through ordinary pathways. Similar comments apply to the other two references cited. More recent work by Mole and Ward [37] showed no concentration of Y 90 in monkey gonads. M. Goldman has strongly refuted the Sternglass hypothesis citing a variety of data including very detailed results of his own group obtained from many years of research with beagle hounds [14]. He emphasizes that Sr 90 is tightly bound in the bone structure, and upon decay the newly formed Y 90 remains trapped in the bones until it has decayed.

The alleged Sr 90 infant mortality correlations have been examined by many investigators and rejected as spurious. A detailed discussion appears in a Public Health Service report by Tompkins and Brown [59]. Shaw and Smith [53] using Sternglass' methodology obtained a negative regression coefficient for Canada which had larger Sr 90 fallout than the U.S. The "Peach Bottom" situation cited by Sternglass is erroneous because the gaseous waste system does not emit the short lived Kr 89 and Xe 138 precursors of Sr 89 and Cs 138 due to the extremely long holdup of the gases before release. Apparently Sternglass misinterpreted the data on pages 57–61 of the Public Health Service report which he cites [26]. The short lived activities cited refer to the main coolant loop and not to the gaseous release.

For these reasons it appears unlikely that Sternglass has identified a hazard from radiostrontium that has gone unrecognized by ICRP or NCRP.

5. Conclusions

Statistical studies which might be capable of demonstrating relationships between health effects and low level radiation pollutants in the environment will be difficult to design, because:

(1) the "pollutants" are small by comparison with natural background radiation and with variations in natural backgrounds;

(2) the radiation burden of medical X-rays are about as large as natural background and thus large compared with the "pollutants," and medical X-ray exposures for which no precise records are kept vary widely among individuals; and,

(3) the biological manifestations of radiation are not uniquely induced by radiation.

The author is indebted to many of his colleagues for illuminating discussions and assistance in obtaining material. Particularly the help of A. P. Hull; C. B. Meinhold; L. D. Hamilton, M.D.; J. L. Bateman, M.D.; and J. S. Robertson, M.D. (all of BNL); F. J. Shore (Queens College); and J. E. McLaughlin (HASL) is gratefully acknowledged.

REFERENCES

[1] J. A. S. Adams and W. M. Lowder (eds.), *The Natural Radiation Environment*, Chicago, University of Chicago Press, 1964.

[2] M. S. Billings, A. Norman, and M. A. Greenfield, "Gonad dose during routine roentgenography," *Radiology*, Vol. 69 (1957), pp. 37–41.

[3] J. O. Blomeke and F. E. Harrington, "Management of radioactive wastes at nuclear power stations," Oak Ridge National Laboratory, Report No. ORNL-4070, January 1968.

[4] V. P. Bond, "Evaluation of potential hazards from tritiated water," *Environmental Aspects of Nuclear Power Stations*, Vienna, International Atomic Energy Agency, 1971, pp. 287–300.

[5] M. L. Brown, et al., *Population Dose from X-rays, U.S. 1964*, U.S. Public Health Service Publication No. 2001, October 1969.

[6] R. F. Brown, J. Heslep, and W. Eads, "Number and distribution of roentgenologic examinations for 100,000 people," *Radiology*, Vol. 74 (1960), pp. 353–362.

[7] P. M. Bryant, "Derivation of working limits for continuous release rates of iodine-131 to atmosphere in a milk producing area," *Health Phys.*, Vol. 10 (1964), pp. 249–257.

[8] P. M. Bryant, "Derivation of working limits for continuous release rates of ^{90}Sr and ^{137}Cs to atmosphere in a milk producing area," *Health Phys.*, Vol. 12 (1966), pp. 1393–1405.

[9] H. Cember, *Introduction to Health Physics*, Oxford, Pergamon Press, 1969.

[10] H. J. DUNSTER, R. J. GARNER, H. HOWELLS, and L. F. U. WIX, "Environmental monitoring associated with the discharge of low activity radioactive waste from Windscale
Works to the Irish Sea," *Health Phys.*, Vol. 10 (1964), pp. 353–362.

[11] M. EISENBUD, "Review of U.S. power reactor operating experience," *Environmental
Aspects of Nuclear Power Stations*, Vienna, International Atomic Energy Agency, 1971,
pp. 861–876.

[12] *Federal Register*, Vol. 36 (9 June 1971), pp. 1113–17.

[13] J. N. GITLIN and P. S. LAWRENCE, *Population Exposures to X-rays, U.S. 1964*, U.S.
Public Health Service Publication No. 1519, 1966.

[14] M. GOLDMAN, testimony of 17 May 1971 before the Atomic Safety and Licensing Board,
U.S. Atomic Energy Commission, Docket No. 50-322, In the matter of Long Island
Lighting Company (Shoreham Nuclear Power Station, Unit 1), AEC Staff Exhibit No.
11, and official transcript, pp. 10562–10603.

[15] E. H. GRAUL and H. HUNDESHAGEN, "Y^{90}-Organverteilungsstudien," *Strahlentherapie*,
Vol. 106 (1958), pp. 405–417.

[16] L. D. HAMILTON, Brookhaven National Laboratory, Upton, N.Y., Personal communication, July 1971.

[17] V. H. HOLTHUSEN, "Genetisch wirksame Belastung durch medizinische Strahlenanwen-
dung in einer groszstädtischen Bevölkerung," *IXth Internat. Congr. Radiology, 23–30
July 1959, München*, Vol. 1 (1959), pp. 127–132.

[18] R. B. HOLTZMAN, "Lead-210 (RaD) and Polonium-210 (RaF) in Potable Waters in
Illinois," *Op. cit.* [1], pp. 227–237.

[19] B. HULTQVIST, "Studies on naturally occuring ionizing radiations," *Kungl. Svenska
Ventenskapsakad. Handl.*, Vol. 6 (1956), pp. 3–124.

[20] C. W. HUVER, "Biological hazards of tritium," Congressional Hearing on Atomic Energy
Plants and Their Effects on the Environment, U.S. Customs Court House, New York,
Feb. 6, 1970, unpublished.

[21] INTERNATIONAL COMMISSION ON RADIOLOGICAL PROTECTION, and INTERNATIONAL COM-
MISSION ON RADIOLOGICAL UNITS AND MEASUREMENTS, "Exposure of man to ionizing
radiation arising from medical procedures: An enquiry into methods of evaluation," *Phys.
Med. Biol.*, Vol. 2 (1957–58), pp. 107–151.

[22] B. KAHN, R. L. BLANCHARD, H. L. KRIEGER, H. E. KOLDE, D. B. SMITH, A. MARTIN,
S. GOLD, W. J. AVERETT, W. L. BRINCK, and G. J. KARCHES, *Radiological Surveillance
Studies at a Boiling Water Nuclear Power Station*, U.S. Dept. of Health, Education and
Welfare (Public Health Service), Report No. BRH/DER 70-1, March 1970.

[23] ———, "Radiological surveillance studies at a boiling water nuclear power reactor,"
Environmental Aspects of Nuclear Power Stations, Vienna, International Atomic Energy
Agency, 1971, pp. 535–548.

[24] A. W. KLEMENT, JR. (ed.), *Radioactive Fallout from Nuclear Weapons Tests, (Conf.—765)*,
Proceedings of a Conference held in Germantown, Md., Nov. 3–6, 1964, USAEC Div. of
Tech, Inf.

[25] J. B. KNOX, "Airborne radiation from the nuclear power industry," *Nuclear News*, Vol.
14 (1971), pp. 27–32.

[26] J. E. LOGSDON and R. I. CHISSLER, *Radioactive Waste Discharges to the Environment from
Nuclear Power Facilities*, U.S. Dept. of Health, Education and Welfare (Public Health
Service), Report No. BRH/DER 70-2, Washington, D.C., March 1970.

[27] W. M. LOWDER, H. L. BECK, and W. CONDON, "Spectrometric determination of dose
rates from natural and fall-out gamma-radiation in the United States, 1962–63," *Nature*,
Vol. 202 (1964), pp. 745–749.

[28] W. LOWDER, USAEC Health and Safety Laboratory, personal communication to M.
Eisenbud, Sept. 1969.

[29] W. M. LOWDER and W. J. CONDON, "Measurement of the exposure of human populations to environmental radiation," *Nature*, Vol. 206 (1965), pp. 658–662.

[30] I. LYON, "Nuclear power and the public interest," *The Bennington Review*, Vol. 3 (1969), pp. 59–63.

[31] A. MAHDAVI, "The thorium, uranium, and potassium contents of Atlantic and Gulf Coast beach sands," *Op. cit.* [1], pp. 87–114.

[32] J. E. MARTIN, E. D. HARWARD, D. L. OAKLEY, J. M. SMITH, and P. H. BEDROSIAN, "Radioactivity from fossil-fuel and nuclear power plants," *Environmental Aspects of Nuclear Power Stations*, Vienna, International Energy Agency, 1971, pp. 325–336.

[33] J. E. MCLAUGHLIN, "Dose rates to man from atmospheric cosmic rays," Unpublished memo to H. D. Bruner, USAEC, Jan. 5, 1971.

[34] ———, "Environmental survey at Grand Central Terminal," Unpublished memo to J. H. Harley, USAEC, Health and Safety Laboratory, March 2 and March 9, 1971.

[35] ———, "Report on preoperational environmental external gamma radiation survey in the vicinity of Millstone Point nuclear generating station," USAEC, Health and Safety Laboratory, New York, Unpublished preliminary report issued Sept. 1970.

[36] R. A. MILLIKAN and G. H. CAMERON, *Phys. Rev.*, Vol. 37 (1931), pp. 235–252.

[37] R. H. MOLE and A. H. WARD, "Yttrium-90 in gonads of monkeys containing strontium-90," *Nature*, Vol. 226 (1970), p. 175.

[38] K. Z. MORGAN, *Bull. Am. Phys. Soc.*, Series II, Vol. 16 (1971), p. 635.

[39] R. MULLER, Personal communication, Aug. 1970.

[40] W. A. MULLER, "Gonad dose in male mice after incorporation of strontium-90," *Nature*, Vol. 214 (1967), pp. 931–933.

[41] NATIONAL COUNCIL ON RADIATION PROTECTION AND MEASUREMENTS, *Maximum Permissible Body Burdens and Maximum Permissible Concentrations of Radionuclides in Air and in Water for Occupational Exposure*, NCRP Report No. 22, Washington, D.C., NCRP Publications, 1959.

[42] H. U. NEHER, "Gamma rays from local radioactive sources," *Science*, Vol. 125 (1957), pp. 1088–1089.

[43] K. O'BRIEN, "The composition of atmospheric cosmic-rays near solar maximum," *Amer. Nuclear Soc. Trans.*, June 1971, pp. 68–69.

[44] H. OHLSEN, "Zur Ermittlung der Bevölkerungsbelastung durch natürliche äußere Strahlung auf dem Gebiet der DDR (Messungen in Häusern)," *Kernenergie*, Vol. 13 (1970), pp. 91–96.

[45] R. L. PENFIL and M. L. BROWN, "Genetically significant dose to the United States population from diagnostic medical roentgenology, 1964," *Radiology*, Vol. 90 (1968), pp. 209–216.

[46] M. L. PETERSON, "Environmental contamination from nuclear reactors," *Scientist and Citizen*, Vol. 8 (November 1965).

[47] G. PHAIR and D. GOTTFRIED, "The Colorado front range, Colorado, U.S.A., as a uranium and thorium province," *Op. cit.* [1], pp. 7-38.

[48] A. PRESTON, "The United Kingdom approach to the application of ICRP standards to controlled disposal of radioactive waste resulting from nuclear power stations," *Environmental Aspects of Nuclear Power Stations*, Vienna, International Atomic Energy Agency, 1971, pp. 147–156.

[49] W. C. ROESCH, R. C. MCCALL, and F. L. RISING, "A pulse reading method for condensor ion chambers," *Health Phys.*, Vol. 1 (1958), pp. 340–344.

[50] W. D. RUCKELSHAUSE, "Environmental Protection Agency," Unpublished letter to Congressman Melvin Price.

[51] L. D. SAMUELS, "A study of environmental exposure to radium in drinking water," *Op. cit.* [1], pp. 239–251.

[52] A. SEGALL, "Radioecology and population exposure to background radiation in New England," *Science*, Vol. 140 (1963), p. 1339.

[53] R. F. SHAW and A. P. SMITH, "Strontium-90 and infant mortality in Canada," *Nature*, Vol. 228 (1970), pp. 667–669.

[54] L. R. SOLON, W. M. LOWDER, A. SHAMBON, and H. BLATZ, "Investigations of natural environmental radiation," *Science*, Vol. 131 (1960), pp. 903–906.

[55] F. W. SPIERS, M. J. McHUGH, and D. B. APPLEBY, "Environmental γ-ray dose to populations: Surveys made with a portable meter," *Op. cit.* [1], pp. 885–905.

[56] V. E. SPODE, "Uber die Verteilung von Radioyttrium und radioaktiven Seltenen Erden im Saugerorganismus," *Zeits f. Naturforshung*, Vol. 13b (1958), pp. 286–291.

[57] E. J. STERNGLASS, "Effects of low-level environmental radiation on infants and children," *Symposium on Biological Consequences of Environmental Radiation*, Univ. of Northern Illinois, March 20, 1971, unpublished.

[58] ———, testimony of 14 April 1971 before the Atomic Safety and Licensing Board, U.S. Atomic Energy Commission Docket No. 50-332, In the matter of Long Island Lighting Company (Shoreham Nuclear Power Station, Unit 1), official transcript, pp. 8849–8850.

[59] E. TOMPKINS and M. L. BROWN, *Evaluation of a Possible Causal Relationship between Fallout Deposition of Strontium-90 and Infant and Fetal Mortality Trends*, U.S. Dept. of Health, Education and Welfare (Public Health Service), Report No. DBE 69-2, Washington, D.C., October 1969.

[60] UNITED NATIONS, *Report of the United Nations Scientific Committee on the Effects of Atomic Radiation, General Assembly, Official Records: 17th Session Supplement No. 16 (A/5216)*, New York, United Nations, 1962.

[61] UNITED NATIONS, *Report of the United Nations Scientific Committee on the Effects of Atomic Radiation, General Assembly, Official Records: 21st Session, Supplement No. 14 (A/6314)*, New York, United Nations, 1966.

[62] J. VILLFORTH, Director of the Bureau of Radiological Health, U.S. Public Health Service, "Elda Anderson Award Acceptance Speech," 15th Annual Meeting of the Health Physics Society, Chicago, Ill., 28 June–2 July, 1970, unpublished.

[63] E. J. B. WILLEY, "Natural levels of radioactivity in Cornwall," *Brit. J. Radiology*, Vol. 31 (1958), pp. 31 and 56.

[64] B. W. WINDEYER, "The contribution of diagnostic radiology to population dose," *IXth Internat. Congr. Radiol., 23–30 July 1959, München*, Vol. 1 (1959), pp. 99–103.

Discussion

Question: J. Neyman, Statistical Laboratory, University of California, Berkeley

Is the radiation from effluents of a nuclear power facility qualitatively similar to or different from what might be called "natural" radiation, that is, approximately the same proportion of α-, β-, γ-radiation, and so forth?

Reply: V. L. Sailor

They are qualitatively similar, although it is necessary, of course, to take into account each individual source of radiation in calculating the dose. The calculation must account for the location of the source (whether internal or external), the energy and type of the radiation, and the duration of the exposure (the half-life for decay, and the residence time in the body). Very voluminous work has been done on dosimetry calculations, and this has been discussed in several of my references (for example, see [7], [8], [9], [10], [41], [60], and [61]). When two different doses have been reduced to units of *rems* they should be directly comparable, within the limits of perfection of the procedure. Note that doses are calculated for *whole body* as well as for *specific organs* of the body.

The detailed procedures for dose calculations were outlined in a publication which is not included in my references (International Commission on Radiological Protection, *Radiation Protection*, ICRP Publication No. 2, New York, Pergamon Press, 1960). Although some of the numbers and recommendations contained in this report have been revised in recent years, the basic procedures are still considered valid, and it gives a good discussion of the calculations.

Question: E. J. Sternglass, School of Medicine, University of Pittsburgh

It is misleading to compare external sources of gamma and X-ray radiation with the internal beta emitters that are inhaled or ingested in the case of nuclear fallout and radioactive isotopes emitted by nuclear facilities for the following reasons.

(1) A given amount of radioactive material on the ground giving rise to gamma rays is far less hazardous or toxic than an equal amount of beta emitters inhaled or ingested, since all the energy from the beta emitters is deposited in a very thin layer of tissue, giving it a much higher amount of damage per unit volume than a gamma emitter. The existing permissible levels of concentration for beta emitters recommended by the International Commission for Radiation Protection recognize differences in toxicity for beta emitters and gamma emitters by many thousands of times.

(2) Unlike all sources of external gamma or X-ray radiation such as cosmic rays or medical X-rays, internal emitters concentrate chemically in certain critical organs of the body. This is especially serious in the case of the early fetus and infant, where the most critical glandular organs are much smaller than in the adult. Measurements have shown that, for instance, iodine 131 gives a dose to a three months fetal thyroid gland some ten times larger than for the full term fetus or young infant, which in turn is known to receive a dose ten times greater than the adult. As a result, the fetal dose for a given amount of iodine intake is typically 100 times greater than for the adult thyroid, which in turn concentrates iodine 100 times more than ordinary muscle tissue!

(3) Medical X-rays are very rarely given to a developing early embryo or fetus during its most critical stage of organ development, while fallout radiation is present throughout the most critical phases of embryonic and fetal life.

Reply: V. L. Sailor

I disagree that there is any misrepresentation. The dose calculations do, in fact, take account of the concentration factor and the beta component, mentioned by Dr. Sternglass. They also set special limits on exposure to particular population groups such as infants, pregnant females, and so forth. One might want to argue about some of the specific details or numbers, since such complex calculations can never be perfect, and this is recognized by ICRP and NCRP who continually strive for more perfect methods. However, allowing for such imperfections, the final doses expressed in units of the rem are directly comparable whether the source was an X-ray machine, a cosmic ray muon, or an ingested radioactive fission product.

Question: A. B. Makhijani, Electrical Engineering and Computer Sciences, University of California, Berkeley

The dose rates from fuel processing and reprocessing plants for nuclear reactors were not discussed, even though the major releases of radioactivity to the environment from nuclear power plant operation occur at these points—and consequently the major radiation exposures will be due to these sources. The radiation released at these points have substantial amounts of high LET components, the carcinogenic effects of which are known to be more severe than those of gamma and beta radiation.

Reply: V. L. Sailor

This is a point of considerable public interest at the present time. Plants for reprocessing nuclear fuel must handle the great bulk of the fission products produced in reactors. These plants are required to meet the same off site dose limitations as any other nuclear facility. So far our experience with such plants is limited because only one commercial reprocessing plant is in operation and that only for a few years. Surveys have shown that the off site effluents from this plant have been only a few per cent of the applicable standards. (For example, see B. Shleien, *An Estimate of Radiation Doses Received by Individuals Living in the Vicinity of a Nuclear Fuel Reprocessing Plant*, U.S. Dept. of Health, Education and Welfare, Report No. BRH/NERHL 70–1, May 1970.)

I assume that when you mention "high LET components" you refer to alpha emitters, particularly transuranic isotopes (Pu, Am, and so forth). I have seen no data which indicate that such isotopes are detectable in the off site effluent.

Question: H. L. Rosenthal, School of Dentistry, Washington University

I think it is necessary to comment about equating the dosage for the medical use of radiation with the permissible dosage in the general population. These two usages are entirely different and must be kept separate. The doses of X-radiation or isotope therapy and diagnosis are prefaced on an entirely different basis for risk benefit. I hope Dr. Sailor is not recommending that doses used for medical purposes are also satisfactory for the general population. These two systems should also be kept separate for statistical and epidemiological purposes.

Reply: V. L. Sailor

I hope that my paper does not give the impression that I *recommend* any dose. Its purpose was to report as accurately as possible what exposures people receive, how these exposures vary and from whence they come. The facts show that medical X-rays are unquestionably the largest man-made radiation burden by a big margin.

Question: Alfred C. Hexter, California Department of Public Health

You and some of the previous speakers have indicated some of the average population exposures. But these are only means. Do not some exposures follow, approximately, a log-normal distribution? If so, this would indicate that a significant portion of the population is receiving substantially greater exposures than these average figures indicate.

Reply: V. L. Sailor

To the best of my knowledge, dose distribution curves have not been constructed with any great precision, but your suggested distribution seems plausible. There is no question that many individuals within the population receive five or ten times the mean dose.

Question: Emanuel Hoffer, California Department of Public Health

You avoided discussing "biological concentration mechanisms." An illustration of this is the "low level waste radiation dumping" in the Columbia River in Washington. This low level radiation went through a number of hosts: plant, fish, bird, and animal, concentrating the initial low level dose 500–1000 times. Man at the end of this food chain could eat the animal. In addition, there is the illustration of clams in the San Francisco Bay concentrating radioactive cobalt from bomb tests.

Reply: V. L. Sailor

In principle, biological concentration of radioisotopes in food chains can enhance the dose individuals receive. We need to exercise care that such situations are not allowed to happen. Some of the measures which have been taken to monitor various facilities are described in [10], [11], [22], and [48]. At the present time, it appears that the contribution of such effects to population exposure is completely negligible.

RADIATION AND RISK— THE SOURCE DATA

H. WADE PATTERSON and RALPH H. THOMAS
LAWRENCE BERKELEY LABORATORY

"A likely impossibility is always preferable to
an unconvincing possibility"
Aristotle—from the *Poetics*

1. Introduction

We have seen evidence in the past several years of a growing concern on the part of the general public over the possible risks to which they may be subjected as a result of man's increasing uses of ionizing radiations.

The specific benefits derived from the uses of ionizing radiations in medicine and industry may be a matter of particular debate, but it seems generally to be accepted that benefits do in fact accrue. Public concern is centered on what risk, if any, is involved in such activities. In the words of the International Commission on Radiological Protection (ICRP), "If the quantitative relationship between dose and the risk of an effect were known, societies or individuals could judge the degree of risk that would be acceptable, taking into account the particular circumstances requiring a radiation exposure. Ideally, such a judgment would involve a balancing of the benefits or necessities of the practice against the risks of the given exposure, which could also be related to that of other risks in the particular society." [1]

With respect to physical and chemical components in the natural environment other than radiation, it would seem that man has, through evolutionary processes, been adapted to function adequately over a rather broad range of exposure. Examples of this are carbon dioxide concentration in air, temperature, and barometric pressure. Observing this, we might be tempted to posit that man's response to radiation exposure would be similar. However, as scientists we must stress that we do not know the effect of small exposures to radiation on human beings. We do not know whether such exposures are deleterious, of no consequence, or beneficial.

It is perhaps true that more is known of man's response to ionizing radiations than to any other self-inflicted pollutant of his environment. This is largely due to the experience of radiation injury resulting from early uses of X-rays and radioactive substances, particularly radium. From these early experiences and

Work done under the auspices of the U.S. Atomic Energy Commission.

313

from studies on certain other groups of individuals subjected to high radiation exposures as a result of radiotherapy, nuclear weapons attack, or radiation accidents, a limited amount of information has been pieced together. Such information is almost entirely about the effects of large exposures and high dose rates. If we are to make any progress in the difficult task of understanding the possible deleterious effects on the health of the population due to small exposures to ionizing radiation at low dose rates it is clear that much greater efforts at interdisciplinary studies are needed. Radiation physicists can measure human exposures to ionizing radiations, physicians can advise on the appropriate indices of health, and statisticians can show us how to analyze available data in the most fruitful manner. It also seems clear that any conclusions we may reach as to the probable risks to human beings of low doses of radiation will almost certainly have been reached by statistical inference. Heretofore much of the analysis of radiation risk data has been performed by nonprofessional statisticians, and we believe that much benefit would derive from a re-evaluation of the existing data by professional statisticians.

Although much of what we say here will be familiar to specialists in the fields of study involved, we do try to draw together what seems to us the relevant threads of the argument involved in setting up an epidemiological study of this nature.

In this paper we first briefly review the source of the studies that have been made of radiation-induced injury for rather large acute exposures. These studies enable one to make some first order approximations on the level of risk involved.

Next we summarize man's natural radiation environment and show that the extreme variations in whole body exposures vary from about 100 mrem/year to an upper limit of a few rem/year. Man-made radiation levels are, with one exception, small compared even with the fluctuations in these natural levels due to geography and personal habits. The one exception will be shown to be due to medical radiology.

2. Size of population needed for an epidemiological study of radiation-induced disease

It seems to us that a most important preparatory step in designing a study to identify the risks of radiation exposure inducing disease is to determine the size of the group needed.

The following simple arguments indicate the size of the population needed to identify the magnitude of risk.

The total number of cases of the disease, N_0, observed in a population, p, over a period of y years is given by

$$(1) \qquad\qquad N_0 = fpy$$

where f is the probability of contracting the disease per year.

Assume that this disease may also be induced by low levels of radiation exposure and further assume that at low doses the dose-effect relationship is linear. At equilibrium an annual dose rate of D rem/year will then produce an additional number of cases of the disease due to radiation, N_R, given by

(2) $$N_R = rDpy$$

where r is the risk per year per rad.

The total number of cases of the disease actually observed, N_T, is then

(3) $$N_T = (f + rD)py$$

and we ask the question, when can we be sure that the difference, Δ, $\Delta = N_T - N_0$ is greater than zero?

(4) $$\Delta = rDpy \pm \epsilon$$

where the error ϵ is given by

(5) $$\epsilon^2 = py(f + rD) + fpy.$$

To be sure of the magnitude of Δ we must demand that $\epsilon \ll rDpy$. Typically, $rDpy$ will be small and this constraint may be difficult to meet. However, let us arbitrarily write

(6) $$\epsilon \approx \frac{rDpy}{2},$$

from which it follows that

(7) $$py \approx \frac{4}{rD}\left(1 + \frac{2f}{rD}\right).$$

This equation enables us to calculate the number of man-years (py) required to form the basis of a study to reveal radiation-induced disease.

As an example, the probability of death in the United States due to malignancies is about 1.5×10^{-3} per year, [2] and one may readily calculate the number of man-years (py) from equation (7) for several dose rates and degrees of radiation-induced risk. Table I summarizes such a calculation.

TABLE I

NUMBER OF MAN-REM YEARS NEEDED FOR AN
EPIDEMIOLOGICAL STUDY OF RADIATION-INDUCED CANCER

Taking "normal" risk of death due to malignancies as 1.5×10^{-3}
per year.

Dose rate (rem/year)	Radiation risk (deaths/year/rad)	Man-years
0.1	10^{-1}	5.2×10^2
0.1	10^{-2}	1.6×10^4
0.1	10^{-3}	1.2×10^6
0.1	10^{-4}	1.2×10^8
0.1	10^{-5}	1.2×10^{10}
1.0	10^{-1}	4.1×10^1
1.0	10^{-2}	5.2×10^2
1.0	10^{-3}	1.6×10^4
1.0	10^{-4}	1.2×10^6
1.0	10^{-5}	1.2×10^8

Professor Neyman has pointed out that the values given in Table I represent an upper limit to the number of man-years required to detect possible effects due to radiation. The actual number is likely to be smaller because the probability, f, of radiation effects is probably not the same for all individuals (as assumed in our model). The more heterogeneous the population studied, the smaller the variance of the number of cases of expected radiation effects and the fewer the number of man years required to obtain the desired precision of the study. Furthermore, the variation of exposures from one individual to another (see paper by V. Sailor) must be incorporated in a precise treatment of this problem. Unfortunately, the actual variability of the probability, f, is unknown and one is compelled to rely on the upper limits given in Table I.

As Sailor has already discussed in this Symposium [3] and we shall show later, it is possible to find differences in radiation exposure rates of substantial populations of up to a few hundred mrem/year. In comparing the death rates due to cancer in groups where radiation exposures have changed with time, studies must extend over periods long compared with the latency of the disease. It would seem mandatory therefore to carry out such investigations over periods of something like 10 to 30 years, and there are those who would suggest even larger periods. If one takes the risk of cancer induction due to radiation as 10^{-4} per rad per year (a conservative upper limit if the interpretation of the pertinent data presented by the International Commission on Radiological Protection (ICRP) is accepted [4]), Table I indicates that populations in excess of 10 million people whose radiation exposures differed by 0.1 rem/year must be studied for extended periods.

There is no chance of finding such large populations within the United States whose environments are so similar and stable over such extended periods—differing only with respect to their radiation exposures. However, much smaller populations are needed to test the hypotheses that the risk of death from radiation-induced disease is much higher than suggested by ICRP.

Gofman and co-workers [5] have suggested that the increase in cancer mortality rates is as high as 2×10^{-2} per rem/year. (This is in fact roughly equivalent to assuming that all cancer mortality is due to radiation exposure, since the "natural" mortality cancer rate is 1.5×10^{-3} deaths per year and the average annual dose rate is about 0.13 rem/year). [3]. One might think this to be an upper limit since chemical carcinogenesis might be suspected to contribute to the death toll.

At levels of risk as high as 10^{-2} per rad, studies with relatively small numbers of people (several hundred) should be capable of revealing significant differences between populations whose radiation exposures differ by a few rads (integrated dose).

One of the populations most frequently exposed to ionizing radiation is atomic energy workers. The USAEC makes annual reports of the exposures for such workers. Using data for 1960, Eisenbud [6] estimated a per capita dose of 0.6 rem to a population of 82,000 workers. Table II summarizes similar data for 1969.

TABLE II

ESTIMATED WHOLE BODY DOSES TO EMPLOYEES OF AEC CONTRACTORS,
AEC LICENSEES, AND AGREEMENT STATE LICENSEES FOR 1969

Annual dose rem	Number of employees		
	AEC contractors	AEC licensees	State licensees
0–1	98,625	59,496	23,082
1–2	2,554	1,489	786
2–3	1,313	ʃ583	321
3–4	335	191	107
4–5	86	109	69
5–6	4	64	56
6–7	0	48	39
7–8	0	36	24
8–9	0	14	6
9–10	0	13	6
10–11	1	3	4
11–12	0	4	0
12+	0	22	19
Total	102,918	62,072	24,519

If we assume, with Eisenbud, that all members receive the mean dose of the dose grouping (probably an overestimate) we can conclude that within the atomic industry the accumulated dose for 1969 was about 110,000 man-rems (at an average per capita dose of 0.58 rem). Failure to find any significant increase in cancer risk in this population should therefore be able to set the risk of cancer induction below about 10^{-3} per year per rad.

3. Radiation and risk studies—a brief review

What has been established "beyond reasonable doubt" thus far?

Fortunately man's experience of radiation-induced injury is nowadays quite infrequent. Nevertheless in the past 70 years a number of persons have been exposed to rather large doses of radiation, and the data obtained from epidemiological and cytogenic studies of them provide some measure of the incidence of radiation-induced diseases. In the main these persons fall into three main groups.

(a) Medical patients undergoing radiotherapy—for example, ankylosing spondylitis patients treated by X-ray irradiation of the spine, radium-therapy and thorium-therapy patients, patients treated for hyperthyroidism, women treated for cervical cancer, or children irradiated for enlarged thymus and tinea capitis. A group of children exposed *in utero* for diagnostic purposes for the mother have also been studied.

(b) Victims of nuclear warfare or testing, for example, those exposed at Hiroshima, Nagasaki, and the Marshall Islands [7].

(c) Occupationally exposed persons, for example, radium-dial painters, radiologists, and uranium miners.

From these three main groups the ankylosing patients, the Hiroshima and Nagasaki victims, and the radium-dial painters have been most extensively studied.

3.1. *Hiroshima and Nagasaki victims.* Perhaps the most thorough and extensive study of the incidence of disease in human populations exposed to ionizing radiations has been performed (and is still in progress) for the victims of the nuclear weapons attacks on Hiroshima and Nagasaki in 1945 [8], [9], [10].

Within about two years from the exposure a significant increase in the incidence of leukemia was observed in the exposed population. Early studies showed the increased frequency of leukemia to be inversely related to distance from the hypocenter. This fact led Lewis [11] to suggest that the incidence of leukemia was linearly related to dose. However, subsequent analyses of the dosimetry have revealed some uncertainties that make such a conclusion uncertain. In his analysis Lewis utilized dose distance curves known by their originators to have substantial errors, but the best available at that time [12].

Auxier and co-workers, [13] in a recent paper on dosimetry, have suggested the probable error in the *air dose* to be ±30 per cent at Hiroshima and ±10 per cent at Nagasaki. Problems of local shielding, spectral distribution, and relative proportions of neutron and γ dose make the assignment of *individual doses* a much more difficult problem. Moloney and Kastenbaum [14] made this distinction when they showed that for persons exposed at the same distance, the incidence of leukemia was higher in those who suffered radiation sickness in the few weeks immediately following the exposure. Milton and Shohoji [15] have reviewed the dose estimates due to Auxier and co-workers and those made by Hashizume and co-workers [16] based on measurements of residual induced activity and thermoluminescence in irradiated material, and concluded that "it is not possible at present to give a quantitative evaluation of either the accuracy or precision of the final (individual dose) estimates."

Inability to assign doses to individuals required that morbidity and mortality data be lumped on the basis of distance. When this is done, even with a distance interval as small as 50 meters, the uncertainty in dose is as large as 30 per cent. And, if the data are lumped in large intervals, as is done in ICRP Publication 8 [17], the dose uncertainty approaches two orders of magnitude. These considerations lead one to conclude that the Hiroshima-Nagasaki data are of insufficient accuracy to test any dose exposure hypotheses. Lewis's analysis of several exposed groups summarized in Table III, assuming a linear dose-effect relationship, suggested the incidence of leukemia to be one to two cases per million person-years at risk per rem.

Recent studies suggest that different types of cancer do not have the same dose incidence relationship [19]. H. Maki and co-workers conclude: "It has been reconfirmed that in both sexes risk of leukemia mortality increases markedly with increase of dose. Also, in both sexes for all sites excluding leukemia, a slight

TABLE III

SUMMARY OF LEWIS'S ESTIMATES OF THE PROBABILITY OF RADIATION-INDUCED LEUKEMIA
PER INDIVIDUAL PER RAD PER YEAR
Source: Lewis, 1957 [11] and Upton [18]

Source of estimate	Type of radiation	Region irradiated	Types of leukemia produced	Probability of leukemia of specified type per individual per rad (or rem) to region irradiated per year		
				Estimated range		
				Lower limit	Upper limit	"Best" estimate
Atom bomb survivors	γ-rays plus neutrons	whole body	all	0.7×10^{-6}	3×10^{-6}	2×10^{-6}
Ankylosing spondylitis patients	X-rays	spine	granulocytic (only?)	0.6×10^{-6}	2×10^{-6}	1×10^{-6}
Thymic enlargement patients	X-rays	chest	lymphocytic (only?)	0.4×10^{-6}	6×10^{-6}	1×10^{-6}
Radiologists	X-rays, radium, etc.	partial to whole body	all(?)	0.4×10^{-6}	11×10^{-6}	1×10^{-6}
Spontaneous incidence of leukemia (Brooklyn, N.Y.)	all natural background sources	whole body	all(?)		10×10^{-6}	2×10^{-6}

trend is noted for the risk to increase with increase in dose. This increment is attributable chiefly to the increase of gastric cancer and lung cancer. Some, for example, uterine cancer, show hardly any effect of exposure."

Studies made during autopsy indicated a slight tendency for higher mortality due to gastric cancer in females and lung cancer in females and lung cancer in both males and females, but the authors note that these trends were not statistically significant. No significant relationship was noted between radiation exposure and *mortality* due to cancer of the liver and biliary ducts and cancer of the uterus (in women).

Studies of the incidence of cancer, however, showed that thyroid cancer, breast cancer, lung cancer, and leukemia all showed increased incidence with increasing exposure. "However, in Nagasaki, while incidence (for leukemia) increased with dose as in Hiroshima for the group exposed to 100 rad or more, no increase was noted under 100 rad." This latter conclusion by Maki and coworkers [19] indicates the difficulties (and possible overestimates) in deriving estimates of cancer incidence in humans at chronic low doses and dose rates from these data on acute high doses.

3.2. *Ankylosing spondylitis patients.* Studies of the subsequent incidence of disease in patients treated with X-rays for ankylosing spondylitis have revealed an elevation in the incidence of leukemia and other cancers (see Table IV).

TABLE IV

Change in Rate of Induced Malignant Disease with Duration of Time
Since Exposure in Irradiated Ankylosing Spondylitics
(Data from Court-Brown and Doll, 1965 [20].)

	Cases per 10,000 man-years at risk	
Years after irradiation	Leukemia + aplastic anemia	Cancers at heavily irradiated sites
0–2	2.5	3.0
3–5	6.0	0.7
6–8	5.2	3.6
9–11	3.6	13
12–14	4.0	17
15–27	0.4	20
Total of expected cases in 10,000 persons in 27 years calculated from the rates given	67	369

Court-Brown and Doll [21] first suggested a correlation between the incidence of leukemia in these patients and radiation exposure. Furthermore, in the dose range studied, the data were consistent with a linear relationship. Court-Brown and Doll, however, excluded those cases in which extraspinal irradiation was given. Brues [22] has noted that this exclusion resulted in a severe bias in the analysis because the cases excluded were predominantly in the high dose range. The complete Court-Brown and Doll data thus indicate not only a curvilinear relationship, but perhaps also a threshold for leukemia induction in the range 50 to 100 R [22] (see Figure 1).

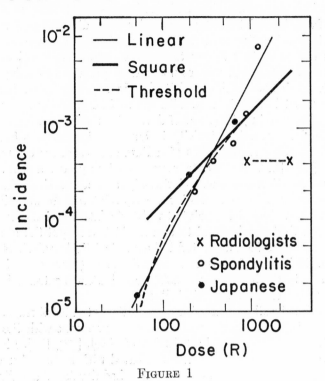

Figure 1

The dose-response relationships for radiation leukemia in radiologists, irradiated spondylitic patients, and Japanese A-bomb survivors. (From Brues, 1959)

Nevertheless, this study clearly demonstrates an almost ten-fold increase in leukemia among irradiated patients and an almost 30-fold increase in the related disease aplastic anemia, whereas cancer of other heavily irradiated sites was increased by a factor of only 1.6. In absolute numbers, 67 cases of leukemia and aplastic anemia were found, 61 cases more than expected as compared with 73 cases of all other cancer beyond the expected. However, there should be some caution in necessarily attributing this increase in cancer (other than leukemia) found in this study to irradiation. The largest contributor to the excess deaths from cancer of patients in the study was contributed by lung cancer, now well known to be caused by smoking and unfortunately the smoking habits of these patients are not known, and it is therefore possible that differences in cigarette smoking may be responsible for part or all of the difference in lung cancer rates between patients and controls. Furthermore, it is not known whether lung cancer may or may not be increased among patients with rheumatoid spondylitis irrespective of radiation. Lung disease is known to occur as part of the primary disease [23]. Still another reason for caution in attributing all these additional cancers to radiation is due to the absence of the typical latent period, peaking, and decline in incidence associated with radiation-induced cancers.

3.3. *Radium-dial painters.* The fate of radium-dial painters who ingested toxic quantities of radium and radium daughters as a direct result of their occupation has been studied over the past 40 years. These painters absorbed radium through the mouth as a result of their practice of tipping their paint brushes with their lips. Radium and its daughters are deposited in bone and in time, if absorbed in sufficient quantities, can lead to skeletal damage, osteosarcoma, and other injury [24]. One of the most extensive and complete analyses of radium and mesothorium toxicity in human beings derives from the MIT group that has followed 604 cases of radium exposure over the past 40 years [25], [26], [27], [28]. These data have been interpreted as showing both a curvilinear dose-effect response relationship and a practical threshold. The time for appearance of bone cancer is inversely related to the quantity of radium absorbed in bone. Thus at the point at which the latent period exceeds probable life span a practical threshold exists, and the MIT data put this at a few tenths of a microgram of radium deposited in bone. Statistical analysis of the data in which some incidence of bone cancer is observed (those cases in which the absorbed dose to the bone exceeds 1200 rads) indicates extreme improbability that the dose-response relationship is linear.

Other studies of radium-dial painters, of patients treated therapeutically with radium, and of animals have shown essential agreement with the conclusions of the MIT group [29], [30], [31], [32], [33], [34], [35], [36]. Finkel and coworkers [37] in a study of 293 patients treated with radium, found no person with a radium body burden below $1.2\mu Ci$ who had developed a malignant tumor ascribable to radium deposition.

Recently Goss [38] has expressed some reservation about the analyses of the data in both these two studies. In the MIT studies it is suggested that the data do not exclude the possibility that the dose response model is linear and with no threshold. In the ANL studies Goss suggests that the higher than expected incidence of tumors of the central nervous system might be significant in an evaluation of risk.

It would seem that here are studies that would benefit from an independent analysis by one or more groups of statisticians.

3.4. *Incidence of lung cancer in uranium miners.* As early as 1500 the high incidence of lung disease amongst miners in the cobalt mines of Saxony and the pitchblende mines of Bohemia was recognized [39]. One component of this disease—colloquially referred to as *"Berg Krankheit"*—was finally identified, at the beginning of the twentieth century, as lung carcinoma. Sikl [40] suggested in 1950 that the one common factor to these mines that seemed primarily responsible for the high incidence of lung cancer was the radiation exposure from the radioactive daughters of uranium, particularly radon and polonium. Several studies of the incidence of lung cancer showed the death rate from lung cancer in these mines to be about 30 times as great as normally expected [39].

Studies of the relationship between the incidence of lung cancer and radiation exposure for uranium miners in the United States have recently been reported [41], [42]. The lowest exposure group studied in 1968 by a National Academy of Sciences Subcommittee [42] had cumulative exposures roughly corresponding to lung doses from radon and its daughter products up to 250 rads. After careful study the subcommittee favored the hypothesis that radiation exposure had probably at least contributed to the higher incidence of lung cancer found in this group of workers than in the general population. However, they were careful to point out that a curvilinear relationship between dose and probability of cancer induction would be expected for lung cancer, which depends on localized tissue damage for its inception. Wagoner and co-workers [43] did in fact find a curvilinear relation between working level months (a rough measure of radiation exposure) and annual incidence of respiratory cancer. Even after correction for the influence of age distribution in the working population, smoking habits, and number of years since onset of cancer, the relationship is still curvilinear.

3.5. *Incidence of leukemia in U.S. radiologists.* Some additional data may be gleaned from a study of the incidence of leukemia in the early U.S. radiologists, who, it is estimated, received doses as high as 2000 rads over a period of many years [44]. Although this cumulative dose resulting from chronic exposure was far in excess of a lethal single dose in man, it resulted in an incidence of leukemia far lower than for either the nuclear bomb victims or the ankylosing spondylitis patients (see Figure 1). This fact suggests that some substantial dose rate effect may be important.

The difficulties in establishing a measure of the risk of radiation-induced disease are evident from this brief review.

In its studies of external radiation effects on humans, ICRP has concentrated on two familiar sets of data: (i) those from a study of victims of the nuclear weapons attacks on Hiroshima and Nagasaki and (ii) those from the study of ankylosing spondylitis patients exposed to high levels of radiation for therapeutic reasons. Neither of these studies provide evidence of an effect with whole body irradiation of less than 100 rads. In order to provide guide lines for the control of radiation exposure, however, ICRP have estimated the risk of the incidence of leukemia and other cancers on the basis of a linear dose effect, no threshold model. This model was not, however, advanced as a scientific hypothesis. Nevertheless, ". . . there must already be many health physicists who believe as a fact that radiation risks are linearly related to dose and independent of dose rate, although this simplification is little more than a convenient simplification from which to derive basic radiation standards" [45].

In discussing its most recent re-examination of the available data, ICRP concluded [46], "In essence this re-examination involved as detailed a subdivision as possible of the category of 'other fatal neoplasms' and the recognition that tissue dose was far from uniform in each of the three chief irradiated human populations—medical radiologists, ankylosing spondylitics and survivors of the atomic bomb explosions in Japan. It had also to be recognized that the time which has elapsed since exposure is still much too short for it to be possible to assess the full tumor incidence in the spondylitics and the Japanese: the following table shows that evidence collected during the first 15 years or so after exposure could be regarded as covering only the beginning of the period in which neoplasms other than leukemia might be expected to appear. If so, relatively small differences in the latent period of neoplasms arising in different tissues could lead to quite erroneous ideas about relative tissue susceptibility.

"The data in the table (Table IV) may also suggest that malignant disease other than leukemia will be 5 to 6 times more frequent than leukemia plus aplastic anemia when the yield is assessed after 27 years of observation. However, in this context the rates cited for 15 to 27 years after irradiation are quantitatively the most important and it should be stressed that these have a considerable statistical uncertainty."

4. Natural background radiation

4.1. *Terrestrial radioactivity*. Those radionuclides which have survived in measurable quantities in the earth's crust are of course those with half-lives comparable with the age of the earth (approximately 5×10^9 years). Three radioactive decay chains account for much of the natural radioactivity to which man is exposed—the familiar uranium series (derived from U^{238}), thorium series (Th^{233}), and the actinium series (Ac^{235}). Of the other naturally occurring radionuclides K^{40} contributes most significantly to the natural background. In addition to these radionuclides of terrestrial origin one must include in this discussion of naturally occurring radioactivity those radionuclides produced by the interaction of cosmic radiation with the earth's atmosphere; of these, the most significant are H^3 and C^{14}. Many extensive studies of terrestrial radioactivity have been made around the world, and the interested reader is referred to excellent summaries prepared by Claus [47], Eisenbud [48], Adams and Lowder [49] and the United Nations [50].

Table V shows the typical concentration of K^{40}, thorium, and uranium in igneous and sedimentary rocks.

These variations in concentration of radionuclides in rock naturally lead to changes in external radiation levels, and Table VI shows estimates of external exposure levels for four regions around the world. We see that natural background levels due to this source may range by more than a factor of ten, principally depending upon the concentration of thorium, uranium, and potassium in the surrounding rocks.

Although there is large variation in external radiation levels from place to place, at a particular location there is little variation with time. Because the contribution to man's external exposure is dominated by the component due to

TABLE V

POTASSIUM 40, THORIUM, AND URANIUM IN IGNEOUS AND SEDIMENTARY ROCKS (IN PPM)

Chemical potassium contains 0.0119 per cent potassium 40.

| | Igneous Rocks | | | Sedimentary Rocks | |
	Basaltic	Granitic	Shales	Sandstones	Carbonates
Potassium 40					
Average	0.8	3.0	2.7	1.1	0.3
Range	0.2–2.0	2.0–6.0	1.6–4.2	0.7–3.8	0.0–2.0
Thorium					
Average	4.0	12.0	12.0	1.7	1.7
Range	0.5–10.0	1.0–25.0	8.0–18.0	0.7–2.0	0.1–7.0
Uranium					
Average	1.0	3.0	3.7	0.5	2.2
Range	0.2–4.0	1.0–7.0	1.5–5.5	0.2–0.6	0.1–9.0

TABLE VI

MEAN DOSE OF IRRADIATION TO GONADS AND BONES FROM NATURAL
EXTERNAL SOURCES IN NORMAL AND MORE ACTIVE REGIONS

Using a shielding factor of 0.63 for γ-rays and a dose rate of 28 mrem/year due to cosmic rays.

Region	Population in millions	Aggregate mean dose (mrem/year)
1. Normal regions	2500	75
2. Granitic regions in France	7	190
3. Monazite region, Kerala in India	0.1	830
4. Monazite region, Brazil	0.05	315

terrestrial radioactivity, it follows that the secular perturbations in the other sources of his external exposure, for example, cosmic radiation, do not have a great influence in the variation of exposure with time.

Considerable variation in radiation exposure from buildings due to the use of differing construction materials is to be expected, however. Studies of the incidence of cancer and leukemia in areas of high terrestrial radioactivity or in areas which utilize building materials of high radioactivity have been suggested as possible sources of information in radiation-induced disease.

Table VII lists some areas of high terrestrial radioactivity, while Table VIII lists areas with high radiation levels in dwelling houses due to the use of special construction materials.

One interesting example of how man may (unwittingly) change his radiation environment due to his use of a naturally radioactive substance has been reported by Jaworowski and co-workers [51]. These authors studied the concentration of Ra^{226} occurring in snow around a coal burning power station in

TABLE VII

SOME DETAILS OF AREAS OF HIGH TERRESTRIAL RADIOACTIVITY

There are also some areas of high natural radiation in the Belgian Congo,
but these are said to be uninhabited.

Area	Population	Demographic information available	Natural radiation received (multiply by 0.63 to get gonad dose)	Possible control populations
Part of Kerala State and adjoining area in Madras State	approx. 80,000	some information on births and deaths: could probably be developed relatively easily	approx. 1300 mR/y (plus about 200 mrad beta rays)	similar ethnic group further along coast
Monazite area in Brazil (States of Espirito Santo and Rio de Janeiro)	approx. 50,000	specially prepared statistics would be required	average 500 mrad/year	?
Mineralized volcanic intrusives in Brazil (States of Minas, Geraes and Goiaz)— 6 km² in a dozen scattered places	pastureland, scattered farms, 1 village with 350 inhabitants	very little	average 1600 mrad/year peak value 12,000 mrad/year	?
Primitive granitic, schistous and sandstone areas of France with slight elevation of natural radiation said to cover about ⅙th of French population (7 million)		specially prepared statistics would be required	180–350 mrem/year	remainder of France estimated at 45–90 mrem/year

Warsaw. Table IX shows their data presented as a function of distance from
the generating plant. Similar data from U.S. coal burning factories and stations
could be developed.

4.2. *Natural radioactivity in the diet.* The natural radioactivity of soil necessarily leads to a transfer of radioactive material to human tissues through ingestion. Much of the α-activity ingested can be directly absorbed to decay
products of the uranium and thorium radioactive series, in particular Ra^{226} and
Ra^{228}, and Pb^{210} (and their decay products).

Table X gives estimates of the total human intake of Ra^{226} and the contribution to the total from different foodstuffs for three different countries. We see
that within the continental United States the average ingestion rate is about
2 pCi/day with some suggestion that the quantity ingested by young people is
somewhat higher.

TABLE VIII

SOME DETAILS OF AREAS WITH HIGH NATURAL RADIATION
IN HOUSES MADE OF SPECIAL MATERIALS

Area	Population	Demographic information available	Natural radiation received (multiply by 0.63 to get gonad dose)	Possible control populations
Sweden—houses made of light-weight concrete containing alum shale	relatively small	special statistics being obtained	158–202 mrad/year (cosmic radiation excluded)	wooden houses 48–75 mrad/year (cosmic radiation excluded)
United Kingdom (Aberdeen)—houses and buildings made of granite	population of Aberdeen approx. 186,000	leukemia statistics being studied	results from a few buildings indicate 102 mrad/year	approx. 78 mrad/year in other cities with brick buildings, for example, Dundee— population 178,000
Austria—granite houses	?	special statistics necessary	granite houses 85–128 mrad/year; brick or concrete houses 75–86 mrad/year	wooden houses 54–64 mrad/year

TABLE IX

CONCENTRATION OF Ra^{226} IN SHOW AROUND
A POWER STATION IN WARSAW

From Jaworowski and co-workers [51].
S is statistical counting error at 0.95 confidence level.

Distance from power plant (km)	pCi/kg ± S
0.6	0.98 ± 0.12
1	0.63 ± 0.07
2	0.45 ± 0.07
4	0.076 ± 0.019
30	0.073 ± 0.033
45	0.019 ± 0.011

TABLE X

ESTIMATES OF TOTAL INTAKE OF Ra^{226} AND OF CONTRIBUTIONS FROM DIFFERENT FOODSTUFF CATEGORIES
(From UNSCEAR Report [50])

Category of foods	United States				Consumers' Union		United Kingdom	India	
	New York, N.Y.	Chicago, Ill.	San Francisco, Cal.	San Juan, P.R.	Five-city study	Teenager twenty-two city study	Country-wide study	Bombay	Kerala State Monazite area
Cereals and grain products	0.56	0.76	0.51				0.17	0.41	1.48
Meat, fish, eggs	0.38	0.37	0.28				0.38		
Milk and dairy products	0.14	0.12	0.13				0.14	0.04	0.19
Green vegetables, fruits and pulses	0.81	0.56	0.48				0.32	0.17	0.81
Root vegetables	0.40	0.22	0.26				0.10	0.02	0.07
Water	~0.02	~0.03	~0.01	~0.7			0.07	0.06	0.29
Total pCi/day	~2.3	~2.1	~1.7		~3 (2.2–4.3)	~5 (2.5–6.5)	~1.2	~0.7	~2.8
pCi Ra^{226}/g Ca	2.2	2.0	1.6	1.3	1.9	2.5	1.1		

It is important to know what quantity of Ra^{226} becomes permanently incorporated in human tissues (principally bone in this case). Table XI shows the quantities of Ra^{226} measured in human bone around the world. It seems that the total quantities of Ra^{226} in the human skeleton correlate with the intake in the diet given in Table X.

TABLE XI

Ra^{226} in Human Bone as Reported after 1962
(from UNSCEAR Report [50])

Skeleton of 7000 g fresh weight yielding 2800 g ash was assumed.
In Illinois, normal areas are those where people are consuming water with "normal" levels of Ra^{226}; high level areas with elevated Ra^{228} concentration.

Location of area	pCi/g ash	pCi/g Ca	Total in the skeleton (pCi)
	Normal Areas		
Central America			
United States			
Puerto Rico	0.006	0.017	17
Europe			
Federal Republic of Germany	0.013	0.040	36
United Kingdom	0.008–0.02		
North America			
United States			
Illinois	0.012		32
New England	0.014		39
New York, N.Y.	0.012	0.032	32
Rochester, N.Y.	0.010; 0.017		28, 48
San Francisco, Calif.	0.0096	0.026	27
	High Level Areas		
Asia			
India			
State of Kerala	0.096		~270
(monazite area)	(0.03–0.14)		
North America			
United States			
Illinois	0.037		~100
Illinois	0.028		78

4.3. *Cosmic rays.* The principal variation in the dose rate from cosmic radiation is with altitude. Table XII shows that the dose rate roughly doubles with an increase in altitude of 5000 feet.

Cosmic radiation contributes only about a third of the total external natural radiation levels and so such a change is not large. Furthermore the relatively small population that lives about 10,000 feet in the United States militates against carrying out a useful epidemiological study. Nevertheless it has been

TABLE XII

COSMIC RAY INTENSITIES AT VARIOUS ALTITUDES
(From S. A. Lough [52])

Altitude, in feet	Cosmic ray intensity (μR/hr)
Sea level	4.0
1,000	4.7
2,000	5.4
3,000	6.2
4,000	7.1
5,000	8.1
6,000	9.1
8,000	11.7
10,000	14.6
12,000	18.0
14,000	21.0

TABLE XIII

DETAILS OF SOME HIGH ALTITUDE AREAS

Populations and altitudes from the Columbia Lippincott Gazeteer of the World (1952).

Area	Population	Demographic information available	Natural radiation received (multiply by 0.63 to get gonad dose)	Possible control populations
La Paz, Bolivia (altitude about 11,909 ft 3630 m); latitude 16° S	approx. 319,600	some statistics available but not comprehensive	approx. 3-fold increase in cosmic rays near equator at 3000–4000 m above sea level cosmic radiation tends to be about a third of total external natural radiation	this might present difficulties as lower oxygen tension at high altitude is a complicating factor

Other high towns in South America—
Quito, Ecuador—altitude 9350 feet (2850 m) lat. 0°; pop. 212,873
Bogota, Colombia—altitude 8660 feet (2640 m) lat. 4° N; pop. 325,658
Cerro de Pasco, Peru—altitude 13,973 feet (4259 m) lat. 10° S; pop. 19,187

Himalayan area: altitude 12,087 feet (3684 m); latitude 30° N; population (Lhasa) about 20,000.

suggested that such studies might be made of populations who live at high altitudes, for example, in La Paz in Bolivia. Table XIII gives details of high cosmic ray intensity areas.

4.4. *Summary.* Table XIV [53] summarizes the exposures to man due to natural background radiation.

TABLE XIV

VARIOUS ESTIMATES OF EXPOSURE OF MAN TO NATURAL BACKGROUND RADIATION (MRAD/YEAR)
(From Morgan and Turner [53])

The values for normal regions were reduced for both sexes by a shielding factor of 0.63.
The totals given in parentheses are in units of mrem/year using an RBE of 10 for alphas.

Type of exposure		Mean dose to gonads (mrad/year)	Mean dose to bone (mrad/year)	Mean dose to lungs (mrad/year)
Internal, from radionuclides	K^{40}	19, 20, 18, 18, 22	10, 11, 7	15
	Ra^{226}		3, 3.8, 6.7, 3	0.5
	Ra^{228}		3	
	Pb^{210}		2	
	C^{14}	1, 0.7, 1.5, 1.3, 1.8	1.6, 1.3	
Internal from $Rn^{220,222}$	Rn^{220} in wooden house			26 ave, 45 max, 2.3 min
	Rn^{220} in brick house			46 ave, 210 max, 2.7 min
	Rn^{220} in concrete house			94 ave, 290 max, 3.9 min
	Rn^{222} in wooden house			19 ave, 39 max, 1.3 min
	Rn^{222} in brick house			58 ave, 120 max, 5.7 min
	Rn^{222} in concrete house			64 ave, 140 max, 4.0 min
External, from U^{238} and Th^{232} series and from K^{40}	normal regions	47, (range 28–82), 39	approximately the same as for gonads	approximately the same as for gonads
	granite regions in France	162		
	monazite regions in India	802		
	monazite regions in Brazil	287		
External, from cosmic radiation	sea level, 0° geo. lat.	23, 24, 35, 28, 30, 33	approximately the same as for gonads	approximately the same as for gonads
	sea level, >50° geo. lat.	26, 27, 41, 37		
	10,000 ft, 0° geo. lat.	56, 50, 89, 80		
	10,000 ft, >50° geo. lat.	84, 66, 128, 120		
Total for normal regions near sea level		100 ave, 150 max, 70 min	100 ave, 150 max, 70 min (180), (250), (140)	200 ave, 570 max, 70 min (1100), (4500), (110)

5. Man-made radiation

There are various sources of man-made radiation which contribute to population exposure. Nuclear reactors are relatively unimportant in terms of the radiation exposure they deliver to the population. This has been estimated by a number of authors to be less than one mrem/year average and no more than a few millirem per year to any individual. As reported at this Symposium, epidemiological studies of populations living near nuclear reactors have shown no evidence of changes in infant mortality due to radiation exposure (the index of health suggested by some as the most sensitive indicator of radiation-induced disease [54].

At the present time there is a dramatic increase in the number of nuclear power plants planned or under construction in the United States, as can be seen by inspecting Figure 2. However, even with this large increase in the number of reactors it seems unlikely that the populations in their immediate vicinity will be suitable for epidemiological studies of radiation-induced disease because of the low exposures involved.

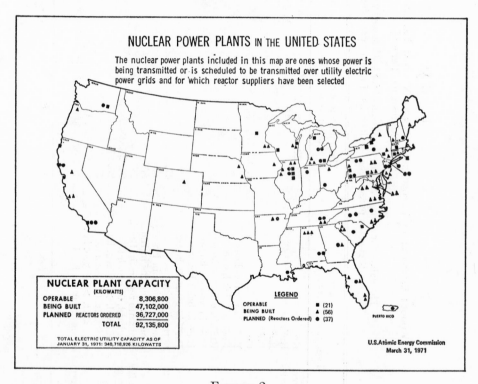

FIGURE 2

Nuclear power plants in the United States.
(From Radiological Health Data and Reports, May 1971)

Fallout from nuclear weapons testing has, in the past, contributed significantly to population exposure. At present, it does not. Table XV gives the dose
commitments from nuclear explosions taking place between 1954 and 1965.

TABLE XV

DOSE COMMITMENTS FROM NUCLEAR EXPLOSIONS
(From UNSCEAR Report [50])

As in the 1964 report, only the doses accumulated up to year 2000 are given for C^{14}; at that
time, the doses from the other nuclides will have essentially been delivered in full. The *total*
dose commitment to the gonads due to C^{14} from tests up to the end of 1965 is about 180
mrads. Totals have been rounded off to two significant figures.

Tissue	Source of radiation	Dose commitments (mrad) for period of testing 1954–1965
Gonads	external, short lived	23
	Cs^{137}	25
	internal, Cs^{137}	15
	C^{14}	13
Total		76
Cells lining bone surfaces	external, short lived	23
	Cs^{137}	25
	internal, Sr^{90}	156
	Cs^{137}	15
	C^{14}	20
	Sr^{89}	0.3
Total		240
Bone marrow	external, short lived	23
	Cs^{137}	25
	internal, Sr^{90}	78
	Cs^{137}	15
	C^{14}	13
	Sr^{89}	0.15
Total		150

5.1. *Radiation exposures resulting from the medical uses of ionizing radiation.*
Several authors, most recently the ICRP [55], have drawn attention to the
increasing medical uses of radiation. The Adrian committee report identified
medical radiology as the dominant component of man-made radiation in the
United Kingdom. Table XVI summarizes typical estimates of the average
genetic dose due to medical radiology in the late 1950's. Morgan [56] estimates
that medical X-ray diagnosis accounts for over 90 per cent of all radiation
exposure from man-made sources. In 1963 the U.S. Public Health Service
reported the genetically significant dose from diagnostic radiology within the
United States was 55 mrem/year. Morgan [56] has estimated that this has
probably increased to 95 mrem/year on the basis of a recent USPHS survey.

TABLE XVI

Average Genetic Dose to Each Member of a Population
from Diagnostic and Therapeutic Use of Ionizing Radiation
(After K. Z. Morgan [53])

Country	Diagnostic (mrem/year)	Therapeutic (mrem/year)	Radioisotopes (mrem/year)
United States	84	12	8
United States	137 ± 100	17	0.25–7
Australia	159	28	—
Hamburg, Germany	17.7	2.2	0.19
France	58.2	5.6	—
Leiden, Netherlands	6.8	4.1–13.1	—
United Kingdom	14.1	5	0.18
Denmark	27.5	1–1.5	—

It is possible to identify single procedures that contribute substantially to these exposures. Thus, for example, Penfil and Brown [57] estimate that nearly half of the genetically significant dose for U.S. males aged 15 to 29 years is due to X-ray examinations of the lower spine (see Figure 3)

"Probably the most important criterion of the somatic damage incurred by a given population is the mean annual bone marrow dose per capita. Surveys have indicated that its magnitude is similar to the per-capita genetically significant dose." This may be seen in Table XVII, which summarizes estimates of the gonadal and bone marrow doses published recently by ICRP.

Great attention has been given to the suggestion first made by Stewart in 1956 [58] that prenatal exposure significantly increases the risk of cancer induction. MacMahon's [59] studies have supported the conclusion of Stewart and co-workers. His data suggested an increase in cancer mortality by 40 per cent among children who were irradiated *in utero*. Gibson and co-workers, [60] however, found no association between *in utero* irradiation *alone* and an increased risk of leukemia. This multivariant study of 13,000,000 children revealed an association between irradiation and an increased risk of leukemia only when other factors were involved.

Most recently Stewart and Kneale [61] have suggested that the leukemia incidence among such children is linearly related to the number of abdominal X-rays taken during pregnancy of the mother.

These studies have led some workers to suggest that infants and the developing embryo are some 100 to 1000 times more sensitive to radiation than the mature adult. [5], [62] Gofman and co-workers [5] in a recent study suggest that *in utero* irradiation will result in a 50 per cent increase in cancer mortality rate per rad.

It is surprising to us (perhaps because we are not statisticians) that there can be such disagreement as to the implications of these studies. It would be of great benefit to have an authoritative study of the mortality rates due to

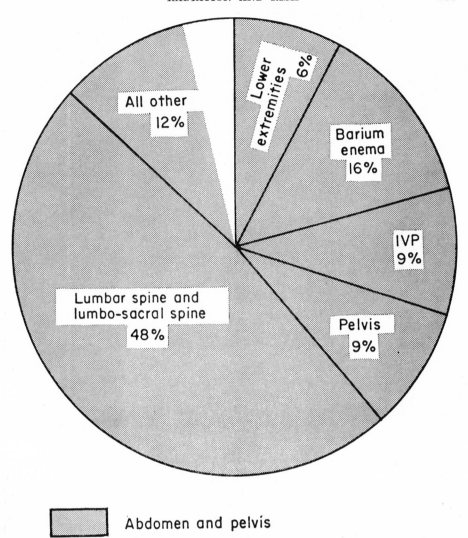

All other
12%

Lower extremities 6%

Barium
enema
16%

IVP
9%

Pelvis
9%

Lumbar spine and
lumbo-sacral spine
48%

Abdomen and pelvis

FIGURE 3

Estimated per cent distribution of genetically significant dose by type of medical roentgenological examination for males aged 15 to 29 years, United States, 1964, indicating that the major contributing examinations are those involving the abdomen and pelvis. (From ICRP Publication 16 [55])

leukemia and cancers in young people over the past 50 years in the United States. If this were coupled with careful measurements of the medical radiation exposure to the individuals in the group studied it should be possible to make some definitive statements. If the risk of cancer induction is indeed as high as suggested by Gofman, Sternglass, and others we can expect to detect substantial

TABLE XVII

Gonad Dose Grouping, Radiological Examination of Adults

The gonad dose values given are composite figures from many measurements in many countries and are to be taken only as an indication of the order of magnitude of dose in the three groups. The mean bone marrow doses included in this table for convenience are similarly composite figures. The mean bone marrow dose is the average dose to the active bone marrow. (From ICRP Publication 16 [55].)

	Gonad dose (mrad) Male	Gonad dose (mrad) Female	Approximate percentage contribution to genetically significant dose	Mean bone marrow dose (mrad)	Approximate percentage contribution to per capita mean bone marrow dose
A. Low gonad dose group					
Head (including cervical spine)	less than 10	less than 10	less than 1	50	3
Dental (full mouth)	"	"	"	20	6
Arm (including forearm and hand)	"	"	"	<10	—
Bony thorax (ribs, sternum, clavical, shoulder)	"	"	"	100	—
Dorsal spine	"	"	"	200	—
Lower leg, foot	"	"	"	<10	—
Chest (heart, lung) including mass miniature radiography	"	"	4	40	35
B. Moderate gonad dose group					
Stomach and upper gastro-intestinal tract	30	150	4	300	15
Cholecystography, cholangiography	5	150	1	100	—
Femur, lower two-thirds	400	50	4	50	—
C. High gonad dose group					
Lumbar spine, lumbosacral	1000	400	18	200	7
Pelvis	700	250	9	100	2
Hip and femur (upper third)	1200	500	8	50	—
Urography	1200	700	12	500	10
Retrograde pyelography	1300	800	4	300	—
Urethrocystography	2000	1500	1	300	—
Lower gastro-intestinal tract	200	800	13	600	8
Abdomen	500	500	4	100	3
Obstetric abdomen	fetal	600	10	100	2
Pelvimetry	fetal	1000	7	800	4
Hysterosalpingography	fetal	1200	<1	2000	—

increase in cancer mortality rates due to medical radiation exposures from studies of fairly small population groups.

6. Conclusion

In reaching our conclusions we should perhaps first indicate our general views as concerned scientists and citizens. Matters concerning the future welfare of mankind are of course, of grave concern to all of us. The fact of man's pollution of his environment is not at debate; the impact of this pollution upon his health is not completely known. It seems to us that one of the first concerns of a symposium such as this should be to order its priorities. Given a limited amount of effort and talent that may be employed on identifying the significantly harmful components of pollution, it would indeed be tragic if this effort were ineptly directed toward trivialities.

We, of course, hope to learn these priorities from symposia such as this, but, while reserving judgment, expect to learn that the risks due to "radiation pollution" do not rate high on the list of urgent priorities.

Nevertheless there are many valuable contributions that independent statistical studies may make to our understanding of the risks of low radiation doses.

At the present time our estimates of radiation risk basically all derive from high dose, acute exposure data. There does not seem to be general satisfaction with the analyses of the data. It would seem to us extremely worthwhile if much of these data were re-examined by fresh minds drawn from all the disciplines necessary for an exhaustive study. Such an authoritative independent study clearly stating what the high dose data tell us about the dose-effect relationship would be invaluable in planning future studies of the induction of disease by low radiation doses.

It does not seem reasonable to expect that we can establish from epidemiological studies that the risk of cancer induction by radiation is less than 10^{-4} per rad per year, since such a study would require a population containing 10 million man rem years at risk. While fairly large differences in radiation exposure from natural sources occur around the world, such differences are at most a few hundred mrem/year within the United States.

Of all man-made sources, medical X-rays are by far the greatest contributor to population exposure and little is known about the individual exposure received by a member of the population. It seems imperative that any statistical study must take both population average exposure and individual exposures into account.

APPENDIX

Radiation concepts and units.

The units and terminology used to quantify exposure to ionizing radiations is a source of confusion to more than laymen. We therefore append some brief definitions of the terms used in this paper, appealing to the knowledgeable reader to forgive us for stating the obvious.

The first attempts to quantify radiation fields began with x and γ radiation. Although the energy absorbed by irradiated material is important in determining the biological response of living organisms, in practice these energies are typically too small to measure directly. Energy absorption in air, however, produces ionization and provides a convenient method of measurement. Therefore the concept of *exposure* was developed [63], [64], [65], which is a measure of the radiation based upon its ability to produce ionization. The special unit of exposure is the roentgen, one roentgen being that exposure that produces one electrostatic unit of charge of both positive and negative signs in one cubic centimeter of air at standard conditions of temperature and pressure.

It should be noted here that in this brief review of radiation units our discussion cannot be of great depth, our purpose being only to paint a broad canvas indicating points of special importance. The reader interested in more detail is referred to texts on radiation dosimetry, for example, that edited by Attix, Roesch, and Tochilin, [66], [67], [68], or the authoritative reports of ICRU.

Despite its great utility, dissatisfaction with the concept of exposure arose because of its exclusiveness—it is, for example, inappropriate for neutron irradiation—and the fact that exposure is not linearly related to energy absorption in tissue. Both disadvantages are due to the basic difference in atomic composition of air and tissue. This difference is most striking for neutrons, since the production of recoil protons is the main mechanism for energy transfer to tissue, but even for photons the different chemical compositions of various tissues—fat, muscle, bone—compared with air become important at low energies [69]. A concept more widely applicable to radiation protection was needed. Since energy absorption seemed to be related to biological response, it was natural to define *absorbed dose*.

Absorbed dose due to any ionizing radiation is the energy imparted to matter by ionizing particles per unit mass of irradiated material at the place of interest. The unit of absorbed dose is the "rad" and is equal to an energy absorption of 100 ergs/g.

Relative biological effectiveness is the ratio of the absorbed dose of reference radiation to the absorbed dose of a different radiation required to produce the same biological effect. An RBE may be specified for any kind of radiation or condition of exposure.

The RBE for radiation of type i is, then,

$$(\text{RBE})_i = D_x/D_i,$$

where D_x, D_i are absorbed doses of 200 keV X-rays and of radiation of type i to produce the same biological effect. Thus the biological effect of irradiation by n different types of radiation would be identical to that from $\sum_{i=1}^{n} (\text{RBE})_i D_i$ rads of 200 keV X-rays. This concept was first known by the term RBE dose, [64] later becoming modified to dose equivalent; [65] its unit is the rem (Roentgen Equivalent Man).

REFERENCES

[1] ICRP, *Recommendations of the International Commission on Radiological Protection*, ICRP Publication 9, Oxford, Pergamon Press, 1966.

[2] U.S. BUREAU OF CENSUS, *Statistical Abstract of the United States*, Washington, D.C., U.S. Dept. of Commerce, 1970, p. 58 (91st ed.).

[3] V. L. SAILOR, "Population exposure to radiation: natural and man made," *Proceedings of the Sixth Berkeley Symposium on Mathematical Statistics and Probability*, Berkeley and Los Angeles, University of California Press, Vol. 6, 1972, pp. 291–311.

[4] ICRP, *Radiosensitivity and Spatial Distribution of Dose*, ICRP Publication 14, Oxford, Pergamon Press, 1969.

[5] J. W. GOFMAN, J. D. GOFMAN, A. R. TAMPLIN, and E. KOVICH, "Radiation as an environmental hazard," paper presented to 1971 Symposium on Fundamental Cancer Research, University of Texas, Houston, Texas, March 3, 1971; *see also* J. W. GOFMAN and A. R. TAMPLIN, "Epidemiological considerations of radiation pollution," *Proceedings of the Sixth Berkeley Symposium on Mathematical Statistics and Probability*, Berkeley and Los Angeles, University of California Press, Vol. 6, 1972, pp. 235–277.

[6] M. EISENBUD, *Environmental Radioactivity*, New York, McGraw-Hill, 1963, p. 395.

[7] R. A. CONARD and A. HICKING, "Medical findings in Marshallese people exposed to fallout radiation: results from a ten-year study," *J. Amer. Med. Assoc.*, Vol. 192 (1965), p. 457.

[8] T. TOMONAGA, M. A. BRILL, I. ITOGA, and R. HEYSSEL, "Leukemia in Nagasaki atomic bomb survivors," Atomic Bomb Casualty Commission Report 11-59, Hiroshima and Nagasaki, 1959.

[9] R. HEYSSEL, A. B. BRILL, L. A. WOODBURY, E. T. NISHIMURA, T. GHOSE, T. HOSHINO, and M. YAMASAKI, "Leukemia in Hiroshima atomic bomb survivors," Atomic Bomb Casualty Commission Report 02-59, Hiroshima and Nagasaki, 1959.

[10] R. E. LANGE, W. C. MOLONEY, and T. YAMAWAKI, "Leukemia in atomic bomb survivors. I. General observations," *Blood*, Vol. 9 (1954), pp. 574–585.

[11] E. B. LEWIS, "Leukemia and ionizing radiation," *Science*, Vol. 125 (1957), pp. 965–972.

[12] J. V. NEEL and W. J. SCHULL, "The effect of exposure to the atomic bomb on pregnancy termination in Hiroshima and Nagasaki," National Academy of Science, National Research Council, Publication 461, Washington, D.C., 1956.

[13] J. A. AUXIER, J. S. CHEKA, F. F. HAYWOOD, T. D. JONES, and J. H. THORNGATE, "Free-field radiation-dose distributions from the Hiroshima and Nagasaki bombings," *Health Phys.*, Vol. 12 (1966), pp. 425–429.

[14] W. C. MOLONEY and M. A. KASTENBAUM, "Leukemogenic effects of ionizing radiation on atomic bomb survivors in Hiroshima city," *Science*, Vol. 121 (1955), pp. 308–309.

[15] R. C. MILTON and T. SHOHOJI, "Tentative 1965 radiation dose estimation for atomic bomb survivors," Atomic Bomb Casualty Commission Technical Report 1-68, 1968.

[16] T. HASHIZUME, T. MARUYAMA, A. SHIRAGI, E. TANAKA, M. IZAWA, S. KAWAMURA, and S. NAGAOKA, "Estimation of the air dose from the atomic bombs in Hiroshima and Nagasaki," *Health Phys.*, Vol. 13 (1967), pp. 149–161.

[17] ICRP, *The Evaluation of Risks from Radiation*, ICRP Publication 8, Oxford, Pergamon Press, 1966.

[18] A. C. UPTON, *Radiation Injury*, Chicago, University of Chicago Press, 1969.

[19] H. MAKI, T. ISHIMARU, H. KATE, and T. WAKABAYASHI, "Carcinogenesis in atomic bomb survivors," Atomic Bomb Casualty Commission Report 24-68, 1968.

[20] W. M. COURT-BROWN and R. DOLL, "Mortality from cancer and other causes after radiotherapy for ankylosing spondylitis," *Brit. Med. J.*, Vol. 2 (1965), pp. 1327–1332.

[21] ———, "Leukemia and aplastic anaemia in patients irradiated for ankylosing spondylitis," *Spec. Rept. Ser. Med. Res. Coun.*, No. 295, London, H. M. Stationery Office, 1957.

[22] A. M. BRUES, "Critique of the linear theory of carcinogenesis," *Science*, Vol. 128 (1958), pp. 693–699.

[23] A. H. CAMPBELL and C. B. MACDONALD, *Brit. J. Dis. Chest*, Vol. 59 (1965), p. 90.

[24] A. S. MARKLAND, "The occurrence of malignancy in radioactive persons" (which is a general view of data gathered in the study of the radium and dial painters, with special reference to the occurrence of osterogenic sarcoma and the interrelationship of certain blood diseases), *Am. J. Cancer*, Vol. 15 (1931), p. 2435.

[25] R. D. EVANS, "The effect of skeletally deposited alpha-ray emitters in man," *Brit. J. Radiol.*, Vol. 39 (1966), pp. 881–895.

[26] ———, "The radium standard for bone seekers—evaluation of the data on radium patients and dial painters," *Health Phys.*, Vol. 13 (1967), pp. 267–278.

[27] R. D. EVANS, A. T. KEANE, R. J. LOENKOW, W. R. NEAL, and M. M. SHANAHAN, "Radiogenic tumor in the radium and mesothorium cases studied at MIT," *Delayed Effects of Bone-Seeking Radionuclides* (edited by C. W. Mays, W. S. S. Jee, R. D. Lloyd, B. J. Stover, J. H. Dougherty, and G. N. Taylor), Salt Lake City, University of Utah Press, 1969.

[28] R. D. EVANS, "Radium and mesothorium poisoning and dosimetry and instrumentation techniques in applied radioactivity," in Annual Progress Report, Physics Department, Massachusetts Institute of Technology, MIT 952-6 (1969), pp. 1–383.

[29] A. J. FINKEL, C. E. MILLER, and R. J. HASTERLIK, "Long term effects of radium deposition in man," in Argonne National Laboratory Health Division, Gamma-Ray Spectroscopy Group Semiannual Report, Report ANL-6839 (1964), pp. 7–11.

[30] M. P. FINKEL, P. B. JINKING, and B. O. BISKIS, "Parameters of radiation dosage that influence production of osteogenic sarcomas in mice," *Nat. Cancer Inst. Monogr.*, No. 14, pp. 243–263.

[31] R. J. HASTERLIK, A. J. FINKEL, and C. E. MILLER, "The cancer hazards of industrial and accidental exposure to radioactive isotopes," *Ann. N.Y. Acad. Sci.*, Vol. 114 (1964), pp. 832–837.

[32] R. J. HASTERLIK and A. J. FINKEL, "Diseases of bones and joints associated with intoxication by radioactive substances principally radium," *Med. Clin. N. Amer.*, Vol. 49 (1965), pp. 285–296.

[33] F. W. SPIERS and P. R. J. BURCH, "Measurements of body radioactivity in a radium worker," *Brit. J. Radiol.*, Suppl. 7 (1957), pp. 81–89.

[34] H. SPIESS, "Über Anwendung and Wirkung des Peteosthor bei pulmonaler und extrapulmonaler Tuberkulose in Kindersalter, augleich eine allgemeine Stellungnahme zur Thorium X- und Peteosthor-Therapie," *Z. Kinderheilk*, Vol. 70 (1951), pp. 213–252.

[35] ———, "Schwere Strahlenschaden nach der Peteosthorbehandlung von Kindern," *Deut. med. Wochschr.*, Vol. 81 (1956), pp. 1053–1054.

[36] ———, "^{224}Ra-induced tumor in children and adults," *Delayed Effects of Bone-Seeking Radionuclides* (edited by C. W. Mays, W. S. S. Jee, R. D. Lloyd, B. J. Stover, J. H. Dougherty, and G. N. Taylor), Salt Lake City, University of Utah Press, 1969, pp. 227–247.

[37] A. J. FINKEL, C. E. MILLER, and R. J. HASTERLIK, "Radium-induced malignant tumors in man," *ibid.*, pp. 195–225.

[38] S. G. GOSS, "The malignant tumour risk from radium body burdens," *Health Phys.*, Vol. 19 (1970), p. 731.

[39] K. Z. MORGAN, "Human experience with man-made sources of ionizing radiation," *Principles of Radiation Protection* (edited by K. Z. Morgan and J. E. Turner), New York, Wiley, 1967, Section 1.3.

[40] H. SIKL, *Acta Union Intern. Contra Cancrum*, Vol. 6, 1950.

[41] *Radiation exposure of uranium miners, hearings before the Subcommittee on Research, Development, and Radiation of the Joint Committee on Atomic Energy, Congress of the United*

States, 90th Congress, first session on Radiation Exposure of Uranium Miners, Part 2,
Washington, D.C., U.S. Government Printing Office, 1967.

[42] *Radiation standards for uranium mining, hearings before the Subcommittee on Research, Development, and Radiation of the Joint Committee on Atomic Energy, Congress of the United States, 91st Congress, first session on Radiation Standards for Uranium Mining, March 17 and 18, 1969,* Washington, D.C., U.S. Government Printing Office, 1969.

[43] J. K. Wagoner, V. E. Archer, F. E. Lundin, D. A. Holaday, and J. W. Lloyd, "Radiation as the cause of lung cancer among uranium miners," *New England J. Med.,* Vol. 273 (1965), pp. 181–188.

[44] C. B. Braestrup, "Past and present radiation exposure to radiologists from the point of view of life expectancy," *Amer. J. Roentgenol., Rad. Therapy, Nucl. Med.,* Vol. 78 (1957), p. 988.

[45] H. J. Dunster, "Safety in numbers or the conservative model," *Health Phys.,* Vol. 16 (1969), p. 248.

[46] ICRP, *Radiosensitivity and Spatial Distribution of Dose,* ICRP Publication 14, Oxford, Pergamon Press, 1969.

[47] W. D. Claus (editor), *Radiation Biology and Medicine,* Reading, Addison-Wesley, 1958.

[48] M. Eisenbud, *Environmental Radioactivity,* New York, McGraw-Hill, 1963.

[49] J. A. S. Adams and W. M. Lowder, *The Natural Radiation Environment,* Chicago, University of Chicago Press, 1964.

[50] United Nations, *Report of the United Nations Scientific Committee on the Effects of Atomic Radiation,* Official Research, 21st Session, Suppl. No. 14 (A/6314), New York, United Nations, 1966.

[51] Z. Jaworowski, J. Bilkiewicz, and E. Zylicz, "^{226}Ra in contemporary and fossil snow," *Health Phys.,* Vol. 20 (1971), pp. 449 and 450.

[52] S. A. Lough, "The Natural Radiation Environment," *Radiation Biology and Medicine* (edited by W. D. Claus), Reading, Addison-Wesley, 1958, Chapter 17.

[53] K. Z. Morgan and J. E. Turner, *Principles of Radiation Protection,* New York, Wiley, 1967.

[54] E. Tompkins, P. Hamilton, and D. Hoffman, "Infant mortality around three nuclear power plants," *Proceedings of the Sixth Berkeley Symposium on Mathematical Statistics and Probability,* Berkeley and Los Angeles, University of California Press, Vol. 6, 1972, pp. 279–290.

[55] ICRP, *Protection of the Patient,* ICRP Publication 16, Oxford, Pergamon Press, 1967.

[56] K. Z. Morgan, "Comments on radiation hazards and risks," Paper KA-3, presented to the American Physical Society, Washington, D.C., 30 April 1971.

[57] R. L. Penfil and M. L. Brown, *Radiology,* Vol. 90 (1968), p. 209.

[58] A. J. Stewart, J. Webb, D. Giles, and D. Hewitt, "Malignant disease in childhood and diagnostic irradiation *in utero*—preliminary communication," *Lancet,* Vol. 271 (1956), p. 447.

[59] MacMahon, "Epidemiologic aspects of cancer," *Ca—a Cancer Journal for Clinicians,* Vol. 19 (1969), pp. 27–35.

[60] R. W. Gibson, I. D. J. Bross, S. Graham, A. M. Lillienfeld, L. M. Schuman, M. L. Levin, and J. E. Dowd, "Leukemia in children exposed to multiple risk factors," *New England J. Med.,* Vol. 279 (1968), pp. 906–909.

[61] A. Stewart and G. W. Kneale, *Lancet,* Vol. 1 (1970), p. 1185.

[62] E. J. Sternglass, "Infant mortality changes near a nuclear fuel reprocessing plant," Unpublished report, November 30, 1970; *see also* E. J. Sternglass, "Environmental radiation and human health," *Proceedings of the Sixth Berkeley Symposium on Mathematical Statistics and Probability,* Berkeley and Los Angeles, University of California Press, Vol. 6, 1972, pp. 145–221.

Appendix

[63] RECOMMENDATIONS OF THE INTERNATIONAL COMMISSION ON RADIOLOGICAL UNITS, Chicago, 1937, *Am. J. Roentgenol., Radium Therapy, and Nucl. Med.*, Vol. 39 (1938), p. 295.

[64] REPORT OF THE ICRU, 1956, *National Bureau of Standards Handbook 62, Washington, D.C.*, United States Department of Commerce, 1957.

[65] REPORT 10A OF ICRU, "Radiation quantities and units," *National Bureau of Standards Handbook 84, Washington, D.C.*, United States Department of Commerce, 1962.

[66] F. H. ATTIX and W. C. ROESCH (editors), "Instrumentation," *Radiation Dosimetry*, New York, Academic Press, Vol. 2, 1966 (2nd ed.).

[67] ———, "Fundamentals," *Radiation Dosimetry*, New York, Academic Press, Vol. 3, 1968 (2nd ed.).

[68] F. H. ATTIX and E. TOCHLIN (editors), "Sources, fields, measurements, and applications," *Radiation Dosimetry*, New York, Academic Press, Vol. 3, 1969 (2nd ed.).

[69] H. E. JOHNS and J. S. LAUGHLIN, "Interactions of radiation with matter," *Radiation Dosimetry* (edited by G. J. Hine and G. L. Brownell), New York, Academic Press, 1956, Chapter 2.

Discussion

Question: Harold L. Rosenthal, School of Dentistry, Washington University

I am somewhat confused by the meaning of the term "genetic dose" and I would appreciate your definition of the term.

Reply: H. W. Patterson and R. H. Thomas

The genetically significant dose was defined in the UNSCEAR 1958 report (Chapter 2 paper) as: ". . . the dose which, if received by every member of the population, would be expected to produce the same total genetic injury to the population as do the actual doses received by the various individuals."

"This definition was based upon the following assumptions and considerations.

(a) The relevant tissue dose is the accumulated dose to the gonads.

(b) The dose effect relation is linear, without a threshold.

(c) The individual gonad dose is weighted with a factor which takes into account the future number of children expected of the irradiated individual compared with an average member of the population (in this connection the fetus is treated as such an irradiated individual and not as a child to be expected)." (Quoted from the UNSCEAR 1962 Report.)

Reply: Alexander Grendon, Donner Laboratory, University of California, Berkeley

The genetically significant dose is not the same as the gonadal dose, which is what you have described. The genetically significant dose is calculated from gonadal dose by weighting for the probability of reproduction, taking into account the ages of those exposed.

Question: R. J. Hickey, Institute for Environmental Studies, University of Pennsylvania, Philadelphia

Based on some material we have heard during this symposium concerning the alleged highly damaging effects of ionizing radiation in the range of "normal" background radiation, could you explain why the geological areas, for example,

Kerala, with very high background radiation are not essentially "denuded" of mammalian life? It is my understanding that man and other mammals live in these areas, and have lived there, in some instances at least, for generations. Would you please comment on this seemingly anomalous situation, or is it anomalous?

Reply: R. H. Thomas

I would only like to say that these facts don't seem to me to be anomalous. As we say in our paper, we do not yet know whether radiation exposures at or about those found in nature are deleterious, of no consequence or even beneficial to man. It is interesting to speculate that man has evolved to operate best at levels of radiation within the range of those found in nature—this seems to be the case with naturally occurring physical and chemical "insults"—perhaps it is true for radiation.

Reply: E. J. Sternglass, School of Medicine, University of Pittsburgh

In connection with the variations in natural background radiation levels in the environment, it is important to note that a number of studies have shown statistically significant effects on man correlated with variations from location to location in the radiation from both external and internal sources.

The most recent of these studies was just reported at the Health Physics Society Meeting (July 11–15) in New York City by M. A. Barcinski and co-workers at the Institute de Biostatistica da U.F.R.J., Rio de Janeiro, Brazil, who found significant differences in chromosome defects in the lymphocytes of individuals living in areas of high thorium content in the soil, and control groups who did not live in the high background areas. At typical exposures of 340 milliroentgens per year, as an example, deletions were found in 90 per cent of the exposed population and only in 19 per cent of the control group. The evidence also favored internal exposures from food grown in the area as a major source of internal exposure.

Another study, carried out to detect possible health effects of naturally occurring radium in drinking water in Illinois sponsored by the P.H.S.'s Bureau of Radiological Health and published in *Public Health Reports*, Vol. 81 (1966), p. 805 (by Peterson, Samuels, Lucas, and Abrahams) indicated a greater incidence of bone tumors in the general population and a higher mortality rate from all causes for children one to nine years old for the exposed population compared with a control population living in areas of low radium concentrations.

Still another study, carried out in upstate New York, showed a correlation between variations in rock content of radioactivity and the incidence of congenital malformations.

As to the comparison of medical diagnostic X-ray doses with those from fallout and nuclear plant emissions, the following points should be recognized.

(1) As far as effects on the most sensitive members of the population are concerned, namely the early embryo, fetus, and infant, the average exposure from diagnostic X-rays is much lower than from measured whole-body doses

near such plants as Humboldt and Dresden, which ranged from 15 to 50 milliroentgens per year from external sources alone. Only the scattered radiation reaches the gonads or the fetus in the case of chest X-rays to the adult, or typically only 1 to 3 milliroentgens per picture, and not the 50 to 100 milliroentgens generally cited as the X-ray exposure to the chest. (See the recent 1964 study of medical exposures published by M. L. Brown at the Bureau of Radiological Health, P.H.S.).

(2) Due to the dominance of internal doses from inhaled or ingested radioactive particles in the case of fission products, concentration effects can increase the dose to critical organs hundreds or thousands of times above the doses calculated for uniform exposure of the soft tissue from X-rays.

Reply: R. H. Thomas

I would like to comment on Professor Sternglass's statement that diagnostic radiology—for example, of the chest—rarely involved irradiation of the gonads. ICRP *Publication 16* reports the finding that about half the genetically significant dose due to males is in fact due to one diagnostic procedure: X-rays of the lower back. This average genetically significant dose is by far the largest man-made contribution to man's radiation exposure.

Question: Alfred C. Hexter, California Department of Public Health

Is there any evidence for a possible favorable effect of very low doses of radiation?

Reply: R. H. Thomas

Yes, there is some evidence but you have to be careful in how you define the term "favorable." One example of an effect that might be termed "favorable" is the observations of a prolongation of life in the experiments of Carlson, Upton, and others—but there are others in the audience better qualified than I to speak on this subject.

Reply: Alexander Grendon

Regarding the question asked about prolongation of life by low levels of radiation, there have been several studies involving chronic irradiation of mice in which the group irradiated at the lowest level did have a longer mean life span than the controls. There have been criticisms of these results directed at the conditions under which controls and irradiated animals were maintained, but I have proposed a hypothesis that might explain a real effect of this kind. In the evolutionary process of developing mechanisms that protect the health of an organism, the response to infectious disease is most significant. It may be that any insult to the system, including radiation at low levels, evokes this response and that it serves to protect the mice from death by infection. Since man has antibiotics, this kind of response does not mean much for him whereas possible increase in tumor incidence does. I don't believe that many in this field think that "a little radiation is good for you," even though some others have said so.

RADIATION AND INFANT MORTALITY—SOME HAZARDS OF METHODOLOGY

EMANUEL LANDAU
ENVIRONMENTAL PROTECTION AGENCY
WASHINGTON, D.C.

A decade ago, in a paper presented at the 1960 State of California Department of Public Health Air Pollution Research Conference, it was suggested by this speaker that an attempt be made to determine whether the distribution of mortality within a city was related in some meaningful way to the geographic distribution of the air pollution within the same geographic area. I said [9],

It may surprise you to have me suggest that one should start with the youngest age group. For many years infant mortality decreased rapidly; very recently this trend has been altered, and no obvious reasons have been found. Could air pollution be the culprit? Therefore, one highly recommended step is the determination of the geographic distribution of infant mortality. The study preferably should include only the postneonatal period; i.e., deaths during the first year of life but excluding the first month to minimize the effect of birth injuries and the like.

Soon after this simplistic view that an environmental factor, air pollution, was implicated in the leveling off in the decline of the infant mortality rate in the United States was presented, the possible hazards of another etiologic agent, radiation, specifically, radioactive fallout from nuclear weapons were cited by Sternglass. He mentioned the need at this time to study the incidence of childhood leukemia and cancer deaths among children born in areas which had received heavy fallout doses six to nine months earlier [15].

In the meantime, some attempt had been made to assess the hazards of extremely low levels of ionizing radiation in the United States, largely by means of the published vital statistics. Grahn and Kratchman in 1963 [8] summarized the studies as follows. "The results of the (bone) tumor and leukemia incidence studies have all been negative. The malformation studies have been suggestive of a radiation effect, though alternative explanations and hidden biases were not entirely accounted for."

In their article, they examined the relationship between the neonatal death rate for the eight-year period, 1950–57 and estimated natural or background radiation exposure in the United States for the white population, by county of residence. The total population of a county was used in place of the white population

345

when the nonwhite portion constituted a small fraction of the total population in the county. The environmental data included radiation estimates based on the distribution of uranium ore deposits obtained from the Atomic Energy Commission and altitude data based primarily on the U.S. Geologic Survey. For some of the analysis, 80 counties with uranium ore reserves, principally in the Mountain States, were compared with 180 counties not containing such reserves in the same states.

The mean county altitude values which were used were in reality mean population-altitude figures. State altitude figures were mean population-altitudes weighted by the live births at risk. To take account of differences in socioeconomic factors, as reflected by median family income, county values were adjusted to a common income. Differences in maternal age were "controlled" by adjusting individual county mortality ratios to a constant maternal age. This adjustment was based on a limited special study of live births conducted in the first quarter of 1950. Since Utah had significantly lower neonatal death rates than the other Mountain States, it was excluded from the subsequent analysis. We will come back to this matter of exclusions later.

A highly significant positive correlation ($r = 0.90$) was found between cosmic ray intensity and neonatal death rates for 11 states in the western portion of the United States, but now excluding Idaho, Montana, and Utah. These death rates were adjusted for age of the mother and family income level. Another highly significant negative correlation ($r = -0.91$) was found when atmospheric pressure was correlated with these death rates.

To ascertain if any relationship would exist between mortality and the presence of uranium ore reserves in the absence of the altitude variable, the authors adjusted the neonatal mortality rates for the 57 counties with ore reserves (excluding Utah) to a constant altitude and then plotted these adjusted death rates against tons of urnaium oxide per 1000 square miles. Since no correlation was found, the authors stated: *"Thus, it can be concluded that the quantity of uranium ore reserves is unassociated with the probability of neonatal mortality"* (authors' emphasis). Yet, they did add the following caution, "However, in the absence of direct measurement of radiation levels in the 57 counties, these results cannot be considered as having entirely eliminated an interpretation based on radiation-induced injury." With reference to altitude, they concluded that historical, experimental and clinical evidence supported the role of the reduced partial pressure of oxygen in causing a reduced fetal growth rate and an increased neonatal death rate.

In his article presented in 1963, published two years later, Grahn made reference to the fact that interest in the possible relationship between infant mortality and environmental radiation had increased due to the testing of nuclear weapons [7]. Also, the increased use of nuclear energy for peaceful purposes had contributed to public concern. He suggested, "Both genetic and somatic endpoints can be studied with equal pertinence to the problem, but genetic endpoints are preferred because they are not confounded by a lifetime's accumulation of environmental experiences." However, the study of infant mortality would not appear to be significantly affected by this consideration.

Later Sternglass expressed his belief that radiation due to nuclear testing had increased the incidence of leukemia in the Albany-Troy area of New York State. He stated that deposition of strontium 90 on the ground in the New York area was associated with excess fetal deaths in New York State. Also, fallout was correlated with excess mortality in the United States for children under one year of age [17]. According to his interpretation, the changing pattern of infant mortality resulted in a large excess of these deaths over that expected. The assumption made is that the rate of infant mortality would have continued to decline without interruption were it not for the effects of fallout due to weapons testing in the United States and overseas. He has noted that the rate of decline of infant mortality, beginning two or three years following the termination of atmospheric testing in 1963, was again approaching that from 1935 to 1950 [18].

Stewart has called our attention to the fact that fetal and infant mortality are the resultant of a number of forces in operation [19]. She said "The most likely explanation of the observed change in trend is that it is a reversion towards normality of a death rate which had, for 20 years, been experiencing booster effects—first from the introduction and dissemination of sulphonomides and then from the introduction and dissemination of antibiotics." She concluded: "In practice, infection deaths are so strongly correlated with sex, age, wealth, climate, density of population, chemotherapy, and so on, that any deviation from normality of related death rates and prevalence rates (for example infant mortality, leukemia mortality and leukemic clusters) can only be regarded as significant *after* these effects have been eliminated" (author's emphasis).

Most recently, Sternglass has postulated that infant mortality has increased around nuclear power plants [18]. I believe that the methodology, and its limitations, if any, will be discussed in greater detail by the next speaker, Mrs. Tompkins.

For those who do not have sufficient exposure to mortality data, it is appropriate to present some definitions of terms and some pitfalls. This will be followed by a discussion of demographic factors affecting infant mortality. Finally, some possible mechanisms of radiation effects will be outlined so that their relevance to possible epidemiologic studies can be examined.

Let us get some perspective on the deceleration in the rate of decline in the infant death rate by looking at selected infant mortality rates. Moriyama had first called our attention to the basic change in the infant mortality trend beginning about 1949 or 1950 [10].

In 1935, there were 55.7 deaths under one year of age per 1000 live births. By 1950, the rate was about half—29.2. By 1964, the rate had declined further but at a much slower pace. It was then 24.8 deaths under one per 1000 live births. Since 1933, the 48 continental states and the District of Columbia have been included in the registration system and so it seemed appropriate to select 1935 as the base year.

Chase has provided in summary fashion an evaluation of the basic data used in infant mortality analysis [2]. She has noted that a distinction between live birth and fetal death is one of the major decisions which has to be made for vital registration statistics. This is significant not only in international comparisons but also in terms of interstate comparisons. A live birth as defined by the World Health Organization (WHO) in 1950 is "the complete expulsion or extraction from its mother of a product of conception, irrespective of the duration of pregnancy, which, after such separation, breathes or shows any other evidence of life, such as beating of the heart, pulsation of the umbilical cord, or definite movement of voluntary muscles, whether or not the umbilical cord has been cut or the placenta is attached; each product of such a birth is considered live born" [22]. Fetal death, as defined by WHO at the same time to complement that of live birth, above, is "death prior to the complete expulsion or extraction from its mother of a product of conception, irrespective of the duration of pregnancy; the death is indicated by the fact that after such separation, the fetus does not breathe or show any other evidence of life, such as beating of the heart, pulsation of the umbilical cord, or definite movement of voluntary muscles" [22]. It should be emphasized that these definitions deliberately omitted any mention of duration of pregnancy or the terms abortion, miscarriage or stillbirth.

Registration of live births, fetal deaths and deaths is required throughout the United States. However, registration requirements for fetal deaths varies widely. Except for Kansas and New York City, registration is based on length of gestation. The minimum period of gestation at which fetal deaths are required to be registered varies. Most national tabulations of fetal deaths include fetal deaths of 20 weeks or more gestation and those with period of gestation unspecified. This is so because this gestation period is common to all States and presumably all States provide complete data for gestations of this period. Nonetheless, any deficiencies in registration data for a state will be reflected in the national data. Special studies have demonstrated there is gross underregistration of fetal deaths in the United States [4], [6], [13]. The difficulties that may be encountered because of varying registration practices of fetal deaths even in a single state are well illustrated by the remarks of Tamplin, Richer, and Longmate [20] on the use of fetal death data for New York State by Sternglass [20].

Tamplin's comments about the differences in reporting procedures for New York City and for the remainder of the state are especially pertinent. In 1939, reporting of early fetal deaths was promoted energetically in New York City and, as a result, an increase in fetal deaths occurred due to the better reporting that resulted.

Attention has already been called to the change in definition of live birth and fetal death which occurred in 1950. These nuances in definition may represent a problem of interpretation as the points in time at which changes may have happened to take place may have been affected by the differences in definition [2]. This is so despite the relatively small effect resulting from the variation in definition.

Also, neonatal deaths were earlier defined as those occurring during the first month of life. Now, they are restricted to those within the first 28 days of life. The difference due to this reason, however, is believed to be relatively slight inasmuch as the risk of death declines rapidly throughout the first month following birth. Postneonatal deaths are those occurring during the remainder of the first year, that is, the 28 days of age to the first anniversary. Death registration is probably more complete for this period than for the neonatal period.

Moriyama has cautioned us to be concerned about the statistical problems involved in trying to correlate changes in trends of infant mortality with presumed etiologic agents [11]. According to him three factors must be taken into account: (1) the mortality level at which the change takes place; (2) the differences in the rate of change; and (3) the time when the change takes place.

Let us now look at the role of demographic characteristics in infant mortality analysis. There are a number of demographic factors which are associated with differences in infant mortality rates in the United States as well as in other countries. Sometimes, however, there are statistical artifacts present. Thus, Norris cites the case where the Japanese in California were stated to have a very low infant death rate [12]. Yet, in actuality, their rate was almost as high as the white population. The explanation was in the coding process. For infants born of mixed marriages, race was coded differently on the death certificate from that on the birth certificate. Births were assigned to the race of the nonwhite parents, while deaths were assigned to the race of the child, which was recorded as white. Therefore, small area comparisons must be made with caution and with detailed consideration being given to possible sources of statistical bias.

What are some of the demographic factors which appear to be important? One is geographical variation and, in the United States, the highest rates for fetal, neonatal and postneonatal deaths are found in the southeastern part of the United States. There are differences between urban and rural rates also. With the advent of the greater availability of hospital and medical facilities, urban experience had tended to become more favorable than rural in the 1930's and 1940's. However, during the 1950's the character of most major cities of the country changed. There had been an inmigration to the central city of low income individuals accompanied by an outmigration to the suburbs of the middle class. Accordingly, it is not surprising to find that neonatal mortality in the largest cities in the United States is higher than among infants living elsewhere in the states containing these cities.

A major concomitant of infant mortality is color or race. For nonwhite infants, fetal death ratios covering 20 weeks or more gestation, per 1000 live births have remained about twice that of white infants since data have been available by color in 1945. Postneonatal death rates are about three times as high for the nonwhite infants while their neonatal death rates are about 1.5 times the mortality experience of white infants.

Maternal age is an important component of relative risk. Thus, fetal and neonatal and postneonatal deaths are lowest when the mother is between 20 to

29 at the time of delivery. Accordingly, the age distribution of women giving birth affects the infant mortality pattern significantly.

Parity or birth order is another significant variable. High birth orders are associated with higher neonatal death rates. But, probably the most important factor is that of birth weight. This, of course, reflects the physical development and maturation of the infant. Neonatal mortality for each sex-color group demonstrates a rate for infants under 2500 grams or less at birth which was at least ten-fold that of heavier infants. Limited data for fetal deaths show that for these as for infant deaths there is an increase in risk for these smaller infants. Any difference in weight distribution for population groups is thus of critical importance.

A replication of the matched birth-death records study in England and Wales which had been carried out initially in 1949-50 was conducted in 1964-65 to determine the changes which had taken place in the 15-year period. The new analysis by Spicer and Lipworth concentrated mainly on regional and social effects on infant mortality [14]. The effects of maternal age and parity mentioned previously were again confirmed in this study. These authors point out that all of the main variables affecting infant mortality are correlated with each other and some process of standardization is required. Moreover, there is interaction in that the effect of one variable is different in the presence of various categories of the others. Thus the relationship of infant mortality to age of mother is different in the lowest social class than in the highest.

One review of this publication, I believe, is particularly germane. In accounting for the differences in infant mortality rates by area, the reviewer said: "We cannot help suspecting that of all possible factors which influence the continuance of the unfavorable rates in the North and in Wales by comparison with other regions that of implementation of the Clean Air Act may still be the most obvious. The other main factor may be assumed to be housing and the greater concentration of Social Classes I and II in the regions with better rates, but here one is entering into the more speculative realm of receptivity to health education" [21].

Before I close this section, I should like to call attention to the findings of trends in late fetal deaths and neonatal mortality as related to birth weight in England and Wales from 1956-65. In this study, Ashford and co-workers found a decrease in the proportion of low birth weight between 1956 and 1965 was associated with a steady reduction of infant and perinatal mortality, in England and Wales during the decade [1]. (Perinatal mortality includes fetal deaths of 20 or more weeks gestation and neonatal deaths.)

They said "Since the changes which have occurred are of multiple causation and since the factors involved are not clearly understood, the use of standard methods of predicting future patterns, such as linear or polynomial interpolation, should be treated with caution. It is unfortunate that information about many of the factors which may be involved is not generally available in England and Wales. A system of monitoring temporal changes in the ethnic, economic, and sociological structure of the population would be an invaluable aid to the interpretation of epidemiological data, including studies of perinatal mortality."

Very recently, a study of perinatal mortality in New York City by Fischler and co-workers examined the use of linear models in the analysis of such perinatal data [5]. They found that additive models, as applied to both the mortality rates and their logit transforms gave only a very rough description of the data. First order interactions, and, in particular, the interactions between age and parity, were found to be important.

Moreover, the issues raised in the present paper are to be supplemented by problems in the environmental parameters which may be even worse. We are considering a new battery of variables such as environmental pollutants not considered in the past. New elements of information regarding pollution are being obtained and we are not certain these are meaningful for correlation with health· Which are the exposures which we should be measuring to see their effect on the first year of life? The problems of the environmental contamination from myriad sources and their measurements need further study.

At this point, I should like to introduce a plea for use of a methodology which avoids dealing with selected states which conform to a chosen hypothesis. As scientists, we should examine with equal interest other states which do not conform. We should not be content to report only those instances which mesh with our hypothesis, nor to overlook or to fail to report what happens in other instances. Thus, I am intrigued by the unusually low infant mortality experience of Utah. As noted previously, this was excluded from the analysis of Grahn and Kratchman [8]. A good study of Utah is needed. Since Chase has reported difficulty in linking infant death and birth records for that state, the registration practices might be looked at carefully [3]. Is it possible that there is a statistical artifact involving infant death rates?

The mechanism by which radiation is presumed to affect infant mortality will help determine the methodology which will be used. Which hypothesis is being tested will point to the use of which portion of the mortality curve should be studied, that is, fetal, neonatal or postneonatal. Thus, the Grahn and Kratchman study evaluated the data in terms of three hypotheses: radiation-induced mutations, radiation-induced injury to the fetus, and hypoxic-induced depression of fetal growth [8]. Only the first two will be considered for our purpose at hand. Neonatal death rates were used because the first hypothesis selected was that increased levels of natural environmental radiation would adversely alter neonatal death rates through genetic effects. It was assumed that an increased mutation rate resulting from radiation would express itself, in part, by an increase in early mortality. The authors estimated the proportion of genetic deaths to total neonatal deaths to be about 20 per cent.

The second hypothesis of radiation-induced injury to the fetus was discarded by the authors because "the present data indicate changes far in excess of any chemical and experimental radiation experience." They said there was no evidence for the extremely high level of radiosensitivity which would be required.

Sternglass hypothesized that a genetic effect was due to a low accumulated gonad dose resulting from fallout [16]. The incorporation of strontium 90 into the genetic material of the parents is identified by him as the mechanism of the change in rate of decline of fetal mortality and infant mortality rates. Thus, following exposure to fallout, postneonatal mortality should be affected earlier and neonatal mortality should be affected later. This delayed genetic reaction could be due to the need to accumulate significant radiation to affect the off-spring *in utero*.

I should like to conclude by stating that, as yet, there is no simple physical law to explain the rate of change in infant mortality rates. Medical, economic and social factors are clearly important and the possible role of environmental, that is, pollution, factors remains to be uncovered.

REFERENCES

[1] J. R. ASHFORD, J. G. FRYER, and F. S. W. BRIMBLECONE, "Secular trends in late foetal deaths, neonatal mortality, and birth weight in England and Wales, (1956–65)," *Brit. J. Prevent. Soc. Med.*, Vol. 23 (1969), pp. 154–162.

[2] H. C. CHASE, "International comparison of perinatal and infant mortality: the United States and six West European countries," *National Center for Health Statistics*, Ser. 3, No. 6, Washington, D.C., 1967.

[3] ———, "A study of infant mortality from linked records: Methods of study and registration aspects, United States," *National Center for Health Statistics*, Ser. 20, No. 7, Washington, D.C., 1970.

[4] C. L. ERHARDT, "Reporting of fetal deaths in New York City," *Pub. Health Rep.*, Vol. 67 (1952), pp. 1161–1167.

[5] B. FISCHLER, E. PERITZ, and J. WINGERD, "On linear models in the study of perinatal mortality," *Demography*, Vol. 8 (1971), pp. 401–410.

[6] F. E. FRENCH and J. M. BIERMAN, "Probabilities of fetal mortality," *Pub. Health Rep.*, Vol. 77 (1962), pp. 835–847.

[7] D. GRAHN, "Methodological problems in the use of standard vital statistical data in the study of neonatal mortality and birth weight," *Genetics and the Epidemiology of Chronic Diseases*, Public Health Service Publication No. 1163, U.S. Government Printing Office, Washington, D.C., 1965, pp. 321–335.

[8] D. GRAHN and J. KRATCHMAN, "Variation in neonatal death rate and birth weight in the United States and possible relations to environmental radiation, geology and altitude," *Amer. J. Hum. Genet.*, Vol. 15 (1963), pp. 329–352.

[9] E. LANDAU, "Methods suitable for utilizing some existing data in air pollution research," presented at a Symposium on Statistical Methods in Air Pollution Research, The Fourth Air Pollution Medical Research Conference, California State Department of Public Health, December 7–9, 1958, pp. 32–43.

[10] I. M. MORIYAMA, "Recent change in infant mortality trend," *Pub. Health Rep.*, Vol. 75 (1960), pp. 391–405.

[11] I. M. MORIYAMA, Personal communication, 1969.

[12] F. NORRIS, "A closer look at race differentials in California's infant mortality, 1965–67," *HSMHA Health Rep.*, in press.

[13] E. R. SCHLESINGER, R. K. BEECROFT, H. F. SILVERMAN, and N. C. ALLAWAY, "Fetal and early neonatal deaths in Onondago County, New York," *Pub. Health Rep.*, Vol. 74 (1959), pp. 1117–1122.

[14] C. C. Spicer and L. Lipworth, "Regional and social factors in infant mortality," General Register Office, Studies on Medical and Population Subjects, No. 19, Her Majesty's Stationery Office, London, 1966.

[15] E. J. Sternglass, "Cancer: Relation of prenatal radiation to development of the disease in childhood," *Science*, Vol. 140 (1963), pp. 1102–1104.

[16] ———, "Infant mortality and nuclear tests," *Bull. Atomic Sci.*, Vol. 25 (1969), pp. 18–20.

[17] ———, "Evidence for low-level radiation effects on the human embryo and fetus," in *Radiation Biology of the Fetal and Juvenile Mammal*, Proceedings of the 9th Annual Hanford Biology Symposium, AEC Symposium Series No. 17, 1969, pp. 693–717.

[18] ———, "Epidemiological study of health effects associated with radiation discharges from nuclear facilities," presented at the 16th Annual Health Physics meeting, July 11–15, 1971, New York.

[19] A. Stewart, "The pitfalls of extrapolation," *New Sci.*, Vol. 41 (1969), p. 181.

[20] A. R. Tamplin, Y. Ricker, and M. F. Longmate, "A criticism of the Sternglass article on fetal and infant mortality," Lawrence Radiation Laboratory Report No. 15506, University of California, Livermore, California, July 1969.

[21] "Regional and social factors in infant mortality," *The Medical Officer*, Vol. 116 (1966), pp. 53–54.

[22] World Health Organization, quoted in "International comparison of perinatal and infant mortality: the United States and six West European countries," *Op. cit.* [2].

Discussion

Question: E. J. Sternglass, School of Medicine, University of Pittsburgh

How is it possible to explain the sharp rise and decline of fetal mortality rates in St. Louis accompanied by a rise and decline of strontium 90 in the fetal jawbone measured by Dr. Harold Rosenthal as an artifact produced by a change in registration requirements?

Reply: E. Landau

Unfortunately, the validity of this apparent association ceases to exist when examined more closely. Although there are serious deficiencies in the presentation of data attempting to correlate strontium 90 levels in the fetal jawbone with fetal mortality, changes in registration practice are not at issue.

On pages 14–19, the 1969 publication of the Bureau of Radiological Health, DBE 69-2, entitled "Evaluation of a possible causal relationship between fallout deposition of strontium 90 and infant and fetal mortality trends" by Tompkins and Brown, discusses the limitations of the analysis by Dr. Sternglass. The authors have noted the use of incomplete data, an error in the estimate for excess fetal deaths for 1963 as well as the use of fetal mortality data for Missouri instead of for St. Louis City and County. They have revised his presentation using his estimating procedure to obtain St. Louis area fetal mortality data and have plotted these revised values against the strontium 90 content of fetal mandibular bone in the St. Louis area for 62 aborted fetuses reported by Dr. Rosenthal. The resulting graph, Figure 16, indicates clearly that the validity of the hypothesis of an association between fetal mortality and fallout as measured by the strontium 90 level of fetal mandibular bone has *not* been substantiated.

MONITORING HUMAN BIRTH DEFECTS: METHODS AND STRATEGIES

ERNEST B. HOOK

NEW YORK STATE DEPARTMENT OF HEALTH

and

ALBANY MEDICAL COLLEGE OF UNION UNIVERSITY

1. Introduction

Since the rationale and goals of a paper in multidisciplinary settings are often unclear to those outside the author's specialty, a statement of purpose may not be out of place. My interest is in the diminution of birth defects and detection of preventable environmental causes of such events.

This paper has been written assuming the reader has no previous knowledge of birth defects. Definition of medical terms not provided in the text are not crucial to the argument, but may be found in any medical dictionary. The record of the New York State Birth Defects Institute's Symposium of October, 1970 will present in much greater detail some of the themes discussed here [10].

Some of the material treated here was originally presented in more condensed form at a National Foundation Symposium in New York City on Environment and Birth Defects, January 27, 1971.

2. Definition and incidence of birth defects

A major human birth defect may be defined as an anatomical structural variant that produces a significant clinical or cosmetic effect. This definition is, of course, a somewhat loose one in that what may be abnormal in one setting may be acceptable in another. For the purposes of monitoring as discussed here, this vagueness will not be a problem. But it should be pointed out that reports of incidence of total birth defects in various groups cannot be compared unless the precise defects scored by the authors and their method of ascertainment are specified [11]. For the purpose of this discussion, the incidence of infants with major defect detectable at birth will be assumed to be about two per cent. The incidence of particular malformations is of course much rarer. Order of magnitude estimates for some of the most frequent major malformations are (in contemporary U.S.A.), anencephaly $1 - 2 \times 10^{-3}$, mongolism 10^{-3}, spina bifida $0.5 - 2.0 \times 10^{-3}$, polydactyly (whites) $0.4 - 0.7 \times 10^{-3}$, polydactyly (blacks) $5 - 10 \times 10^{-3}$. Some might call the latter a minor malformation. (The definition of a minor birth defect is given in Section 8.) But most types of major

355

malformations are probably very rare in liveborns having an incidence of no greater than $3 - 5 \times 10^{-5}$.

3. Defects by likelihood of ascertainment

For the purposes of ascertainment we can divide defects by their likelihood to be diagnosed at birth. (a) Externally dramatic defects include such malformations as anencephaly, cleft lip, and phocomelia. (b) Externally detectable malformations likely to be noted in superficial physical examination at birth but not likely to be starting on initial observation include such defects as syndactyly and imperforate anus. (c) Cryptic but diagnosable defects include among others, structural abnormalities of the genitourinary system (for example, polycystic kidney) and some congenital heart defects (for example, those producing cyanotic heart disease). (d) Cryptic defects not likely to be diagnosed or diagnosable in infancy may include occult tumors, Merkel's diverticulum, and so forth. Some in this category may only be found at autopsy.

4. Defects by causal mechanism

At least five classes may be distinguished, (a) defects genetically inherited such as simple Mendelian recessive or dominant traits, (b) defects associated with inherited chromosomal translocations (which may be included with some of class (a), (c) defects occurring as a result of a germinal mutation (in the previous generation) resulting in the appearance of either a simple Mendelian trait or a fresh chromosome abnormality, (d) defects produced by some agent to which the mother is exposed during gestation, and (e) defects produced by some unknown concatenation of environmental and genetic events, each of presumably relatively small effect but which interact to produce a defect (the multifactorial hypothesis). These categories are obviously not mutually exclusive. Among other things, the effect of any environmental agent may be modified by the genetic background and the effects of genetic factors may be influenced by environmental events during or before pregnancy. This has been well documented in nonhuman species and probably accounts for at least some of the fluctuation in expression of known single genetic or environmental causes of human birth defects. Thus even the boundaries between these cateogries of causes are not sharp.

From the viewpoint of the environment, an external factor may produce defects by a number of different mechanisms. It would be hazardous to suggest all possible routes by which structural development can go awry. The list below is of only very general categories but may still be incomplete. An agent may induce defects by (1) inducing a chromosomal translocation or chromosomal nondisjunction or a simple point mutation in germinal cells which then eventually results in a gamete carrying an abnormal genotype; (2) producing somatic mutations in cells of the developing fetus; (3) inducing structural disorganizational events during gestation, directly, or (4) affecting the mother's physiological metabolic pathways in such a manner that the effects are dysmorphogenetic in the fetus she is or will be carrying. Agents producing the first two types of events may be said to be *mutagens*, those producing the latter two types, *teratogens*. The effect of a known teratogen varies markedly with the time of exposure of the fetus. For most teratogens there is a critical period, usually relatively early in gestation, when the most dramatic effects are produced. In humans, this is almost always at some time in the first two or three months. The literature seems fairly consistent in suggesting that any agent which is mutagenic is also terato-genic. That is, any agent which will induce changes in the genetic constitution of cells will also be dysmorphogenetic if administered at the right time. This may be because accumulation of sufficient mutations during (early) development may eventually result in structural abnormality. The converse however, is not true. Thus, to my knowledge, rubella and thalidomide have not yet been shown to be mutagenic in humans. In any event it seems likely that the magnitude of terato-genic effect is much greater than any mutagenic effect they might possess. In the discussion below we will use the term teratogen to denote any environmental agent that may produce a birth defect. Since toxic effects of environmental agents upon the fetus are not necessarily limited to production of malformation (see Section 8), the term *embryotoxin* is used for an agent with any deleterious effect upon the conceptus [21].

From the viewpoint of the defect, any single malformation could be produced by any combination of mechanisms mentioned above. But given a *single* malforma-in a *particular* infant it is likely that no single causal event can be identified, so such occurrences are usually attributed to category (e), the multifactorial hy-pothesis. A very rough estimate would be that perhaps 90 to 95 per cent of all human defects fall into this latter category which is simply a reflection of our ignorance about the exact pathogenetic events resulting in malformation in such cases. Some may be caused by relatively few "strong" environmental causes. An abrupt increase in a particular defect could be simply a chance event due to concatenation of background events or result from introduction or increase of some environmental insult. (See Section 9 for further discussion of this point.)

5. Detecting causes—the rationale for monitoring

The best way to detect environmental causes of birth defects would be to find some experimental model in which all extraneous background events could be manipulated and controlled and test specific agents. With the possible exception of higher primates there are, however, no such models generally applicable to the human situation. Recently it has been suggested that with liberalization of abortion laws, women about to undergo an elective abortion might volunteer to take a teratogen experimentally. While many factors will be uncontrolled here and a large number of social and ethical questions are unresolved, there is the ad-ditional problem that having taken a teratogen, a woman might then change her mind about the abortion. Many obstetricians are thus reluctant to wait more than a brief period after exposure, which may not be long enough for an effect to manifest. Barring such approaches resort must be made to large scale studies comparing outcomes of pregnancies where the extent of maternal exposure to an agent in question is known. But this is difficult since the exposed and unexposed groups are unlikely ever to be completely identical with regard to other factors. Nevertheless, a good deal of information may be gained thereby, particularly if the effect of the suspected agent is relatively strong compared to the background events.

All the above approaches assume that we have already been able to identify an agent as suspicious on one grounds or another. Of course an optimal social strategy might require testing *all* compounds and agents to which we are exposed no matter how apparently innocuous. The scale of such an attempt, even given a suitable animal model, is such that at least in the initial approach we would have to limit ourselves to the most ubiquitous compounds and/or those which were already highly suspicious for other reasons. On the other hand by monitoring the incidence of birth defects we hope to have some way of quickly detecting the effect of still unknown teratogens that have, presumably, been recently introduced in the environment (or a recent increase in exposure to already present teratogens). This approach does not necessarily help us to identify the causes of the background defects produced by already existing factors. Nevertheless, since it may help prevent the rate of defects from increasing, it seems at least one reasonable strategy if it is not too demanding, can provide information of interest and possible utility, and is not unduly costly. But it must be remembered that such an attempt is close to being research without hypothesis and subject to all the perils of such an attempt.

6. Aspects of a useful monitoring scheme

These are fairly obvious but are worth specifying. (a) We should have confidence both in the accuracy of the diagnosis of markers used and in our ability to ascertain a high proportion. (b) We should ascertain our markers as close as possible in time to the presumed environmental causal event. (c) The more frequent the background rate of our marker the more likely we are to detect a significant increase in the rate. (d) The larger the population base the greater the absolute number of markers that can be detected and the better the opportunity to study rare defects. (e) The more intense scrutiny we can give to a population, the more markers that can be ascertained. (f) Given the opportunity to monitor rates of markers, we must also have the opportunity to correlate our observations with the environmental exposure of those in whom the marker is detected. (g) The cheaper the scheme the better.

7. Direct monitoring of major malformations

Complete ascertainment of all birth defects is probably impractical. The cryptic category of birth defects may actually be more frequent than those externally evident at birth. One study showed that about two per cent of all infants at birth had detectable significant malformations [13]. But by age one year perhaps two or three times this number will have had at least one defect diagnosed [14]. Thus it is clear that a monitoring scheme for malformations must essentially limit itself to defects that are likely to be readily detectable. But it should not be overlooked that we still may be missing a significant proportion of malformations in so doing.

One approach to monitoring defects is to use information provided on birth certificates. Almost all states in this country currently request those filling out the certificates (not always the physician delivering the baby) to note whether or not a malformation is present. Thus, there is at least a partial handle on surveillance in the data collected by many health departments.

There are however at least six drawbacks to the use of such certificates for monitoring. (a) As already noted only the externally detectable malformations are likely to be recorded by the individual filling out the birth certificates. (b) Of those diagnosable by gross inspection at birth, a large proportion are overlooked or underreported by the physician or whoever else fills out the birth certificate. As might be suspected the externally dramatic defects, for example, anencephaly and cleft lip which are the most frequent in this category, are relatively accurately reported. The externally detectable less dramatic defects such as imperforate anus or syndactyly are not recorded as accurately and thus the data on such are less useful [5]. Occasionally over compulsive physicians may report anatomical variants that are likely not to be classified as defects by other physicians [15]. A single such individual may create a pseudoepidemic that may be difficult to track down. Furthermore, the particular psychological set of those noting defects in any reporting system may change with time, particularly if recent publicity has called attention to the medical significance of defects that might otherwise have gone overlooked. It is suggested that such an effect occurred with a pseudoepidemic of congenital hip dislocation in England [7]. (c) Diagnostic reports are particularly inaccurate when multiple malformations are present in a baby. The more malformations that are present in an individual the less likely any particular malformation is to be recorded on the certificate. It is likely that only the most externally dramatic ones will appear. Occasionally "multiple malformations" may be noted without any further specification. (d) Even if the birth certificate is filled out as completely and accurately as possible, the coding of the defects that are reported upon such certificates is incomplete in many jurisdictions. Thus, in some instances if two or three defects are present only what may be regarded as the most serious will be coded. (This raises another point concerning clustering of major defects in individuals. An individual with a single specific major defect is more likely than an individual chosen randomly to have another major defect elsewhere.) Another problem with coding is that defects are often grouped together in the rubric of so called "other miscellaneous defects" particularly if they are rare and apparently of lesser significance. Thus a ten fold increase in a particular rare defect may be masked if hidden in a large miscellaneous category. (e) There is an inertia in any reporting system working from vital records that probably introduces a two month delay in pulling together all the relevant information. The interval between environmental insult and detection of a defect using birth certificates may thus be at least over a year. (f) As already noted the total incidence of external defects visible at birth is about two per cent but the incidence of particular defects is much less frequent. Teratogens may affect only one or just a few organ systems so one cannot lump together all defects but must separate malformations meticulously. Since the incidence of most particular malformations is quite rare and even the most frequent serious external defects have an incidence of no more than one to two per thousand, establishing a doubling or tripling in such rates will be very difficult without an enormous population base.

Nevertheless, despite the difficulties these approaches are already in use in some areas and can be progressively refined perhaps with relatively little time and money. For instance one can include questions on birth certificates relating to specific malformations, thus increasing the likelihood of ascertainment. Secondly, such records can be supplemented by using other sources as well. Large data files exist in many jurisdictions for purposes of justifying hospital insurance payments, characterizing medical care, and for other purposes. These may also provide useful information concerning the incidence of birth defects. Infants with malformations are much more likely, of course, to be hospitalized during the first year of life and to undergo extensive diagnostic evaluation and corrective surgery. A certain proportion will, unfortunately, miss these benefits and some may even die before receiving medical attention. For the latter, information from death certificates may still be of some value. By using such supplemental records one is considering information derived well after birth and thus is even further from the time of the presumed environmental event. The ascertainment with such a system however, is likely to be much better than using birth certificates alone. For instance, Banister in Ottawa using medical insurance records increased the ascertainment of all major defects by three fold up to 3.8 per cent, and similarly Gittelsohn using computerized hospital abstracts had a similar magnitude of effect, compared to that observed using birth certificates alone [1], [6].

One problem that presents itself in the use of such data, as well as birth certificates, is having identified an increase in malformation rate how does one then track down the presumed environmental cause of the increase? The computerized hospital abstracts used by Gittelsohn could be matched with other hospital records but they were still anonymous. At some point someone will have to contact the original family involved to inquire about exposure (unless the records themselves already have such complete data, which is unlikely). The records used in this context have been acquired for other reasons. Individuals who have access to them are thus extraordinarily sensitive to the question of confidentiality, with good reason, but just this problem alone may interfere with attempts to investigate etiology. At least the report of an increase of some defect in a jurisdiction, however, would lead to focusing of attention upon this particular malformation and would probably lead many individuals in particular regions to look for possible environmental events, without investigation of the central monitoring agency.

While birth certificate data is quite incomplete, newborn hospital records concerning the externally dramatic and externally detectable defects are usually fairly accurate in medically competent settings. Most babies stay in the hospital three or four days before discharge. By that time a number of significant defects are likely to be picked up that were not noted on the birth certificate. Simply by visiting nurseries and delivery rooms regularly in a circumscribed area in which there are a number of hospitals one can get fairly good counts of at least the major defects diagnosed neonatally. In fact, in Atlanta the CDC is currently doing just this [3]. But there is no requirement that an outside agency do the monitoring. Hospitals at teaching centers in large metropolitan areas could cooperate to report monthly the incidence of defects detected on their newborn services.

Information on the incidence of birth defects can also be derived from study of fetal wastage. The proportion of malformations in human spontaneous abortuses is at least twice as high as in human newborns [16], [17], [18]. Thus, the changes in the incidence of malformations here may be helpful in a monitoring scheme. This would require that in particular areas one would have to organize centers for the collection and the examination of fetal wastage. The difficulties with this approach to monitoring are discussed in the next section.

Since teratogens vary widely in the spectrum of effect, at least as far as major malformation is concerned, one cannot monitor for just a few major defects in the hope that the effects of all possible teratogens will be detected thereby. Mutagens, however, are unlikely to have such specific effects in that they are unlikely to have a strong predilection for a *particular* gene locus. (For the purposes of *monitoring* mutations, this statement is probably acceptable in that if an agent induces a mutation at one gene locus it will *probably* do so at many other loci. But differential base pair composition or base sequences in loci may affect DNA reactivity with base analogues or other mutagens, and chromosomal proteins and RNA in one region of a chromosome may also be more likely to bind some specific mutagen.) Since some particular inherited defects (for example, achondroplasia and Apert's syndrome) are likely to reflect recent dominant mutations, monitoring for these *particular* defects may provide an index of what is happening to the mutation rate at some other loci. But there are a number of problems with this approach. While a trained diagnostician could probably accurately confirm the diagnosis in suspected cases it would be perilous to rely simply upon outside reporting of these defects. Since the incidence of such defects may be only one in 50,000 births just a few inaccurate diagnoses could produce a pseudoepidemic of mutations. However, as Smith has suggested, all reported instances of such cases could be sought out by knowledgeable diagnosticians in a particular jurisdiction and at least confirmed [19].

It is occasionally stated that the presumed environmental events responsible for a rise in mutation rate are not likely to have a strong effect on the total number of birth defects, but rather upon the genetic load of the population, the effects of which may not be manifest for some generations. Assuming that mutagens also are teratogens however, this may not necessarily be the case.

8. Indirect monitoring of major malformations

If agents which induce major defects produce other effects as well, then monitoring for these secondary or indirect markers may also be helpful. There seems to be good evidence that in the rhesus monkey, a much better model for the human than the rodent, the main pattern of response to an embryotoxin is not the induction of malformation in a liveborn, but production of abortion [21]. This makes biological "sense", in that if a fetus has a defect that makes survival to the time of reproduction unlikely it is considerably more efficient for the species that this organism be lost at the very beginning of (fetal) life. It seems at least plausible to suppose that strong selective factors have produced this as a general category of response to many environmental fetal insults. (One can also imagine why such response might not be as strong in other species with many animals in a litter and greater relative fecundity.) Of course whether abortion or a defect is produced may heavily depend on the dose and the timing of the embryotoxin. If the same pattern occurs in man as well as monkey it is probable that the birth defects we observe reflect only the "upper tip of the iceberg" of human embryotoxicity. The first and the most ubiquitous effects of compounds which have teratogenic effect might thus be fetal loss. It is interesting that the drug amethopterin which is known to be teratogenic in humans, was once used specifically as an pharmacologic abortificant. The rare cases of amethoperin embryopathy in liveborns may represent only a small fraction of the exposed aborted fetuses.

The outstanding problem with monitoring the *rate* of fetal wastage however, is that the denominator from which the samples are drawn is simply unknown. (This is also a problem with monitoring the proportion of fetal wastage with detectable abnormality.) Without monitoring the entire female population of reproductive age one cannot tell how many are pregnant at any time and how many have lost a fetus very early in pregnancy without reporting it to a physician. The rate of fetal wastage seems likely to be as high as 25 per cent of all conceptions. But in a unit devoted to collecting fetal wastage, even in the most cooperative hospital the proportion of specimens was no more than six per cent of all live births [17], [18].

Another type of indirect marker that is at least worth considering is low birth weight. One must distinguish two categories: (a) that occurring because of a shortened gestation (SG), and (b) that occurring because of intrauterine growth retardation (IUGR). Of course IUGR and SG may occur together in which case the birth weight will be even lower than that appropriate for the shortened length of gestation. IUGR has been noted in association with many teratogens. Monitoring the incidence of IUGR requires that data on gestation length as well as birth weight be available. This is usually calculated from the first day of the last menstrual period (LMP) of the mother and the birth date. But often bleeding in the first month or so of pregnancy may be confused with a menstrual period so that the length of time may not be estimated correctly. Frequently the obstetrician makes an estimate of the LMP (or the expected date of confinement, EDC) based upon the size of the fetus, which is significantly different from date the mother recalls. But my suspicion is that the mother is more likely to be correct because an obstetrician is judging the size of the fetus against some normal standard, and if IUGR is present he may underestimate the length of gestation.

Data on the length of gestation in most hospital charts or birth certificates is useless since many medical personnel routinely write "term" or "40 weeks" for all infants not obviously pre- or postmature. (The assumption is that any discrepancy between the interval between LMP and the birth date and "40 weeks" is due to maternal error in memory.) Usually the LMP as stated by the mother is relatively accurately recorded however. In those jurisdictions where "LMP" and not "gestation length" is recorded on birth certificates, one has a handle on monitoring IUGR about as accurate as the monitoring of dramatic defects with vital records. I suspect that such data may be among the most useful that can be derived from vital records in monitoring for environmental hazards. It should be relatively easy to derive these data and furthermore one can measure the extent of IUGR as a continuous variable and thus derive additional power from such an analysis. (The actual variable analyzed would be the "percent of mean appropriate weight for gestational age.") One last point should be made. By timing the length of gestation from the LMP one is adding on two weeks, since ovulation does not occur until about two weeks later (usually). Thus the true length of human gestation is closer to 38 weeks, although most of the medical literature refers to it as 40 weeks.

Another indirect approach to seeking environmental agents that produce birth defects is to use anatomical variants which are of no significance in themselves but which are still correlated with the presence of major defects. Such variants may be called minor birth defects. Relatively trivial abnormalities in flexion creases of the palm, dermatoglyphic ridge patterns, and anatomical shape of the external ear and the eye are included. None of these have any significance per se but analysis has shown some to occur often in association with major malformation or with syndromes of malformations that are produced by the few human teratogens which have been discovered. Obtaining this type information however will require that infants receive systematic physical examinations by trained observers who use precise diagnostic criteria. The main advantage of using such minor defects is that they are very common compared to major malformations. The incidence of all minor malformations taken together varies from perhaps 15 to 30 per cent of the newborn population depending upon what variants are taken to be minor defects, the defects scored, and the population examined. And the incidence of one particular minor defect is over five per cent in at least one population.

It is still uncertain however, whether minor malformations are likely to be produced by significant teratogens in the absence of major malformations. Thus it is unknown whether an increase in minor malformations might be taken as evidence for a smaller increase in significant cryptic defects in a particular population. Indirect evidence suggests this may be likely but there is not yet definitive proof [8], [9], [13], [19].

9. Strategies in study of causes

It seems unlikely that any particular method discussed above could be recommended to the exclusion of any other. On the other hand, many methods may be used simultaneously in the same setting to provide additional information not necessarily gained from a single approach.

The use of data acquired in surveillance deserves comment. First of all it will provide information on the incidence of particular human major malformations. Data on the indirect markers are also of intrinsic interest. Even if there should be no change at all over a period of time this information is important and it is worth trying to get, if the attempt is not too expensive. Certainly in a negative sense it would be reassuring to know that rates were not increasing. But if rates were found to be increasing, one cannot specify in advance the exact approach to tracking down the cause.

It should be mentioned in this context that extensive epidemiological data are already available on a number of major malformations. Anencephaly for instance, has shown remarkable association with a variety of demographic factors in the population such as low socioeconomic status, elevated maternal age, and so forth [4]. There is a cline in the British Isles that is striking and may even be related to softness of water [20]. Use of old hospital records has revealed that an epidemic of this and related malformations occurred in New England earlier in this century and then subsided [12]. But despite all the data accumulated we still do not yet know what specific environmental factors are responsible for these variations in incidence. Although it seems likely that genetic factors contribute at least something to differences between ethnic groups, it seems unlikely they account for all of it.

Many studies of this and other defects have involved comparisons of "background" rates in different populations. In investigation of the causes of such differences every single variable that differs between the two populations must be considered as possibly significant. On the other hand when following the same population (that is, monitoring) and observing an abrupt increase, one is in a better strategic position because one has at least a particular time to focus upon and a specific population in which to look for the introduction of a new cause. While investigation may be no more productive here it may be easier to develop and investigate plausible hypotheses. In fact when abrupt increases of birth defects have been observed in particular jurisdictions even practicing physicians have been able to suggest the likely cause. It was in such an anecdotal way that rubella was first suspected as a cause of cataract. With refined statistical techniques in use for systematic monitoring, hopefully we will not have to depend upon personal anecdotal impressions to document a future increase or detect its cause. (But I suspect that inspired guesses have contributed more to our knowledge of single major environmental causes of human birth defects than any methodical discipline.)

Given the presumed "background" causes of major defects a certain number of clusters in time and space are to be expected on probabilistic grounds alone. Thus any single observed rise in a relatively small population may only reflect coincidental occurrence of many "background" causes together. While investigation may disclose no new environmental insult as a likely cause, it may reveal some of the "background" causes. But if there is an increase in the number of clusters in time and space, and if we presume that the increase is caused by some newly introduced environmental factor, we will not of course be able to say *a priori* which observed cluster is due to the environmental event and which to a concordance of background events, or even sort out such factors in any particular setting. Thus it is clear that all such clusters should be investigated but that we should not be discouraged if unable to identify a new environmental cause in any particular case.

Lastly, there is one point which is perhaps obvious but which is often overlooked. This is that the greatest collection of talent and knowledge from different fields that we can bring to bear on problems of this sort the more likely we will be to reach socially useful conclusions.

REFERENCES

[1] P. BANISTER, "Evaluation of vital record usage for congenital anomaly surveillance," op. cit. [10].

[2] H. EKELUND, S. KULLANDER, and B. KALLEN, "Major and minor malformations in newborns and infants up to one year of age," *Acta Pediatrica Scand.*, Vol. 59 (1970), pp. 297–302.

[3] J. W. FLYNT, JR., A. T. EBBIN, G. P. OAKLEY, A. FALEK, and C. W. HEATH, JR., "Metropolitan Atlanta congenital defects program," op. cit. [10].

[4] F. C. FRASER, "The epidemiology of common major malformations as related to environmental monitoring," op. cit. [10].

[5] A. M. GITTELSOHN and S. MILHAM, JR., "Vital record incidence of congenital malformations in New York State," *Genetics and Epidemiology of Chronic Diseases* (edited by J. V. Neel, M. W. Shaw, and W. J. Shull), U.S. Department of Health, Education and Welfare—Public Health Service Publication No. 1163, 1965, pp. 305–320.

[6] A. M. GITTELSOHN, "Feasibility of vital and hospital record linkage in monitoring birth defects," op. cit. [10].

[7] G. B. HILL, C. C. SPICER, and J. A. C. WEATHERALL, "The computer surveillance of congenital malformations," *Brit. Med. Bull.*, Vol. 24 (1968), pp. 215–218.

[8] E. B. HOOK, "The possible value of dermatoglyphic ridge patterns and asymmetry as teratologic markers," op. cit. [10].

[9] ———, "Some general considerations concerning monitoring: Application to utility of minor birth defects as markers," op. cit. [10].

[10] E. B. HOOK, D. T. JANERICH, and I. H. PORTER (editors), Monitoring, Birth Defects, and Environment—The Problem of Surveillance, Academic Press, New York, in press.

[11] W. P. KENNEDY, "Epidemiologic aspects of the problem of congenital malformations," Birth Defects Original Article Series of the National Foundation, Vol. III, No. 2 (1967), pp. 1–18.

[12] B. MacMAHON and S. YEN, "Epidemic of neural tube deformities," Lancet, Vol. 1 (1971), pp. 31–33.

[13] P. M. MARDEN, D. W. SMITH, and M. J. McDONALD, "Congenital anomalies in the new-born infant, including minor variations," J. Pediatrics, Vol. 64 (1964), pp. 357–371.

[14] R. McINTOSH, K. K. MERRITT, M. R. RICHARDS, M. H. SAMUELS, and M. T. BELLOWS, "The incidence of congenital malformation: A study of 5,964 pregnancies," Pediatrics, Vol. 13 (1954), pp. 501–521.

[15] S. MILHAM, JR., "Experience with malformation surveillance," op. cit. [10].

[16] J. R. MILLER and B. J. POLAND, "Monitoring of human embryonic and fetal wastage," op. cit. [10].

[17] T. NELSON, G. P. OAKLEY, JR., and T. H. SHEPARD, "Collection of human embryos and fetuses. II. Classification and tabulation of conceptual wastage," op. cit. [10].

[18] T. H. SHEPARD, T. NELSON, G. P. OAKLEY, and R. J. LEMIRE, "Collection of human embryos and fetuses. I. Methods," op. cit. [10].

[19] D. W. SMITH, Discussion, op. cit. [10].

[20] P. STOCKS, "Incidence of congenital malformations in the regions of England and Wales," Brit. J. Prevent. Soc. Med., Vol. 24 (1970), pp. 67–77.

[21] J. G. WILSON, "Use of Rhesus monkeys in teratological studies," Fed. Proc., Vol. 30 (1971), pp. 104–109.

A METHOD FOR MONITORING
ADVERSE DRUG REACTIONS

GARY D. FRIEDMAN
and
MORRIS F. COLLEN
PERMANENTE MEDICAL GROUP, OAKLAND

1. Introduction

A recent international conference on Adverse Reactions Reporting Systems [6] has again underscored the great need for continuing surveillance of therapeutic drugs after they are marketed. Formal clinical drug trials and other premarketing studies are usually too small in scale and too formally structured to detect all of the problems that a drug may cause when it is employed in the varied and complex setting of actual patient care. These drug-caused problems, known medically as *adverse drug reactions*, consist of a wide variety of untoward effects, some of which occur quite rarely.

In the Kaiser-Permanente Department of Medical Methods Research, a computerized medical data system [9] is being developed which now records the essential medical data for patients seen in the Kaiser-Permanente outpatient department in San Francisco. In attempting to minotor the risks of untoward events due to therapeutic drugs, we have employed an analytic method that delves into a relatively unstructured situation in an effort to bring out some orderly and useful observations.

After describing the method we shall discuss the relationship of drug monitoring to studies of the effects on health of environmental pollution, the theme of this part of the Symposium.

2. Data currently available

In contrast to most other drug monitoring programs, we have been working with outpatient data, that is, information about what takes place in outpatient clinics rather than in a hospital ward [3]. Analytic methods applied to inpatient data have been described by others [1], [5], [7], [8].

Because outpatients are not under continuous observation as are hospitalized patients, outpatient data are necessarily imprecise and less complete than inpatient data. Regarding drug usage, in an outpatient setting, one can ascertain

The research activities of this program have been supported in part by the Kaiser Foundation Research Institute, Food and Drug Administration Contract 68–38) and National Center for Health Services Research and Development (Contract HS00288).

that a drug has been prescribed, or even, as in the Kaiser-Permanente system, that a drug has been dispensed. However, one cannot be sure that the drug was taken, without special interviews or tests which are impractical in a monitoring situation. Furthermore, to detect the development of untoward events related to a drug it is necessary in an outpatient setting that the patient report his problem to the physician. Despite these relative deficiencies in the completeness of outpatient data, outpatient surveillance is a necessary component of drug monitoring. Only in this setting can patients receiving chronically administered drugs, such as antihypertensives, antidiabetics, and oral contraceptives, be followed up for the occurrence of long term or delayed side effects.

Our basic outpatient data come from a sequence of visits made by patients to the pharmacy and to various clinics. We have some identifying information about each patient: medical record number, name, sex and month and year of birth. When each clinic visit occurs, the date, the time, the identity of the clinic and the attending physician and the diagnosis made by the physician are recorded and entered into the computer record. Necessary modifiers of the diagnosis are also entered, that is, "new" (meaning that the condition is new or recurrent), "old" (the condition continues from a previous visit) or "worse" (the condition is pre-existing but has worsened since last seen). Certain clinics provide for the entry of procedures such as minor operations, injections and dispensing of drug samples by the doctor. For the clinic visit information to be stored, the patient identifying information must agree with that in the patient computer record, and the date and a diagnosis (or procedure) must be recorded. However, a missing visit time or physician identifying number will not prevent the storage of the visit information.

When a pharmacy visit occurs the dispensing of a prescription is recorded and verified "on-line" by the pharmacist. The stored data include the date and time of the visit, the identity of the doctor who wrote the prescription, the sequential prescription number used by the pharmacy, the name of the drug dispensed (usually the trade name rather than the generic), the form of the drug (for example, tablets, syrup, eye drops, and so on), the strength of the drug (for example, 50 mg. of drug per tablet, or 2 per cent concentration of the drug in the ointment, and so forth), the "sig" or instructions to the patient as to how the drug is to be used, the amount dispensed and the total amount left to be dispensed by subsequent refills.

The various visits are arranged in chronological sequence in the patient's computer record. Such a sequence for an actual patient during three months is summarized in Table I. An example of the greater detail that is available about a prescription is shown in Table II.

We believe that drug reaction studies covering a broad group of drugs should be undertaken in at least two stages. The first stage is a monitoring and screening procedure. This search for drug-event associations can be applied on a large scale to a variety of drugs. It should indicate the presence of statistically significant associations between drugs and subsequent untoward events and provide a meas-

TABLE I

Summary of a Patient's Visit Data During July 1 Through September 30, 1969

Patient Number: 1234567 (fictitious number)   Sex: Female   Birth Date: 03/1895

Date	Time	Visit location	Drug	Diagnosis
July 15, 1969	11:36 a.m.	Pharmacy	pyridoxine	
July 17, 1969	1:30 p.m.	Medical clinic		diabetes mellitus, old arteriosclerotic heart disease, old edema, peripheral, new
July 17, 1969	3:00 p.m.	Pharmacy	lasix	
August 5, 1969	1:36 p.m.	Pharmacy	folic acid orinase	
August 21, 1969	—	Medical clinic		arteriosclerotic heart disease, old arrhythmia, paroxysmal tachycardia, worsening
Sept. 5, 1969	5:00 p.m.	Pharmacy	pyridoxine	
Sept. 16, 1969	11:42 a.m.	Pharmacy	digoxin	
Sept. 19, 1969	4:12 p.m.	Pharmacy	phenobarbital	

TABLE II

Actual Stored Data Concerning an Individual Prescription

(all identifying numbers are fictitious)

Patient number: 7654321	Physician number: 54321
Date: September 3, 1969	Time: 5:06 p.m.
Prescription number :	174626
Drug name :	ferrous sulfate
Drug form :	enteric-coated tablets
Strength :	3.5 grains
Sig :	take 1 tablet 3 times a day, before meals as directed
Amount dispensed :	200
Amount remaining for refills :	600

ure of the magnitude of the associations. The second stage is more of an *ad hoc* in depth study that is applied to individual drug-event associations to determine the likelihood that the drug actually causes the event and to define better the patient and drug characteristics and other circumstances that foster the adverse reaction.

In order to be broadly applicable, our first stage monitoring uses only the summary data shown in Table I. In Finney's [2] classification of drug monitoring records, these data correspond to class II records, which require patient identification, drugs, events and diagnoses leading to drug prescriptions. To some extent they also meet the criteria for class I records which contain, in addition, patient attributes and past medical history. Some patients have had extensive multiphasic examinations the results of which are stored in the computer record. These patients have very good class I records, which can be used for detailed second stage studies as can the drug data illustrated in Table II.

3. Monitoring analysis for a particular drug

The first step in bringing some order out of the varying picture of drug usage by outpatients was to establish a *time frame*. The beginning point in chronological time that was imposed on us by technology was the end of June, 1969, when pharmacy data were first entered into the computer. Diagnostic data from most clinics were being entered by that time or soon after. The first large set of data that we worked with was retrieved in January, 1970 and covered the six month period July through December, 1969. We are now beginning to analyze data for the one year period that ended on June 30, 1970.

We define a time period called the "selection interval," during which we identify the patients who received the drug to be studied. Then each patient is followed up for the development of untoward events for a defined period of time called the "follow-up interval." This latter interval begins at the time the patient first receives the drug. Observation of events before the drug is received, extends from the beginning of the selection interval until the time the patient first receives the drug. These intervals are illustrated in Figure 1.

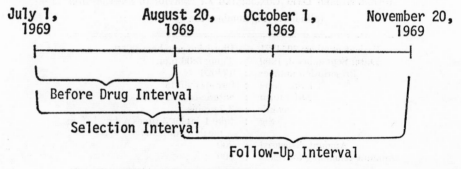

FIGURE 1

Example of "selection" and "follow-up" intervals. For this study, the selection interval is chosen as July through September, 1969. The follow-up interval is chosen as three months. A patient first receiving the drug on August 20 is followed until November 20. Calculation of event incident rate after the drug is based on this follow-up interval. The before drug incidence rate is based on the before drug interval, July 1 through August 20, 1969.

This has been the time frame that we have most often used so far. The selection interval can be made as long as desired within the constraints of the duration covered by the available data and the minimum duration of follow-up desired. The follow-up interval is less easily selected. For chronically used drugs one would desire as long a follow-up as the data would allow. For drugs used for a short duration (about one day to two weeks) or for a medium duration (up to a few months) one view is that the period of usage of the drug should be computed from the detailed dispensing data and patients should be followed for that time only. However, we feel that this would be too cumbersome from a computational point of view, particularly since patients often do not follow instructions exactly and since some adverse reactions may occur after the drug is stopped. It seems more prudent to select a reasonable follow-up interval based on the type of drug. For short term antibiotics we used a one month period. For long term drugs we used three months to follow-up. As more data become available we will use longer and longer periods of follow-up but after three or six months have elapsed it would seem desirable in many cases, to require continuing evidence that the patient is still using the drug.

4. Incidence of untoward events

The occurrence of any "new" or "worse" diagnosis is considered an untoward event that is worth at least some initial scrutiny as to whether it is or is not related to the drug. In this way we are open to the discovery of previously unsuspected reactions. Any patient receiving the drug who develops the event at least once during follow-up is counted. The incidence rate for each event is the proportion of drug users who develop the event per unit of time of follow-up.

While we have generally used a fixed after-drug follow-up period, this is not a requirement. Where follow-up duration varies from person to person the incidence rate can be expressed in terms of person-days of observation.

In traditional prospective epidemiologic studies the presence or absence of a disease under investigation is ascertained at the onset by special study. Only those certified to be free of the disease are considered at risk for the development of the disease and only they are used in the denominator of an incidence rate. In a large scale monitoring program concerned with a wide variety of untoward events this approach obviously is not practical. Instead, we have depended on the doctors' determinations as to whether conditions are new or worsening. Naturally, errors may occasionally occur, particularly in situations when a patient changes physicians or when the old chart is not readily available at the time the patient is seen.

5. Comparison groups

5.1. *Nonusers.* To evaluate the observed incidence rates of events in users of the drug one needs some basis of comparison. Our primary comparison group to date has been those persons who came to the clinic or pharmacy during the selection interval but who did not receive the drug. The incidence of the untoward events observed in these nonusers is computed in an analogous fashion to that in the users.

We considered using the entire Kaiser Health Plan population residing in San Francisco as our basic study group and source of nonuser incidence rates. However, there was some doubt as to whether all of these persons consistently used the San Francisco facility for their care since Health Plan members are free to use facilities in other locations or to utilize non-Kaiser physicians. It seemed more prudent to require that nonusers show evidence of visiting the data collecting facility as do users. This would provide some assurance that untoward events developing in users and nonusers have a reasonably equal chance of being detected by the monitoring system.

The users' follow-up begins when they first were known to have received the drug. We begin the nonusers' follow-up when they first were seen at the pharmacy or any clinic during the selection interval.

There are, of course, a number of possible sources of bias that should be considered when users of a drug are compared with nonusers. For example, it occurred to us that there might be a substantial difference in the distribution of starting times between users and nonusers of a particular drug. In the study of events with a seasonal variation substantial differences between users and nonusers as to the timing of follow up might introduce artificial differences that have nothing to do with the drug. An example of a seasonal condition in our data is Acute Bronchitis which had a relatively low frequency during the summer months but increases through autumn and winter (Table III).

TABLE III

MONTHLY FREQUENCIES OF MEDICAL CLINIC DIAGNOSES
OF "ACUTE BRONCHITIS—NEW"

Month	Number of diagnoses	
	Total	Per 1000 visits
July	64	8.5
August	58	8.3
September	84	10.9
October	112	14.6
November	122	15.4
December	142	18.3

We examined the distributions of starting times for users and nonusers of various drugs. Anticipating longer periods of data for analysis we assumed that the entire six month period, July 1 through December 31, 1969, was a selection interval and computed the percentage of users and nonusers that began follow-up each week. The nonuser distribution curves for all drugs were quite similar since the users of even the most popular drugs comprise only a small proportion of patients, leaving the vast bulk of the population as nonusers. Nonuser starting times tended to be most frequent early in the period with a gradual decrease as time passed. This is what would be expected since patients with more than one visit during the period would be started at their earliest visit. Incidentally, there were the expected dips in the distribution curve for the weeks with holidays in them (Figures 2 and 3).

For chronically used drugs such as oral contraceptives (Figure 2) the distributions of starting times for users and nonusers were quite similar. This was not the case for antibiotics, such as penicillin, used on a short term basis (Figure 3). These showed a gradual rise in percentage starting each week until a peak was reached at the end of October. With drugs of this nature we will have to be concerned about spurious associations due to seasonal trends.

Even more basic than starting times, however, is the question of whether users are really users and nonusers are really nonusers. The limitations of outpatient data for determining whether patients actually use drugs, have been mentioned above. Regarding nonuser status we carried out some validation studies using

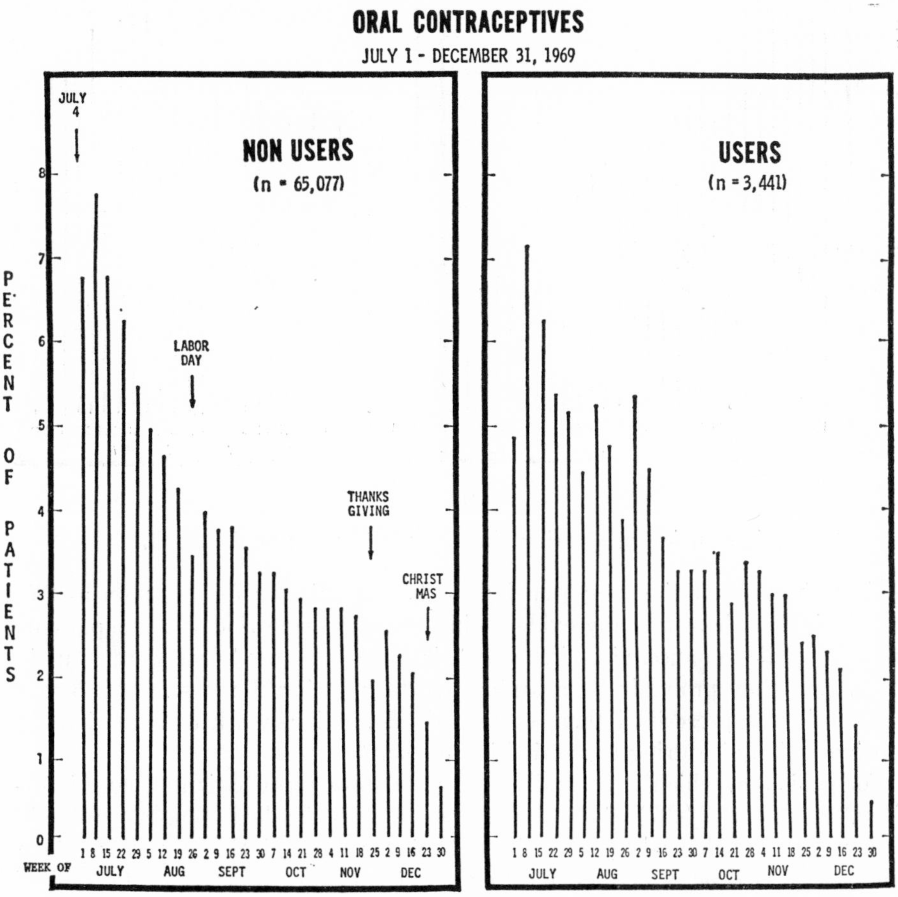

ORAL CONTRACEPTIVES
JULY 1 - DECEMBER 31, 1969

FIGURE 2

Distribution of starting times for follow-up: oral contraceptives.

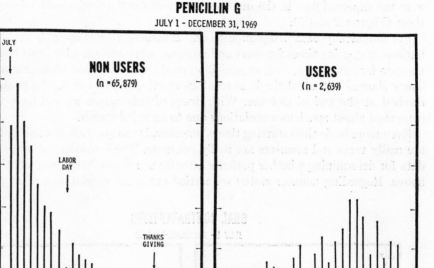

FIGURE 3

Distribution of starting times for follow-up: penicillin G.

patient charts after the computer data had been collected for two months. We asked whether there was any evidence in the chart suggesting that the patients whom we called nonusers by virtue of their computer-stored data were actually users. Had they probably been taking the drug during the two month period because it was prescribed then, or prescribed previously with instructions to continue taking it during the period in question?

For penicillin, which is generally used on a short term basis for acute infections, 2 per cent (7/297) of the persons whom we had called nonusers probably were users as evidenced by notes in their charts. For thiazide diuretics, which are usually used on a medium term or long term basis, 6 per cent (14/225) of "nonusers" aged 15 and over probably were users. For oral contraceptives, which are usually used on a long term basis, 11 per cent (11/97) of "nonuser" women aged 15–54 probably were users. So, it appeared that our computer records for a two month period were reasonably good at identifying nonusers. They were less accurate for oral contraceptives, but prescriptions for these drugs are usually given every three or six months, so that longer interval computer records would be expected greatly to increase the accuracy.

5.2. *Users before receiving the drug.* Another comparison group that should be suitable for monitoring comprises the drug users themselves, looked at before they received the drug. A distinct advantage in using this group is that many of the important characteristics that might differ between users and nonusers, such as age, sex and socioeconomic status are automatically matched. However, people do change over time and the fact that they have just obtained a drug suggests that they have developed a new disease or have arrived at a new stage in their disease or have changed doctors, all of which can change the likelihood of new untoward events being reported. Furthermore the problem of seasonal changes in event rates previously mentioned would be a very important consideration in the before-after comparison.

Because we have dealt primarily with the earliest data collected, there is considerable doubt as to what has occurred before the first recorded issuance of a drug for a patient. Therefore, to say that before the first recorded issuance is really before the drug was given, would not be appropriate. In the chart review studies for the first two month period, we also checked to see whether there was evidence that patients who were called users were also taking the drug during the period before the first recorded dispensing. For penicillin, 6 per cent (1/16) of users had received the drug before the first recorded dispensing. For thiazides, the drug was used before by 53 per cent (10/19) of users and for oral contraceptives, by 77 per cent (10/13) users. Thus the before-after comparison would be quite unsatisfactory in our early data for medium term and long term drugs.

However as more data are collected we will be able to look farther back from the time of issuance of a drug to determine that the patient had indeed not received it before. With this in mind we have begun to use a "search-back" procedure on data collected over a one year period or more. This places the "selection" interval later in the available time period, to be preceded by a "search-back" interval, during which period the use of the drug being studied is again checked. Users and nonusers are defined during the "selection" interval as before. However, any evidence of use of the same drug during the "search-back" interval will result in the exclusion of users from the before-after comparison. Furthermore any nonusers who received the drug during the "search-back" interval will be excluded from the nonuser comparison group. This modified study design is illustrated in Figure 4 where the first three months of a one year period are allotted for "search-back," the second three months for "selection" and the remainder for follow-up.

We have formalized our comparison of incidence rates in users *versus* nonusers, and after *versus* before the drug, by looking at the ratio of two rates (commonly known in epidemiology as the *relative risk*) and by looking at the difference between two rates (commonly known as the *attributable risk*). The 95

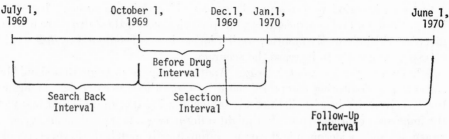

FIGURE 4

Example of time intervals resulting from the use of a "search-back" procedure. For this study the selection interval is chosen as October through December, 1969. The follow-up interval is chosen as six months. The search-back interval is July through September, 1969. A patient first receiving the drug during the selection interval on December 1, 1969 is followed until June 1, 1970. If he also received the drug during the search-back interval he is excluded from the before-after comparison but not from the user-nonuser comparison. Calculation of the event incidence rate after the drug is based on the length of the follow-up interval. The incidence rate before the drug is based on the length of the before-drug interval, October 1 through December 1, 1969. A nonuser during the selection interval, is excluded from the nonuser comparison group if he received the drug during the search-back interval.

per cent and 99 per cent confidence limits of the difference were examined to see if they included zero difference. If not, then one or two asterisks were placed on the computer output next to the event, indicating a significant association deserving further scrutiny.

As an example of our screening analyses, Table IV shows our current findings concerning a particular drug-event relationship, that is, between oral contraceptives and anxiety reaction. While the results varied among age groups, the over all six month incidence of physician-diagnosed anxiety reaction in oral contraceptive users was similar to the age adjusted incidence in nonusers (12.9/1000 and 15.4/1000 respectively). The apparent lack of association between anxiety reaction and oral contraceptive usage illustrates another value of population based drug monitoring. In addition to detecting adverse drug reactions, monitoring may provide valuable negative results. Some conditions are not brought about by a drug, despite their reported occurrence in some users of the drug.

The use of the relative risk, attributable risk, statistical significance tests and medical judgment to determine which drug-event associations should be pursued further have been discussed in detail elsewhere [4].

6. Refining user versus nonuser comparisons

Since users and nonusers may differ greatly in characteristics that influence the event rates it is important to control for these characteristics in the analysis

TABLE IV

EXAMPLE OF A USER—NONUSER COMPARISON
Oral Contraceptives and Anxiety Reaction

(Women aged 15–54; drug use and nonuse ascertained during October–December, 1969;
follow up for six months; search-back during July–September, 1969)
* Age adjusted. The overall proportion with an event for nonusers is a weighted average of the
proportions for individual age-sex subgroups. The weighting is according to the proportion
of *users* who fall into each age-sex subgroup.

Age group	Total number		No. developing anxiety reaction		Incidence/1000 of anxiety reaction		Relative risk	Attributable risk per 1000
	Users	Non-users	Users	Non-users	Users	Non-users	(User/nonuser)	(User − nonuser)
15–24	905	3,798	7	30	7.7	7.9	.98	−0.2
25–34	1,261	3,816	9	64	7.1	16.8	.43	−9.7
35–44	539	3,265	13	76	24.1	23.3	1.04	0.8
45–54	241	4,165	9	80	37.3	19.2	1.94	18.1
15–54	2,946	15,044	38	250	12.9	15.4*	.84	−2.5

whenever possible. Control by randomization is not possible in the observational monitoring situation. We control for age and sex differences by adjustment, that is, in computing the overall nonuser rate, weighting the rates for nonuser subgroups according to the proportions of users that fall into each corresponding subgroup. However, more control of extraneous variables is often needed to make a fair comparison. Ideally one would like to control for many different variables with each separate drug-event association but this is not feasible in the first screening stage of monitoring. That is why we favor applying a second stage detailed study to selected drug-event associations.

We have had some experience controlling for a third variable in our screening analyses, that is, the disease being treated. For example, we have restricted some of our analyses concerning antidiabetic drugs to patients with diabetes mellitus. In this way users and nonusers are matched for an important variable.

When we compared the results of restricting the analysis in this manner to using an unselected nonuser comparison group certain differences were noted as expected. There were reductions in the total number of users available for study since not all users had had the disease in question recorded during the six month study period. This problem will be largely solved by longer periods of data collection. Corresponding to the decrease in the number of users there were fewer different events in users to be studied, but the decrease in the number of different events was proportionally much less than the decrease in the number of users. The main discernible benefit that resulted from restricting the analysis to users and nonusers with a specific disease was a reduction in the number of events which appeared to be associated with the drug but were actually related to the basic disease being treated by the drug.

7. Relationship of drug monitoring to studies of the effects on health of environmental pollution

Medically prescribed drugs are not customarily thought of as environmental pollutants. However, there are a number of analogies between drugs and pollutants as well as similarities in the requirements for their study.

Many environmental pollutants are chemicals and so are drugs. Perhaps if drugs or medicines were referred to as "chemicals" they would not be so eagerly sought by some patients.

The harmful consequences of environmental pollutants, are, like those of drugs, unwanted byproducts of socially useful or desirable inventions. Just as the physician must consider the potential benefits *versus* risks of each drug he prescribes, decisions resulting from studies of the effects of pollution on health will involve weighing benefits and undesired side effects.

Both in the study of environmental pollution and in drug monitoring, one will encounter a very complex situation in attempting to measure degree of exposure. Quantities of pollutants or drugs and the time patterns of exposure are quite variable. Reasonable systems for classifying exposure, will have to be developed in order to compare population subgroups. While the presence or absence of gross exposures may be relatively easy to determine, some individuals may be exposed to small quantities of drugs or pollutants which go undetected or unquantified. For example, very little is known about the exposure of humans to antibiotics or other drugs given to the animals they eat; and what, if any, role these exposures might play in promoting or preventing adverse reactions to the same drugs prescribed medically.

Just as it is wished to have objective measures of health and disease it is also important to have objective measures of exposure. Interview or questionnaire data about drug usage are often inadequate because of the unfamiliarity of some people with what medications they are taking. Furthermore many people are unwilling to admit that they are not taking the drugs prescribed for fear of displeasing their physician. Periodic health examinations may provide the means of measuring blood levels or other objective indices of exposure of a population to drugs and environmental contaminants such as trace metals.

8. Interpretation and evaluation of findings

Monitoring of drug reactions and of the health effects of environmental pollutants involves primarily observational rather than experimental study. The investigator cannot control the situation and randomly allocate persons to exposed and nonexposed groups. Thus the problem of self-selection, already discussed in this symposium, must be reckoned with. We must be concerned with all of the ways that users of a particular drug might differ from nonusers and that those exposed to a particular pollutant might differ from those not exposed. Occasionally there will occur a "natural experiment" in which an exposure is so universal or haphazard that the influence of self-selection or other important confounding variables seems quite remote. Usually, however, this will not be the case.

The critic of an observational study can usually think of additional reasons why an observed association might have been fortuitous or explained by a third variable. Ultimately it becomes a matter of judgment as to how far this skepticism should be allowed to proceed. While the investigator has the major responsibility for testing and defending his findings the critic should be able to provide some evidence that his questions and doubts are reasonable. This is particularly true in the case of findings affecting human health and safety. Often it will be most prudent for the decision maker to act on the basis of incomplete evidence from observational studies even before some of the reasonable questions have been answered.

The difficulty and expense of long term observational studies of human populations should also be kept in mind. The classical approach in laboratory science of other investigators verifying findings by repeating experiments may be impractical for the kinds of studies we are talking about. Other means of confirmation may have to be sought. Existing vital statistics data can often be explored at little expense, to test hypotheses that are generated by findings of large scale population studies. For example, if a pollutant appears associated with a particular fatal disease, known time trends or geographic differences in exposure to this pollutant should be correlated with similar trends and differences in the mortality rates for this disease.

Large scale population monitoring studies may well involve follow-up periods measured in decades. Many of the adverse health effects of drugs or environmental pollutants may take long periods to develop even after exposures of short duration. The offspring of the exposed should also be studied.

It would be unfortunate if the value of a large scale population surveillance program were judged solely by the number of hazards that were detected. Just as with medical checkups to detect early disease, it is also desirable to find out that there is nothing wrong. The study should be judged according to its ability to detect problems if they exist. If it has this ability and no problems are detected, this reassuring information is well worth the expense and effort in our changing environment.

9. Conclusion

An epidemiologic method for monitoring adverse drug reactions has been described. It is clear that some of the decisions as to how the data are to be analyzed are arbitrary and will be modified by future experience. The conditions that are imposed by a real medical care program are less than ideal for formal scientific study. However, the medical care situation is a crucial setting for monitoring the harmful effects of drugs.

Drug monitoring is, in many ways, similar to studying a population for the effects of environmental pollutants. In both instances the methods are observational rather than experimental, and exposures may be difficult to detect and classify.

REFERENCES

[1] L. E. CLUFF, G. THORNTON, L. SEIDL, and J. SMITH, "Epidemiological study of adverse drug reactions," *Trans. Assoc. Amer. Phys.*, Vol. 78 (1965), pp. 255–268.
[2] D. J. FINNEY, "The design and logic of a monitor of drug use," *J. Chron. Dis.*, Vol. 18 (1965), pp. 77–98.
[3] G. D. FRIEDMAN, M. F. COLLEN, L. E. HARRIS, E. E. VAN BRUNT, and L. S. DAVIS, "Experience in monitoring drug reactions in outpatients: the Kaiser-Permanente drug reaction monitoring system," *J. Amer. Med. Assoc.*, Vol. 217 (1971), pp. 567–572.
[4] G. D. FRIEDMAN, "Screening criteria for drug monitoring: The Kaiser-Permanente drug reaction monitoring system," *J. Chron. Dis.*, in press.
[5] N. HURWITZ and O. L. WADE, "Intensive hospital monitoring of adverse reactions to drugs," *Brit. Med. J.*, Vol. 1 (1969), pp. 531–536.
[6] International Conference on *Adverse Reactions Reporting Systems*, sponsored by the National Research Council, Washington, D.C. (October 23–24, 1970), in press.
[7] H. JICK, O. S. MIETTINEN, S. SHAPIRO, G. P. LEWIS, V. SISKIND, and D. SLONE, "Comprehensive drug surveillance," *J. Amer. Med. Assoc.*, Vol. 213 (1970), pp. 1455–1460.
[8] D. KODLIN and J. STANDISH, "A response time model for drug surveillance," *Computers and Biomed. Res.*, Vol. 3 (1971), pp. 620–636.
[9] E. E. VAN BRUNT, M. F. COLLEN, L. S. DAVIS, E. BESAG, and S. J. SINGER, "A pilot data system for a medical center," *Proc. of the IEEE*, Vol. 57 (1969), pp. 1934–1940.

Discussion

Question: John R. Goldsmith, Environmental Epidemiology, California Department of Public Health

Time trends in needs or demands for drugs used to treat bronchospasm can be of great value in environmental epidemiology since they may be the most sensitive index of the buildup of pollutants causing respiratory irritation. Many patients would not be aware of such a possible relationship.

Epidemiological study of effects of drugs of abuse requires more than the customary examination of vital statistics to find evidence of increased fatality among young people due to overdose or increased prevalence of hepatitis. What is more difficult and more urgent is the need for longitudinal studies of such possible effects as fetal loss or birth defects. The empaneling of populations of drug abusers is, therefore, also necessary for a program of drug monitoring which is comprehensive.

Another type of study is based on the retrospective examination of drugs used by individuals with causes of death commonly related to drug ingestion (such as blood dyscrasias, and certain renal or skin conditions). Unfortunately, many serious drug reactions are only beginning to be reflected in morbidity and systems of medical records. Those at Kaiser have an unusually important role to play in defining the parameters in studying morbidity reactions in follow-back investigations.

Reply: G. Friedman

Regarding the study of time trends in the usage of bronchodilators, our pharmacy data might be a very good source of this kind of information. This is because the dispensing of prescription refills is recorded even if not immediately preceded by a physician visit. Patients in distress might go directly to the pharmacy for a refill of their medicines.

At this time our monitoring system only deals with prescription drugs and not with the drugs of abuse such as marijuana, heroin, and so on. We thought that by looking for persons at the upper ends of the distributions of numbers of prescriptions dispensed, the monitoring system might be able to detect previously unsuspected abusers of drugs such as tranquillizers and analgesics. However a preliminary review of the records of some of the patients who had received many prescriptions for these drugs indicated that they were using these drugs for good medical reasons and that their physicians were well aware of their heavy drug usage.

We agree that the retrospective method of investigation may play an important role in drug monitoring, particularly when one is dealing with rare events.

Question: Alexander Grendon, Donner Laboratory, University of California, Berkeley

You were in a position to use a control group which, unlike the environmental pollutant case, was not exposed to *any* "pollutant" drug; yet your "nonusers" apparently might have been using other drugs, so long as they were not using the drug you were investigating. Would it not have been better to select as controls nonusers of *all* drugs, who came to your clinic for medical care not requiring treatment with drugs?

Reply: G. Friedman

Your suggestion is an interesting one and perhaps deserves to be tried out. However, there has been much concern that users of a drug might receive more intensive follow-up medical care and might therefore be more likely to have untoward events detected, than might those who are not receiving any drug at all. The comparison would then be biased in the direction of finding higher event rates in users than in nonusers.

Question: E. B. Hook, Birth Defects Institute, Albany Medical College

In a conference such as this devoted to planning a comprehensive program, it is worth emphasizing that the drugs dispensed by a hospital pharmacy represent only a small proportion of chemicals ingested by the population, and about which we should have some concern. In a sense they are relatively easy to study compared to such agents as:

(1) Over the counter, self administered preparations: aspirin, vitamin C (in high doses), and so on.

(2) Illicit drugs: LSD, amphetamines, and so on, as well as their multitudinous contaminants.

(3) Food additives and preservatives: monosodium glutamate, sodium benzoate, and so on.

This list of categories is not exhaustive, but it illustrates that whatever the population studied, very detailed scrutiny of these and other "occult" agents which may represent "cocarcinogens," "coteratogens," or just "cotoxins," would be required.

Reply: G. Friedman

So far our outpatient monitoring involves only prescription drugs. It would certainly be desirable to be able to study as well the items that you have mentioned.

Question: Colin White, Department of Public Health, Yale School of Medicine

What are the unique contributions of a monitoring program? In particular, when should a designed study be preferred to monitoring? Monitoring may fall into disrepute if it is used for investigations that ought to be carefully planned rather than made on an observational basis.

Reply: G. Friedman

We regard monitoring as an initial screening process to provide clues and hypotheses that can later be studied more carefully. By covering a large number of drugs and events, even though superficially, monitoring provides a means of detecting previously unsuspected reactions. We wish to emphasize the limitations of the data we have presented. They certainly are not meant as a substitute for well controlled studies.

CHEMICAL INDUCTION OF MUTAGENESIS AND CARCINOGENESIS

ALEC D. KEITH

UNIVERSITY OF CALIFORNIA, BERKELEY

1. Introduction

It is taken for granted that man strives to control his environment by the extensive use of energy sources. Many of these energy sources are chemical and result in unstable intermediates in side products. Areas which require large energy outputs for industrial purposes, automotive uses and a variety of other needs may be expected to contain harmful concentrations of chemical side products.

Most industrialized cities and cities with extensive automotive traffic have large areas containing variable amounts of chemical impurities, at all times. More rural areas may have seasonal or transitory periods where field burning or some equivalent action results in high concentrations of air or water borne impurities. Local zones having very high concentrations of chemical impurities may also exist for variable periods. Notably, kitchens and bathrooms may be exposed to a variety of potentially hostile chemicals such as an almost infinite variety of aerosols, aromatics and "germ-killing" agents most of which have unknown long range effects on man and other organisms.

Fossil fuels are being used at a rapid rate such that we may expect a world-wide and ever increasing amount of atmospheric impurities over the next several years. The major sources of chemical impurities comes from hydrocarbon combustion, insecticides and herbicides, cosmetics and cleaning agents including dyes, food additives and perhaps, indirectly, the extensive use of inorganic fertilizers.

Before proceeding, I wish to state working definitions of the basic genetic conditions, mutagenesis and carcinogenesis. DNA is localized in the chromosomes of organisms and is composed of four small molecules arranged in triplet information bits. A set of triplets comprises a basic information unit called a gene or cistron. In cellular function, the information contained in a cistron is transcribed onto a complementary RNA segment which, in turn, is translated into one amino acid. The amino acids are assembled into the same sequence as was contained in the cistron. If the triplet code is intact, then the condition is referred to as *wild type*. On the other hand, if the triplet sequence, composition

Supported in part by USPHS (Am-12939) and Project Agreement 194 from the U.S. Atomic Energy Commission.

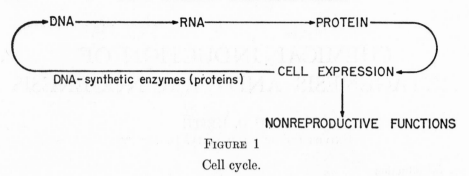

FIGURE 1

Cell cycle.

of each triplet, or number of triplets is altered, then the net information in a given cistron is altered and the condition is referred to as a mutation. There is usually a correspondence between a functional protein unit (enzyme) and a cistron; however, some proteins, such as hemoglobin, may be comprised of peptides emanating from more than one cistron. In such cases, a mutation in either peptide results in an altered protein and the relationship between protein and DNA is still functionally the same.

2. Mutagenesis

Mutagenesis results from the genetic transferable material in an individual undergoing a chemical alteration such that the genetic information in the altered cell contains some different information. In higher organisms we are only concerned with the reproductive cells; however, in lower forms such as one celled organisms generally all the cells are reproductive.

3. Carcinogenesis

Carcinogenesis results from an alteration in a cell or tissue such that the growth and functional properties of the cells are altered. An alteration of the genetic information in a cell is not a requirement. It is generally recognized by developmental biologists that every cell in higher organisms has the same genetic information in a given animal and that specific tissues have cells whose genetic expression is modified (reduced) such that a specific tissue performs limited functions. For example, brain cells do not normally divide in adult mammals; however, if some of the brain cells in an individual do start to divide (grow) a brain tumor is the result. It is generally believed that these cells have become de-repressed and in so doing have become more primitive such that they can now divide. Cell division is apparently inhibited by the repression of the appropriate genetic information in mature mammalian brain cells. Several mechanisms may account for the causes of cancer. I merely wish to establish that mutagenesis and carcinogenesis are not mechanistically the same, although some carcinomas may be caused by mutagenic events. Certainly, many chemicals will cause either or both.

4. Chemicals which are reactive with nucleic acids

I define these as either causing changes in covalent bonding (Type I) or as having some noncovalent interaction (Type II). Type I would include base analogs, some chemicals which attack functional groups on nucleic acids and the chemical alkylating agents. Type II would be dyes and other chemicals which intercalate into DNA or in some way interfere with nucleic acid enzyme functions.

Some known mutagens have an incredible degree of specificity for a given cell type. A small chemical alkylating agent, ethylmethane sulfonate (EMS) has been used extensively in chemical and mutagenic studies and in the fruitfly, *Drosophila melanogaster*, it acts primarily on mature sperm [1]. EMS has very little effect on other stages of sperm development, somatic cells, or on female *Drosophila*, in general, at equivalent concentrations. Normally, *Drosophila* males are placed in vials containing a small piece of porous paper saturated with a solution of 0.003M EMS and exposure is for 24 hours. By contrast, haploid yeast is exposed to about 0.3M of EMS for an hour or less in solution and this concentration results in a relatively high mutation rate. The extent of contact is hard to compare but certainly much larger effective doses are used for yeast. Probably any cell is sensitive to EMS at some concentration. Therefore, the seemingly unique sensitivity of sperm (mammalian sperm, as well) may relate to permeability processes. The scheme shown in Figure 2 shows the

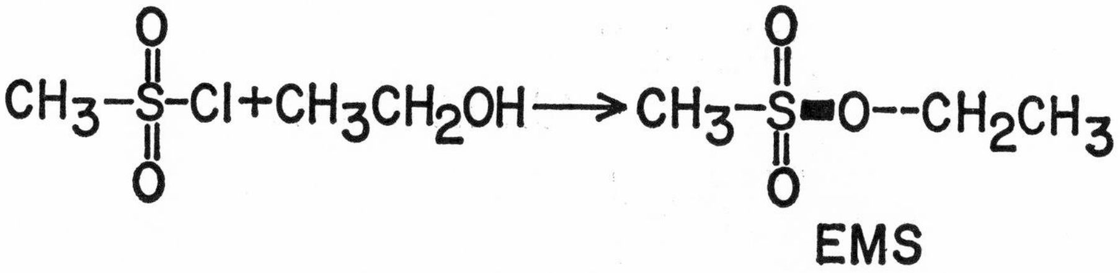

FIGURE 2

Synthesis of ethylmethane sulfonate (EMS).

synthesis of EMS and illustrates that the bond shown by the heavy line is formed during the synthesis. Conversely, the decomposition of EMS proceeds by breaking the bond shown by the dotted line (Figure 3). The alkyl group has a high affinity for nucleophilic sites such as those on the N_7 position of guanine or else is hydrolyzed by water to yield the original starting materials, methane sulfonic acid and ethanol (Figure 4). This reaction proceeds by the familiar S_N2 mechanism. Many known mutagenic agents are alkylating agents which alkylate by the S_N2 mechanism.

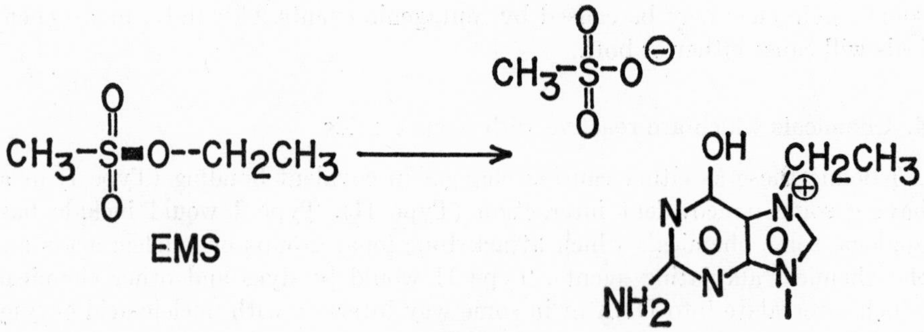

FIGURE 3

EMS alkylates guanacine.

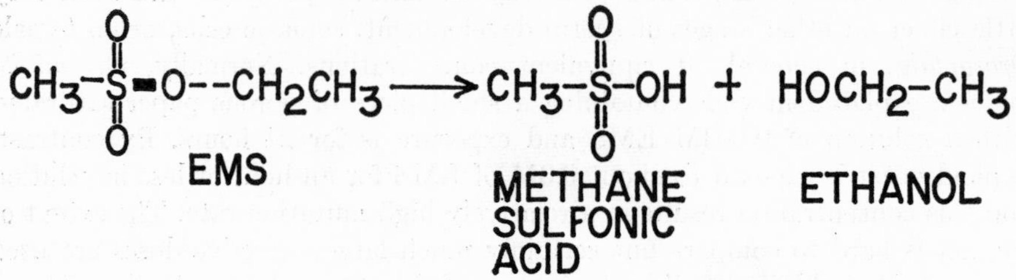

FIGURE 4

EMS decomposes in water.

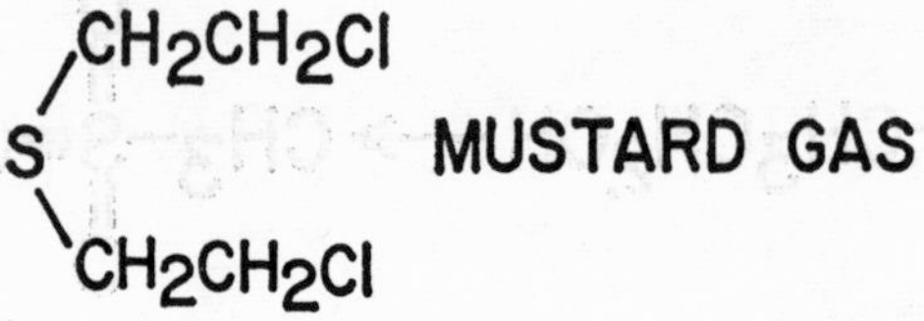

MUSTARD GAS

FIGURE 5

Another chemical mutagen.

Mustard gas is also a mutagenic and alkylating agent (see Figure 5). This compound has a different cellular specificity than EMS in that *Drosophila* late spermatogonia are sensitive to mutation while the mature sperm are resistant [2]. This is determined by observing that broods from virgin females mated on sequential days to the exposed male do not show appreciable mutations until about the fifth or sixth day, while treatment with EMS results in high mutation rates the first three days and then demonstrates a drastic reduction of mutation rate in subsequent days.

There is a large body of literature dealing with induced chemical mutagenesis. The two chemicals mentioned and a few other alkylating agents have had extensive research carried out on them and demonstrate that chemically unstable species are potentially extraordinarily dangerous. The myriad of chemicals human populations come in contact with, by and large, have not been characterized with respect to mutagenesis and carcinogenesis, especially on human subjects.

Many chemicals also cause carcinogenesis. Some are natural products and some are synthetic. For present purposes these fall into two groups. Those that are carcinogenic in their native state without requiring metabolism of the host organism. The other group must be acted upon by the host organism; therefore, the second group are detoxification products or products of enzyme-mediated reactions. In either case the carcinogen has some action on a given tissue such that the tissue becomes abnormal in its growth and/or function.

In the early 60's in England 100,000 turkeys died of cancer which had been caused by grain infected with the mold *Aspergillus flavus* which produce chemicals known as Aflatoxins. The Aflatoxins are among the most carcinogenic

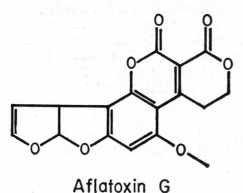

Aflatoxin G

FIGURE 6

A fungal metabolic product which is carcinogenic.

substances known. These chemicals in μg/kg body weight quantities cause cancer with an extremely high rate. Whether a given chemical is carcinogenic is often not predictable and must be determined by experimental means on the organism in question. Therefore, while we know such chemicals as Aflatoxins cause cancer it was not possible to determine this by inspection of the chemical formula but only through experimental evidence.

The majority of research carried out on chemical carcinogenesis has been phenomenological; however, some work in recent years has been more analytical and allows some generalizations. Huggins and co-workers [3] showed how small chemical alterations on substituted benzanthracene molecules drastically altered the frequency of induced carcinogenesis in rats. Methyl groups on carbons 7

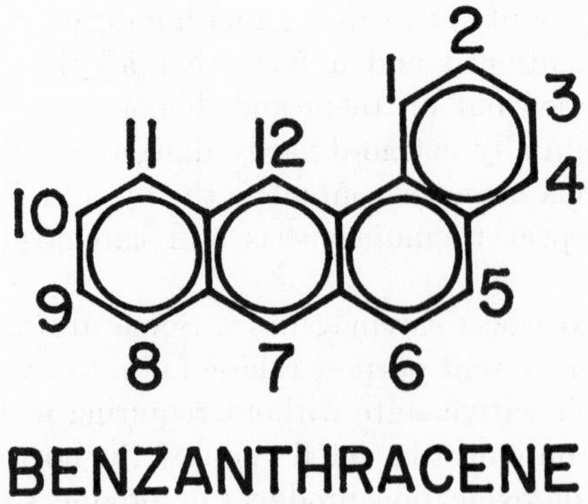

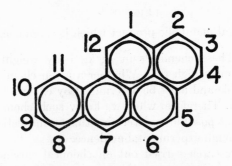

BENZANTHRACENE

FIGURE 7

A polycyclic aromatic hydrocarbon carcinogenic.

and 12 result in a compound which is among the most carcinogenic of all chemicals known. Ethyl groups on carbons 7 and 12 result in a compound which is comparatively harmless. These authors tested a variety of substituted benzanthracene compounds at different ring positions and concluded that the positions 6, 7, 8, and 12 were the most active sites. The exact molecular geometry is critical as to whether a given polycyclicaromatic hydrocarbon is carcinogenic. The consideration is also important as to whether the compound itself is carcinogenic or a metabolic product. If the latter is true, then we expect considerable variation based on small molecular geometry changes due to enzyme specificity as a function of molecular geometry. Cavalieri and Calvin [4], some years later, proposed that 7, 12-dimethyl-benzanthracene was a carcinogen due to an enzyme-mediated electrophilic attack on either carbons 6 or 8. They elaborated somewhat more on a related substance, benzapyrene. They proposed that benzapyrene is possibly mutagenic in rats due to enzyme-mediated electrophilic

BENZAPYRENE

FIGURE 8

A polycyclic aromatic hydrocarbon carcinogenic.

attack on carbon 6 by an oxygen atom causing positions 1 or 3 to become a reactive electrophilic center. The electrophilic zone would then be expected to react with a cellular site which would result in cellular transformation into a tumor type. The implications of these last two papers are that, an animal, in trying to protect itself from harmful chemicals may, in fact, produce new chemical species which are carcinogenic. Mammalian organs and tissues are capable of a great many enzyme-mediated oxidative reactions. The closest molecular analogs to the two polycyclicaromatic hydrocarbons just mentioned are probably sterol molecules of which cholesterol is an example. Mammals are

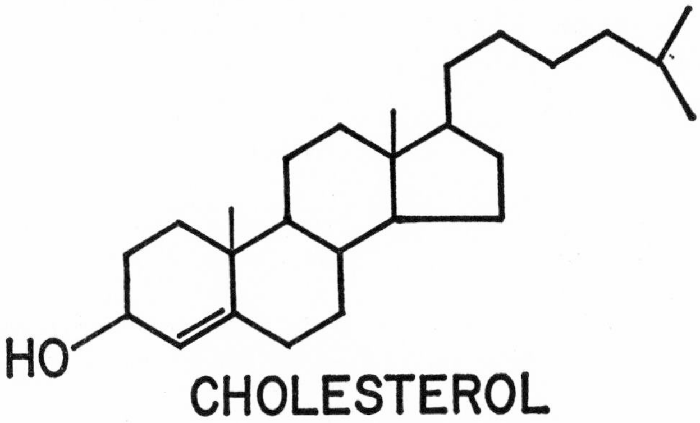

CHOLESTEROL

FIGURE 9

A natural product cyclic hydrocarbon in animals.

capable of both complete chemical synthesis and oxidative degradation of sterols. A hydrocarbon of similar structure could well serve as a pseudosubstrate. It is important to understand these metabolic relationships in consideration of the control of cancer.

Studies dealing with either mutagenesis or carcinogenesis are difficult to apply to humans or human populations. Some chemicals probably are mutagenic or carcinogenic to virtually all cells of all organisms. Among these would be the unstable chemical species such as alkylating agents and other reactive chemicals. A wide variety of other chemicals have wide variability dependent upon the species, age, or metabolic condition of a given organism. For example, compound A may be mutagenic to *E. coli*, bladder cancer inducing to man, cause liver cancer in the rat, and have no observable effect on the guinea pig. An ensemble of such compounds exists which have variable and species-dependent effects.

The statistician works with systems where partial ignorance exists. The amount of ignorance is great in consideration of present day cancer or mutation induction by chemical means; consequently, efforts must be made to reduce the number of variables on these systems. I also wish to emphasize that cancer frequency may be trivial to future generations, however frightening it appears to the current generation. Mutation frequency, on the other hand, may seem (and truly be) unimportant to the present generation since germ cells are un-important to the owner's personal physiology; however, the long range survival and quality of survival of man depends more on mutation frequencies.

Probably the only feasible way to monitor the possible effects of chemical impurities on human populations is to use microbes and higher organisms with shorter generation times than those of man. Appropriate sampling of impurities and controlled experimentation will allow high probability extrapolation to determine the effect on humans. Agencies having control over allowable limits of air and water impurities should realize that as the frequency of deleterious genes increases in human populations an increase in the frequency of monster births and defective offspring is not far behind. Some upper limit exists for the genetic load beyond which a species cannot survive.

REFERENCES

[1] O. G. FAHMY and M. J. FAHMY, "Mutagenic response to alkyl methane sulfonates during spermatogenesis in *Drosophila melanogaster*," *Nature*, Vol. 180 (1957), pp. 31–34.
[2] C. AUERBACH, "Sensitivity of *Drosophila* germ cells to mutagens," *Heredity*, Vol. 6 (1953), pp. 247–257.
[3] C. B. HUGGINS, J. PATAKI, and R. G. HARVEY, "Geometry of carcinogenic polycyclicaro-matic hydrocarbons," *Proc. Nat. Acad. Sci.*, Vol. 58 (1967), pp. 2253–2260.
[4] E. CAVALIERI and M. CALVIN, "Molecular characteristics of some carcinogenic hydro-carbons," *Proc. Nat. Acad. Sci.*, Vol. 68 (1971), pp. 1251–1253.

Discussion

Question: R. J. Hickey, Institute for Environmental Studies, University of Pennsylvania, Philadelphia

I believe you stated that there are chemicals which are carcinogenic but not mutagenic. Would you take the position that this will be a true statement for all time, and that it will never be demonstrated that such carcinogenic chemicals are not, in fact, also mutagenic?

Regarding the de-repressor concept, is it not possible that this phenotypic character of the cell could originate in part, and perhaps in large part, with the cell genotype? Could not modification of the cell genotype of some particular pattern effect repressor activity or function? I confess little knowledge in this area. I am merely curious.

Reply: A. Keith

It is simply a fact that some chemicals are carcinogenic in a given species and are not mutagenic in the same species. The way you have asked your question invites uncertainty. Of course, some chemicals which meet the criteria I have just mentioned may be mutagenic on some other species. The main point is that there is not a one to one correspondence between chemicals which are carcinogenic and ones which are mutagenic.

In answer to your second question, yes.

Question: Alexander Grendon, Donner Laboratory, University of California, Berkeley

Even though, as you pointed out, some chemical agents produce cancers and have not been shown to produce mutations and though your hypothesis as to their carcinogenic mechanism seems very plausible, have there, in fact, been any experiments that rule out the possibility that the process occurring during the formation of such cancers involves a change in the DNA of the affected cells?

Reply: A. Keith

I know of no such experiments.

Harold L. Rosenthal, School of Dentistry, Washington University

Your paper shows so beautifully what man can accomplish using his ingenuity and intelligence to develop that which is useful and to correct those things that may be harmful. At the same time, you make the statement that man cannot return to the land or give up the technology that may endanger our environment. I must take issue with the statement that we "can't" return. It may be necessary for man to return to the land if we don't use our intelligence and ingenuity to understand and combat the dangers of man's activities.

THE BIOCHEMICAL APPROACH TO MUTATION MONITORING IN MAN

M. C. CLARK, D. GOODMAN, and A. C. WILSON
UNIVERSITY OF CALIFORNIA, BERKELEY

1. Introduction

At present, there is no simple, inexpensive, direct way to determine whether an increase in the mutation rate in human populations has occurred or is occurring. The problem is increasingly serious with the ever more widespread use in our society of many potentially mutagenic agents. A widespread, highly mutagenic agent could have serious long term effects on the overall viability and fertility of a population, as well as increasing the incidence of specific genetic defects and diseases.

In order to protect human populations from the effects of potential mutagens, we must be able to make two kinds of measurements. First, we must determine the mutagenicity of specific environmental agents, preferably prior to their wide scale use, on a variety of experimental organisms. Second, we must be able to measure accurately the overall mutation rate in human populations, as a last check against a genetic emergency caused by previously undetected mutagens or mixtures of individually innocuous substances.

2. Tests on organisms other than man

Most work on the measurement of mutation rates has been done with nonhuman systems, both because of the greater number of organisms that can conveniently be studied, and because obviously it would be unethical to use humans as laboratory animals for the preliminary screening of potentially very dangerous materials. It is clear that the results of mutagenicity determinations on nonhuman systems cannot be blindly extrapolated to man. Organisms differ both in their inherent sensitivity to mutational damage, in their ability to repair mutation effects, and the ability to detoxify mutagens or to convert innocuous material into mutagens. However, from the standpoint of safety it would seem obvious that no material significantly mutagenic in a mammal should even be tested on man much less used in the environment, and that no material significantly mutagenic in *any* biosystem should be used in the environ-

Preparation of this paper has been supported by Grant GB-13119 from the National Science Foundation and Grant GM-18578 from the National Institutes of Health to A. C. Wilson. M. C. Clark is a Predoctoral Fellow of the National Science Foundation, and D. Goodman is a Postdoctoral Fellow of the National Institutes of Health.

ment as a whole unless it can be clearly shown to be nonmutagenic on other mammals.

With this basic limitation in mind, we will now consider briefly some of the most widely used methods for detecting specific mutagenic effects in lower organisms. Probably the most sensitive and economical test is the bacterial one developed by Ames [1]. Specific strains of bacteria are exposed to a suspected compound under conditions such that only bacteria mutagenized by the compound will form colonies. Various bacterial strains can be used to test for different kinds of mutations: base pair substitutions, insertions, or deletions. This method is simple and extremely sensitive, since 10^9 cells can be tested on a single petri dish; but it suffers from the limitation that both the metabolism and the genetics of the organism tested are unlike those of people.

A related method, the host-mediated assay (Legator and Malling, [6]) involves the injection of test bacteria into a mammal. The mammal is then treated with the potential mutagen, and the bacteria subsequently removed and tested for mutations. This test indicates whether the mammal can detoxify the examined compound, or metabolize it to form mutagenic products. This system is extremely useful in that it combines the sensitivity and range of the bacterial test with the metabolic (although not the genetic) functions of the host mammal.

The choice of a test mammal for any mutation detecting system depends on three factors: ability to extrapolate experimental results to man, generation time, and maintenance cost. Mice are frequently used because their short generation time allows the effects of a potential mutagen to be measured for several generations in a relatively brief period. Also, mice are perhaps the least expensive mammals to maintain in a laboratory colony. These advantages have led investigators to question the efficiency of using larger mammals, such as monkeys and chimpanzees, for mutation detection tests. Just how much better can we predict human response using these primates? Comparative studies with proteins and nucleic acids (Sarich and Wilson, unpublished) indicate that the genetic difference between chimpanzees and man is 10 to 20 times less than the mouse-man difference, and the rhesus monkey-man difference is 3 to 4 times less than the mouse-man difference. The quantitative estimates of genetic relatedness may enable investigators to more effectively balance generation time, cost, and predictive value.

There are a number of ways to detect directly an increase in extremely deleterious mutations in mammals. We will mention just one, the dominant lethal test (Bateman and Epstein, [2]). Generally, a male mammal is treated with a potential mutagen, and mated with an untreated female. The pregnant female is dissected and the embryos examined for morphological or chromosomal aberrations which would have caused abortion. This test has the disadvantage of being much more laborious and expensive than the bacterial methods, and of only detecting one class of mutants, namely, those with a dominant effect strong enough to cause visible defects in the embryo. These mutants are probably only a small class of the deleterious mutations possible.

3. Monitoring human populations

Let us now turn to the main point, the methods available for the detection of increased mutation rates in human populations. The classical method is to determine the frequency of certain dominant mutant phenotypes in the population, Crow [4]. To be useful in such a system, a mutant phenotype must meet the following criteria: (1) dominant, so as to appear in the first generation after a mutation occurs, (2) expressed at birth or early childhood, (3) serious enough to cause death before reproduction or reproductive failure so that all cases are due to new mutations, (4) not mimicked by phenocopies, and (5) so easily diagnosed that all such individuals will be recognized and reported. There are no mutant phenotypes now known that meet all these criteria. At best, this system is expensive and somewhat ambiguous, since a rise in congenital defects may have many causes other than mutation. Of course, the interaction of mutational events with other environmental insults is also of interest, so such an ambiguity is not necessarily a disadvantage. A more serious difficulty is that we are measuring only one class of mutants, as was the case with the dominant lethal test. In addition, of course, the detection of all individuals in a large population with a particular serious genetic abnormality is a task of tremendous scope.

A method of detecting a greater variety of mutations, including point mutations, is to examine a number of proteins electrophoretically. There are at least ten proteins which seem suitable (see Neel and Bloom [9]). These proteins show little or no polymorphism in the population and are detectable by electrophoresis with sufficient sensitivity and precision. The frequency of such mutants in man is not well known. Crow [4] suggests that the frequency of such mutations is on the order of 10^{-5} per gamete. In that case, since there are about 3×10^6 individuals born per year in the U.S., about 60 new mutants would be detected at each locus in a year, assuming complete detection. As the standard deviation is about 8, only an increase of one-third in the mutation rate, to 80 new mutants in the population, would be significant at the five per cent level. If we examine ten proteins, we would need to screen 300,000 infants a year to detect an increase of one-third in the mutation rate.

The frequency of electrophoretically detectable mutants, however, may be an order of magnitude higher than Crow estimates. Neel [8] estimated that certain rare, highly deleterious mutant phenotypes appeared at a frequency of about 10^{-5} in human populations. In *Drosophila*, there are about 30 times more mutations showing only a small decrease in viability than there are lethals (Mukai, [7]), and in *Salmonella*, only 10^{-1} of the mutants are lethals (Whitfield, Martin, and Ames, [10]). Thus, it may be reasonable to guess that the overall mutation rate in man could be as high as $10^{-5} \times 10 = 10^{-4}$. In this case, a one-third increase in the mutation rate could be detected by screening 30,000 infants a year for ten proteins. This is probably within current capabilities, as it involves examining 100 to 150 samples a day. This implies that, although electrophoresis is probably not capable of detecting small increases in the mutation rate, it might be useful for detecting large increases. Obviously, the more individuals or the more proteins examined, the smaller the detectable mutation rate increase. If the mutation rate in human populations is as high as 10^{-4}, the problems with electrophoresis are probably financial rather than technical.

Electrophoretic methods detect only those mutations causing a change in the charge of the protein. On the basis of the genetic code, these are about one-third of all new point mutations. The immunochemical method of microcomplement fixation can detect a substantial proportion of these electrophoretically "silent" mutations. Microcomplement fixation can detect single amino acid substitutions in proteins (Cocks and Wilson, [3]) and can also distinguish between hetero-zygotes and homozygotes of protein variants. For example, blood from wild type individuals and blood from individuals heterozygous for the sickle cell trait react differently when tested against antihemoglobin serum, as shown in Figure 1. Since most point mutations will appear in heterozygous form this is a particularly valuable asset. This method probably measures the true mutation rate most realistically, since it is sensitive to all classes of mutations. The main difficulty is that in its present manual form it is by far too laborious—much more so than electrophoresis. Therefore, the only hope of using microcomple-ment fixation for the detection of mutation rate increases is through automa-tion, which is now under investigation in our laboratory but has not yet been accomplished.

4. Tests for somatic mutations in man

A major problem in screening human populations is that each person tested provides only one item of data, thus requiring large sample sizes and the attend-ant problems of sample handling. Probably in large scale electrophoretic screen-ing, the expense of sample procurement and preparation would considerably exceed that of the actual biochemical testing. One way of greatly amplifying the mutation detecting power of any monitoring method is by measuring the rate of somatic mutations, rather than of those occurring only in the germ line. Of course we are basically concerned with the effect of mutagens on the germinal mutation rate, but since the two sorts of mutations probably occur by the same underlying mechanisms, and are subject to similar physiological factors, the correlation between the two types of mutations is undoubtedly very high.

Two approaches have been proposed to make use of somatic mutations. The first involves the cytological examination of blood cells, usually from cord blood. Unfortunately this method detects only chromosomal abnormalities and not point mutations, which are probably of much greater importance. In addition, the effort necessary to examine cytologically large numbers of cells is prohib-itive. Much effort has been spent on the automation of this procedure, however, and it is possible that it may eventually be useful as a screening method.

Another somatic cell procedure involves the detection of biochemical mutants

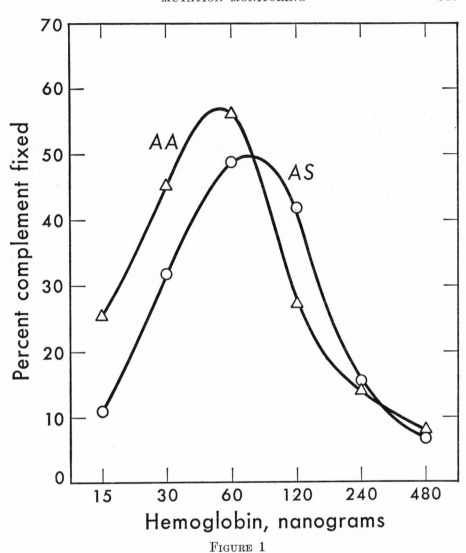

FIGURE 1

Microcomplement fixation test conducted with an antiserum prepared by
immunizing a rabbit with pure hemoglobin A. The antiserum was
reacted with hemolysates from a homozygous (AA) individual and from a
heterozygous (AS) individual. (Goodman and Wilson, unpublished data.)

in human leukocytes using cytochemical techniques. Sutton has suggested
testing for point mutations at the glucose-6-phosphate dehydrogenase locus in
leukocytes from peripheral human blood (Hook, [5]). Over 70 variants of this
enzyme are known in human populations. A substantial number of the variant
enzymes can use 2-deoxyglucose-6-phosphate as a substrate in place of glucose-

6-phosphate. Sutton suggests developing a staining technique which will detect an enzyme at this locus with an altered substrate specificity. Such alterations might be shown to be due to somatic mutations in the stem cells of the leukocytes.

A cytochemical approach, using this or other enzymes, offers several unique possibilities. For one thing, meaningful data can be obtained from the cells of a single individual. This would permit us to correlate exposure to a particular environmental factor with its mutagenicity, and would enable us to determine the actual mutational damage to any particular person exposed to high risks. A further advantage lies in the possible use of this method as a final screening procedure for compounds that have been found to be not appreciably mutagenic in animals. Since humans will be exposed to such compounds anyway, it is surely better to test them on one or two individuals, measuring any increase in the somatic mutation rate in their leukocytes, than on the population at large, which is the only alternative and the one currently practiced.

To summarize, the technically possible methods of determining meaningful mutation rates in man are at present limited to monitoring for rare dominant deleterious phenotypes, and to electrophoretic screening. In each case the problems are those of efficient sampling and finance, not of genetics or biochemistry. The more sensitive methods of microcomplement fixation and detection of somatic cell mutations, in spite of their great potential, do not yet seem technologically feasible for large scale use. Because of this potential, though, we urge, along with Crow and others, that major effort be given to their development.

Note added in proof.   We have tried various modifications of Sutton's method for staining white blood cells for glucose-6-phosphate dehydrogenase activity, with glucose-6-phosphate and 2-deoxyglucose-6-phosphate as substrates. In our hands this method does not give reproducible results (Clark and Wilson, unpublished work).

REFERENCES

[1] B. N. Ames, "The detection of chemical mutagens with enteric bacteria," *Chemical Mutagens: Principles and Methods for Their Detection* (edited by Alexander Hollaender), New York, Plenum Press, 1971, Vol. 1, pp. 267–282.
[2] A. J. Bateman and S. S. Epstein, "Dominant lethal mutations in mammals," *Chemical Mutagens, op. cit.*, Vol. 2, pp. 541–568.
[3] G. T. Cocks and A. C. Wilson, "Immunological detection of single amino acid substitutions in alkaline phosphatase," *Science*, Vol. 164 (1969), pp. 188–189.
[4] J. F. Crow, "Human population monitoring," *Chemical Mutagens, op. cit.*, Vol. 2, pp. 591–605.
[5] E. B. Hook, "Monitoring human birth defects and mutations to detect environmental effects," *Science*, Vol. 172 (1971), pp. 1363–1366.
[6] M. S. Legator and H. V. Malling, "The host-mediated assay, a practical procedure for evaluating potential mutagenic agents in mammals," *Chemical Mutagens, op. cit.*, Vol. 2, pp. 569–589.

[7] T. Mukai, "The genetic structure of natural populations of *Drosophila melanogaster*, I. Spontaneous mutation rate of polygenes controlling viability," *Genetics*, Vol. 50 (1964), pp. 1–19.

[8] J. V. Neel, "Mutations in the human population," *Methodology in Human Genetics* (edited by W. J. Burdette), San Francisco, Holden Day, 1962, pp. 203–224.

[9] J. V. Neel and A. D. Bloom, "The detection of environmental mutagens," *Med. Clin. N. Amer.*, Vol. 53 (1969), pp. 1243–1256.

[10] H. J. Whitfield, R. G. Martin, and B. N. Ames, "Classification of amino-transferase (C gene) mutants in the histidine operson," *J. Mol. Biol.*, Vol. 21 (1966), pp. 335–355.

Discussion

Question: John R. Goldsmith, Environmental Epidemiology, California Department of Public Health

Bateman has emphasized the value of spontaneous abortion rate and of cytogenetic abnormalities as an index of mutagenesis (and to a different extent teratogenesis). As therapeutic abortion becomes more frequent, cannot the cytogenetic analysis and also biochemical analysis of abortuses yield a greater amount of data on prevalence of mutations, since by definition some of them would not yield a live birth? Do you know of any laboratories studying biochemical mutagenesis in abortuses?

Reply: M. C. Clark, D. Goodman, and A. C. Wilson

We believe this question deserves serious attention, but we suspect that the mutation rate in fetuses from therapeutic abortion will not be as high as the rate in fetuses from spontaneous abortions. In addition, it would be difficult to obtain a representative sample of either spontaneous or induced abortions. We know of no laboratories studying biochemical mutagenesis in abortuses; several groups, including T. Shepard at the University of Washington and J. Miller at the University of British Columbia, are studying the frequency of gross abnormalities in abortuses.

Warren Winkelstein, Jr., Epidemiology, University of California, Berkeley

Since potential mutagens are probably unequally distributed between urban, rural, and regional areas, the number of samples required may be very substantial to provide adequate nationwide surveillance.

X-RAY FLUORESCENCE—AN IMPROVED ANALYTICAL TOOL FOR TRACE ELEMENT STUDIES

FREDERICK S. GOULDING

LAWRENCE BERKELEY LABORATORY

1. Introduction

This conference is dominated by discussion of the hazards associated with radiation; the present author feels that this exaggerates the importance of radiation levels as a hazard to health, when compared to other environmental insults to which we are subjected. One such insult is that produced by toxic metals introduced into living systems via many routes from industrial and natural sources. The purpose of this paper is to describe the use of X-ray fluorescence analysis with semiconductor detector spectrometers as a tool to permit fast analysis of specimens for a broad range of chemical elements present in trace quantities (that is, < 1 ppm by weight).

Before discussing trace element analysis, it may be useful to examine some possible reasons for the emphasis on radiation seen at this meeting, and to relate these to the situation seen in regard to trace elements.

1.1 Sources of radiation exposure to living things are well defined both in location and time. Nuclear explosions, reactors, X-ray sources and other radiation sources are constantly scrutinized by local, national and international agencies. Contrasting with this situation, the release of toxic elements by natural and industrial sources is subject to virtually no control or monitoring.

1.2 Public and governmental sensitivity to the hazardous nature of radiation has resulted in large programs to improve radiation measurements, and to evaluate radiation effects. Minor parallel steps are only now being taken to establish similar parameters for trace elements.

1.3 Indices of radiation effects are fairly well established. While authorities may differ in their interpretation of such studies, incidence of cancer and leukemia, longevity, and infant mortality rates have all been used as indicators of radiation effects. Although relationships between trace elements and certain diseases are known to exist, and others are suspected, few large scale statistical studies have been made to define the range of possible connections. We should also note that the wide variety of trace elements, and of their effects, make

This work was done under the auspices of the U.S. Atomic Energy Commission Contract No. W-7405-eng-48.

studies equivalent to those of radiation effects much more complex and difficult in this area.

1.4 Many trace elements are known to be essential to life, but at higher concentrations they become toxic. Despite occasional murmurs to the contrary, it seems that any increase in radiation above natural background must be considered harmful. It is, therefore, much easier to define the objectives of radiation measurement and control programs than the equivalent objectives for trace element studies.

1.5 Synergistic effects are known to be extremely important in determining the behavior of trace elements in living systems, and any study of the effects of a trace element must take account of the presence of other elements. While synergistic effects are also observed in radiation studies it appears that they are much less important than in trace element studies.

In summary, we might conclude that the complexity of trace element work, the lack of funding, and the resulting shortage of suitable measurement techniques, weighed against the relatively better situation in radiation studies, makes the latter a more attractive field of research. *However, it does not necessarily follow that such research is more valuable or more important than research on other forms of insult to life.*

I am in the fortunate position, in this talk, of being able to point to a possible important contribution of one nuclear research program to the problems of trace element studies. The value of the technique lies in its potential to permit large scale statistical studies of trace element distributions in our environment, and in living things, and thereby to provide the basic data needed to understand their effect. Furthermore, the technique promises to provide the fast monitoring method required for control of certain environmental contaminants. Before illustrating some applications, we will discuss the method and recent improvements in it that have greatly enhanced its usefulness.

2. Physics of the X-ray fluorescence method

Every scientist is familiar with the brilliant yellow sodium light produced when salt falls into a flame, and with the fact that our knowledge of the constitution of stars derives from studies of the light they emit. Optical spectroscopy, thanks to our fine natural light detectors, and also to Newton's use of prisms to disperse light according to its color, is now a commonplace tool in chemical analysis. However, a casual observer is overwhelmed by the complexity of optical spectra, caused mostly by the abundance of levels in the outer shells of atoms involved in light production. By observing X-rays, we can do elemental analysis much more easily, for the inner atomic shell structure producing X-rays is quite simple, and only a few X-ray wavelengths are emitted. X-ray spectroscopy requires both detection of X-rays and measurement of their energy; both functions are performed conveniently and well by semiconductor detectors—hence, the power of these devices as analytical tools.

The simple design of a semiconductor X-ray fluorescence spectrometer is shown in Figure 1; the physical mechanisms involved in the fluorescence process are also illustrated. Spectroscopy of photoelectrons and Auger electrons can also be used for analysis purposes, and the outstanding energy resolution of electron spectrometers permits measurement of changes in energy levels due to chemical bonding effects. However, the short range of electrons in materials limits the use of electron (compared with X-ray) spectroscopy. A wide range of methods of exciting characteristic X-rays is available. We discuss here only X-ray excitation of the sample, as our objective is to illustrate a relatively simple instrument; other, more complex, methods of excitation may be employed and are favored for some purposes.

The atomic shell transitions generating the X-rays of interest in this paper

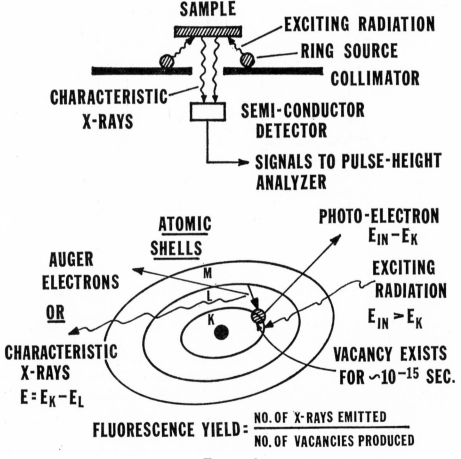

$$\text{FLUORESCENCE YIELD} = \frac{\text{NO. OF X-RAYS EMITTED}}{\text{NO. OF VACANCIES PRODUCED}}$$

FIGURE 1

Basic X-ray fluorescence spectrometer and the atomic process.

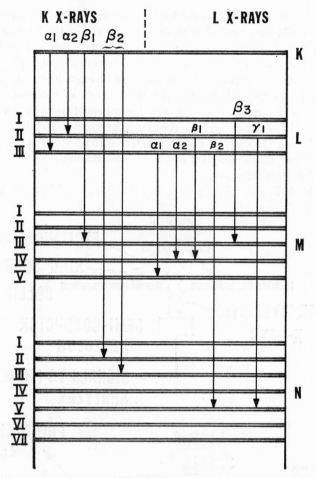

FIGURE 2

Atomic energy levels involved in the emission of X-rays of interest in this paper.

are shown in Figure 2. In the energy range of interest here, the energy resolution of semiconductor-detector spectrometers is such that the $K\alpha_1$ and $K\alpha_2$ lines are observed as a single line, as are the ($K\beta_1$, $K\beta_2$) and the ($L\beta_1$, $L\beta_2$, $L\beta_3$) lines. The intensity of the $K\beta$ peak is always much smaller than that of the $K\alpha$ line, while the $L\alpha$ and $L\beta$ lines are of similar intensity to each other, and are large compared with the other L X-rays.

Energies of primary interest lie below 30 keV, where the K X-rays of elements with $Z < 55$ are observed, and the L X-rays of the heavy elements also appear. Figure 3 shows the energies of these X-rays. The L X-ray energy of a heavy element may be the same as the K X-ray energy of a light element causing confusion in analysis; but most heavy elements are relatively rare in biological

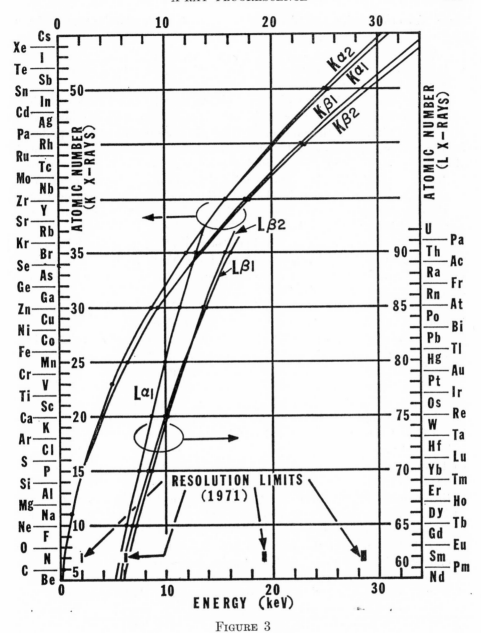

FIGURE 3

Energies of K and L X-rays up to 30 keV.

material, so the resulting problem is of less importance in practice than perhaps might be expected. Often, the presence of two strong L lines is useful—for example, Pb Lα coincides with the common As Kβ, but Pb Lβ only clashes with Kr, an uncommon contaminant. This fact permits determination of lead even when arsenic is present.

As shown in Figure 1, Auger electron emission provides an alternative method of filling vacancies in the atomic shells, thereby reducing the yield of fluorescent X-rays. Unfortunately, the emission of Auger electrons becomes highly probable for low energy transitions, so the fluorescent yield becomes very small for light elements, as shown in Figure 4. A similar behavior is seen for L X-rays emitted by heavy elements. The low fluorescent yield makes X-ray fluorescence spectrometers less sensitive for light elements; the problem is further exaggerated by the self-absorption of the fluorescent X-rays in the sample itself. This factor restricts analysis to a vanishingly small surface layer of the sample for low Z elements. Figure 5 shows the thickness of typical organic material required to attenuate different X-rays by a factor of $1/e$—this curve shows that samples less than 1 mm thick must be used to avoid severe attenuation even of elements as high as iron in the periodic table.

Two further efficiency factors are important in the design of an X-ray spectrometer. The efficiency of any detector falls at high energies causing a loss of counts. Lithium-drifted silicon detectors 5 mm thick were used in these studies; their efficiency is essentially unity in the energy range of interest here. Choosing the energy of the exciting radiation is a major step in the design of an X-ray fluorescence experiment; the probability of creating a vacancy is a maximum when the energy of the exciting radiation just exceeds the binding energy of the atomic shell involved. Figure 6 shows how the probability of vacancy creation in the K and L shells of various atoms depends on the energy of the exciting radiation. As an example of the use of these data, consider the case of a sample containing equal parts by weight of lead and bromine excited by Mo Kα radiation (17.4 keV). The lead L-shell vacancies will be excited 3.5 times as efficiently as the bromine K-shell vacancies; but since lead atoms are three times as heavy as bromine, only one-third as many lead atoms are present. Moreover, the fluorescent yield for Pb L X-rays is about 25 per cent less than that for Br K X-rays (Figure 4). The total number of lead X-rays from the sample will, therefore, be slightly smaller than that of bromine K X-rays.

We have so far considered only the fluorescent X-rays produced in the sample, their excitation and transmission to the detector, and their absorption in the detector. The main mechanism for absorption in the detector in this energy range is the photoelectric process; even where Compton scattering occurs in the detector, the scattered photon will nearly always be absorbed photoelectrically to produce a total charge in the detector proportional to the total energy of the original incoming X-ray. This situation is a pleasant contrast to the dominance of the Compton effect in spectra observed when detecting higher energy γ-rays. However, scattering processes are still a major factor in X-ray fluorescence

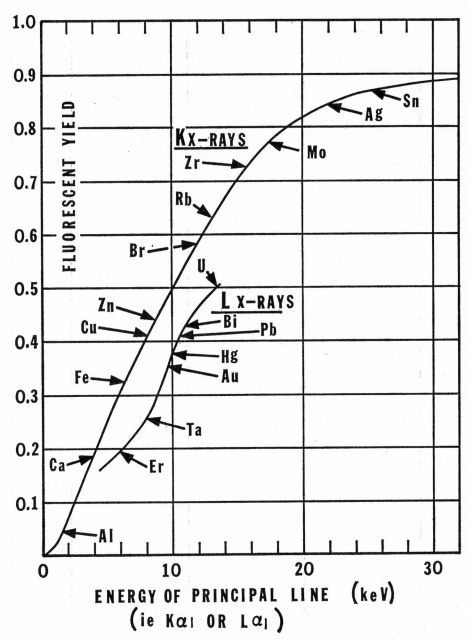

FIGURE 4

Variation of fluorescent yield with energy for K and L X-rays.

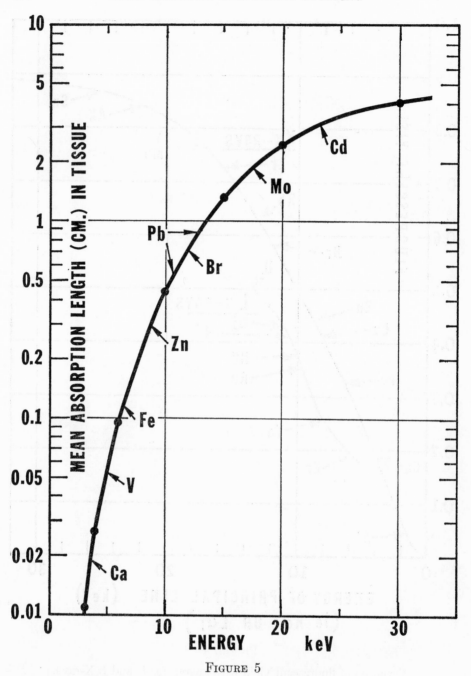

FIGURE 5

Absorption length of X-rays in typical organic material as a function of energy.

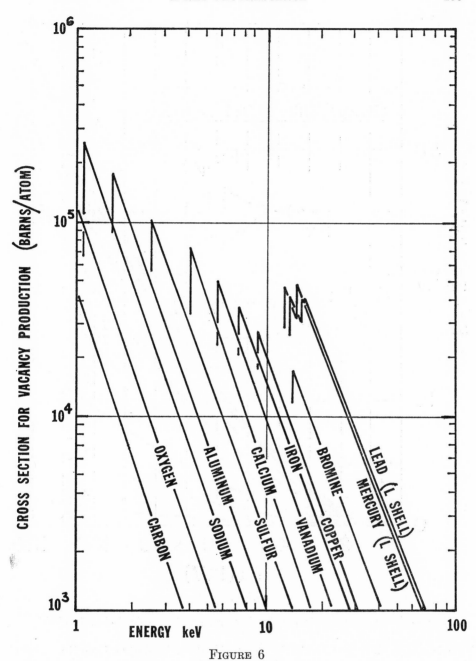

FIGURE 6

Cross section for K- (or L-) shell vacancy formation in several elements as a function of energy.

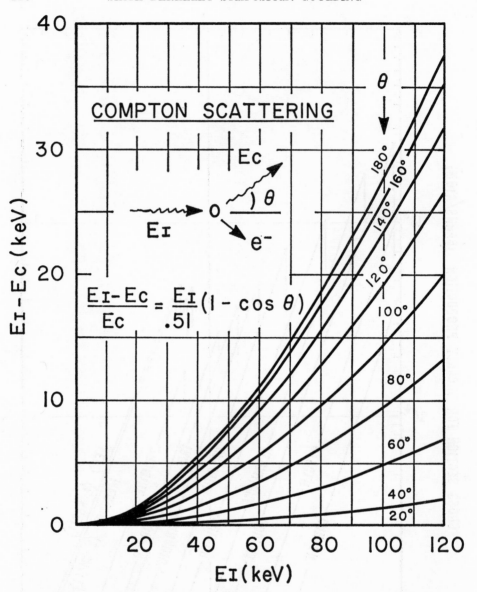

FIGURE 7

Energetics of the Compton-scattering process at energies below 120 keV.

spectroscopy—this time due to the effect of scattered radiation from the sample into the detector. Both coherent (elastic) and incoherent (Compton) scattering from the sample matrix, the organic base containing the trace elements, are present, the relative proportions depending on the ratio of light to heavy nuclei present. Figure 7 shows the energy exchanges involved in Compton scattering at these energies. Even 180° scattering of 20-keV photons involves a loss of energy of only about 1.5 keV to the photon involved in the scattering process.

We are now in a position to appreciate the general form of the spectrum (Figure 8) observed by a X-ray fluorescence spectrometer. The dominant fea-

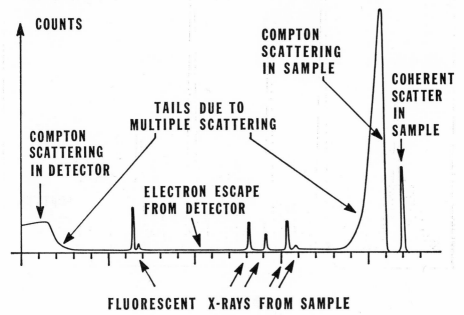

FIGURE 8

Idealized spectrum observed by an X-ray fluorescence spectrometer.

ture of the spectrum is the large scatter peaks that may constitute 99 per cent of the total counts observed. At the very low energy end of the spectrum we see the effect of scattered photons from the sample which happen to Compton-scatter from electrons in the detector and escape, leaving only the knock-on electron energy in the detector. The central region of the spectrum contains the interesting information on fluorescent X-rays emitted by the sample—unfortunately, superimposed on a background that, in an ideal case, is due only to photoelectrons from the detector escaping from its surface. Since the photons producing these electrons are primarily those scattered from the sample, and only part of the photon energy is converted into ionization in the detector, a continuum of pulse heights below the scatter peaks is generated by this process. As we will see later, other processes in the detector degrade events that should appear in the scatter peaks, resulting in increased background in the region of interest, thereby limiting our ability to see minute traces of impurities. The results given in this paper are made possible only by the methods of background reduction described herein.

The validity of this general picture is illustrated in an X-ray fluorescence

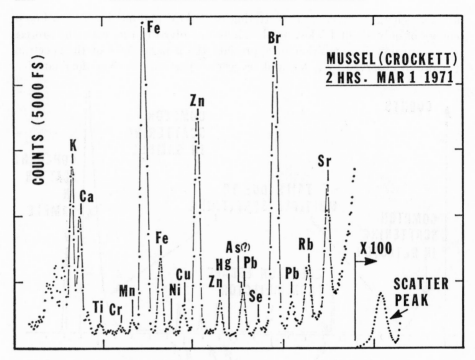

ENERGY

FIGURE 9

A practical spectrum obtained with Mo X-ray excitation of a freeze dried mussel sample.

spectrum obtained recently, shown in Figure 9. The large incoherent scatter peak is shown, while the coherent scatter peak is just beyond the right margin of the figure. The Compton edge due to the detector, which would normally appear at the left of this figure, has been removed by a discriminator in the electronic pulse pileup rejection system employed in this experiment. The background level seen in this figure, while low, is at least ten times as high as can be explained by photoelectrons escaping from the detector surface.

In the next section we will discuss the origins of background in X-ray spectrometers and the factors that determine the sensitivity of the spectrometers for analysis of trace elements.

3. Background and sensitivity factors

3.1 *Energy resolution and counting-rate performance.* The ability of an X-ray spectrometer to identify elements clearly depends on the width of the peaks produced in a spectrum like that shown in Figure 9—if lines due to adjacent elements are not separated, then separate identification of these elements is impossible. It is only in the last six years that electronic noise sources have been reduced sufficiently to make semiconductor detector X-ray fluorescence spectrometers a practical tool [1]. Recent work [2] has improved energy resolution to the point where even lines due to elements as light as carbon are well resolved [3].

A further benefit results from improvements in resolution—narrower peaks stand higher above the flat background observed in a spectrum like that in Figure 9. Therefore, better resolution results not only in the ability to resolve peaks due to adjacent elements, but also in lower detection limits for these elements.

The ultimate lower detection limit for an element is determined by comparing the statistical variations in background below a peak with the counts produced in a peak by a trace element contained in a sample. The number of counts in a peak is obviously proportional to the integrated dose of X-rays on the sample in the counting time. The background level is also proportional to the dose, though we will see later (Section 3.3) that the constant of proportionality is very dependent on the actual spectrometer design, and on the type of source of exciting radiation. Statistical fluctuations in the background are proportional to the square root of the dose. The detection limit for a peak is, therefore, proportional to $1/(\text{dose})^{1/2}$ (that is the sensitivity is proportional to $(\text{dose})^{1/2}$).

A consequence of these considerations is that fast determination of trace elements demands both intense X-ray excitation sources (see Section 3.2), and also electronics capable of handling high counting rates. Recent work [4] has resulted in electronics capable of handling adequate counting rates (2×10^4 to 10^5 c/sec) with virtually no degradation of resolution.

3.2 *Choice of excitation source.* In the work described in this paper we have chosen to use X-ray excitation of the fluorescent X-rays of elements in the sample. Possible alternative methods of excitation include charged particles [5] and electrons. The first alternative is eliminated partly because it requires a rather elaborate piece of equipment in the particle accelerator; also because comparisons between our own results and these achieved with charged-particle excitation have indicated that the sensitivity achieved in a short time is generally better with X-ray excitation of the right type. Direct-electron excitation of the sample is only useful for examination of very thin samples, and is, therefore, unsuitable for analysis of a broad range of samples. Furthermore, the Bremsstrahlung background produced by electron bombardment masks lines produced by trace elements, thereby seriously limiting sensitivity of the spectrometer.

Having chosen to use X-ray excitation, choice of the actual method of generation of X-rays must be considered. Radioactive sources such as I 125 (producing Te X-rays) are commonly employed, but such sources (of reasonable size) provide only relatively weak excitation. Consequently, long counting periods must be used to provide good trace element sensitivity. An attractive alternative is to use a small X-ray tube for excitation—intensities equivalent to tens of curies of radioisotopes are produced by X-ray tubes dissipating only a watt or two of electrical power. Furthermore, the emission of X-ray tubes can be cut to zero when not required—a contrast to radioactive sources which must be considered potentially hazardous to the general public when used on a large scale.

Unfortunately, the output of conventional X-ray tubes consists mostly of continuous Bremsstrahlung X-rays with the characteristic anode material X-rays superimposed on the continuum. When such a tube is used for excitation in a X-ray fluorescence spectrometer, the continuous X-ray spectrum is scattered from the sample to produce a large background in the spectral region where the characteristic fluorescent X-ray peaks should be observed. The flat background seen in Figure 8 is made very large in this case, and sensitivity to trace elements is very poor.

To overcome this limitation we have adopted two new designs of X-ray tube [6].

3.2.1 *Transmission anode tube.* In this tube a thin foil is used for the anode of a small X-ray tube, and the X-ray output is taken from the side of the anode opposite that bombarded by electrons. The filtering effect produced by passage of the X-rays through the thin anode foil cuts out most of the Bremsstrahlung spectrum, so that the output spectrum consists of almost monoenergetic X-rays characteristic of the anode material. A diagram of a tube of this type is shown in Figure 10.

3.2.2 *Secondary fluorescence anode tube.* In this tube a reflection X-ray output geometry is used for the anode, with the X-rays produced by the anode exciting fluorescent X-rays from a secondary target. Collimation of the output permits only the secondary-target X-rays to leave the tube. To obtain high output intensities from the tube it is necessary to have very close coupling between the primary anode X-rays and the secondary fluorescence target. Figure 11 shows an X-ray tube of this type. Its output is almost monochromatic, consisting of the X-rays characteristic of the secondary fluorescence target, with almost no background at lower energies.

These tubes have been demonstrated to be ideal excitation sources for fluorescence X-ray spectrometers, permitting fast analysis of samples for trace elements (for example, 0.1 ppm sensitivity in 15 minutes in some cases).

3.3 *Conventional detectors.* Figure 12 shows three silicon detector configurations used in X-ray spectrometers. The first type, originally used by Miller [7] is referred to as the "top hat" geometry, and is characterized by low leakage current and excellent high voltage behavior. Both characteristics are desirable in high resolution spectrometers, so this configuration is commonly used in these applications. The second type, the "grooved" detector, was used originally by E. Woo, and now employed in "Kevex" X-ray systems, possesses the same advantages as the "top hat" geometry. The third geometry, generally referred to as "planar" exhibits higher leakage current and capacity than the other two

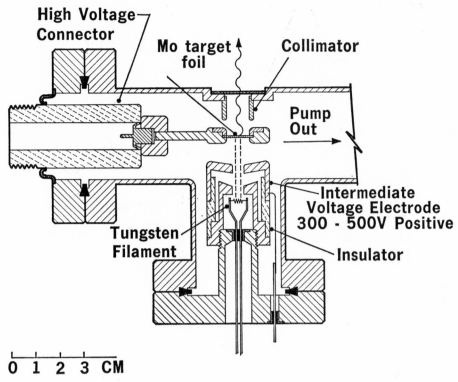

FIGURE 10

Schematic of transmission X-ray tube.

types, and is, therefore, rarely used in X-ray spectrometers. However, its background properties probably deserve investigation. Llacer [8] analyzed the behavior of these geometries with regard to their ability to sustain high voltage operation, and to produce low leakage current. His results, and those obtained in our laboratory, indicate that an n-type surface channel normally exists on the surface of silicon detectors. This channel acts as an extension of the n-type lithium-diffused region, and since it represents a poor junction to the bulk material, it contributes most of the leakage current, and sets the voltage limitation on detector operation. The fact that such channels can be "pinched off" by internal electric fields normal to the surface explains the difference in behavior between the structures (a) and (b) of Figure 12 and that of structure (c) of Figure 12.

These arguments fail to take into account the collection of charge produced by radiation in the bulk of the detector. The presence of n-type surface layers distorts the internal electric field pattern in the detector in such a way that collection of charge produced by X-rays interacting in some parts (shown by horizontal line shading in Figure 12) is via the surface layers. This causes a loss

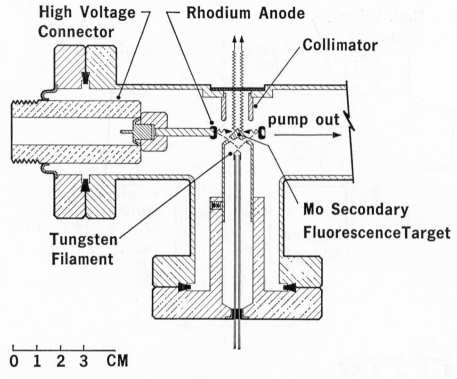

High Voltage
Connector   Rhodium Anode

Collimator

pump out

Mo Secondary
FluorescenceTarget

Tungsten
Filament

0 1 2 3 CM

FIGURE 11

Schematic of secondary fluorescence tube with a rhodium anode and molybdenum secondary target.

of charge, so that signals that should appear in the backscatter peaks appear
in the general background in spectra. The tests presented in this paper show
that this is the predominant source of background in existing spectrometers.

At first sight, it may appear that collimation of X-rays to prevent their interaction in the poor field regions might reduce background, and, indeed, tests
show that some improvement can be achieved by this method. It is also obvious
that improvement results from increasing the detector area while collimating
to a small central region, but the large consequent increase in detector capacity
seriously degrades the system resolution—an intolerable price to pay. The
degree of collimation that can be used is determined by the requirement for
good sample detector geometry; as shown in the typical geometry shown in
Figure 13, this implies a wide divergence of X-rays hitting the detector. Despite
the possible auxiliary collimator shown in this figure, mounted on the detector
face—an expedient rarely adopted as it is difficult to change this collimator to
suit the energy range of interest—X-rays like that travelling from A to B still
interact in regions of poor charge collection. On the other hand, the X-ray from

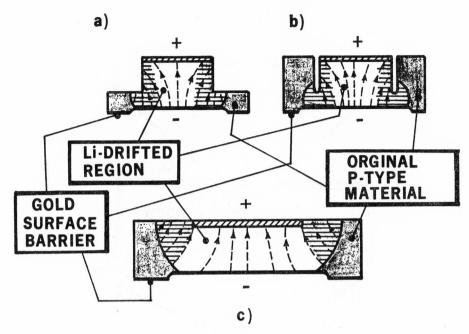

FIGURE 12

Types of detector configuration used in X-ray spectrometers. (a) Top hat detector, (b) grooved detector, (c) planar detector.

C to D interacts in a region of good charge collection, and produces the correct signal.

As shown in Figure 14, the background due to degraded pulses increases as the energy of the X-rays impinging on the detector increases. For cadmium X-rays, the total integrated background count approaches the total number of counts in the main X-ray peak. The proportion of counts in the background decreases considerably in the case of zirconium, and still further for lower energy X-rays.

3.4 *Guard ring detectors.* Guard rings have been employed for many years to overcome fringing field effects in standard capacitors, and have also found application in semiconductor detectors as a device to reduce edge leakage [9]. It, therefore, seems an obvious step to use a guard ring to define the boundary of the sensitive volume of a detector by internal electric field lines rather than by a physical surface with its unknown charge-trapping characteristics. This can also be considered as an electronic collimation technique. Figure 15(a) shows the simple implementation of the idea; note that the output signal is derived only from the central region, while the guard ring and the central region are maintained at the same dc potential (ground).

Even this configuration suffers from a signal degradation problem at the edge

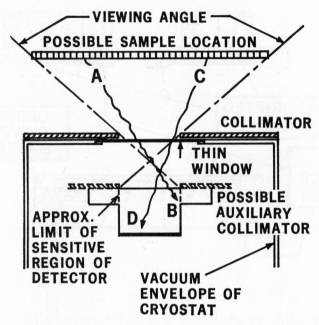

FIGURE 13

Preferred collimation geometry for an X-ray spectrometer.

of the central region. The initial X-ray interaction in this peripheral region produces a dense cloud of charge ~5 microns in diameter; in the electric field, holes and electrons are separated, drifting toward their appropriate electrode. The internal repulsive fields existing within the hole and electron clouds are very large compared with the drift field in the detector—therefore, the cloud dimensions rapidly increase until the internal repulsive field approaches the same value as the drift field. This means that the cloud dimensions reach about 100 microns during the charge collection process. Consequently, a peripheral region of 100 microns thickness exists around the sensitive region from which only part of the charge due to an event is collected in the central region—this means that many of the backscattered events appear in general background. Our measurements show that the background present with a simple guard ring detector is from 2 to 10 times smaller than that with a top hat detector, the exact factor depending on the energy of the backscatter peak.

A further reduction in background is achieved by sensing coincident signals between the guard ring and central regions, and rejecting the central region signal when such a coincidence is registered. This "guard ring reject" system

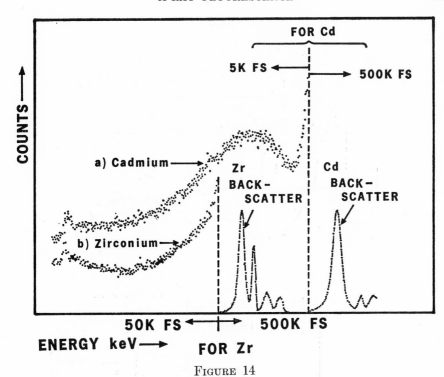

FIGURE 14

The variation in background of a top hat detector as a function of the energy
of radiation striking the detector. (a) Cd X-rays scattered by lucite, (b) Zr
X-rays scattered by lucite.

effectively eliminates the partial collection from the peripheral region of the
sensitive volume of the detector. With such an arrangement, we approach the
background level expected due to electron escape from the detector surface.
In our actual detector, shown in Figure 15(b), a double guard ring is used, the
outer ring serving to reduce edge leakage in the inner ring, and thereby improving its noise properties so that the inner guard ring signal discriminator
can be set low to detect very small signals.

The improvement in background resulting from use of the guard ring reject
method is shown in Figure 16. Using exactly the same geometry, cadmium
X-rays scattered from lucite were used to irradiate a standard top hat detector
(a), and the guard ring reject detector system (b). For the same number in the
cadmium peak, total counts recorded in the background is 40 times smaller for
the guard ring reject system than for the top hat detector. Larger factors are
obtained when the detectors are irradiated by radiation of higher energy than
cadmium X-rays.

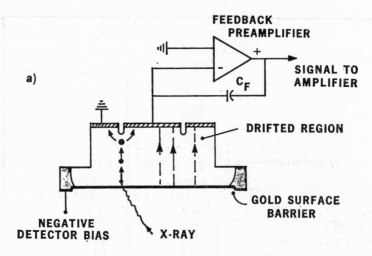

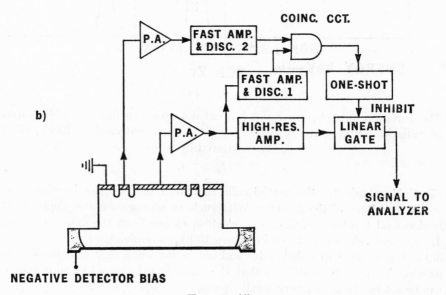

FIGURE 15

Guard ring detectors. (a) Simple guard ring approach showing the mechanism
for degraded pulses, (b) double guard ring detector with pulse-reject circuitry
to remove degraded pulses.

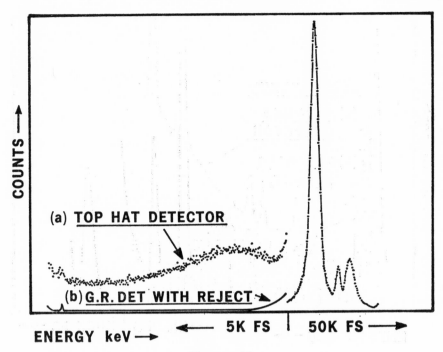

FIGURE 16

Background produced in detector systems by de X-rays. (a) Top hat detector,
(b) guard ring detector with reject circuitry.

Figure 17 illustrates the improvement achieved by using this technique on
a typical sample. A blood serum sample was examined by a system using a top
hat detector, then by one using a guard ring detector, and the same total number
of counts were accumulated in the molybdenum X-ray backscatter peak. The
reduction in background seen in the second spectrum, averaged over the full
energy range, is about a factor of 15. Much larger factors, ranging up to about
60, have been observed for higher energy excitation. Comparison of the two
results in Figure 17 shows the improvement in ability to see small traces of
elements, such as nickel, present in the specimen at a level near 0.1 ppm. Better
statistics realized by a longer count, or with more intense excitation, would
further reduce the detection limit.

4. Some experimental results and applications

Semiconductor X-ray spectrometers are distinguished by their ability to
simultaneously survey a whole spectrum of trace elements present in samples
at levels less than 1 ppm. More sensitive methods can be devised for particular
elements; for example, atomic absorption can be used for such elements as

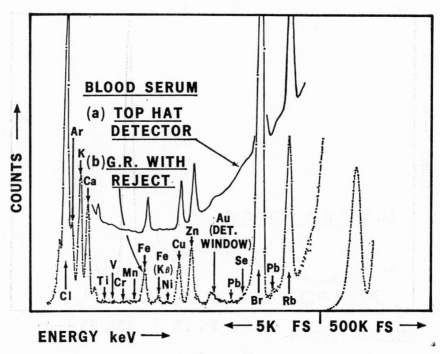

FIGURE 17

Fluorescence X-ray spectrum obtained on a blood serum specimen using a spectrometer equipped with guard ring detector and reject circuitry (b). For comparison, the spectrum obtained on the same sample with the same geometry, and the same total counts in the scatter peak, but with a simple top hat detector, is also shown (a).

mercury. Neutron activation can exhibit greater sensitivity than X-ray fluorescence for some elements, but its sensitivity is poor for many elements, and the activation process is slow and costly. The activation of sodium in biological samples necessitates a long "cooling off" period (often a month) before a sample can be analyzed. Other methods, including atomic absorption, require extensive sample preparation, while X-ray fluorescence requires very little, though removal of water by freeze drying is useful in many cases. For these reasons, it seems that X-ray fluorescence analysis may find a wide range of applications. A few of these are discussed in this section, and some early results are presented.

4.1 *Analysis of foods.* Public attention has recently been focused on the adverse effects of certain trace elements in foods. Mercury and cadmium, in particular, have received attention. The toxicity of these elements results from their accumulating in the body and inhibiting important organic functions. This is to be distinguished from the toxicity of chemicals that cause immediate effects on body functions. Furthermore, the toxicity of an element depends a great deal on its chemical form. In the well-known case of mercury, organic mercury compounds are more toxic than are most of its inorganic salts. X-ray fluorescence gives no information on the chemical form of elements, so it can only be used to flag possible hazards.

Since some trace elements are toxic while others are essential to life, detailed maps of trace elements in a wide range of foods should be of value in nutritional studies.

With these points in mind, we will examine a few measurements on foods made with Mo X-ray excitation. In practice, Mo X-ray excitation restricts the range of elements measured with good sensitivity to the heavy elements (L X-rays), and to the light elements from iron through rubidium. Higher energy excitation must be used for elements heavier than rubidium. Measurements of elements below iron demand lower energy excitation, and also the use of thin samples. In both cases, a slight loss of sensitivity will result.

A test on a freeze dried tuna sample, depicted in Figure 18, shows the presence of mercury (~0.5 ppm). However, the sample was concentrated by freeze drying; the original tuna would have assayed at 0.1 ppm of Hg—far below the FDA tolerance (0.5 ppm).

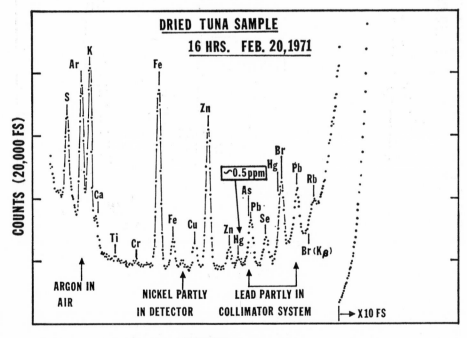

FIGURE 18

Dried tuna spectrum. Features due to the collimator system and other extraneous materials are indicated.

Several features in this spectrum are due to instrumental artifacts. The lead peaks are due mainly to lead contamination of the collimator, while the nickel peak is partly due to the nickel surface barrier used in this particular detector. These factors are not present in the remaining spectra in this paper. The argon peak is due to excitation of argon present at a level of 1 per cent in the air surrounding the system. We also note that krypton—present to only 1 ppm in air— is observed in some spectra. Helium can be used to displace air when necessary.

A word about calibration is in order at this point. A calibration procedure devised by J. Jaklevic and R. Giauque uses thin metal layers of known weight to determine geometry factors, excitation efficiency, and sample absorption [10]. This method permits rapid quantitative assessment of the concentrations of impurities observed in spectra.

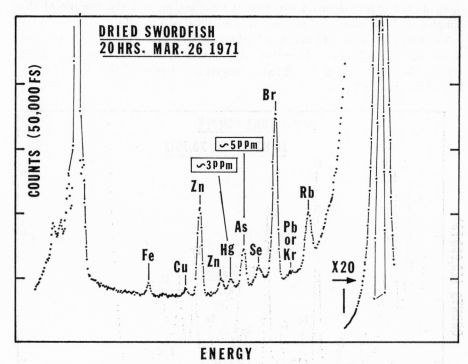

FIGURE 19

Dried swordfish spectrum.

A test on dried swordfish (see Figure 19) is to be compared with that on tuna shown in Figure 18. A very different distribution of elements is seen: iron is almost absent in swordfish, while much more mercury and arsenic are present. Arsenic is probably not a serious problem, but this sample of swordfish (recently purchased) certainly contains more mercury than the FDA limit.

The results of a test on liver sausage (freeze dried) are seen in Figure 20.

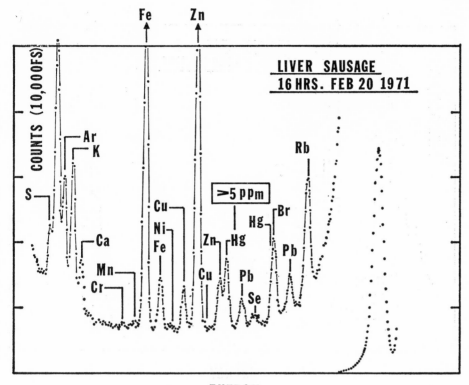

FIGURE 20

Liver sausage spectrum.

Both mercury and lead are present in higher concentrations than in Figure 18, but the mercury in this case might well be largely in the form of an inorganic salt. Although the FDA limit of 0.5 ppm contains no qualification as to chemical form, eating liver sausage might not be a hazardous pursuit!

4.2 *Plants.* Trace elements are known to be an important factor in the growth of plants, and examination of plant material and soils may provide useful information in agriculture. Perhaps the most surprising result of our analysis of plant material has been the high concentration of relatively uncommon elements.

Figure 21 shows the spectrum obtained with a camellia leaf. We show it as a contrast to the other spectra presented here. The large quantities of manganese and strontium were surprising. The ubiquitous lead peak is seen although this particular plant grows far from freeways or city traffic.

4.3 *Medical uses.* Medical applications of this technique could turn out to be the most important. Studies have shown that trace elements in blood may correlate with disease states, indicating a possible diagnostic technique [11].

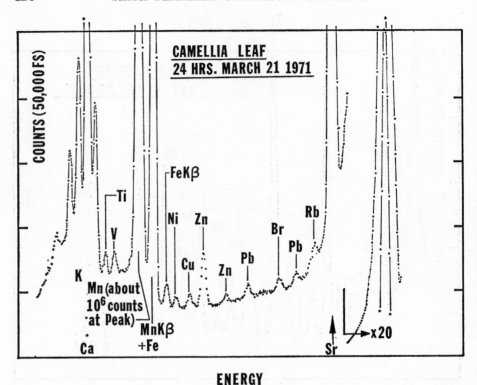

ENERGY

FIGURE 21

Camellia leaf spectrum.

Other readily available samples include urine, hair, fingernails, and so forth. The statistical correlations between trace elements in these various samples and disease have hardly been studied at all—mostly due to the lack of a suitable fast, convenient, and accurate method of measurement. Semiconductor detector X-ray fluorescence spectrometers may well stimulate new interest in this field of study.

More specific types of problem also arise in medicine. The examination of organs for trace elements may be an important tool in studying the accumulation of elements in organs and its relationship to diseases, including cancer. One example is the analysis of blood samples for lead, as an indicator of lead poisoning in infants. The X-ray fluorescence method is sensitive enough for this purpose as illustrated in Figure 22, where lead is clearly determined in a freeze dried sample (about 3 ml) of blood obtained from an infant with symptoms of lead poisoning. The sensitivity is adequate to permit use of very small samples (~0.25 milliliters).

Two other features should be noted in Figure 22. The large iron peak results in an increase in background at energies below that of the peak. This is due to

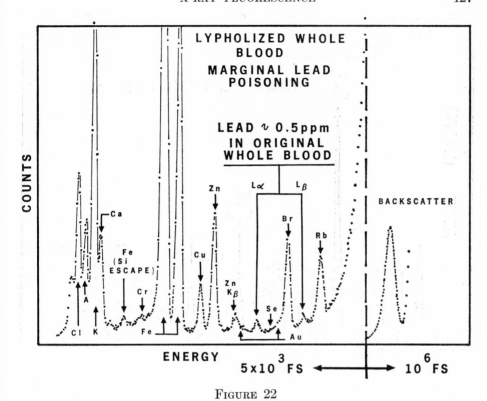

FIGURE 22

Freeze dried whole blood—original blood volume = 3 ml.

a small dead layer in this particular detector. We also note a small peak resulting from escape of silicon K X-rays excited by the iron X-rays. This indicates the care required in interpretation of results, as the escape peak is quite close to the expected location of a vanadium peak (but resolvable by the system).

4.4 *Air pollution studies.* Elemental analysis of particulate deposits on air pollution filters presents a simple problem for this technique. With short sample collection times, the mass of the particulate deposit becomes a substantial fraction of the air filter material. Consequently, when the filter is subjected to analysis, background due to backscatter in the filter material is quite small. An example of air filter analysis is shown in Figure 23. The quantity of lead measured in this sample (from Detroit) corresponds to almost 2 μg/meter^3 of air, a typical number for air in an industrial area.

Figure 23 also illustrates the use of an X-ray tube to speed up analysis. The transmission-anode tube of Figure 10 was used to expose the sample for only ten minutes to produce this result—opening up the possibility of on-line analysis of pollution with the capability of seeing fluctuations in particulate pollutants on a time scale much shorter than an hour.

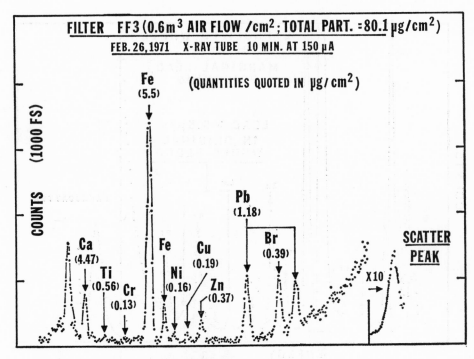

FIGURE 23

Air filter spectrum.

5. Conclusions—relationship to epidemiological study

The instrument described here provides the potential for study, on a large scale, of a broad range of elements present at the <1 ppm in many types of sample. Sample preparation can be very simple at this level of sensitivity, and the measurement time can be short. It is possible to envisage a broad range trace element analysis facility capable of analyzing 50 or more samples per day down to a level of about 0.1 ppm by weight at a cost in the region of $100,000. A limited range of elements could be examined at much lower cost. When this is combined with a computer analysis system, direct readout of trace element concentrations for this large number of samples would be possible.

The samples analyzed with the instrument may cover the whole range required for epidemiological studies. Foods, water, air, particulates, plants, and other environmental factors can be subjected to analysis, and studies of the deposition of trace elements in the human body may also be facilitated. The broad range capability of the method will permit studies of synergistic relationships between the elements.

A necessary factor in any epidemiological study would seem to be the data base available on the problem to be studied. We believe that the use of X-ray fluorescence analysis represents a significant new technique in acquiring such a data base for trace element studies.

It is a pleasure to acknowledge the cooperation and advice of J. Jaklevic, B. Jarrett, R. Pehl, D. Landis and J. Walton throughout this program. The assistance of R. Giauque in some of the analytical chemistry aspects of the work is also recognized.

REFERENCES

[1] H. R. Bowman, E. K. Hyde, S. G. Thompson, and R. C. Jared, "Application of high resolution semiconductor detector in x-ray emission spectrography," *Science*, Vol. 151 (1966), pp. 562–572.

[2] F. S. Goulding, J. Walton, and D. F. Malone, "An opto-electronic feedback pre-amplifier for high-resolution nuclear spectroscopy," *Nucl. Instr. Methods*, Vol. 71 (1969), pp. 273–279.

[3] J. M. Jaklevic and F. S. Goulding, "Detection of low energy x-rays with Si(Li) detectors," *IEEE Trans. Nucl. Sci.*, Vol. NS-18 (1971), pp. 187–191.

[4] D. Landis, F. S. Goulding, R. H. Pehl, and J. T. Walton, "Pulsed feedback techniques for semiconductor detector radiation spectrometers," *IEEE Trans. Nucl. Sci.*, Vol. NS-18 (1971), pp. 115–124.

[5] T. B. Johanssen, R. Akselsson, and S. A. E. Johanssen, "X-ray analysis: elemental trace analysis at the 10^{-12}g level," *Nucl. Inst. Methods*, Vol. 84 (1970), pp. 141–143.

[6] J. M. Jaklevic, R. D. Giauque, D. F. Malone, and W. L. Searles, "Small x-ray tubes for energy dispersive analysis using semiconductor spectrometers," *Advan. X-ray Anal.*, Vol. 15 (1971); also, Lawrence Berkeley Laboratory Report No. LBL-10.

[7] G. L. Miller, private communication referenced in Llacer's paper (Ref. [8]), p. 99.

[8] J. Llacer, "Geometric control of surface leakage current and noise in lithium drifted silicon detectors," *IEEE Trans. Nucl. Sci.*, Vol. NS-13 (1966), pp. 93–103.

[9] W. L. Hansen and F. S. Goulding, "Leakage, noise, guard rings and resolution in detectors," *Proc. Asheville Conf.*, NAS-NRC Report No. 32 (1961), pp. 202–209.

[10] R. D. Giauque and J. M. Jaklevic, "Rapid quantitative analysis by x-ray spectrometery," *Advan. X-ray Anal.*, Vol. 15 (1971), to be published; also, Lawrence Berkeley Laboratory Report No. LBL-204.

[11] J. W. Gofman, "Chemical elements in the blood and health," *Advan. Biol. Med. Phys.*, Vol. 8 (1962); *also*, UCRL-10211 p. 62, Semiannual Report, *Biology and Medicine*, Donner Laboratory.

Discussion

Question: E. Tompkins, Human Studies Branch, Environmental Protection Agency

Would you please relate the level of lead shown in the blood of the child to that which would be seen in a child with acute lead poisoning?

Reply: F. S. Goulding

Based on published data, the 0.5 ppm lead (in whole blood), seen in this case, produces minimum clinical symptoms. Acute cases involve levels many times higher than seen in this case.

Question: Emanuel Hoffer, California Department of Public Health

You said we didn't have any information or work on trace elements in human tissues. What about the work of Isabel Tipton in spectroscopy at Oak Ridge?

You also made a statement that we do not have any information on the distribution of these elements in human tissues. There is the work of a statistician at the University of Cincinnati who studied lead, chromium, mercury, and others, and found that these trace elements followed a log normal distribution.

And finally, you made the statement that the level of detection for your fluorescence method would detect the lowest levels of lead where you would get clinical symptoms. What about the delta-aminolevulinic acid test which can detect subclinical levels of symptomology?

Reply: F. S. Goulding

I plead guilty to the charge of minimizing work already done in trace elements. I think, however, that it is fair to say that there is a lack of sufficient data on trace element distributions in human organs, on the normal fluctuations in the levels, on the levels in the environment, foods and other species, and their effect on human beings. I try to relate the few good studies in this area to the vast amount of work on radiation effects, bearing in mind that trace elements represent a much more complex problem.

As seen in the spectrum of Figure 22 the level of lead in this specimen is easily seen, though it is characteristic of just detectable poisoning symptoms. The method is clearly capable of observing levels 5 to 10 times below the clinically observable amount. The ALA in urine test does not appear to be as sensitive since very little change in ALA in urine occurs until lead approaches the clinically detectable level. A nonlinear relationship exists in a plot of ALA in urine against the ALA activity in blood.

Question: Thomas F. Budinger, Donner Laboratory, University of California, Berkeley

We have some data on the distribution of trace elements in tissues of normal and diseased individuals. I would guess that between Professor Isabel Tipton, Dr. Schroeder, and others, even including Dr. John Gofman who worked in this field before he took up his chromosome and linear radiation work, there are over 2000 cadaver studies. These studies include measurements of quantities of about 40 elements above the 1 to 2 ppm lower sensitivity range. Furthermore, we have measures of the variance of concentration in organs and the change in content with age, for example, cadmium. I agree the possible threat from trace elements which accumulate irreversibly in body organs such as kidney and liver is perhaps far greater than anticipated radiation pollution. However, I plead that we avoid trace element fishing trips.

For example, we have collected about 2000 food measurements of Cd and well over 300 references, some of which would seduce us to conclude Cd causes hypertension and tumors. Perhaps it has an important role as perhaps chromium deficiency is involved in diabetes. However, without critical experiments aimed at elucidating mechanisms, I do not see how we are going to establish cause effect. One element is a lifetime work for one man. Or study of many trace elements in one plant or animal system is a lifetime work. Not both.

Reply: F. S. Goulding

Thank you for detailing the scope of existing studies. While I agree that the purely statistical approach to trace elements will be time consuming, and the problem might better be approached in terms of specific mechanisms, we should recognize that accumulation of statistical data often points to mechanisms that might not be predicted.

George B. Morgan, Monitoring Support Division, Environmental Protection Agency

The X-ray fluorescent instrument is excellent for biological tissue. It definitely lacks sufficient sensitivity for measuring concentrations in the environment. To determine transport of pollutants through the media of air, water and food, sensitivities 50 to 100 times the X-ray instrument are needed.

EPA is using the X-ray instrument described for source samples. For ambient samples the following instruments are necessary because some are particularly sensitive for certain elements: (1) emission spectrometer (computerized)—40 elements simultaneously; bad for zinc, arsenic, lithium, mercury, and so forth; (2) spark and plasma TET; good for Pb, Cr, U, Cu, Be, Co, Ti, Fe, Mn, Sr, and so forth; (3) activation analysis—very good for selenium, mercury, arsenic, (4) atomic absorption spectrometer—100 samples per hour for two elements.

There is no one instrument that can do the entire job for environmental measurements. To select the optimum instrument, the researcher must consider the pollutant and the type of sample. The cost for a comprehensive laboratory is about 700 to 800K which includes data acquisition.

AVERAGING TIME AND MAXIMA FOR AIR POLLUTION CONCENTRATIONS

RICHARD E. BARLOW

UNIVERSITY OF CALIFORNIA, BERKELEY

1. Introduction

The Public Health Service, and now the Environmental Protection Agency (EPA), has operated a Continuous Air Monitoring Program (CAMP) since January 1962 (see Larsen [7]). Under CAMP, air pollutant concentrations are punched automatically into a computer tape every five minutes. Air pollutants which are being monitored include carbon monoxide, various hydrocarbons, nitric oxide, nitrogen dioxide, total oxidants (chiefly ozone), and sulfur dioxide. Monitoring stations are located in Chicago, Cincinnati, Los Angeles, New Orleans, Philadelphia, San Francisco, and Washington, D.C. Measurements are recorded in parts per million (ppm), parts per hundred million (pphm), and parts per billion (ppb), and micrograms per cubic meter ($\mu g/m^3$). For example, oxidant, a chief constituent of smog, is considered undesirable if its concentration reaches or exceeds 0.1 ppm.

In San Francisco, the Bay Area Air Pollution Control District (BAAPCD) publishes, on a monthly basis, daily average high hour oxidant values as well as daily peak oxidant values. Carbon monoxide values are similarly recorded. However, sulfur dioxide values (in ppb) and particulate values ($\mu g/m^3$) are recorded only as 24 hour averages. Averaging times vary widely because of the nature of the pollutant and the monitoring system used. For example, particulate matter is measured by the high volume sampler. In this device, air is blown through a filter which is then weighed after 24 hours. In the San Francisco Bay Area, particulate readings tend to be made (for a 24 hour period) every other day and occasionally every third day. Particulate readings are recorded at nine locations in the Bay Area and there are wide variations in the data due to location.

Table I, taken from the pamphlet "Air Pollution and the San Francisco Bay Area" provides a summary of the main air pollutants, the 1969 California state standards for these pollutants and reasons for controlling their concentrations.

For purposes of evaluating air quality, it is important to know the probability of maximum pollutant concentrations exceeding state standards which are

This research has been partially supported by the Office of Naval Research under Contract N00014-69-A-0200-1036 and the National Science Foundation under Grants GP-29123 and GK-23153 with the University of California. Reproduction in whole or in part is permitted for any purpose of the United States Government.

TABLE I

AIR QUALITY STANDARDS—BAY AREA 1969

Substance	State standard	Objective
Oxidant	0.1 ppm for 1 hour	To prevent eye irritation and possible impairment of lung function in people with respiratory problems. Also to prevent damage to vegetation
Carbon monoxide	20 ppm for 8 hours	To prevent carboxyhemoglobin levels greater than 2%
Sulfur dioxide	0.04 ppm for 24 hrs. (particulate $>100 \, \mu g/m^3$)	To prevent possible increase in chronic respiratory disease and damage to vegetation
	0.5 ppm for 1 hour (regardless of particulate)	To prevent possible alteration in lung function; also odor prevention
Particulate matter	60 $\mu g/m^3$ ann. geom. mean No single 24 hour sample to exceed 100 $\mu g/m^3$	To improve visibility and prevent acute illness when present with about 0.05 ppm sulfur dioxide.
Visibility reducing particles	Visibility of not less than 10 mi. when relative humidity is less than 70%	To improve visibility
Nitrogen dioxide	0.25 ppm for 1 hour	To prevent possible risk to public health and atmospheric discoloration
Hydrogen sulfide	0.03 ppm for 1 hour	To prevent odor

stated for various averaging times. We use extreme value theory to determine the limiting distribution of maximum air pollutant concentrations as a function of averaging time. Bounds on the location parameter of the corresponding extreme value distribution are used to evaluate air quality. These are then used to evaluate suspended particulate data.

2. Larsen's results

R. I. Larsen and co-workers in a series of papers [6], [7] analyzed three years of gaseous air pollutant data, from December 1961 to December 1964, for the seven cities previously mentioned. They stated [6] two main conclusions:

(1) Concentrations are approximately lognormally distributed for all pollutants in all cities for all averaging times.

(2) The median concentration (50th percentile) is proportional to averaging time to an exponent (and thus plots as a straight line on logarithmic paper).

Figure 1, [6], is a plot of three years' data for Washington, D.C. illustrating empirically Larsen's second assertion.

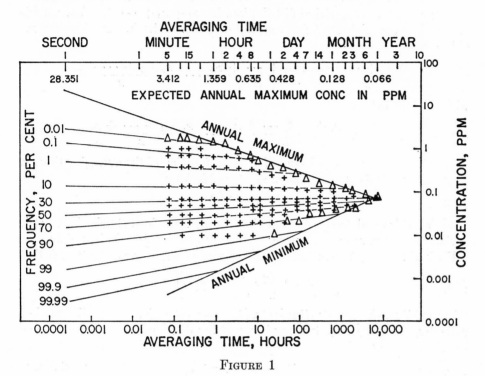

FIGURE 1

Concentration *versus* averaging time and frequency for nitrogen oxides in Washington from 12/1/61 to 12/1/64.

Since sums of (independent) lognormal distributed random variables are not distributed as lognormal random variables, Larsen's first result might be considered suspect. Histograms of air pollution concentration data are highly skewed, much as are life test data plots. One suspects that the data might as easily be fit with gamma or Weibull distributions as with a lognormal distribution. The randomness in air pollutant concentrations results mainly from meteorological phenomena. For this reason, the observations will not be independent. However, observations averaged over long time periods and within a given season of the year may be considered to be statistically independent and identically distributed.

Recently, N. D. Singpurwalla has interpreted Larsen's results using extreme value theory [11]. However, he again approximates sums of lognormal random variables by lognormal random variables.

3. A mathematical model based on extreme value theory

Suppose n observations $x_1, x_2, \cdots, x_k, x_{k+1}, \cdots, x_n$ are taken, say over a year or a season. We assume, for now, that observations are independent with distribution F. For Larsen [7], F is the lognormal distribution. He estimates parameters from the empirical distribution. Let $x_1 + x_2 + \cdots + x_k$ have distribution F_k so that F_k is the k-fold convolution of F with itself. Consider averages of length k

$$(3.1) \qquad \frac{x_1 + x_2 + \cdots + x_k}{k}, \frac{x_{k+1} + \cdots + x_{2k}}{k}, \cdots, \frac{x_{n-k+1} + \cdots + x_n}{k}$$

where $k \ll n$. Let

$$(3.2) \qquad \eta_{k,n} = \max \left\{ \frac{x_1 + x_2 + \cdots + x_k}{k}, \cdots, \frac{x_{n-k+1} + \cdots + x_n}{k} \right\}.$$

We are interested in the behavior of $\eta_{k,n}$ as a function of the averaging time, k. Larsen [6] estimates the median of $\eta_{k,n}$ (say $M_{k,n}$) by

$$(3.3) \qquad M_{k,n} = Ck^{-b}$$

where $b > 0$ is tabulated [6] as a function of various one hour standard geometric deviations. These values were apparently computed empirically from data.

Fix k and let $\alpha_{k,n} > 0$ and $\beta_{k,n}$ be a sequence of norming constants such that

$$(3.4) \qquad \lim_{n \to \infty} P \left[\frac{\eta_{k,n} - \beta_{k,n}}{\alpha_{k,n}} \leq x \right]$$

exists and is nondegenerate. Gnedenko [4] showed that, for nonnegative random variables, there are only two possible limiting distributions (up to scale and location) namely

$$(3.5) \qquad \Lambda(x) = \exp\{-e^{-x}\} \qquad\qquad -\infty < x < \infty$$

and

$$(3.6) \qquad \Phi_\alpha(x) = \begin{cases} 0 & x \leq 0 \\ \exp\{-x^{-\alpha}\} & x > 0 \end{cases} \qquad \alpha > 0.$$

The lognormal, Weibull, gamma and most other commonly used distributions lie in the domain of attraction of Λ; that is, there exist constants $\alpha_{k,n} > 0$, $\beta_{k,n}$ for these distributions such that (3.4) equals $\Lambda(x)$. Marcus and Pinsky [9] give necessary and sufficient conditions for a distribution to lie in the domain of attraction of Λ. For some $\alpha > 0$, F belongs to the domain of attraction of Φ_α if and only if $1 - F(x) = x^{-\alpha}L(x)$ where

$$(3.7) \qquad \lim_{t \to \infty} \frac{L(tx)}{L(t)} = 1$$

for each $x > 0$. (See [3], pp. 270–272). Intuitively, Λ seems the more reasonable limiting distribution for air pollution concentrations and we make that assumption henceforth. In particular, we assume that F_k belongs in the domain of attraction of Λ.

Let $G(x) = 1 - e^{-x}$ for $x \geq 0$ and $R_k(x) = G^{-1}F_k(x)$. Gnedenko [4] (see [9]) showed that the norming constants could be expressed as

$$(3.8) \qquad \beta_{k,n} = \frac{R_k^{-1}[\log n]}{k}$$

and

$$(3.9) \qquad \alpha_{k,n} = \frac{R_k^{-1}[1 + \log n] - R_k^{-1}[\log n]}{k}$$

Hence for large n,

$$(3.10) \qquad P[\eta_{k,n} \leq x] \sim \Lambda \left[\frac{x - \beta_{k,n}}{\alpha_{k,n}} \right]$$

where $\beta_{k,n}$ and $\alpha_{k,n}$ are given by (3.8) and (3.9). $\beta_{k,n}$ is the location parameter and also approximately the 37th percentile of $\Lambda[(x - \beta_{k,n})/\alpha_{k,n}]$. Since $\alpha_{k,n}$ is typically small relative to $\beta_{k,n}$, the latter provides a convenient way of summarizing $\eta_{k,n}$. The distribution of Λ is tabulated by Owen [10].

The main difficulty in using $\beta_{k,n}$ occurs in computing the convolution F_k. In the case where

$$(3.11) \qquad F(x) = \int_0^x \frac{u^{\lambda-1}e^{-u/\theta}}{\theta^\lambda \Gamma(\lambda)} du$$

that is, the gamma distribution, then, of course, F_k is again a gamma distribution and there is no problem in computing R_k. For n large and $k \ll n$, Gurland [5] has approximated $\beta_{k,n}$ as

$$(3.12) \quad \beta_{k,n} \sim \frac{1}{k} \left[\theta \log \left(\frac{n}{\left(\frac{\theta}{k} \right)^{\lambda k-1} \Gamma(\lambda k)} \right) + \theta(k\lambda - 1) \log \left(\frac{\theta}{k} \log n \right) \right]$$

where $\Gamma(\cdot)$ is the gamma function. Hence,

$$(3.13) \qquad \beta_{k,n} \sim \frac{\theta}{k} \log n = C_n k^{-1}$$

for large n. Since the right tail of the gamma distribution behaves like the exponential distribution, (3.13) is not surprising. If we let

$$(3.14) \qquad F(x) = 1 - \exp \left\{ - \left(\frac{x}{\delta} \right)^{1/b} \right\} \qquad \text{for} \quad x \geq 0$$

then $\beta_{k,n}$ behaves like k^{-b} in a sense to be made precise. Of course, $b = 1$ corresponds to the exponential distribution. The lognormal distribution has an $R(x)$ which is first convex and then concave over adjacent intervals.

4. Bounds on $\beta_{k,n}$

We wish to obtain bounds on $\beta_{k,n}$ for distributions other than the gamma distribution. To motivate this discussion, consider the Weibull distribution

(3.14). Unfortunately, numerical methods are necessary to compute convolutions of Weibull distributions. For this distribution,

$$(4.1) \qquad\qquad R(x) = G^{-1}F(x) = \left(\frac{x}{\delta}\right)^{1/b}.$$

Note that for $0 < b < 1$, R is convex while for $b > 1$, R is concave. Many other distributions useful in life testing, especially, have the property that R is convex for certain parameter values and concave for others. [A distribution F for which R is convex (concave) is called IFR (DFR) in Barlow and Proschan [2].] It is easily seen that if $R(0) = 0$ and R is convex, then R is *superadditive*, that is,

$$(4.2) \qquad\qquad R(x + y) \geq R(x) + R(y) \qquad\qquad \text{for} \quad x, y \geq 0.$$

This weaker property is sufficient to provide one sided bounds on $R_k(x)$ and hence on $\beta_{k,n}$.

THEOREM 4.1. *If F is continuous, $F(0) = 0$ and R is superadditive (subadditive), then*

$$(4.3) \qquad\qquad R_k(x) \leq (\geq)G^{-1}\Gamma_k[R(x)] \qquad\qquad x \geq 0,$$

where $\Gamma_k(x) = 1 - e^{-x}\left[\sum_{j=0}^{k-1} \frac{x^j}{j!}\right]$ for $x \geq 0$ is the gamma distribution and $G \equiv \Gamma_1$.

PROOF.   Assume R is superadditive so that R^{-1} is subadditive, that is,

$$(4.4) \qquad\qquad R^{-1}(x + y) \leq R^{-1}(x) + R^{-1}(y) \qquad\qquad \text{for} \quad x, y \geq 0.$$

Let $Y_j, j = 1, 2, \cdots, n$, be i.i.d. random variables with distribution $G(x) = 1 - e^{-x}$ for $x \geq 0$. Then $X_j = R^{-1}(Y_j), j = 1, 2, \cdots n$, are i.i.d. with distribution F. Since R^{-1} is subadditive,

$$(4.5) \qquad\qquad \sum_{j=1}^{n} R^{-1}(Y_j) \geq R^{-1}\left(\sum_{j=1}^{n} Y_j\right).$$

Then

$$(4.6) \qquad 1 - F_n(x) = P\left[\sum_{j=1}^{n} X_j > x\right] \geq P\left[R^{-1}\left(\sum_{j=1}^{n} Y_j\right) > x\right]$$

$$= P\left[\sum_{j=1}^{n} Y_j > R(x)\right]$$

$$= e^{-R(x)}\sum_{j=0}^{n-1} \frac{[R(x)]^j}{j!} = 1 - \Gamma_k[R(x)].$$

Hence, $R_k(x) \leq G^{-1}\Gamma_k[R(x)]$ as claimed. The proof for the subadditive case is similar. Q.E.D.

The proof for R convex was first noted by Erwin Straub [12]. Note that (4.3) is an equality if $k = 1$ or if $R(x) = ax$ for some $a > 0$.

The following theorem provides additional bounds.

THEOREM 4.2. *If F is continuous, $F(0) = 0$ and R is convex (concave), then*

$$(4.7) \qquad\qquad R_k(x) \geq (\leq)G^{-1}\Gamma_k\left[kR\left(\frac{x}{k}\right)\right] \qquad\qquad \text{for} \quad x \geq 0.$$

The proof is due to Straub [12].

We can now state useful bounds on the extreme value location parameter, $\beta_{k,n}$.

COROLLARY 4.3. *If F is continuous, $F(0) = 0$ and R is convex (concave), then*

$$(4.8) \qquad \frac{1}{k} R^{-1} \Gamma_k^{-1} G(\log n) \leqq (\geqq) \beta_{k,n} \leqq (\geqq) R^{-1} \left[\frac{1}{k} \Gamma_k^{-1} G(\log n) \right].$$

PROOF. Equation (4.8) follows from (4.3), (4.7) and repeated use of the fact that $(AB)^{-1}(x) = B^{-1} A^{-1}(x)$ where $AB(x)$ means $A[B(x)]$; that is, functional composition. Q.E.D.

From (3.12), we see that for large n and $k \ll n$

$$(4.9) \qquad \Gamma_k^{-1} G(\log n) \sim \left[\log \left(\frac{n}{\Gamma(k)} \right) + (k-1) \log \log n \right] = c_{k,n}$$

so that for R convex (concave)

$$(4.10) \qquad \frac{1}{k} R^{-1}[c_{k,n}] \leqq (\geqq) \beta_{k,n} \leqq (\geqq) R^{-1} \left[\frac{c_{k,n}}{k} \right].$$

EXAMPLE. If $F(x) = 1 - \exp\{-(x/\delta)^{1/b}\}$, then $R^{-1}(y) = \delta y^b$ and

$$(4.11) \qquad \begin{aligned} \beta_{k,n} &\leqq \delta(c_{k,n})^b k^{-b} \sim \delta[\log n]^b k^{-b} \\ \beta_{k,n} &\geqq \delta(c_{k,n})^b k^{b-1} \sim \delta[\log n]^b k^{b-1} \end{aligned}$$

if $0 < b < 1$. For the Weibull distribution, we see the connection between the power of the averaging time and the shape parameter of the distribution.

5. Bounds on $\alpha_{k,n}$

We could also provide bounds on

$$(5.1) \qquad \alpha_{k,n} = \frac{R_k^{-1}(1 + \log n) - R_k^{-1}(\log n)}{k}$$

by using Theorems 4.1 and 4.2. However, the bounds will be less elegant than the bounds on $\beta_{k,n}$. Typically, $\alpha_{k,n}$ will be small and probability will tend to be concentrated around the location parameter $\beta_{k,n}$. When R is convex, an upper bound on $\alpha_{k,n}$ is available.

THEOREM 5.1. *If F is continuous with mean θ, $F(0) = 0$ and R is convex, then*

$$(5.2) \qquad \alpha_{k,n} \leqq \frac{1}{k^2 \theta} \qquad \text{for large } n.$$

PROOF. If R is convex, then R_k is also convex by Theorem 5.1 on p. 36 of [2]. Since $R_k^{-1}(x)$ is concave, it crosses the ray $x/k\theta$ at most once, and from above. $R_k(x)$ crosses $x/k\theta$ exactly once since F_k and $G(x) = 1 - \exp\{-x/k\theta\}$ have the same mean, namely $k\theta$. Hence for large values of x, the slope of R_k^{-1} is less than the slope of $x/k\theta$. It follows that

$$(5.3) \qquad \alpha_{k,n} = \frac{R_k^{-1}[1 + \log n] - R^{-1}[\log n]}{k} \leqq \frac{1}{k^2 \theta}. \qquad \text{Q.E.D,}$$

6. An example using suspended particulate data

As was mentioned in the introduction, suspended particulates are averaged over a 24 hour period. The California state standard for particulate matter is 60 $\mu g/m^3$ annual geometric mean. (The geometric mean is used because of the lognormal distribution assumption.) The state standard also specifies that no single 24 hour sample is to exceed 100 $\mu g/m^3$. A severe pollution episode occurs if average particulate values remain high for several days in succession. Hence, an interesting question to ask is, what is the probability that the maximum of, say, three day averages over the course of several seasons will exceed any specified amount? An alternative approach is to ask for upper bounds on $\beta_{k,n}$, the location parameter of $\eta_{k,n}$. Table II shows 24 hour average particulate measurements recorded in $\mu g/m^3$ for San Jose, California during November, 1969.

TABLE II

24 HOUR PARTICULATE MEASUREMENTS
FOR SAN JOSE, CALIFORNIA
NOVEMBER, 1969

Day in November 1969		Suspended particulates ($\mu g/m^3$)
S	1	140
Sunday	2	.
M	3	.
T	4	122
W	5	.
T	6	39
F	7	.
S	8	.
Sunday	9	.
M	10	.
T	11	129
W	12	.
T	13	147
F	14	.
S	15	31
Sunday	16	.
M	17	.
T	18	.
W	19	132
T	20	124
F	21	.
S	22	.
Sunday	23	.
M	24	.
T	25	158
W	26	.
T	27	105
F	28	.
S	29	140
Sunday	30	.

They were taken from the BAAPCD Contaminant and Weather Summary. The particulate "season" in the San Francisco Bay Area is roughly September through December, and November, 1969, was unusually high. Notice that only 11 out of 30 days were actually recorded. The mean value for this month was 115 $\mu g/m^3$.

Assuming a Weibull distribution for particulate values during this month, we found the linear invariant estimates for δ and b using tables computed by Nancy Mann [8]. For this particular data, we found $\hat{\delta} = 127.74$ and $\hat{b} = .2727$. In this case,

$$(6.1) \qquad R(x) = \left(\frac{x}{\delta}\right)^{1/b}$$

is convex. There is no *a priori* reason, however, why b should lie in $[0,1]$. Letting $n = 270$ days corresponding to three seasons of 90 days each, we computed the upper bounds on $\beta_{k,n}$ for $k = 1, 2, \cdots, 7$, shown in Table III. The lower bounds on $\beta_{k,n}$, however, were unreasonably low in view of the data at hand.

TABLE III

UPPER BOUNDS ON $\beta_{k,n}$ $(n = 270)$

k	$(\mu g/m^3)$
1	176.25
2	145.89
3	130.68
4	120.76
5	113.61
6	108.07
7	103.62

◇ ◇ ◇ ◇ ◇

The author would like to acknowledge Professor Nozer Singpurwalla for bringing this air pollution problem to his attention and for making available to him preprints of his papers.

REFERENCES

[1] "Air pollution and the San Francisco bay area," Bay Area Air Pollution Control District Publications, San Francisco, California, 1970.
[2] R. E. BARLOW and F. PROSCHAN, *Mathematical Theory of Reliability*, New York, Wiley, 1965.
[3] W. FELLER, *An Introduction to Probability Theory and its Applications*, Vol. 2, New York, Wiley, 1966.
[4] B. V. GNEDENKO, "Sur la distribution limite du terme maximum d'une série aléatoire," *Annals of Mathematics*, Vol. 44 (1943), pp. 423–453.
[5] J. GURLAND, "Distribution of the maximum of the arithmetic mean of correlated random variables," *Ann. Math. Statist.*, Vol. 26 (1955), pp. 294–300.

[6] R. I. LARSEN, "A new mathematical model of air pollutant concentration averaging time and frequency," *J. Air Poll. Cont. Assoc.*, Vol. 19 (1969), pp. 24–30.

[7] R. I. LARSEN and C. E. ZIMMER, "Calculating air quality and its control," *J. Air Poll. Cont. Assoc.*, Vol. 15 (1965), pp. 565–572.

[8] N. R. MANN, "Tables for obtaining the best linear invariant estimates of parameters of the Weibull distribution," *Technometrics*, Vol. 9 (1967), pp. 629–645.

[9] M. MARCUS and M. PINSKY, "On the domain of attraction of $e^{-e^{-x}}$," *J. Math. Anal. Appl.*, Vol. 28 (1969), pp. 440–449.

[10] D. B. OWEN, *Handbook of Statistical Tables*, Addison-Wesley, p. 962.

[11] N. D. SINGPURWALLA, "Extreme values from a lognormal law with applications to air pollution problems," *Technometrics*, to appear.

[12] E. STRAUB, "Application of reliability theory to insurance," ASTIN Colloquium, Randers, Denmark (1970).

EFFECTS OF ENVIRONMENTAL POLLUTANTS UPON ANIMALS OTHER THAN MAN

ROBERT W. RISEBROUGH
UNIVERSITY OF CALIFORNIA, BERKELEY

All pollutants are waste products. They include both "natural" compounds that are present in undesirable concentrations in local ecosystems and chemical species foreign to the environment. What distinguishes them from other chemical wastes is, by definition, a potential capacity to inflict harm upon one or more species of an ecosystem. Many populations of wildlife are currently affected by pollutants that decrease the life span of adults or lower their reproductive capacity. Since these species breathe the same air as does man and consume some of the same food, they constitute an early warning system for the future health of man. In some cases it is not immediately evident which pollutants are producing the observed effects and an increasing amount of research is being devoted to these problems. The present paper will attempt to summarize our current knowledge in the field of pollutant ecology that relates to the effect of pollutants upon wildlife populations. Hopefully it will provide a useful background to the formulation of programs that will look for effects of these same pollutants upon human health.

1. Dimensions of the system

It is evident that the capacity of the earth to support life is finite. A convenient parameter with which to discuss the dimensions of the global ecosystem is the amount of organic carbon synthesized per year by the photosynthetic activity of plants. Production of organic carbon in the sea has been estimated to be in the order of 2×10^{16} grams per year [52]. Photosynthetic processes on land produce in the order of 6×10^{16} grams of organic carbon per year [38]. The sum of these numbers, 8×10^{16} grams, is therefore a useful number with which to compare such parameters as annual petroleum production, the annual U.S. production of organic chemicals, global mineral production, and the total amount of waste material formed by the sum total of global technology. The latter might be considered a measure of the current level of human activity.

2. Waste products as pollutants

Carbon monoxide is a waste product that has become a locally dangerous pollutant in urban areas. Worldwide emissions from major industrial sources, excluding fuel consumption, amounted to 2.6×10^{13} grams in 1968 [38]. The atmospheric concentrations of carbon monoxide, however, do not appear to be increasing [16], [30] in spite of the considerable input. In sinks such as soil [29] carbon monoxide may be converted to either carbon dioxide or methane. The ecosystem appears therefore to be able to absorb the increased input.

In contrast, the atmospheric concentrations of carbon dioxide are increasing. The burning of fossil fuels currently releases in the order of 14 billion metric tons of carbon dioxide into the atmosphere per year or 1.4×10^{16} grams. This figure is close to the activity level of the natural ecosystem. Approximately one-half of this input remains in the atmosphere, resulting in an increase in CO_2 concentrations at the rate of 0.2 per cent per year [38], [34], [42].

Several of the heavy metals may become pollutants when environmental levels are increased significantly above background. Zinc, copper and iron are essential components of enzymes or other proteins and are not toxic at lower concentrations. Lead, mercury and cadmium, however, may be highly toxic to biological systems. All are natural components of the earth's crust and are transported to the sea upon weathering of the rocks by water or wind. Ultimately they are deposited in sediments which in turn are uplifted to form new mountains.

The present rate of input of lead into the oceans is approximately ten times greater than the rate of introduction by natural weathering [13]. Concentrations of lead in surface waters are higher than in deeper waters. Moreover, the isotope composition of the lead in surface waters and in recent precipitation is more similar to that of mined ore leads than to that in marine sediments [12]. There are almost no data, however, that would suggest that the higher concentrations of lead in surface seawater derived from the lead transported through the atmosphere have resulted in higher concentrations in marine wildlife. Lead poisoning is frequently encountered in waterfowl that have ingested lead shot from the bottom mud of marshes, but there are very few data that would indicate what should be "natural" levels in terrestrial and fresh water wildlife. Lead concentrations that have been measured in liver and bone of selected species of birds have been compiled by Bagley and Locke [5]. Annual global production of lead is in the order of 3×10^{12} grams [38].

Global production of cadmium ranged from 1.2 to 1.4×10^{10} grams per year between 1963 and 1968 [38] and in addition significant quantities are released into the environment as by-products of zinc mining operations. These quantities are sufficiently high to indicate that cadmium could be a significant pollutant in local areas. Extensive poisoning of human populations living downstream from a zinc mine in Japan has been documented [64]. Cadmium levels in tissues of the Ashy Petrel (*Oceanodroma homochroa*), an oceanic species resident in the coastal waters of California, were approximately twice as high as concentrations in tissues of two populations of the Wilson's Petrel (*Oceanites oceanicus*) that breed in the Antarctic but summer in the Atlantic and Australian regions respectively. Cadmium levels in tissues of the Snow Petrel (*Pagodroma nivea*), a species that never leaves the Antarctic pack ice, were in the same order of magnitude as those in the Wilson's Petrel. Cadmium levels in eggs of the Common Tern (*Sterna hirundo*) from Long Island Sound were in the order of 0.2 ppm dry weight, not appreciably higher than those in the Antarctic Tern (*Sterna vittata*) from the Antarctic with levels in the order of 0.1 ppm. These limited data do not suggest that cadmium has become a significant marine pollutant [2].

An estimated 5,000 tons of mercury, or 5×10^9 grams, are transferred per year from the continents to the oceans as a result of continental weathering [35]. Global production of mercury is currently about twice as high, in the order of 9×10^9 grams per year [23]. In addition, the burning of petroleum releases in the order of 1.6×10^9 grams into the atmosphere per year [9]. A conservative estimate of the amount of mercury released per year into the global environment from the burning of coal is in the order of 3×10^9 grams [33]. Thus the amount of mercury mobilized by man is considerably higher than the amounts released by natural weathering. Nevertheless the amount of mercury estimated to be in the oceans is in the order of 10^{14} grams, approximately three orders of magnitude higher than the amount of mercury consumed in the United States since 1900. Mercury in marine organisms is therefore most likely of natural origin. Thus, mercury concentrations in the tissues of the Ashy Petrel from the coastal waters of California, the site of most of the mercury mines in the United States, were in the same order of magnitude as the mercury concentrations in tissues of the Snow Petrel from Antarctica. Mercury concentrations in nine eggs of the Common Tern from Long Island Sound were only slightly higher than in nine eggs of the Antarctic Tern from Antarctica [2].

Environmental residues of mercury in Sweden, as measured by concentrations of mercury in feathers of several species of birds, rose dramatically in the years following 1940. The source was considered to be the alkyl-mercury compounds used as seed dressings [8]. The use of mercury compounds as seed dressings ha also resulted in higher environmental levels of mercury in Alberta [19]. Local pollution of lakes and rivers in North America has resulted from the discharge of the wastes from factories using mercury, notably chlorine-caustic soda plants.

Pollution of the environment by heavy metals might therefore be considered equivalent to an accelerated weathering process, resulting in local concentrations that are higher than background levels and in some cases higher than a threshold of damage to one or more species. Local pollution by other inorganic compounds such as salts are also instances of concentrations of naturally occurring compounds exceeding those normally encountered by wildlife.

In contrast, several of the organic compounds that have become pollutants are synthetic, unknown to exist in the environment before the development of chemical technology and represent a new evolutionary factor in ecosystems. Total U.S. production of synthetic organic chemicals in 1968 was 120,000 million pounds or approximately 5×10^{13} grams, a 15 per cent increase over 1967 [59]. Growth over the past decade has been comparable, and it can be anticipated that a comparable rate of growth will continue, not only in the

United States but in the remainder of the world. Among the groups of chemicals showing high rates of growth have been the plastics, pesticides, and resins. U.S. production of DDT in 1968 was 140 million pounds or about 6×10^{10} grams; production of the dioctyl phthalates, one of the principal groups of plasticizers, was 440 million pounds [59]. Production figures of the polychlorinated biphenyls (PCB), industrial compounds that like the DDT compounds have become widespread pollutants in the global environment [32], [50] are not available because of current U.S. practice that protects proprietary information.

Eventually all of the organic chemicals synthesized become waste products and their potential threat to the environment is evidently a function of their input, their chemical and biological stability, mobility, toxicity, and of the properties of breakdown products. Carcinogens and mutagens clearly pose a much more significant threat to human populations, which value each individual life, than to wildlife populations which can much more readily withstand the loss of individuals as long as the reproductive capacity is not impaired. Although wildlife populations might serve to monitor increases in environmental levels of chemical carcinogens or mutagens, damage to a species from such chemicals has not been documented. Environmental chemicals that lower the reproductive capacity of a population pose a much more serious threat to wildlife.

3. The thin eggshell phenomenon

In 1967 Ratcliffe [43] published a paper that showed for the first time that physiological changes in wildlife species were correlated with geographical and temporal patterns of environmental pollution. Eggs of the Peregrine Falcon (*Falco peregrinus*), Sparrowhawk (*Accipiter nisus*) and Golden Eagle (*Aquila chrysaetos*) laid in several regions of Great Britain after 1947 tended to have lower amounts of calcium carbonate in the shells than did eggs laid by the same species in the past. No changes were detected in the Central and Eastern Highlands of Scotland which are comparatively remote from pollution sources. Hickey and Anderson [28] subsequently showed that many populations of species of fish-eating and raptorial birds in North America were also suffering from the same syndrome and, as in Britain, the changes were first detectable in eggs laid in 1947. Ratcliffe [44] has documented the species for which shell thinning has been shown in Britain. In North America the phenomenon has now been documented in the following families of birds: the pelicans, *Pelecanidae* [4], [48]: cormorants, *Phalacrocoracidae* [4]; hawks, *Accipitridae*, [28]; falcons, *Falconidae* [28]; ospreys, *Pandionidae* [28]; herons, *Ardeidae* [18]; gulls, *Laridae* [28]; auks, *Alcidae* [22]; and petrels, *Procellariidae* [14].

Possible causes of the phenomenon were first discussed by Ratcliffe [43]. In North America the scientific considerations have frequently been obscured by political and economic factors [15]. Of the species affected, the Brown Pelican (*Pelecanus occidentalis*) is perhaps best known to the general public and has been frequently mentioned in public hearings [15], [58]. The available pollutant data and relevant biological information on this species are discussed in detail below.

4. The Brown Pelican: a species endangered by pollution

The Brown Pelican is a coastal species and is not found on inland waters. It feeds primarily upon fish and is therefore exposed to pollutants by eating contaminated fish. In the United States breeding colonies have been found in Central and Southern California, along the Gulf States, and on the Atlantic coast north to North Carolina [1]. The present status of the Brown Pelicans in the United States has recently been reviewed by Schreiber and Risebrough [54]. Although it is the state bird of Louisiana, no wild birds have nested there since 1961. The reasons for their sudden disappearance and for the death of adult birds has not been documented, but factors other than a decrease in reproductive capacity were evidently responsible.

Anderson and Hickey [3], in comparing eggs of the Brown Pelican obtained after 1949 with those obtained before 1943, found that recent eggs from Florida, Texas, and California were thin-shelled. Field observations carried out in California and northern Baja California in 1968 showed that reproduction was abnormally low [53]. More extensive field studies in 1969 on Anacapa Island, the only breeding site in California, showed that the birds could no longer reproduce because virtually all of the eggs laid by the pelicans were so thin-shelled that they collapsed during incubation [51]. The eggs laid in 1970 and 1971 were also thin-shelled and very few young were hatched [48].

Following the discovery of the breeding failures on Anacapa Island in 1969 J. R. Jehl visited Mexican islands along the western shore of Baja California that had been traditional nesting sites of Brown Pelicans. On the Islas Coronados near San Diego, California, the pelicans were experiencing the same kind of reproductive failures that had been observed on Anacapa. Most of the eggs laid by the birds were crushed; the remains were scattered about the colony. To the south, on Isla San Martin, some shell thinning was evident, and discarded, broken eggs were found. Nevertheless, some survived to permit hatching of young birds. On the Islas San Benitos still further to the south, most of the eggs did not appear to be thin-shelled upon superficial examination [31].

Studies of the breeding biology of the Brown Pelicans in Florida had been begun in 1968 by R. W. Schreiber. In 1969 and 1970 a total of 87 eggs were obtained from four different colonies for thickness measurement and pollutant analysis. No reduction in population numbers of the Florida pelicans was evident and apparently normal numbers of young birds were being fledged [54].

The methodology of measuring concentrations of chlorinated hydrocarbons in the eggs has been described elsewhere [45], [47]. The eggshell thickness measurements were made with a micrometer at the girth of the egg, and represent the means of at least four measurements of each egg. The thickness value includes the proteinaceous membrane attached to the shell. The mean thickness of this membrane was 0.11 mm and this value should therefore be subtracted from the measurements presented to obtain the thickness of the calcium carbonate layer.

TABLE I

DDT and PCB Residues in Yolk Lipids of Brown Pelicans,
DDE-PCB Relationships, and Associated Eggshell Changes
Concentrations in parts per million of the lipid weight. Total DDT is the sum of p,p'-DDE,
p,p'-DDD and p,p'-DDT.

From Risebrough, Gress, Baptista, Anderson and Schreiber [48].

$* p < 0.05; ** p < 0.01; *** p < 0.001.$

| General breeding area Colony | Sample size | Residue means | | | PCB *versus* DDE | Mean eggshell thickness (mm) |
		Total DDT	p,p'-DDE	PCB	r value	
Pacific Coast						
Anacapa, California	65	1,223	1,176	210	0.520***	0.32
Los Coronados	28	1,158	1,109	266	0.495**	0.34
San Martin	6	429	411	72	0.970**	0.45
San Benitos	10	128	121	39	0.973***	0.51
Gulf of California	4	13	11	4	0.952	0.56
Atlantic and Gulf Coasts						
Tampa Bay, Florida, 1969	14	56	37	120	0.837***	0.51
Tampa Bay, Florida, 1970	21	36	26	69	0.346	0.51
Cocoa Beach, Florida, 1970	22	32	28	64	0.837***	0.50
Hemp Key, Florida, 1970	20	24	18	45	0.830***	0.52
Vero Beach, Florida, 1970	10	26	21	77	0.523	0.50
All Florida	87	34	26	71	0.701***	0.51

A summary of results is presented in Table I. Because distribution of pollutant residues in these samples was not Gaussian [6], confidence limits of the arithmetic means of pollutant concentrations are not given.

Concentrations of the DDT compounds are evidently much higher in the west than in Florida. The arithmetic mean of the concentrations of total DDT residues in the yolk lipids of the 87 Florida eggs was 34 parts per million, whereas the arithmetic mean of the DDT concentrations of 65 broken eggs from Anacapa Island was 1,223 ppm. Furthermore, a north-south gradient is evident along the west coast, with concentrations highest in southern California. PCB concentrations are also higher in southern California than in Florida. In addition, many of the eggs were analyzed for dieldrin and endrin, two highly toxic organochlorine compounds used as insecticides that have also become marine pollutants. Five eggs obtained on Anacapa in 1971 were analyzed for mercury, lead, cadmium, chromium and several other heavy metals, but to date the metal analyses of eggs from other regions have not yet been finished.

Nevertheless, it is possible to determine the nature of the relationships between the magnitude of the thinning effect and the concentrations in the eggs of the DDT and PCB compounds, the two pollutant groups suspected to be major contributors to the shell thinning phenomenon. Table II lists for all eggs analyzed the shell thickness, the concentrations in the yolk lipids of the principal DDT compound, p, p'-DDE, the total concentrations of the three DDT compounds, p, p'-DDE, p, p'-DDD and p, p'-DDT, and the concentrations of PCB.

The thinning cannot be considered a direct consequence of the amount or kind of pollutant present in the egg, since pollutants are laid down with the yolk or albumin before the shell is formed. Rather, the degree of thinning must be a result of the physiological condition of the shell gland where the shell is formed. Since the shell gland is highly vascular, the sensitive sites are exposed to pollutants in the blood. It is therefore assumed that the thinning effect is a function of the pollutant levels in the blood. Since the blood carries both the yolk lipids and associated pollutants to the ovary where they are deposited in the forming yolk, it is further assumed that the pollutant concentration in the yolk lipid is also a function of its concentration in the blood. The relationships found between the thinning and the pollutant concentrations in the egg are therefore indirect.

It is also necessary to make an assumption about the comparability of the east coast and west coast eggs. The Florida Brown Pelicans belong to a different geographical race, *Pelecanus occidentalis carolinensis* than do the western birds, which are *Pelecanus occidentalis californicus*. The eastern birds are smaller, with slightly different body coloration [62]. As a result of their smaller body size, the Florida birds lay slightly smaller eggs than do the western pelicans. Thus, the mean thickness of 172 eggs obtained in Florida prior to 1943 and now preserved in museums was 0.557 ± 0.004 mm (95 per cent confidence limits) whereas the mean thickness of 111 eggs obtained on the west coast prior to 1943 was 0.571 mm [3]. The measured thickness of all Florida eggs obtained in 1969 and 1970 was therefore multiplied by the constant factor 1.03 to make the east coast eggs comparable to those from the west coast for the consideration of pollutant effects. Thus the thickness data for the Florida eggs of Table I represent the original values, whereas those of Table II have been multiplied by the correction factor.

From examination of Table I it is apparent that there is a marked difference between the distributions of the PCB and DDT compounds on the east and west coasts. In the east, PCB is more abundant than is DDE, whereas in the west DDE is more abundant than PCB. Moreover, within most colonies there is a highly significant linear relationship between the concentrations of DDE and PCB in the pelican eggs (Table I). Both compounds are very resistant to chemical and biological degradation, both are very insoluble in water and highly soluble in fats, and both move through ecosystems in comparable ways. Comparable relationships have been observed in other ecosystems [50]. Other pollutants with similar biological and chemical properties but that are undetected by the methodology that measures chlorinated hydrocarbons might therefore show a comparable correlation with the concentrations of DDE. Correlations between

TABLE II

DDE, TOTAL DDT, AND PCB CONCENTRATIONS IN BROWN PELICAN EGGS
PARTS PER MILLION OF THE YOLK LIPIDS

Locality	Sample no.	Thickness (mm)	DDE	Total DDT	PCB
Anacapa	15	0.14	2500	2571	452
	85	0.19	1337	1382	184
	14	0.20	1003	1055	115
	65	0.20	1780	1875	315
	64	0.21	1780	1862	139
	58	0.22	617	647	177
	59	0.22	1310	1376	214
	66	0.22	2145	2219	356
	101	0.23	1623	1660	166
	43	0.23	1638	1694	246
	1	0.23	1760	1764	177
	41	0.23	1925	2020	289
	84	0.24	1410	1454	175
	3	0.25	1458	1525	296
	19	0.25	1495	1586	205
	50	0.26	860	965	324
	72	0.26	1220	1296	260
	31	0.26	1290	1353	188
	27	0.26	1380	1425	208
	86	0.27	950	981	109
	55	0.28	600	648	204
	46	0.28	641	663	89
	52	0.28	1205	1280	320
	70	0.29	569	569	265
	2	0.29	862	898	138
	12	0.29	1170	1215	198
	70	0.29	1373	1442	191
	48	0.29	1940	1956	193
	32	0.29	2100	2192	316
	17	0.30	613	632	122
	102	0.30	800	850	305
	24	0.30	1060	1096	203
	93	0.30	1418	1478	396
	103	0.30	1430	1490	250
	124	0.30	2010	2031	230
	104	0.30	2379	2440	214
	105	0.31	301	305	46
	49	0.31	1560	1622	256
	63	0.32	1680	1758	204
	21	0.34	748	780	150
	42	0.34	1405	1473	218
	69	0.34	1435	1490	261
	16	0.35	969	999	143
	54	0.35	1080	1132	229
	37	0.36	949	975	173
	67	0.36	1275	1316	132
	22	0.36	1355	1418	175
	74	0.37	703	734	236
	30	0.37	1055	1091	220
	29	0.37	1140	1176	212

TABLE II (continued)

Locality	Sample no.	Thickness (mm)	DDE	Total DDT	PCB
Anacapa	90	0.39	390	403	119
(continued)	88	0.39	441	462	144
	79	0.39	882	928	147
	96	0.40	700	733	171
	91	0.41	895	943	216
	11	0.41	1430	1513	232
	94	0.42	745	783	216
	87	0.42	850	879	164
	92	0.42	890	921	185
	89	0.44	416	438	87
	98	0.46	625	645	216
	95	0.46	785	824	199
	97	0.47	755	768	236
	100	0.49	657	672	237
	99	0.49	686	708	206
Los Coronados	2137	0.23	2610	2620	464
	2136	0.25	1598	1686	590
	2140	0.25	1135	1185	162
	2288	0.25	448	472	202
	2128	0.27	1730	1807	206
	2130	0.27	1410	1499	171
	2133	0.27	1732	1823	230
	2139	0.28	2221	2299	302
	2135	0.29	1470	1475	270
	2131	0.31	1422	1527	178
	2129	0.32	1598	1710	338
	2078	0.33	1330	1393	412
	2074	0.34	1140	1206	213
	3138	0.34	622	650	85
	2072	0.34	720	770	218
	2289	0.37	691	730	204
	2132	0.37	1500	1579	167
	2286	0.38	685	712	481
	2134	0.39	1985	2050	530
	2076	0.39	335	351	130
	2287	0.40	246	258	102
	2285	0.40	321	334	110
	2075	0.42	1420	1493	1060
	2077	0.42	340	370	77
	2284	0.44	98	104	41
	2294	0.44	1035	1055	191
	2073	0.44	450	486	157
	2079	0.45	748	792	154
San Martin	2184	0.27	1710	1779	194
	2082	0.37	384	397	73
	2080	0.49	103	114	70
	2083	0.49	118	122	30
	2081	0.52	59	65	37
	2151	0.54	91	95	28
San Benito	2067	0.39	636	679	234
	2066	0.40	247	252	39
	2070	0.44	85	90	35
	2297	0.49	25	28	11

TABLE II (continued)

Locality	Sample no.	Thickness (mm)	DDE	Total DDT	PCB
San Benito	2298	0.53	25	27	12
(continued)	2071	0.54	65	73	17
	2069	0.55	36	38	14
	2068	0.58	17	19	7
	2020	0.59	40	43	11
	2065	0.59	30	33	8
Gulf of California	9000	0.50	9.3	11.5	1.8
	2182	0.58	12.9	15.8	5.1
	9001	0.59	6.6	7.5	2.0
Florida	2346	0.403	84.7	124	61.6
	2123	0.433	63.0	105	340
	2328	0.433	26.5	39.7	92.5
	2110	0.464	51.2	65.5	200
	2330	0.464	45.8	53.5	70.5
	2338	0.474	23.2	31.4	95.5
	2122	0.484	55.4	82.5	214
	2340	0.494	24.6	36.6	48.3
	2341	0.495	15.9	18.8	48.8
	2117	0.505	45.5	65.1	163
	2121	0.505	58.0	91.5	118
	2329	0.505	16.9	28.2	42.3
	2336	0.515	15.8	21.7	51.3
	2337	0.515	36.0	54.1	99.5
	2111	0.525	39.6	50.2	58.0
	2124	0.525	44.0	74.5	153
	2109	0.536	38.8	55.2	96.0
	2114	0.536	31.4	53.5	100
	2343	0.536	43.9	56.1	163
	2112	0.546	21.4	30.1	75.0
	2125	0.546	28.8	45.2	59.0
	2326	0.547	23.6	33.1	79.2
	2331	0.547	28.0	41.1	121
	2333	0.547	15.5	20.5	36.3
	2334	0.547	18.5	22.0	70.2
	2113	0.556	23.2	35.7	48.3
	2335	0.556	29.3	38.2	59.3
	2339	0.556	17.8	22.4	50.0
	2119	0.567	10.2	14.0	29.9
	2120	0.567	10.7	15.8	27.5
	2327	0.567	19.0	29.8	64.8
	2332	0.567	15.3	20.6	38.0
	2344	0.567	14.6	17.9	34.8
	2342	0.577	19.8	22.4	83.0
	2345	0.598	19.7	27.6	36.2
	2348	0.474	26.6	31.6	60.2
	2349	0.474	35.7	43.5	114
	2357	0.474	18.4	23.2	78.2
	2352	0.485	32.7	38.3	88.4
	2360	0.485	16.6	19.2	26.8
	2354	0.495	38.6	45.5	96.8
	2356	0.495	40.1	45.6	89.6
	2366	0.495	39.7	43.6	68.8

TABLE II (continued)

Locality	Sample no.	Thickness (mm)	DDE	Total DDT	PCB
Florida	2350	0.505	22.8	30.1	56.8
(continued)	2363	0.505	22.0	24.7	34.7
	2368	0.505	65.1	74.3	134
	2358	0.526	22.0	25.1	45.7
	2362	0.526	22.7	24.9	35.3
	2364	0.526	19.9	22.8	49.5
	2367	0.526	23.1	27.7	66.2
	2351	0.536	41.0	44.6	69.8
	2353	0.536	31.2	34.7	62.5
	2355	0.547	29.6	33.9	66.3
	2361	0.547	13.3	15.0	21.7
	2365	0.547	21.7	23.9	50.7
	2359	0.556	19.9	22.3	72.9
	2347	0.577	8.4	10.6	15.1
	2372	0.443	42.3	59.9	87.6
	2380	0.443	41.0	47.9	180
	2371	0.474	22.6	28.1	40.7
	2379	0.485	30.1	39.2	122
	2382	0.485	13.8	17.6	52.8
	2376	0.495	10.1	12.9	32.7
	2387	0.505	33.0	42.2	77.7
	2386	0.515	14.2	18.5	50.3
	2369	0.536	10.5	12.3	13.3
	2374	0.547	16.0	20.4	10.5
	2375	0.547	20.6	30.2	43.2
	2378	0.547	13.2	19.0	20.9
	2373	0.567	9.9	13.5	17.2
	2377	0.567	4.4	5.3	15.2
	2381	0.567	12.4	15.1	31.3
	2384	0.567	20.2	24.1	19.1
	2385	0.567	18.1	29.5	38.0
	2370	0.577	5.0	6.1	3.7
	2383	0.597	19.0	25.0	19.4
	2388	0.629	9.5	10.8	16.8
	2389	0.474	20.8	24.1	57.0
	2398	0.485	22.9	29.9	129
	2390	0.495	18.1	22.0	64.8
	2392	0.505	19.6	24.1	43.1
	2397	0.526	11.1	11.7	32.7
	2394	0.526	37.1	45.9	105
	2391	0.536	22.1	25.2	154
	2393	0.536	20.9	30.2	71.2
	2396	0.536	24.2	31.6	64.3
	2395	0.547	9.2	11.8	53.3

concentrations of a pollutant and the environmental effect in only one locality must therefore be interpreted with caution.

Table III lists the simple correlation coefficients, r, found to exist between thickness and the pollutants for each locality [20]. In all cases there is a significant negative correlation between thickness and the concentrations of either DDE or total DDT. In two cases there was no significant correlation between

TABLE III

SIMPLE LINEAR CORRELATION (r) BETWEEN SHELL THICKNESS AND CONCENTRATIONS
OF DDE, TOTAL DDT AND PCB IN BROWN PELICAN EGGS

From Risebrough, Gress, Baptista, Anderson and Schreiber [48].
*$p < 0.05$; **$p < 0.01$; ***$p < 0.001$.

Sample	N	DDE	Total DDT	PCB
All eggs	199	−0.8512***	−0.8534***	−0.6128***
All Florida	87	−0.5794***	−0.5509***	−0.4987***
Anacapa	65	−0.5605***	−0.5659***	−0.2527*
Los Coronados	28	−0.5994***	−0.6003***	−0.0987
All eggs, DDE < 100 ppm	101	−0.5059***	Not determined	−0.4746***
All eggs, DDE > 99 ppm	98	−0.5985***	−0.6016***	−0.1545
West Coast, DDE < 100	14	−0.5961*	−0.5946*	−0.6736**

thickness and PCB; DDE concentrations were high in each. In only one category, the west coast eggs with concentrations of DDE less than 100 ppm, was the correlation coefficient for PCB higher than for DDE. Because of the relationships between DDE and PCB (Table I) this approach does not permit any conclusion about the contribution of PCB to shell thinning, especially when DDE concentrations are low.

The plot of the curve between thickness and DDE [48] shows a very steep drop in thickness with small concentrations of DDE, with a decreasing slope as DDE concentrations increase, suggesting a logarithmic rather than a linear relationship. A logarithmic relationship is consistent with a plausible physiological model, consisting of a finite number of sensitive sites in the shell gland. The number of pollutant molecules required to block each site would then be a function of the number of sites already blocked.

The DDE and PCB concentrations of Table II were therefore transformed to their logarithms, as well as to their square roots, squares, exponentials, and cross products. Of these, including the original variables, the logarithm of DDE concentrations showed the highest linear correlation with thickness, ($r = -0.8906$, $N = 199$). The logarithm of DDE was therefore selected as the initial independent variable of a regression equation relating pollutant concentrations with thickness. In order to determine whether PCB might be contributing to the mathematical relationship between ln DDE and thickness, the data of Table II were analyzed with a program that determines stepwise regression (BMD 02R, Biomedical Computer Programs (edited by Dixon, 1970) on the Control Data Corporation 6400 computer of the University of California, Berkeley. The other potential independent variables consisted of DDE, ln PCB, PCB, ln (DDE × PCB) and (DDE × PCB). The program first selected the parameter that is most highly correlated with the dependent variable, in this case the logarithm of the DDE concentration, with $F = 754$, and degrees of freedom 1,197. The program then entered DDE with a partial F of 14.4155, degrees of freedom (1,197), ($p < 0.005$). The other variables did not contribute significantly to the regression ($p > 0.10$). The program then added DDE to ln DDE, with $F = 409$, d.f. (1,196). The part al F of the cross product DDE $\times$ PCB was 4.54, $0.02 < p < 0.05$. In these and subsequent operations none of the other variables contributed significantly to the regression.

The mathematical relationship between thickness and the added linear DDE term is also consistent with the biological model, since increasing amounts of DDE in the lipo-protein membranes of the shell gland could adversely affect ion transport across them, in addition to the blocking of sensitive sites. It is more difficult to interpret the contribution of the cross-product DDE $\times$ PCB, especially when no significant contribution of this term appeared in the first operation of the program. Synergistic effects between PCB and DDT compounds have been observed in other biological systems [37], but the magnitude of the effect here is not sufficiently convincing to conclude that PCB is modifying the DDE effect on pelican eggs. This is clearly a borderline case in the attempts to determine causal relationships between pollutants and a pollutant effect. An experimental approach is clearly preferable.

Unfortunately an experimental approach is not always feasible in the evaluation of pollutant effects upon ecosystems, especially when the ecosystems are unique. Species such as pelicans are extremely difficult to keep in captivity, especially under conditions that would approximate those encountered in their natural habitat. Because different species of birds differ widely in their physiological responses as a result of different food habits, different habitat selection, the necessity of some species to eliminate excess salt, and of many other factors, extreme caution must be made in extrapolating conclusions about the real world without recourse to adequate experimentation. Thus, Jukes [15] has written ". . . without conducting controlled experiments, DDE has been blamed for the thin shells of eggs laid by pelicans on Anacapa Island."

The conclusions about the shell thinning effects of DDE and PCB on pelican eggs based upon field data might, however, be compared with the results of an experiment designed to test the combined effects of DDE and PCB upon the thinning of eggs of mallard ducks (*Anas platyrhynchos*), [46]. Fifty mallard hens were divided at random into eight groups of five birds and one group of ten. The latter served as spare birds in case of need, but were not used. Each group of five was placed in a cage with two drakes. Two groups served as control, two groups were fed DDE at a concentration of 40 ppm, dry weight, of their diet, two were fed 40 ppm PCB (Aroclor 1254) and the remaining two received a combination of 40 ppm DDE + 40 ppm PCB. Instead of thickness, a thickness index, derived by dividing the weight of the dried shell in grams by the product of the length and breadth in cm² was used [43]. The data are summarized in Table IV. The distribution of shell thickness indices in each group was approximately normal [46], permitting determination of the 95 per cent confidence limits of the mean by means of the t test [20].

Thus PCB increased slightly the shell thickness index and DDE caused a significant reduction. Shell thickness index of birds receiving the combination

TABLE IV

SHELL THICKNESS INDEX CHANGES OF MALLARD DUCK EGGS

From Risebrough and Anderson [46].

Diet group	N	Thickness index $\pm$ 95% C.L.
Control	500	0.2098 ± 0.0014
DDE, 40 ppm	473	0.1745 ± 0.0017
DDE, 40 ppm and PCB, 40 ppm	264	0.1700 ± 0.0019
PCB, 40 ppm	388	0.2143 ± 0.0020

was slightly lower than that of birds receiving DDE alone. One of the birds on the DDE diet, however, began to lay thick-shelled eggs toward the end of the experiment [46]. If these eggs, representing the top eight values, are omitted from the calculations, the index becomes 0.1729 ± 0.0013, not significantly different from combination birds. The situation is again borderline, but the combination of DDE and PCB had other effects upon reproduction [46].

Other experiments have also shown that PCB alone has no effect upon the shell thickness of eggs of mallard ducks [26] or of ring doves, *Streptopelia risoria* [40].

More difficult is the evaluation of possible contributions by other pollutants, alone or in combination with DDE, which were not measured by the techniques used. The relationship between thickness and the logarithm of DDE fitted well the data from all colony sites, including both the Atlantic and Gulf coasts of Florida, California and western Mexico, except that at higher concentrations of DDE, thickness decreased more rapidly than predicted. These eggs were all from southern California and adjacent areas of Mexico. As indicated above, addition of a linear DDE term significantly improved the fit, and was further-more consistent with a biological model. The possibility can not be eliminated however, that other pollutants are contributing slightly to the DDE effect. No correlation was found between thickness and concentrations of either dieldrin, which shows a slight shell-thinning effect in mallard eggs [36], or endrin [48]. The mean arithmetic concentrations of total mercury in the wet weight contents of the five eggs obtained in 1971 was 0.083 ppm (standard deviation 0.036 ppm) [2], lower than the mercury concentrations found in 85 eggs of 13 species of fish-eating birds from the Great Lakes region, all of which showed lesser amounts of thinning than the Brown Pelicans [17]. Mercury residues in a fish-eating species from Antarctica were only slightly lower [2]. Cadmium levels in these eggs were in the order of 0.1 ppm, dry weight, in the same range as the Antarctic species [2]. Lead levels were not above background (maximum concentration 0.1 ppm dry weight).

DDE appears therefore to be the major cause of the thinning of Brown Pelican eggshells. The amount of data presently available is not sufficient to show whether pollutants other than the DDT and PCB compounds are contrib-uting to a minor extent.

5. Pollutant effects on other wildlife species

Among species of fish-eating and raptorial birds, patterns of regional decline and reproductive failures have almost always been associated with thinning of eggshells [28], [43], [44], [48]. The available evidence indicates that DDE is the pollutant primarily responsible for this phenomenon. Other potential contributors such as dieldrin [36] or metallic mercury [56] are usually much less abundant in environmental samples than is DDE.

A statistically significant inverse correlation between shell thickness and the concentrations of DDE in the egg have been recorded for: Herring Gull (*Larus argentatus*) [28]; Double-crested Cormorant (*Phalacrocorax auritus*) [4]; White Pelican (*Pelecanus erythrorhynchos*) [4]; and the Peregrine Falcon [11]. Low concentrations of DDE in the diet comparable to those found in fish have produced shell thinning under controlled experimental conditions in Mallard Ducks [25], [46], American Kestrel (*Falco sparverius*) [63] and Japanese Quail (*Coturnix*) [55]. Shell thinning has also been induced experimentally with DDE in Ring Doves [41].

Like the DDT compounds, dieldrin may be dispersed through the atmosphere [49], [57]. The greatest hazard of dieldrin exists to fish-eating birds such as the Bald Eagle (*Haliaeetus leucocephalus*) [39] and Common Egret (*Casmerodius albus*) [18] and to the Peregrine Falcon [44] which may accumulate lethal levels from fish or birds which are not themselves harmed.

Long term effects of PCB upon wildlife and on the environment in general are not known. PCB may increase susceptibility to infectious agents such as virus diseases [21]. Like other chlorinated hydrocarbons PCB increases the activity of liver enzymes that degrade sex hormones [50]. Highly toxic byproducts may also be associated with PCB [60], [61].

Illness of zoo animals in New York City caused by lead poisoning, presumably by breathing lead contaminated air, is an indication of the hazards of lead additives in gasoline to both wildlife and man [7].

The use of alkyl-mercury compounds as seed dressings in Sweden caused the death of numbers of seed-eating birds [10]. In Finland mercury may have contributed to the decline of the White-tailed Sea Eagle (*Haliaeetus albicilla*) in regions where the species feeds upon marine fish and marine birds [27].

In all cases the compounds suspected to cause damage have been the heavy metals or relatively persistent chlorinated hydrocarbons. No damage to a population has been linked to date with a compound acting primarily as a mutagen or teratogen.

6. Monitoring environmental levels of mutagens and teratogens

The number of individuals in wildlife populations that would be affected by environmental mutagens or teratogens is inevitably small. Without careful monitoring they would not be observed in most species. Because of high rates of natural mortality it is unlikely that deaths of comparatively few individuals would affect population numbers. If a wildlife population were to be used, however, to monitor the levels of environmental chemicals that cause birth defects, the following criteria might apply:

(1) a species that occupies a relatively high position in the food web, such as a fish-eating bird, can be expected to accumulate greater amounts and varieties of pollutants that might cause birth defects than would a seed-eating species;

(2) the population studied should occupy a relatively polluted habitat, but other populations of the species should occupy relatively clean environments;

(3) it should be possible to examine large numbers of young at the time of birth.

These conditions are met in a colony of Common Terns on Great Gull Island in Long Island Sound, which has been studied intensively by ornithologists from the American Museum of Natural History. The terns prey upon small fish, and Long Island Sound is one of the areas of the sea that are heavily polluted with such industrial chemicals as PCB [24]. Nests are marked when the eggs are first laid and examined at the time of hatching. The young terns are marked with a unique combination of bands, some of which are colored. In 1970, 33 or 1.5 per cent of the 2,316 young Common Terns possessed a gross abnormality. Most frequently observed was a loss of feathers when the birds were two to three weeks old. This phenomenon appears to be pollutant-induced [24]. Other abnormalities included beak, eye and foot deformities. At the present time data from control colonies are not sufficient to permit conclusions that the incidence of abnormalities is significantly above the expected. If levels of a chemical that causes birth defects are increasing, however, in the environment, the effects might first be noted in a polluted area adjacent to a highly industrialized, densely populated region such as Long Island Sound. Increasing levels of deformities in the tern colony on Great Gull Island would serve as an "early warning" to the human populations which share the environment with the terns.

Research was supported by NSF Grant GB 11649 to the Institute of Marine Resources, University of California, H. S. Olcott, principal investigator. I thank Joyce K. Baptista, Daniel W. and Irene T. Anderson, Ralph W. Schreiber, and Elizabeth L. Scott for their assistance.

REFERENCES

[1] AMERICAN ORNITHOLOGISTS UNION, *Check list of North American Birds*, Baltimore, Lord Baltimore Press, 1957 (5th ed.).

[2] V. C. ANDERLINI, P. G. CONNORS, R. W. RISEBROUGH, and J. H. MARTIN, "Heavy metal concentrations in some Antartic and Northern Hemisphere sea birds," *Proc. Coll. Conserv. Prob. Antarctica Circumpolar Waters*, in press (1971).

[3] D. W. ANDERSON and J. J. HICKEY, "Oological data on eggs and breeding characteristics of Brown Pelicans," *Wilson Bull.*, Vol. 82 (1970), pp. 14–28.

[4] D. W. ANDERSON, J. J. HICKEY, R. W. RISEBROUGH, D. F. HUGHES, and R. E. CHRISTEN-

SEN, "Significance of chlorinated hydrocarbon residues to breeding pelicans and cormorants," *Canad. Field-Natur.*, Vol. 83 (1969), pp. 91–112.

[5] G. E. BAGLEY and L. N. LOCKE, "The occurrence of lead in tissues of wild birds," *Bull. Environ. Contam. Toxicol.*, Vol. 2 (1967), pp. 297–305.

[6] J. K. BAPTISTA and R. W. RISEBROUGH, unpublished observations.

[7] R. J. BAZELL, "Lead poisoning: zoo animals may be the first victims," *Science*, Vol. 173 (1971), pp. 130–131.

[8] W. BERG, A. JOHNELS, B. SJOSTRAND, and T. WESTERMARK, "Mercury content in feathers of Swedish birds from the past 100 years," *Oikos*, Vol. 17 (1966), pp. 71–83.

[9] K. K. BERTINE and E. D. GOLDBERG, "Fossil fuel combustion and the major sedimentary cycle," *Science*, Vol. 173 (1971), pp. 233–235.

[10] K. BORG, H. WANNTORP, K. ERNE, and E. HANKO, "Alkyl mercury poisoning in terrestrial Swedish wildlife," *Viltrevy*, Vol. 6 (1969), pp. 301–379.

[11] T. J. CADE, J. L. LINCER, C. M. WHITE, D. G. ROSENEAU, and L. G. SWARTZ, "DDE residues and eggshell changes in Alaskan falcons and hawks," *Science*, Vol. 172 (1971), pp. 955–957.

[12] T. J. CHOW, "Isotope analysis of seawater by mass spectrometry," *J. Water Poll. Cont. Fed.*, Vol. 40 (1968), pp. 399–411.

[13] T. J. CHOW and C. PATTERSON, "The occurrence and significance of lead isotopes in pelagic sediments," *Acta Geochim. Cosmochim.*, Vol. 26 (1962), pp. 263–293.

[14] M. C. COULTER and R. W. RISEBROUGH, "Shell thinning of eggs of the Ashy Petrel, *Oceanodroma homochroa*," unpublished manuscript.

[15] M. SOBLEMAN (compiler), *DDT: Selected statements from State of Washington DDT Hearings*, Torrance, Montrose Chemical Company, 1970.

[16] B. DIMITRIADES and M. WHISMAN, "Carbon monoxide in lower atmosphere reactions," *Environ. Sci. Technol.*, Vol. 5 (1971), pp. 129–222.

[17] R. A. FABER and J. J. HICKEY, "Insecticides, PCB's and mercury in inland aquatic bird eggs: Progress report on Contract 14-16-0008-515 to Bureau of Sport Fisheries and Wildlife," Department of Wildlife Ecology, University of Wisconsin, 1970.

[18] R. A. FABER, R. W. RISEBROUGH, and H. PRATT, "Organochlorines and mercury in Common Egrets and Great Blue Herons," *Envir. Poll.*, in press.

[19] N. FIMREITE, R. W. FYFE, and J. A. KEITH, "Mercury contamination of Canadian prairie seed-eaters and their avian predators," *Canad. Field-Natur.*, Vol. 84 (1970), pp. 269–276.

[20] R. A. FISHER and F. YATES, *Statistical Tables for Biological and Medical Research, 3rd Edition*, Edinburgh, Oliver and Boyd, 1948.

[21] M. FRIEND and D. O. TRAINER, "Polychlorinated biphenyl: interaction with duck hepatitis virus," *Science*, Vol. 170 (1970), pp. 1314–1316.

[22] F. GRESS, R. W. RISEBROUGH, and F. C. SIBLEY, "Shell thinning in eggs of Common Murres, *Uria aalge*, from the Farallon Islands, California," *Condor*, Vol. 73 (1971), pp. 368–369.

[23] A. L. HAMMOND, "Mercury in the environment: Natural and human factors," *Science*, Vol. 171 (1971), pp. 788–789.

[24] H. HAYS and R. W. RISEBROUGH, "Pollutant concentrations in abnormal young terns from Long Island Sound," *Auk*, in press.

[25] R. G. HEATH, J. W. SPANN, and J. F. KREITZER, "Marked DDE impairment of Mallard reproduction in controlled studies," *Nature*, Vol. 224 (1969), pp. 47–48.

[26] R. G. HEATH, J. W. SPANN, J. F. KREITZER, and C. VANCE, "Effects of polychlorinated biphenyls on birds," *Proc. XV Internat. Congr.*, in press.

[27] K. HENRIKSSON, E. KARPPANEN, and M. HEIMINEN, "High residue of mercury in Finnish White-tailed Eagles," *Ornis Fennica*, Vol. 43 (1966), pp. 38–45.

[28] J. J. HICKEY and D. W. ANDERSON, "Chlorinated hydrocarbons and eggshell changes in raptorial and fish-eating birds," *Science*, Vol. 162 (1968), pp. 271–273.

[29] R. E. INMAN, R. B. INGERSOLL, and E. A. LEVY, "Soil: a natural sink for carbon monoxide," *Science*, Vol. 172 (1971), pp. 1229–1231.

[30] L. S. JAFFE, "Ambient carbon monoxide and its fate in the atmosphere," *J. Air Poll. Cont. Assoc.*, Vol. 18 (1968), pp. 534–540.

[31] J. R. JEHL, personal communication.

[32] S. JENSEN, A. G. JOHNELS, M. OLSSON, and G. OTTERLIND, "DDT and PCB in marine animals from Swedish waters," *Nature*, Vol. 224 (1969), pp. 247–250.

[33] O. I. JOENSUU, "Fossil fuels as a source of mercury pollution," *Science*, Vol. 172 (1971), pp. 1027–1028.

[34] F. S. JOHNSON, "The oxygen and carbon dioxide balance in the earth's atmosphere," in *Global Effect of Environmental Pollution* (edited by S. F. Singer), New York, Springer-Verlag, 1970.

[35] D. H. KLEIN and E. D. GOLDBERG, "Mercury in the marine environment," *Environ. Sci. Technol.*, Vol. 4 (1970), pp. 765–768.

[36] P. LEHNER and A. EGBERT, "Dieldrin and eggshell thickness in ducks," *Nature*, Vol. 224 (1969), pp. 1218–1219.

[37] E. P. LICHTENSTEIN, K. R. SCHULZ, T. W. FUHREMANN, and T. T. LIANG, "Biological interaction between plasticizers and insecticides," *J. Econ. Entomology*, Vol. 62 (1969), pp. 761–765.

[38] *Man's Impact on the Global Environment: Report of the Study of Critical Environmental Problems*, Cambridge, M.I.T. Press, 1970.

[39] B. M. MULHERN, W. L. REICHEL, L. N. LOCKE, T. G. LAMONT, A. BELISLE, E. CROMARTIE, G. E. BAGLEY, and R. M. PROUTY, "Organochlorine residues and autopsy data from Bald Eagles, 1966–1968," *Pesticides Monitoring J.*, Vol. 4 (1970), pp. 141–144.

[40] D. B. PEAKALL, "Effect of polychlorinated biphenyls on the egg shells of Ring Doves," *Bull. Environ. Contam. Toxicol.*, Vol. 6 (1971), pp. 100–101.

[41] D. B. PEAKALL, "p,p'-DDT: Effect on calcium metabolism and concentration of estradiol in the blood," *Science*, Vol. 168 (1970), pp. 592–594.

[42] E. K. PETERSON, "Carbon dioxide affects global ecology," *Environ. Sci. Technol.*, Vol. 3 (1969), pp. 1162–1169.

[43] D. A. RATCLIFFE, "Decrease in eggshell weight in certain birds of prey," *Nature*, Vol. 215 (1967), pp. 208–210.

[44] ———, "Changes attributable to pesticides in egg breakage frequency and eggshell thickness in some British birds," *J. Appl. Ecol.*, Vol. 7 (1970), pp. 67–115.

[45] R. W. RISEBROUGH, "Determination of polychlorinated biphenyls in environmental samples," *Proc. Internat. Symp. Identif. Measur. Environ. Poll.*, Ottawa, in press.

[46] R. W. RISEBROUGH and D. W. ANDERSON, "Synergistic effects of DDE and PCB upon the reproduction of Mallards," unpublished manuscript.

[47] R. W. RISEBROUGH, G. L. FLORANT, and D. D. BERGER, "Organochlorine pollutants in Peregrines and Merlins migrating through Wisconsin," *Canad. Field-Natur.*, Vol. 84 (1970), pp. 247–253.

[48] R. W. RISEBROUGH, F. GRESS, J. K. BAPTISTA, D. W. ANDERSON, and R. W. SCHREIBER, "Oceanic pollution: Effects on the reproduction of Brown Pelicans *Pelecanus occidentalis*," unpublished manuscript.

[49] R. W. RISEBROUGH, R. J. HUGGETT, J. J. GRIFFIN, and E. D. GOLDBERG, "Pesticides: Transatlantic movements in the North-east Trades," *Science*, Vol. 159 (1968), pp. 1233–1236.

[50] R. W. RISEBROUGH, P. REICHE, D. B. PEAKALL, S. G. HERMAN, and M. N. KIRVEN, "Polychlorinated biphenyls in the global ecosystem," *Nature*, Vol. 220 (1968), pp. 1098–1102.

[51] R. W. RISEBROUGH, F. C. SIBLEY, and M. N. KIRVEN, "Reproductive failure of the Brown Pelican on Anacapa Island in 1969," *American Birds*, Vol. 25 (1971), pp. 8–9.

[52] J. H. RYTHER, "Photosynthesis and fish production in the sea," *Science*, Vol. 166 (1969), pp. 72–76.

[53] R. W. SCHREIBER and R. L. DELONG, "Brown Pelican status in California," *Audubon Field Notes*, Vol. 23 (1969), pp. 57–59.

[54] R. W. SCHREIBER and R. W. RISEBROUGH, "Studies of the Brown Pelican *Pelecanus occidentalis*," *Wilson Bull.*, in press.

[55] L. F. STICKEL and L. I. RHODES, "The thin eggshell problem," *Biological Impact of Pesticides in the Environment* (edited by J. W. Gillett), Corvallis, Oregon, Environmental Health Series No. 1, 1970, pp. 31–35.

[56] G. S. STOEWSAND, J. L. ANDERSON, W. H. GUTENMANN, C. A. BACHE, and D. J. LISK, "Eggshell thinning in Japanese Quail fed mercuric chloride," *Science*, Vol. 173 (1971), pp. 1030–1031.

[57] K. R. TARRANT and J. O'G. TATTON, "Organochlorine pesticides in rainwater in the British Isles," *Nature*, Vol. 219 (1968), pp. 725–727.

[58] United States Senate: Hearings before the Subcommittee on Agricultural Research and General Legislation of the Committee on Agriculture and Forestry, 1971.

[59] United States Tariff Commission: *Synthetic Organic Chemicals, United States Production and Sales, 1968*, Tariff Commission Production 327, Washington, U.S. Government Printing Office, 1970.

[60] J. G. VOS and J. H. KOEMAN, "Comparative toxicologic study with polychlorinated biphenyls in chickens with special reference to porphyria, edema formation, liver necrosis and tissue residues," *Toxicol. Applied Pharmacol.*, Vol. 17 (1970), pp. 656–668.

[61] J. G. VOS, J. H. KOEMAN, H. L. VAN DER MAAS, M. C. TEN NOEVER DE BRAUW, and R. H. DE VOS, "Identification and toxicological evaluation of chlorinated dibenzofuran and chlorinated naphthalene in two commerical polychlorinated biphenyls," *Fd. Cosmet. Toxicol.*, Vol. 8 (1970), pp. 625–633.

[62] A. WETMORE, "A review of the forms of the Brown Pelican," *Auk*, Vol. 62 (1945), pp. 577–586.

[63] S. WIEMEYER and R. PORTER, "DDE thins eggshells of captive American Kestrels," *Nature*, Vol. 227 (1970), pp. 737–738.

[64] N. YAMAGATA and I. SHIGEMATSU, "Cadmium pollution in perspective," *Bull. Inst. Publ. Health*, Vol. 19 (1970), pp. 1–27.

Discussion

Question: E. B. Hook, Birth Defects Institute, Albany Medical College

Regarding the observation of congenital defects in the terns in Long Island Sound:

(1) Over how long a period of time have the birds been systematically examined?

(2) What is the observed incidence rate of these defects?

(3) Where else do these birds breed and/or feed?

(4) Have these defects been observed in other species of birds and/or have any other birds been systematically examined?

(5) Are there internal defects found on autopsy of those birds with external limb defects in higher frequency than in birds without external defects?

Reply: R. Risebrough

(1) Detailed studies were begun in 1969 and have continued through 1971.

(2) In 1969 only three of 3,160 young terns showed abnormalities. Both

Roseate Terns and Common Terns were studied. In 1970, 3,122 young terns of both species were examined; 38 showed abnormalities. Data for 1971 have not yet been compiled, but several abnormalities have been noted to date.

(3) At least some of the birds winter in the Caribbean, as shown from the banding returns. The Common Tern also breeds in many localities along the east coast and in eastern and middle North America.

(4) To my knowledge they have not been looked for in other species.

(5) These were not looked for, but the specimens have been preserved for future studies, as required.

Question: E. Tompkins, Human Studies Branch, Environmental Protection Agency

What is your explanation for the sharp dichotomy between the thickness of the egg shells in museums in England between 1944 and 1945? It seems to be a sharp drop to associate with the gradual buildup of an environmental pollutant.

Reply: R. W. Risebrough

The sharpest drop occurred between 1946 and 1947. The year 1947 was also the year when thin shelled eggs of the Peregrine were first observed in Massachusetts and California. The curve relating shell thickness to DDE concentrations in the Brown Pelican eggs, and in eggs of other species as well, shows a very sharp initial drop with low concentrations of DDE. In a given ecosystem, the Peregrine always has the highest concentrations of DDE. The relatively sharp drop in thickness is, therefore, compatible with other observations.

Question: Alexander Grendon, Donner Laboratory, University of California, Berkeley

I understood you to say that the eggs with low levels of DDE were museum specimens. Since these would be empty shells, how did you determine their DDE content?

Reply: R. W. Risebrough

No, the eggs with low levels of DDE were from Florida or from the Gulf of California. Chlorinated hydrocarbons were, therefore, measured in the egg contents. I am not aware of any attempts to measure DDE in egg shells and I doubt that they would be worthwhile.

Question: Burton E. Vaughan, Ecosystems Department, Battelle Memorial Institute, Richland, Washington

Would you say something about DDT levels in open ocean as compared to estuarine and other coastal water samplings? Are open ocean levels high in relation to the levels necessary for egg shell thinning?

Reply: R. W. Risebrough

California coastal waters are a special case, since they have received huge amounts of DDT compounds from the effluent of a factory in Los Angeles. The Peregrine Falcons of Amchitka Island in the North Pacific show slight but significant amounts of shell thinning and DDE levels in the fat are in the order of 100–300 ppm. On the east coast, the gannets which feed on fish in the coastal waters show significant thinning of shells. Of the true sea birds, the petrels, shearwaters, and albatrosses, shell thinning has been looked for and found only in the Ashy Petrel of California. But levels in some of the other species appear sufficiently high to cause a slight amount of thinning.

Question: E. J. Sternglass, School of Medicine, University of Pittsburgh

Have you examined the possible effect of radioactive fallout on animal reproduction, for instance as observed by Dr. Norman B. French at U.C.L.A. for a natural population of desert animals exposed to very low levels of daily doses?

Reply: R. W. Risebrough

Ratcliffe's initial studies of shell thinning showed that the geographical patterns of thinning corresponded to pollution patterns. Remote areas in Scotland, which presumably received as much fallout as southern England, showed relatively slight thinning of the birds' eggs. In North America, the pattern of shell thinning has corresponded closely to pollution patterns rather than to one shown by fallout. The species first affected have invariably been those highest in the food webs. Related species inhabiting the same areas have either not shown thinning or have shown it several years later than the most sensitive species. The few available data on radioactivity in birds show that body burdens are higher in mountain areas than along the coast. But coastal species have shown many more symptoms of reproductive failures than have mountain species.

Question: B. G. Greenberg, School of Public Health, University of North Carolina, Chapel Hill

I wonder if Dr. Risebrough can help us to transfer or extrapolate the findings about thickness of eggshells in bird studies to human populations. Is the formation of bone structure or teeth likely to be affected and are there other organs affected?

Reply: R. W. Risebrough

The shell thinning phenomenon appears to be associated with the membranes of the shell gland. Most likely the transport of calcium ions is inhibited, perhaps by inactivation of an ATPase. Another potentially sensitive process is the transport of bicarbonate or carbonate across the membrane and the maintenance of the pH gradient. The deposition of calcium carbonate in the eggshell is, therefore, a very different process from the deposition of calcium phosphate crystals in bone. If a human system were sensitive to DDE it would, therefore, most likely have membranes across which ions are transported. But even the relatively uncontaminated pelicans from Florida have much more DDE than does the North American human population.

IMPLICATIONS OF ENVIRONMENTAL STRONTIUM 90 ACCUMULATION IN TEETH AND BONE OF CHILDREN

HAROLD L. ROSENTHAL

WASHINGTON UNIVERSITY SCHOOL OF DENTISTRY

1. Introduction

In order to define and document possible hazards of environmental radioactive pollutants that occur from fallout or nuclear reaction emissions, my laboratory has been studying the accumulation of strontium 90 in teeth and bone of children as related to dietary consumption of the nuclide. Following the suggestion of Kalckar [1], we selected to study strontium 90 for the following reasons:

(1) Strontium 90, deposited during formation of stable calcified tissues, such as teeth, represents a marker atom that indicates the maximum amount of nuclide deposited during formation of the teeth.

(2) Strontium is biochemically similar to calcium and is permanently deposited in calcified tissues.

(3) Strontium 90 is a potentially hazardous nuclide because of its long physical half-life ($T/2 = 28$ yrs.) and its very slow biological turnover time, ranging between no turnover for teeth to less than eight per cent per year in vertebral bone [2].

We have concentrated our studies on deciduous and permanent teeth because deposition of alkaline earth radionuclides is only minimally affected by such factors as mineral turnover, exchange, accretion, and remodeling during the time the tooth crown is formed. Thus, the concentration of radionuclide in the tooth crown represents the equilibrium established between the crown and the diet at the time the crown is mineralized. Once the crown is complete, the nuclide concentration becomes a permanent record and is representative of the total mineralization process when mineralizing tissues are in their most active metabolic state.

Our previous data [3], [4], [5], [6] demonstrated that the accumulation of strontium 90 in the deciduous and permanent teeth of children was adequately described by a linear equation of the form $C_T = KC_D$ where C_T and C_D represent tooth crown and diet strontium 90 concentrations respectively, and K is a constant. The constant K differs for each specific kind of tooth formed *in utero* and after birth (Table I), chronological age for tooth development, attendant discrimination factors, and other factors as far as they are known.

465

TABLE I

FRACTION OF TOOTH CROWN DEPOSITED DURING TOOTH DEVELOPMENT

Tooth	Prenatal (X)	Postnatal (Y)
Fetal buds	1.00	0.00
Incisor (deciduous)	0.32	0.68
2nd molar (deciduous)	0.05	0.95
1st bicuspid (permanent)	0.00	1.00

A theoretical expression for the Sr 90/g calcium of teeth (C_t) was derived by expansion of the basic equation of Reiss [7]

(1) $C_t = X$ prenatal Sr 90/g Ca $+ Y$ postnatal Sr 90/g Ca.

The factors X and Y represent the fraction of tooth crown calcium deposited during pre- and postnatal periods respectively. These factors vary with the kind of tooth as shown in Table I.

In order to account for dietary intake and discrimination factors, equation 1 becomes:

(2) $$C_t = (X)(A)C_a^m D_m + (Y)(A)C_a^I D_I,$$

where C_a^m is the mother's dietary intake of Sr 90/g calcium from commercial milk, D_m is the discrimination factor against strontium 90 between dietary intake and calcified tissue. The variable A relates the concentration of Sr 90/g calcium in milk to that of the total diet, and varies with the dietary habits of the individual. For pregnant mothers, A is estimated to be equal to 1.6. For bottle-fed infants during the first year of life the value is 1.0 and for pre-adolescent children between one and 14 years of age, the value is about 1.2.

In order to solve equation 2, values for D_m and D_I must be experimentally determined. A discrimination factor of 0.18 for D_m appears reasonable (see end of Section 3). A value of 0.8 for D_I for bottle-fed infants has been selected as an intermediate value on the basis that children under 60 days of age do not discriminate against strontium 90 [5], [8], and strontium 90 discrimination in children under one year of age is probably less than 0.5 [8], [9], and 0.35 for D_I of pre-adolescents [5] appears to be a satisfactory estimate.

With these estimates, equation 2 may be solved for each appropriate age group to become:

(3) $$C_T = KC_D,$$

where C_T and C_D are the Sr 90/g calcium for tooth crown and commerical bottle milk respectively and K is a constant.

We prefer to relate our data to cow's milk because the strontium 90 content of commercial milk is readily available for past and current years and cow's milk represents the primary source of dietary calcium in the American diet.

2. Strontium 90 content of milk and diet

In St. Louis, the strontium 90 level in milk increased from negligible levels prior to 1950 to an average maximum level of 20 pCi/g Ca for the first six months of 1964 and declined rapidly thereafter until 1967. For the past four years since 1967, the strontium 90 level in milk appears to be quite steady, and averages about 9 pCi/g Ca (Figure 1). A maximum monthly average of 38 pCi/g Ca occurred during June 1963.

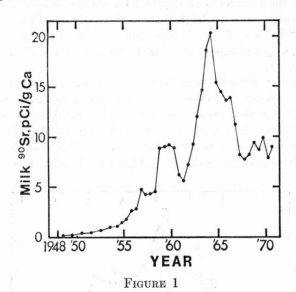

FIGURE 1

Average yearly milk levels of strontium 90 in St. Louis.

The strontium 90 level in milk appears to be the best parameter of dietary intake because measurements are easily obtained, milk calcium represents the greatest source of dietary calcium in the American diet, and because milk is a relatively stable foodstuff with respect to chemical composition. Nonetheless, the total dietary strontium 90 intake is the important datum and numerous measurements to relate the contribution of milk strontium 90 to diet strontium 90 have been made [10]. Although the dietary contribution varies with individuals, with geographic area, with seasonal variation and rainfall, and with cultural or habitual dietary habits, the Federal Radiation Commission has accepted a factor of 1.2 to 1.6 times the milk strontium 90 level as the best average value for the adult American diet [11]. For bottle-fed American children during the first post-natal year, the factor is essentially 1.0 and the factor gradually increases as the dietary habits change and approach that of the adult society.

It must be recognized, however, that the conversion factor is probably valid only for the United States and European countries in which milk and dairy products contribute the major portion of calcium and strontium to the total diet.

In countries where grains and cereals represent the major source of alkaline-earth elements, this factor is probably not valid and measurements of strontium 90 and calcium in the total diet are required.

3. Accumulation of strontium 90 in fetal bone and teeth

The relative strontium 90 content of fetal tooth buds and mandibular bone obtained from the same fetus averaged 0.99 ± 0.18 (S.D.) for 62 fetal samples. In 56 comparisons between fetal femur and mandibular bone, an average of 0.94 ± 0.22 (S.D.) was obtained. It is apparent therefore that strontium 90 is distributed equally throughout the various hard tissues of the fetus during *in utero* development.

During the years 1961 through 1970, a period encompassing both increasing and decreasing strontium 90 fallout, strontium 90 concentrations in fetal mandibular bone followed that present in market milk. Fetal bone reached average maximal strontium 90 concentrations of 6.1 pCi/g Ca during the latter half of 1964. However, the peak strontium 90 content of milk occurs about six months earlier than that occurring in the fetus (Figure 2). This lag period appears to be

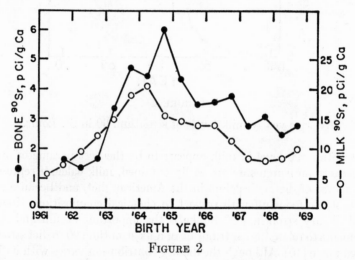

FIGURE 2

Strontium 90 content of fetal mandibular bone (●) and of commerical cow's milk (○) vs. the first and last half of the year of birth. Each point represents average values for 2 to 18 fetuses.

due to at least two factors that are difficult to evaluate by direct analysis. In the first instance, it needs to be recognized that fetal calcium (and strontium) is drawn from the mother's body mineral pool in addition to the mother's dietary intake. Consequently the fetal calcium and strontium represents, in part, a contribution from the mother's mineral stores previously deposited before fetal bone mineralization occurs. In the second instance, part of the mother's dietary intake of calcium and strontium includes dairy products (cheese, powdered skim milk, and so forth) and other foods that have an appreciable shelf life and are consumed at some time after processing. Under steady state conditions, when the dietary intake of strontium 90 is constant or changing very slowly, correction for the lag period will be of no consequence. This situation appears to be occurring for the period after 1965.

The data for fetal mandibular bone, when plotted for six month intervals against the milk concentration existing six months previous to abortion, adequately fits a linear equation for both increasing and decreasing strontium 90 concentrations of milk, with a slope K of 0.28 (Figure 3). If the total diet strontium 90 during pregnancy is considered to be 1.6 times the milk concentration, then the slope K becomes 0.18. This value is D_m in equation 2.

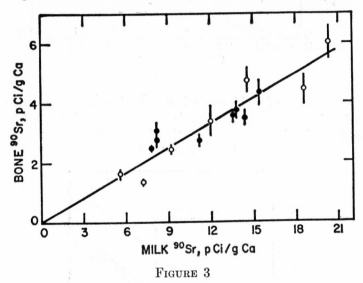

FIGURE 3

Strontium 90 content of fetal mandibular bone during increasing (ϕ) and decreasing fallout ($\blacklozenge$) *versus* strontium 90 in cow's milk 6 months prior to birth. The vertical bars represent $\pm$S.E. for samples shown in Figure 2.

4. Accumulation of strontium 90 in deciduous teeth of infants

The concentration of strontium 90 in the tooth crown of sound incisors and carious second molars of children born between 1951 through 1963 (Figure 4), and who were bottle-fed from birth, is also adequately described by a linear equation (Figure 5). The strontium 90 content in deciduous tooth crowns increased from negligible amounts prior to 1950 to maximum average values of 10.6 pCi/g Ca for incisors and 14.6 pCi/g Ca for 2nd molars during late 1963 and early 1964 when milk strontium 90 was maximal. The values of K for in-

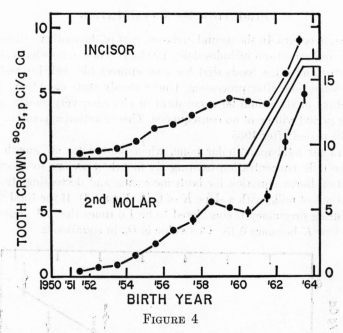

FIGURE 4

Strontium 90 content of deciduous incisor and 2nd molar crowns *versus* year of birth. Each point represents average values for 5 to 21 pooled samples. The S.E. are given as vertical bars except where the diameter of the points is equal to or larger than ±1 S.E.

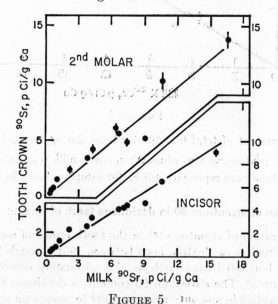

FIGURE 5

Strontium 90 content of deciduous incisor and 2nd molar crowns *versus* strontium 90 content of cow's milk. See Figure 4 for details.

cisors (0.63) and 2nd molars (0.77) are consistent with results previously reported for limited data. For this age group, the strontium 90 content of milk and total diet is approximately equal.

5. Accumulation of strontium 90 in permanent teeth of adolescents

Calcification and crown formation of 1st bicuspids begins at about $1\frac{1}{2}$ to 2 years of age and is completed at about 6 years of age, while the root forms between ages 6 to 14 years. The midpoint for calcification appears to occur at 4 years of age for crowns and 9 years of age for roots. In the tooth crowns, the concentration of strontium 90 represents the dietary contribution and is not influenced by maternal contributions. The root is less stable than the crown, is more like bone and reflects calcification during early adolescence. Maximum strontium 90 levels of 5 pCi/g Ca were found in the tooth crowns of children during high environmental strontium 90 levels that peaked in 1963–1964 (Figure 6). For roots calcifying during these peak years, maximum levels of 4.5 pCi/g Ca

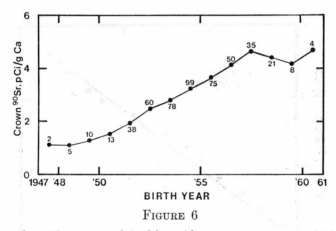

FIGURE 6

Strontium 90 content of 1st bicuspid crowns *versus* year of birth.

were found (Figure 7). Correlations between crown (Figure 8) and root (Figure 9) strontium 90 levels during the time of calcification with that found in the diet is essentially linear. The constant, K, is 0.31 for crown and 0.24 for root. Consideration of variables of turnover, exchange, remodeling and radioactive decay do not alter the equations.

6. Regional variation of strontium 90 in deciduous incisors

The strontium 90 content of teeth for children born between 1956–1958 in various geographic areas increases as the latitude decreases from Toronto to New Orleans (Table II). This relationship follows the same general pattern for atmos-

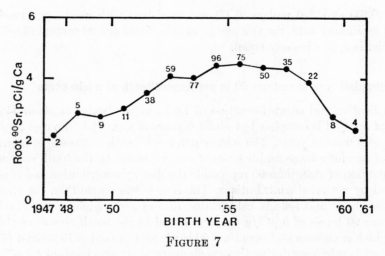

FIGURE 7

Strontium 90 content of 1st bicuspid roots *versus* year of birth.

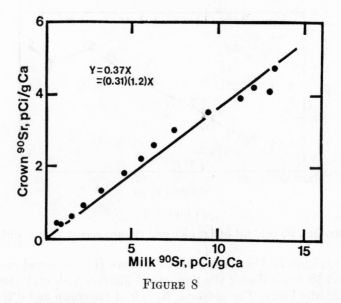

FIGURE 8

Strontium 90 content of 1st biscupid crowns *versus* strontium 90 content of cow's milk for samples shown in Figure 5.

pheric radioactivity as determined by the surface air sampling program of the AEC-HASL installation in which contamination is lower in northern latitudes (Toronto) and higher in southern latitudes (New Orleans). Teeth from children born in California contain approximately 50 per cent less strontium 90 than teeth of St. Louis children. Two samples of deciduous teeth of children born in Japan [13] during 1956 contained 1.28 and 1.54 pCi/g Ca—values that are

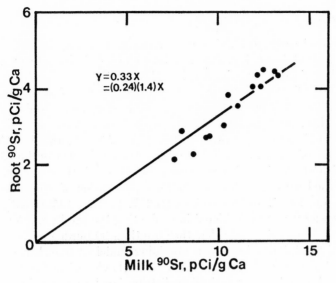

FIGURE 9

Strontium 90 content of 1st bicuspid roots *versus* strontium 90 content of cow's milk for samples shown in Figure 7.

TABLE II

STRONTIUM 90 IN DECIDUOUS INCISORS OF AMERICAN CHILDREN

Area	Birth year	No. of samples	pCi/g Ca	% of St. Louis values
Toronto	1956	6	1.82 ± 0.05	83
	1957	5	1.96 ± 0.09	70
Michigan	1957	12	2.47 ± 0.07	89
Indianapolis and Chicago	1956	5	2.46 ± 0.21	112
	1957	3	2.77 ± 0.21	99
St. Louis	1956	19	2.19 ± 0.08	100
	1957	21	2.79 ± 0.08	100
	1958	18	3.46 ± 0.12	100
East Texas and New Orleans	1957	7	3.43 ± 0.14	123
California	1956	2	1.21 ± 0.35	55
	1957	5	1.53 ± 0.13	55
	1958	3	1.74 ± 0.25	50

similar to that found in California for comparable years. Furthermore, the deciduous teeth of German children [14] and St. Louis children born during 1952–1954 contained comparable amounts of strontium 90 that averaged 0.45 pCi/g Ca. Although the data are sparse, they indicate the worldwide dissemination of strontium 90 in the teeth and calcified tissues of children.

7. Conclusions

It is apparent from our studies that strontium 90 accumulation into calcified tissues of pre-natal and juvenile American children is linearly related to the concentration of the nuclide in the diet. The equations that we have developed appear to be valid for increasing and decreasing levels of environmental contamination. These equations become even more useful when only one variable is known. Thus, measurement of tooth strontium 90 levels may describe the dietary level of ingested strontium 90 at the time of tooth formation. Conversely, measurements of dietary strontium 90 define the maximal strontium 90 calcified tissue burden at the time of tooth formation. Although these studies have only been concerned with strontium 90 burdens, the burden of other nuclides with both short and long half-lives such as Cs 137, I 131, I 125, strontium 89 and many others need to be considered in evaluating the total radiation burden of body tissues. It would appear that the strontium 90 burden might serve as an indicator of the total body burden providing sufficient knowledge is obtained concerning the biochemical distribution and behavior of the various nuclides, their half-lives, and the quantitative distributional relationship between the various nuclides in the environment.

The rate of decrease of strontium 90 in milk from the St. Louis watershed during the past four years is somewhat less than the physical decay of the isotope. We interpret this situation to mean that additional strontium 90 is being injected into our environment in quantities sufficient to obviate natural decay. If this situation continues, children born at present levels of about 9 pCi Sr/gm Ca in milk will continue to develop, mature and bear their children at about the same level of environmental contamination. The question must then be asked, "What immediate or future biological effects, if any, are to be expected from such levels of radiation burden?" The answer to this question is unknown at present, although some effects on infant mortality have recently been suggested by Sternglass for radiation near the vicinity of nuclear reactors [15].

The accumulation of strontium 90 in bone and calcified tissues and its relationship to osteogenic tumor induction has occupied the attention of most investigators for many years [16]. With the exception of the proximity of bone strontium 90 to rapidly proliferating bone marrow, biochemical effects of low levels of strontium 90 on soft tissues such as testes, ovary, pituitary and other glands have been essentially ignored. Furthermore, the yttrium 90 daughter of strontium 90 would be expected to leach out of bone and to be in equilibrium with soft tissues. Preliminary studies in my laboratory [17] show that strontium is bound to testicular and ovarian homogenates and cell fractions more than to liver and kidney tissue groups. These binding groups appear to be largely the phosphate groups of DNA and RNA although other protein groups also bind strontium 90 and cannot be ignored. Because DNA is predominantly present in the nucleus and is the major sensitive component for genetic transmission and mutation, it is conceivable that continuous low levels of radiation for long periods of time would tend to increase DNA modification and mutation rates. It would appear, therefore, that the strontium 90 bone stores and its yttrium 90 daughter could yield such a continuous radiation to sensitive soft tissue components throughout the growth period and reproductive life of the individual.

We have previously implied that the strontium 90 body burden might result in a statistical appearance of increased mutation rates in 70 to 90 years equivalent to perhaps three or four generations. However, developments in medical knowledge during the past fifteen years with respect to genetic engineering, chromosome mapping, and biochemical function may obviate the need to wait such a long time. A number of genetic defects, such as hemoglobinopathies and inborn errors of metabolism are known to be unusually common in man and can be readily characterized in the population [18]. For example, phenylketonuria is known to occur once in each 10,000 births. The heterozygote occurs about once in 100 people and can readily be determined by phenylalanine loading tests. In some populations of the world ranging from Africa to Southeast Asia, as many as 30 per cent of the population will have some form of readily detectable hemoglobin variant. Of the many mutations amenable for study, some will be more fruitful than others. The proper choices can only be made by an interdisciplinary consortium of knowledgeable manpower.

It appears possible, therefore, to determine and characterize the existing mutation rates for many human and animal pathopathies and to study any changes in these rates during the next few years. Such studies will require a concerted effort on behalf of the scientific community such as specialists in medicine, public health, environmental engineers, statisticians and biological scientists. It is conceivable, of course, that many biological effects may result from synergistic relationships between multiple mutagens. A thorough discussion of molecular evolutionary events [19] and an annual symposium report on birth defects and mutations [20] have recently appeared. Until such studies are initiated and relevant knowledge has been obtained, it appears desirable to re-evaluate the consequences of and the need to proliferate environmental contamination by nuclear reactors and devices.

Although most mutations in plants and animals are deleterious, many mutations are either innocuous or contribute some benefit to the individual. Thus, sickle cell anemia is advantageous to individuals in tropical areas because it contributes some immunity to malaria. Because an increase in radiation from whatever source—nuclear reactors or natural cosmic radiation for space travelers —will speed up the mutation rate and evolutionary development, some human selection for the kind of acceptable mutation (Orwell's 1984?) must be made. The answer to this question lies in philosophical considerations beyond the limits of this report. Nonetheless, it appears desirable to thoroughly understand the genetic implications of low level radiation on human and animal evolutionary mechanisms in order to make such important philosophical judgments.

This study was supported, in part, by Grant RH-00461 and EP-00102 from the U.S. National Center for Radiological Health and the U.S. Environmental Protection Agency. Special thanks are due to Mrs. Sylvia Raymond and Mrs. Marian Bueler for their aid in making this study possible.

REFERENCES

[1] H. M. KALCKAR, "An international milk tooth radiation census," *Nature*, Vol. 182 (1958), p. 983.
[2] J. RIVERA, "Strontium turnover rates in human bones," *Radiol. Health Data*, Vol. 5 (1964), pp. 98–99.
[3] H. L. ROSENTHAL, "Accumulation of environmental 90Strontium in teeth of children," *Radiation Biology of the Fetal and Juvenile Mammal* (edited by M. R. Sikov and D. D. Mahlum), *USAEC Symposium Series*, No. 17 (1969), pp. 163–171.
[4] ———, "Discussion of effectiveness of monitoring systems," *Pediatrics*, Vol. 41 (1968), pp. 202–206.
[5] H. L. ROSENTHAL, J. T. BIRD, and J. E. GILSTER, "Strontium-90 content of first bicuspids," *Nature*, Vol. 210 (1966), pp. 210–212.
[6] H. L. ROSENTHAL, S. A. AUSTIN, J. E. GILSTER, and J. T. BIRD, "Accumulation of Strontium-90 into human fetal teeth and bone," *Proc. Soc. Exper. Biol. Med.*, Vol. 125 (1967), pp. 493–495.
[7] L. F. REISS, "Strontium-90 absorption by deciduous teeth," *Science*, Vol. 134 (1961), pp. 1669–1673.
[8] S. LOUGH, J. RIVERA, and C. L. COMAR, "Retention of strontium, calcium and phosphorus in human infants," *Proc. Soc. Exper. Biol. Med.*, Vol. 112 (1963), pp. 631–636.
[9] F. J. BRYANT and J. F. LOUTIT, "Human bone metabolism deduced from strontium assays," British Report AERE-R 3718, 1961.
[10] B. G. BENNETT, *HASL Report 242*, USAEC, April, 1971, Part I, pp. 108–120.
[11] FEDERAL RADIATION COUNCIL, *Report No. 6*, October, 1964.
[12] H. L. ROSENTHAL, J. E. GILSTER, J. T. BIRD, and P. V. C. PINTO, "Regional variation of strontium-90 content in human deciduous incisors," *Arch. Oral. Biol.*, Vol. 11 (1966), pp. 135–137.
[13] E. ONISHI, Personal communication.
[14] R. KEIL, F. RANDOW, and H. A. SCHULZE, "Strontium-90 in Menschenzahnen," *Strahlentherapie*, Vol. 122 (1963), pp. 117–122.
[15] E. J. STERNGLASS, "Evidence for low-level radiation effects on the human embryo and fetus," *Radiation Biology of the Fetal and Juvenile Mammal* (edited by M. R. Sikov and D. D. Mahlum), *USAEC Symposium Series*, No. 17 (1969), pp. 693–717.
[16] A. ENGSTROM, R. BJORNERSTEDT, C. CLEMEDSON, and A. NELSON, *Bone and radiostrontium*, New York, Wiley, 1958.
[17] H. L. ROSENTHAL and O. A. COCHRAN, unpublished results.
[18] R. W. MCGILVERY, *Biochemistry*, Philadelphia, Saunders, 1970, pp. 387–388.
[19] T. UZZELL and K. W. CORBIN, "Fitting discrete probability distributions to evolutionary events," *Science*, Vol. 172 (1971), pp. 1089–1096.
[20] E. B. HOOK, "Monitoring human birth defects and mutations to detect environmental effects," *Science*, Vol. 172 (1971), pp. 1363–1366.

Discussion

Question: Alexander Grendon, Donner Laboratory, University of California, Berkeley

Since you commented on the possibility of genetic effects from such small amounts of Sr 90, I want to add to the record the note that the probability of such effects is extremely small. The hypothesis to which you evidently refer is that Sr 90 may replace some of the few calcium atoms that serve as binding elements in DNA, since other Sr 90 atoms are almost all in bone and remote from the gonads. One must then consider atom ratios, and 10 pCi Sr 90/gCa represents about one Sr 90 atom per 10^{13} to 10^{14} atoms of calcium. Even if there is no discrimination against strontium, the probability of an Sr 90 atom replacing a Ca atom in a cell that forms a child is extremely small.

Reply: Harold L. Rosenthal

We have studied Sr 90 deposition primarily as a long range indicator atom. However, as I mentioned earlier, strontium 90 body burdens might serve as an indicator of the total body burden if all factors concerning the biochemical distribution and physical behavior of all the various nuclides (their half lives, and so forth) and the quantitative distributional relationship between the various nuclides in the environment are considered. For example, a knowledge of the fission yield of nuclides relative to Sr 90 could be used to estimate the exposure of humans and animals to all of the radionuclides produced. I believe that upwards of 1600 nuclides have now been described so that summation and cumulative effects of all the exposures, Cs 137, I 131, Po 210, and so forth, certainly adds up to more than from just Sr 90. It would be very desirable for this conference to do such a summation recognizing all of the chemical, physical and biological factors.

Question: E. J. Sternglass, School of Medicine, University of Pittsburgh

I believe it is important to point out the great value of the data Dr. Rosenthal has so carefully gathered over the years, since it is the only truly meaningful and accurately recorded measure of the actual amounts of radioactivity in the developing human fetus and infant we now have available for future epidemiological studies.

In view of the very serious fact that the concentrations of strontium 90 did not continue to decline at the rate at which they decreased after their peak from atmospheric testing by the U.S., U.S.S.R. and U.K. in 1964 but, instead, have actually shown renewed rises since 1968, it would seem urgent not only to continue this unique monitoring technique of fallout in the developing human infant, but to expand it so as to include other geographical areas.

The fact that there exists a high correlation between the excess fetal mortality rate and the amounts of strontium 90-yttrium 90 in the fetal tooth buds strongly suggests the importance of further research in the area of the endocrinological aspects of these isotopes and other rare earth elements so as to explain the biological mechanisms whereby the premature birth of the infant and the attendant higher mortality rates are produced.

Above all, these data showing renewed rises in the bone concentrations of these isotopes in the newborn in recent years would suggest the need for a halt to all further introduction of fission products into the environment as a matter of prudence and concern for the health of our children.

Reply: Harold L. Rosenthal

We have made some measurements on regional variations in deciduous teeth of children as we showed in Table II. In general, teeth of children from Toronto contain about 25 per cent less than that for St. Louis children and New Orleans shows about 25 per cent more. This seems to follow the same north-south patterns of atmospheric levels. It is interesting to note that teeth of northern California children contain about half of the St. Louis value—presumably because the major testing program is east of California and the prevailing winds carry the fission products in an easterly direction. We don't know if the California children are contaminated from Russian or Pacific Island tests or if the Nevada test products completely circled the globe.

PROBLEMS IN DETERMINING IF A COMMONLY USED HERBICIDE (2,4,5-T) HAS AN EFFECT ON HUMAN HEALTH

THEODOR D. STERLING

WASHINGTON UNIVERSITY

One has to be reminded occasionally that the roots of statistics lie in problems of inference, especially in the study of efficient and useful experimental designs from which conclusions can be drawn. I start with this reminder because many of the observations and conclusions concerning the effects of pollutants derive from experiments which "happened" more than they were designed and which "presented" their data rather than analyzed them. However, there is yet one other reason for this reminder. Problems of inference raised by studies on pollution seem to show that what is needed is not so much a study of the design of experiments but a study of the strategies of acquiring and analyzing wide ranges of observations and that the end products are not simple inferences about "states of nature" but formulations of "public policies."

Keeping these reminders in mind, we shall next turn to a review of the issues and problems surrounding the question if a commonly used herbicide, 2,4,5-T, has effects that are of concern to the large community of this country.

1. 2,4,5-T is a general pollutant

The widespread (albeit inadvertent) consequences of present practices to control animal and plant pests by use of chemical agents have been recognized only recently as a general pollution problem. In view of the known toxicity of many of the agents, industrial physicians and engineers have been concerned with the manufacture and distribution of these toxic materials and with instituting proper warning procedures so as to avoid what has been commonly called "accidents." It is recognized now that while herbicides and pesticides are designed to affect only a specific target species, their indiscriminate and widespread use creates very general pollution problems. First, all herbicides and pesticides have inadvertent effects on nontarget species. The toxic agent may be extremely widely spread by wind and water to places where it was never intended to show up. For instance, 2,4,5-T applied as a spray can be transported in the atmosphere as a drop of spray, as a gaseous state of 2,4,5-T, or adsorbed on dust or other particulate matters in the air. In this way 2,4,5-T was found adsorbed on dust in a trace of rain in Cincinnati, Ohio, presumably from applications in Texas [32].

Secondly, traces of the chemical may show up in the food and water consumed by human users. Heretofore concern has been mainly for residues that may appear in foods that have been grown with the help of pesticides. In most circumstances this residue can be kept to a very minimum (although this is not always possible). Much more serious, however, is the danger that the pesticide becomes concentrated in the ecological food chain. Man, being on the apex of a pyramid in this food chain, might thus be exposed to large doses of the chemical in a wide variety of foods, especially meats. Thus, herbicides and pesticides used for specific purposes, even in rather isolated rural areas, might find their way in large doses to man wherever he resides. Finally, because the active chemical agents may be also mutagenic and teratogenic, their effect may not be limited to the present population exposed to that toxicity but to future generations through mutations and introduction of malformations. (We should note in passing that until now the mutagenic and teratogenic dangers inherent in some of the commonly seen air and water pollutants have received very little attention.)

Pesticides are of immeasurable benefit. They make it possible for man to inhabit his planet in large numbers by insuring an ample food supply and eliminating or controlling carriers of disease and discomfort. Yet, the dangers created by these same pesticides to human health and survival have to be examined very carefully and weighed against the benefits derived from their use.

2. Patterns of use of 2,4,5-T

The herbicide, 2,4,5-T (2,4,5-trichlorophenoxyacetic acid), is used primarily to control woody plants and a few herbaceous species against which it is more effective than other herbicides. 2,4,5-T is never "pure." It has not been possible so far to manufacture 2,4,5-T without one specific impurity, 2,3,7,8-tetrachloro-dibenzo-p-dioxin (commonly referred to as TCDD). While even pure 2,4,5-T has a known (albeit small) toxic and teratogenic effect, the toxicity and teratogenicity of TCDD can be described only as "virulent." While most species tested (with the exception of the dog) can survive a single oral dose of 2,4,5-T in excess of 100 mg/kg, and several can survive daily treatments for a number of days at this high level [12], embryo toxicity appears in litters of females given TCDD in doses as low as .000125 mg/kg per day, and a dose of .008 mg/kg per day caused pallor and debilitation. The LD_{50} for TCDD has been found to be .022 to .045 mg/kg in the rat and .0006 mg/kg in the guinea pig [18]. Manufacturers of 2,4,5-T had paid only limited attention to the amount of impurities present. Both Dow Chemical Company and Hercules Incorporated developed procedures in which the contaminant, dioxin, was held to less than 1 ppm, but other companies manufactured a product that contained as much as 40 ppm of the dioxin, TCDD.

Because of its effectiveness, the chlorophenoxy-herbicides, 2,4-D and 2,4,5-T, have been widely used for over twenty years to control broad leafed weeds. Since a number of studies conducted in the 1950's were interpreted to indicate that the herbicide was not toxic to man at the doses at which man is exposed,

2,4,5-T has been widely used in forestry, agriculture, by the park service, and around the home. One of the major uses of 2,4,5-T has been for rights of way (approximately 50 per cent) and for nonfarm forests (approximately 10 per cent). However, roughly 20 per cent of the produced 2,4,5-T goes into hay, pasture, and rangeland and other farm uses where residues of the herbicide, in one form or the other, may show up in food. The remainder of the 2,4,5-T is used for parks, especially lawn and turf care, gardens, and other purposes where it may come into direct contact with human users in a variety of ways [30]. The pattern of use of the herbicide is especially important for its pollution potential. Its use in grazing and forestry areas may make it subject to entering the food chain and being concentrated in a large number of ways. For instance, it may be concentrated through grazing animals [15] or, by being washed into the lakes and oceans where it is picked up by plankton, it may contaminate fish, crustacea, and mollusks [3]. 2,4,5-T may come into contact with humans through residues on foodstuffs, especially on domestic rice. Finally, 2,4,5-T is a favorite herbicide around the home and garden. It is especially effective in controlling poison ivy. Its unsafe use by a home user may contaminate the family's food in a large number of ways.

One of the big problems, therefore, is whether or not 2,4,5-T can be accumulated in the food chain and stored in tissues. Unfortunately, not much is known about the fate of 2,4,5-T after it is applied, and even less is known about the fate of its active teratogenic and toxic impurity, TCDD. 2,4,5-T is immediately subject to physical and chemical actions that continually reduce the amount remaining at a site of application (by degradation by soil microorganisms, leaching and surface movement in water, volatilization, movement by wind, and photochemical decomposition). Chemically detectable amounts of two pounds per acre have been found in soil after three to seven months after application, but no detectable amount was found in the same soil after one year [1]. The half-life of 2,4,5-T has been estimated as 40 days, at least in forests [22]. Thus, although 2,4,5-T disappears from the soil relatively quickly, it remains long enough to appear as a possible residue in foodstuffs, and certainly its half-life is long enough to enable it to enter the ecological food chain. TCDD is not as easily eliminated from the environment as is 2,4,5-T. Because of its low water solubility (only .2 ppb), TCDD does not move through the soil nor is it easily leached out [30]. Using radiolabeled TCDD, it was found that the radioactive material (probably TCDD) in the soil decreased only 15 to 20 per cent in 160 days, indicating that this compound was very slowly degraded and would persist for more than a year [19]. Thus, there is ample evidence that TCDD remains in the environment for a relatively protracted period.

3. The recent 2,4,5-T controversy

Although a number of studies were done some twenty years ago to evaluate the toxicity of 2,4,5-T, and although the possible teratogenicity of this agent has been under careful investigation during the last two years, there is disagreement on whether or not 2,4,5-T is toxic and teratogenic to humans exposed to low doses. Scientists who are familiar with existing studies are divided in how they interpret the data. For statisticians it is important to notice such divisions, especially if the disagreement is not at what level of a probability (computed or subjective) a hypothesis ought to be accepted or abandoned. What makes the study of the controversy surrounding 2,4,5-T so valuable is that it gives us the opportunity to see the scientific process in operation and, perhaps, enables us to evaluate the usefulness of present methods of designing and analyzing experiments. After all, 2,4,5-T originally was evaluated by a number of conventionally designed toxicity studies, and the recent public controversy surrounding its possible teratogenicity has motivated a number of experiments by both government and industry that were designed specifically to clarify, once and for all, if 2,4,5-T was teratogenic—at least in the experimental animals.

The review of 2,4,5-T came about quite inadvertently through a screening study which the National Cancer Institute contracted with Bionetics Research Laboratories. In this study Bionetics Research Laboratories performed screening studies for carcinogenicity and teratogenicity on a number of pesticides and industrial chemicals. The results released in October, 1969, indicated that 2,4,5-T showed embryo toxicity in two stocks of mice at doses of 113 mg/kg per day when given for several days during organogenesis. As it turned out, the sample of 2,4,5-T used in the Bionetics study was known to have been contaminated with 27 ± 8 ppm of TCDD, and the results were no longer considered valid indication of the teratogenicity of the herbicide. (The experiment has been summarized in a number of forms [8], [23], [31].)

The findings of the Bionetics Laboratories, together with resports by South Vietnamese newspapers of an increased occurrence of birth defects during June and July of 1969 (2,4,5-T had been the major defoliant used in Vietnam), elicited far reaching reactions from governmental agencies and segments of the scientific and concerned lay communities. A number of animal experiments performed early in 1970 confirmed that even the pure samples of 2,4,5-T did, indeed, result in delivery of malformed offspring. In April, 1970, the Secretaries of Agriculture, of Health, Education, and Welfare, and of the Interior jointly announced the suspension of the registration of 2,4,5-T.

Next, the Dow Chemical Company and Hercules Incorporated exercised their right under Section 4.c. of the Federal Insecticide, Fungicide, and Rodenticide Act (U.S.C. Vol. 7, p. 135 ff.) to petition for referral of the matter to an advisory committee. Such a committee was then formed from a list supplied by the National Academy of Sciences and met during the early five months of 1971. The committee submitted its report in May, 1971. Although the committee was in substantial agreement on the facts of the case (as far as the evidence was palpable enough to supply the facts), its recommendations on the restoration of registration were divided [6], [14].

At the same time, the United States Military in South Vietnam undertook a survey to evaluate the human evidence of birth defects possibly due to defoliation practices. An Army report was issued in December, 1970 [10]. The Army studies surveyed obstetrical records for the years 1960 to 1969 of 22 provincial, district, and maternity hospitals in 18 cities and other areas in various geographical localities. The findings of the Army studies, in the main, were that no differences in stillbirth rates were observed geographically that could not be attributed to better maternal and neonatal care or to more competent or thorough examination for congenital malformations in the capitol area, and that the rates of stillbirth declined and of congenital malformations remained unchanged during this ten year period despite the heavy spraying between 1966 and 1969. At the same time, the American Association for the Advancement of Science appointed a commission led by Dr. Meselson to investigate the Vietnam charges. This report (HAC) noted that the Army report had been heavily influenced by data from the capitol area which generally experienced little or no exposure to 2,4,5-T. By deducting the capitol area data and considering only data from other parts of the country, the declining trend was reversed, giving considerably increased stillbirth and malformation rates in 1966–69 (heavy spraying period) over 1960–65 (no or light spraying period). Also there was an increase in the incidence of cleft palates and spina bifida (in some regions) that was in need of an explanation. In addition, the HAC report pointed out that in the Tay Ninh Provincial Hospital, which serviced a population that was directly exposed to heavy defoliation or lived near rivers draining areas of defoliation and serving as the source for fish, the stillbirth rate was much higher than in any of the other hospitals surveyed by the Army [21]. (There are two other reported incidents where spraying was thought to be followed by malformed human offspring. One came after a spraying project near Globe, Arizona, and the other from a Swedish defoliation project in Lapland. Claims for both areas were investigated by teams of experts and found to be, most likely, not associated with the spraying incidents.)

4. Sources of uncertainty

It would be a mistake to view the present attitude toward the use of 2,4,5-T as indicating a deep division in the scientific or concerned lay communities. There is general agreement that 2,4,5-T (with or without its impurities) is toxic. The disagreements reflect the burden of responsibilities or of special interests. Governmental agencies tend toward the restriction and more stringent regulation of 2,4,5-T, and agricultural and manufacturing industry and representatives of the forest service tend toward the restoration of the use of 2,4,5-T to its prior status. However, the division does reflect the uncertainty about the dangers to human health. Despite a large number of experiments, some of them directly designed to assess the toxicity and teratogenicity of 2,4,5-T, a substantial uncertainty remains about almost every important question that needs to be answered about the toxic and teratogenic dangers entailed in human exposure.

The problem which causes uncertainty consists of two parts.

(1) The determination of toxicity and teratogenicity of a substance is not a simple matter. Sometimes the experimental solution to questions of fact concerning toxicity cannot be answered by any known experimental procedure, and sometimes they can be answered only partially.

(2) The types and kinds of experimental designs required and the sort of analyses necessary in order to extrapolate from present findings to questions of human health require a high degree of sophistication in experimental design, mathematical analysis (especially in extrapolation techniques), and in information processing. Unfortunately, these skills appear to be very sparsely distributed among toxicological or teratogenic researchers, and proper help from statistics and data processing sources appear to be either unavailable or unused.

As a consequence, a difficult scientific problem is infinitely compounded because proper tools are not brought to bear on its solution. These two facets create and compound the uncertainty about the danger inherent in the use of 2,4,5-T to such an extent that we shall treat them separately and refer to them, respectively, as first and second order uncertainty.

4.1. *First order uncertainty: the experimental model.* All toxicological and teratogenic studies suffer from three major shortcomings.

(1) Animal experiments are, by and large, inadequate;

(2) the dose and response relationship is difficult to determine; and

(3) information on the effects on humans is almost impossible to obtain.

4.1.1. *The proper animal "model."* It is naively assumed that the antecedents of disease can be easily studied on animals and that a demonstration that certain conditions lead to disease in animals may invariably teach us how disease is caused in humans. Unfortunately, this is not always true. Although we can learn a great deal from animal experiments, toxicological and teratological information from animal experiments turns out to be much less useful than is commonly thought. Animals react with a wide range of physical and behavioral responses to the presence or absence of chemical stimuli. There is not only a wide difference in the reaction of different animal species, but the difference becomes even more pronounced when we compare animal to man. Difficulties in inferring human reactions from animal studies are further compounded by the use of rodents as a favorite experimental animal because of their convenience and low cost as laboratory animals. However, rodents are much further removed phylogenetically from the human animal than are dogs or monkeys. On the other hand, experiments with dogs and monkeys are inordinately expensive. The problem raised by the use of rodents is clearly seen in 2,4,5-T toxicology studies, where it was found that rats may be able to maintain on doses of 100 mg/kg per day for a number of days while dogs may succumb from a single dose of 20 mg/kg [12]. Yet, most inferences about toxicity of 2,4,5-T to man were based on work with rodents. It should be noted that the problems of choosing a proper model (that is, an animal system that reacts in the same way as a human subject) have had marked effects on such recent affairs as thalidomide, riboflavin deficiencies, vitaminoses, and others.

4.1.2. *Relationship between dose and response.* As a corollary to the problem of finding the right model of animal stands the observation that every known substance (including water and oxygen) is harmful and may have toxic, teratogenic, mutagenic, and carcinogenic effects if it is given in a large enough dose [4], [7], [29]. As a consequence, the meaning of a toxic response in an animal system to a very high dose of an agent may be unclear. True, the kind of physiological or anatomical reaction occurring offers an important clue to how human tissue may respond to the same agent. But the mere fact that an agent is toxic at some high dose is not considered surprising, since a toxic reaction is to be expected for some dose. What is important is to see the rate at which a reaction disappears as a function of lower and lower doses.

Unfortunately, the investigation of the dose response curves in animals is beset by two extremely difficult problems.

(1) A low exposure does not necessarily result in a smaller reaction in an animal but, rather, in fewer animals in which this reaction may be observed. Thus, with a large enough exposure to kill all animals, the question of whether or not a particular dose is toxic creates no great experimental problems. However, an immensely large number of animals would be needed to determine doses with the toxicity to affect 1 per cent, 0.5 per cent, or 0.01 per cent of the animal population.

(2) The response of biological systems to any agent is not uniformly good or bad. It is true that biological systems do respond to a large variety of chemical agents in different ways at different doses. For instance, copper, which is a very toxic substance at high doses, is not toxic at all in low doses and is, in fact, necessary to sustain life, so that its complete absence is a definite hazard. The same is, of course, also true with many other substances, including water and oxygen.

Thus, the major utility of an animal experiment is to see whether or not the reactions to relatively high doses can serve for extrapolating a dose response function that will tell if a zero or nontoxic response to some dose will occur or whether or not the agent may be toxic at any dose.

4.1.3. *Clinical (human) studies.* The best measure of toxic and teratogenic effects is man. Unfortunately, man is also the most unusable experimental subject. Three major difficulties meet the attempts to study effects of pollution on man (including that of herbicides and pesticides).

(1) Toxic agents that have dramatic effects are easily spotted and eliminated. Pollutants that affect the occasional individual, individuals who are ill from other causes, or that cause small reactions in individuals tend to go unnoticed.

(2) Even when noticeable reactions occur, the information that they do occur may be almost impossible to obtain and evaluate. Abnormal human reactions usually are recorded only when they motivate the diseased individual to seek medical attention. Thus, processes of self selection emanate which make it almost impossible to find proper control groups for prospective or retrospective studies, even if these records become available. However, the fact that records exist somewhere about a reaction to a pollutant is, in most instances, immaterial

since these recorded instances are difficult to find or, once located, to concentrate for review and analysis.

(3) There are restraints to either withdrawing or exerting a specific treatment on humans, even if an agent or treatment is deemed to be harmful (or its opposite).

4.2. *Second order uncertainty: the experimenter.* Experiments that may yield answers to such questions as those asked here require sophistication in the design, the analysis of multivariate data, the mathematical techniques for extrapolation, and data processing facilities.

Yet, what do we find? The literature contains less than two dozen key reports of studies on the toxicity of 2,4,5-T and on its teratogenicity. While most of these experiments were performed using some variation in combination of different dose levels, different concentrations at which dose levels were applied, different vehicles which carried the doses, different amounts of impurities (especially TCDD), and some were performed on different species or used different products coming from different manufacturers (and so ideally suited to factorial designs), the reports of some 22 experiments analyzed by the 2,4,5-T Advisory Committee did not contain a *single* experiment that was designed to tell something about the effect of 2,4,5-T at very low doses and attempted, by statistical analysis or mathematical techniques, to milk the available data (although inadequate) for whatever information could have been obtained about reactions of animals to very small doses.

The relevant design features and approaches to the analysis of data of ten toxicological and ten teratological experiments are summarized in Tables I and II. In view of the lack of statistical sophistication practiced in this important field, much is to be learned from their study.

All experiments were designed to test the effect of 2,4,5-T at high doses on relatively small numbers of animals. No provisions were built into the experimental design to permit picking up the effect of 2,4,5-T at a low dose. In fact, some of the conclusions reached by the scientists who reviewed these data were that 2,4,5-T has a toxic and teratogenic effect only at high doses. This impression may have been created by the fact that no adequate experimental design existed to assess the effect of 2,4,5-T at low doses.

Very serious, in view of the wealth of available statistical and mathematical techniques, is the relative naiveté with which these studies were analyzed. Many of these reports did not subject their data to any statistical analysis whatsoever. Simple summary figures, such as the arithmetic average, were in most instances not accompanied by measures of dispersion. The most sophisticated statistical analyses were multiple application of Student's t tests comparing sets of measurements from each individual group (resulting from combinations of doses and other factors) with a control group. The numbers so analyzed often were discrete rather than continuous and came from obvious nonnormal distributions. Those values of "t" that met the famous criteria of statistical significance at 0.05 were duly starred with an asterisk, adding perhaps insult to injury [28]. Thus, not a

TABLE I

Summary of Relevant Design Features and Analyses of Data Obtained from Key Experiments Investigating the Toxicity of 2,4,5-T

Reference	Kind of animal	Number of animals	Dose levels (and other conditions)	(Statistical) analyses
Rowe [24]	Rats Mice Guinea pigs Chicks	10 10 10 3	Different doses but amounts not given	LD$_{50}$ (by method of Lichtfield). No statistical analyses.
Drill [12]	Dogs	48	4 dose levels each for 2 separate experiments (acute and chronic), for 2,4-D and 2,4,5-T and for males and females	Tabular and descriptive summaries (given for each animal) of body weight, symptoms, autopsy findings, blood counts, organ weights, microscopic changes. LD$_{50}$ is estimated but no indication given on how it was done. No statistical analyses.
Rowe [25]	Rats Guinea pigs Rabbits Mice Chicks Dogs	Unclear how many	15 different herbicidal materials and 12 different herbicidal formulations	LD$_{50}$ (by method of Lichtfield). Various qualitative observations listed. No statistical analyses.
Dow [11]	Rats	100	5 dose levels each for males and females	Tabular summary of mortality, autopsy findings, averages for food consumption, body and organ weights, blood and urine analyses. Statistical analyses limited to comparing some groups to controls using Student's t tests.
Hazelton [16]	Rats Rabbits Rabbits Rabbits	30 3 16 12	For eye irritation phase—6 dose levels in 3 concentrations. For first dermal irritation phase—4 dose levels. For second dermal irritation phase—3 dose levels.	Tabular summaries and qualitative descriptions of weekly body weights, signs of dermal irritation, blood and urine analyses. No statistical analyses.
Butler [3]	Crustacea Mollusks Fish Phyloplankton	Not given either by volume or number	Different pesticides at different water conditions as salinity and temperature and time intervals	Tables of LD$_{50}$ values. Qualitative description of a variety of observations. No statistical analyses.
Hegyi [17]	Rabbits	32	10 different solutions	Pictures and qualitative description of a variety of observations. No statistical analyses.
McCollister [20]	Rats	50	5 dose levels each for males and females	Tables giving blood and urine counts (for each individual animal), individual body and organ weights and averages, listing of gross histological results. No statistical analyses.

TABLE II

SUMMARY OF RELEVANT DESIGN FEATURES AND ANALYSES OF DATA OBTAINED
FROM KEY EXPERIMENTS INVESTIGATING THE TERATOGENICITY OF 2,4,5-T

Reference	Kind of animal	Number of animals	Dose levels (and other conditions)	(Statistical) analyses
Emerson [13]	Rats	175	6	Qualitative description of observations. Tabular summaries of organ and body weights, number of pregnancies, implantations, corpora lutea, viable pups, resorptions, average and range of pup weights, skeletal and visceral abnormalities. No statistical analyses.
Courtney [8]	Mice (two different strains) Rats	Not given. However, number of litters is.	4, 2, and 3 for different strains or time periods.	Tabular summaries of per cent (of live litters) of live fetuses, fetal mortality, abnormal litters, abnormal fetuses, cleft palate, cystic kidney, fetal and maternal average body and liver weights. Used "2 × 2 Chi Square tests" to compare porportion of abnormal fetuses with controls. (It is worthy of note that there were no concurrent controls but controls used were the cumulated experience over preceding three years.) No further statistical analyses.
Sparschu [26]	Rats	75	6 dose levels and concentrations	Qualitative description and tabular summaries of average maternal body weights, means of number of pregnancies, corpora lutea, viable fetuses, resorption sites, sexes, numbers of gross, visceral, and skeletal abnormalities. Student's t test used for comparing test groups to controls. No further statistical analyses.
Emerson [13]	Rabbits	80	4	Qualitative description of observations and tabular summaries of average maternal body weights, total and average number of implantations, corpora lutea, resorptions, viable kits, average and range of kit weights, sex and number of skeletal and visceral abnormalities. No statistical analyses.

TABLE II, (continued)

Reference	Kind of animal	Number of animals	Dose levels (and other conditions)	(Statistical) analyses
Bionetics [2]	Mice	Not given	1 dose of each of 2 compounds given via 2 vehicles and controls in each vehicle	Tabular description of observations on each individual animal and litter and tabular summary. No statistical analyses.
Collins [5]	Hamsters	6 per group	Up to 4 dose levels each for 7 different sources of 2,4,5-T and 3 sources of 2,4-D and one control group.	Tabular summaries of number and averages and per cents. No statistical analyses.
Sparschu [27]	Rats	75	2 doses and control	Tabular summaries of numbers, averages, and per cents. Student's t and Chi Square tests—individual test groups vs. control.
Bionetics [2]	Mice	Not given directly	1 dose of 2 compounds in 2 vehicles and controls for each vehicle	Tabular description of observations on each individual litter and summaries of totals, per cents, and averages. No statistical analyses.
Courtney [9]	Mice (3 strains) Rats	Not given, but number of litters is.	Up to 4 dose levels (varying) of 2 compounds each	Tabular summaries of number, per cent, averages. No statistical analyses.
Wilson [33]	Rats	Not given, but number of litters is.	4 dose levels, 3 of these given on day 9, 1 given on various days (7 to 13)	Tabular summary of total implants, per cent dead or resorbed, mean weight of survivors, per cent of survivors malformed. It is noteworthy that there are no concurrent controls. No statistical analyses.

single one of these experiments subjected its data to the robust and generally available factorial techniques (not to say anything about taking out the effect of confounding variables). Not a single one of these experiments made any attempt to extrapolate their data toward low doses.

5. Conclusion

It is not easy to understand why so much inadequate data had been generated. After all, these toxicological problems are not new and solutions exist for most of them. One simple answer might be that these practices have grown up around the need to *license* or *register* possibly toxic products and not around their widespread *use*. Another explanation might be that statisticians and mathematicians have failed to *educate* the community of scientists, especially those in medical areas. Undoubtedly, immediate interests in licensing and lack of proper preparation of scientists are influential factors. But it would be negligent not to look further. After all, a simple explanation would hardly fit a case involving so many investigators from so many different sources. Surely, lack of education or surrender to special interests are not *that* widespread in our scientific community.

There is an alternative answer to this question. Perhaps proper statistical and experimental techniques do not exist that would enable us to deal with the issues created by pollution and community health! Perhaps to derive a statement of public policy is quite a different problem from making an inference about a state of nature. For it is true that the end product of the process of inference in instances of pollution and community health is *administrative policy decisions* and not judgments about possible states of nature. These policy decisions may be to build or not to build nuclear power plants, to build them at particular sites, or to choose other strategies to deal with a power shortage. Similarly, the policy decision might be to register or license the use of a herbicide, restrict its use to certain instances, restrict its use with certain limitations on impurities, or examine alternative ways of pest control. Fortunately, the field of statistics is prepared to deal with a broad experimental strategy of the kind required to serve as a foundation for public health policy. Statisticians have brought the science of making inferences in the face of uncertainty to a high plateau. What may be needed as a next step is to assess problems created by pollutants for community health and to make decisions about community action and to develop the kinds of experiment to decision strategies that are best suited to derive rational and sensible public policies.

REFERENCES

[1] W. L. BAMESBERGER and D. R. ADAMS, "Organic pesticides in the environment," *Adv. Chem.* Ser. 60 (1966), Washington, D.C., ACS Publication.
[2] BIONETICS RESEARCH LABORATORIES, "Teratogenic effects of 2,4,5-T in mice, report submitted to Hercules Incorporated, December, 1970," Report submitted to Hercules Incorporated, March, 1971.

[3] P. A. BUTLER, "Pesticide-wildlife studies; a review of Fish and Wildlife Service investigations; commercial fisheries investigation," Circular 167 (1963); Circular 199 (1964); Circular 266 (1965).

[4] W. CARRUTHERS, "Carcinogens related to the aetiology of bronchial carcinoma," *Physiotherapy*, Vol. 44 (1958), pp. 307–312.

[5] T. F. X. COLLINS and C. H. WILLIAMS, "Teratogenic studies with 2,4,5-T and 2,4-D in the hamster" (prepublication copy to 2,4,5-T Advisory Committee, AE-9), Department of Health, Education, and Welfare, 1971.

[6] "Conflicting philosophies over 2,4,5-T," Editorial in *Nature*, Vol. 231 (1971), pp. 483–485.

[7] J. CORNFIELD, W. HAENSZEL, E. C. HAMMOND, A. M. LILIENFELD, M. B. SHIMKIN, and E. L. WYNDER, "Smoking and lung cancer: recent evidence and a discussion of some questions," *J. Nat. Cancer Inst.*, Vol. 22 (1959), pp. 173–203.

[8] K. D. COURTNEY, D. W. GAYLOR, M. D. HOGAN, H. L. FALK, R. R. BATES, and I. MITCHELL, "Teratogenic evaluation of 2,4,5-T," *Science*, Vol. 168 (1970), pp. 864–866.

[9] D. K. COURTNEY and J. A. MOORE, "Teratology studies with 2,4,5-T and tetrachlorodioxin" (manuscript for 2,4,5-T Advisory Committee), National Institute of Environmental Health Sciences, 1971.

[10] R. T. CUTTING, T. H. PHUOC, J. M. BLLO, M. W. BENENSON, and C. H. EVANS, *Congenital Malformations, Hydatidiform Moles, and Stillbirth in the Republic of Vietnam, 1960–1969*, Washington, D.C., U.S. Government Printing Office, 1970.

[11] DOW CHEMICAL COMPANY, "Results of 90-day dietary feeding studies of Dowanol, 97B Ester 2,4,5-T in rats," Biochemical Research Laboratory Report, November 27, 1961.

[12] V. A. DRILL and T. HIRATZKA, "Toxicity of 2,4-dichlorophenoxy acetic acid and 2,4,5-trichlorophenoxy acetic acid, A report on the acute and chronic toxicity in dogs," *AMA Arch. Indust. Hygiene Occupat. Med.*, Vol. 7 (1953), pp. 61–67.

[13] J. L. EMERSON, D. J. THOMPSON, C. G. GERBIE, and V. B. ROBINSON, "Results of teratogenic studies of 2,4,5-trichlorophenoxy acetic acid in rats," Dow Chemical Company, Human Health Research and Development Laboratory Report, March 30, 1970.

[14] ENVIRONMENTAL PROTECTION AGENCY, Report of the Advisory Committee on 2,4,5-T to the Administrator, May 7, 1971, to appear.

[15] B. H. GRISBY and E. D. FARVELL, "Some effects of herbicides on pasture and grazing livestock," *Quart. Bull.*, Vol. 32 (1950), pp. 378–385.

[16] HAZELTON LABORATORIES INCORPORATED, Report to the Diamond Alkali Company, February, 1962, mimeograph.

[17] E. HEGYI, Z. ST'OTA, and A. LUPTAKOVA, "Die Rolle des Alkali im Hinblick auf die keratogene Wirkung des Technischen 2,4,5-Trichlorophenol," *Derufs-dermatosen*, Vol. 6 (1969), pp. 327–337.

[18] J. E. JOHNSON, "Symposium on possible public health implications of widespread use of herbicides," AIBS Meeting, August 26, 1970.

[19] T. C. KERNEY, "Chlorinated dioxin research," presented before a Joint Meeting on Pesticides, United Kingdom, Canada, United States, November 5, 1970.

[20] S. B. McCOLLISTER and R. J. KOCIBA, "Results of 90-day dietary feeding study on 2,4,5-trichlorophenoxy acetic acid (2,4,5-T) in rats," Biomedical Research Laboratory, The Dow Chemical Company, September 18, 1970.

[21] M. S. MESELSON, A. H. WESTING, and J. D. CONSTABLE, "Background material relevant to presentation of the 1970 annual meeting of the AAAS," Herbicide Assessment Commission of the American Association for the Advancement of Science (revised January 14, 1971).

[22] L. A. NORRIS, "Degradation of herbicides in the forest floor," *Tree Growth and Forest Soils* (by C. T. Youngberg and C. B. Davey), Corvallis, Oregon State University Press, 1970, pp. 397–411.

[23] OFFICE OF SCIENCE AND TECHNOLOGY, *Report on 2,4,5-T of the Panel on Herbicides*, April, 1971.

[24] V. K. Rowe, D. D. McCollister, and H. C. Spencer, "The acute oral toxicity of 2,4,5-trichlorophenoxy acetic acid to rats, mice, guinea pigs, and chicks," Report of the Biochemical Research Laboratory, Dow Chemical Company, 1950.

[25] V. K. Rowe and T. A. Hymas, "Summary of toxicological information on 2,4-D and 2,4,5-T type herbicides and an evaluation of the hazards to livestock associated with their use," Amer. J. Veterinary Res., Vol. 15 (1954), pp. 622–629.

[26] G. L. Sparschu, F. L. Dunn, and V. K. Rowe, "Teratogenic study of 2,3,7,8-tetrachloro-diobenzo-p-dioxin in the rat," Biochemical Research Laboratory Report, Dow Chemical Company, 1970.

[27] G. L. Sparschu, F. L. Dunn, R. W. Lisowe, and V. K. Rowe, "Study of the effect of high levels of 2,4,5-trichlorophenoxy acetic acids (2,4,5-T) on rat fetal development," Chemical Biological Research Report, Dow Chemical Company, 1971.

[28] T. D. Sterling, "Publication decisions and their possible effects on inferences drawn from tests of significance—or vice versa," J. Amer. Statist. Assoc., Vol. 54 (1959), pp. 30–34.

[29] T. D. Sterling, "Epidemiology of disease associated with lead," Arch. Environ. Health, Vol. 8 (1964), pp. 333–348.

[30] United States Department of Agriculture, "Progress report on dioxin research IV," March 25, 1970, Unpublished report.

[31] U.S. Department of Health, Education, and Welfare, "Report of the Secretary's Commission on Pesticides and their Relationship to Environmental Health," Washington, D.C., 1969.

[32] S. R. Weibel, R. B. Weidner, J. M. Cohen, and A. G. Christianson, "Pesticides and other contaminants in rainfall and runoff," J. Amer. Waterworks Assoc., Vol. 58 (1966), pp. 1075–1084.

[33] J. G. Wilson, experiment reported to the 2,4,5-T Advisory Committee, 1971; this experiment is part of the Final Report of the 2,4,5-T Advisory Committee [23].

Discussion

Question: E. B. Hook, Birth Defects Institute, Albany Medical College

Would you specify the nature of the increase in congenital malformations in the AAAS Vietnamese study (that is, which specific defects were increased), and what order of magnitude of increase there was, and what was the "denominator" of population studied?

Also, is there evidence that the other civilian ravages of war in Vietnam were associated with exposure to herbicide spraying?

Reply: T. Sterling

The HAC study found in some geographic areas an increase in spina bifida and cleft palates. These are the same malformations also noted in the teratogenicity animal studies. The increase was noticeable rather than impressive. However, this does not mean much because these data are full of rather large errors. The denominator of the population studied, incidentally, was, of course, all the children recorded in a given hospital.

Question: Dominick Mendola, Ecology, San Diego State University

In your paper you referred to a report utilizing radioactive tracer techniques to trace the fate of dioxin in soils. You said that there was a 15 per cent to 20 per cent decrease of dioxin applied to soils after 160 days. Do you know how long this experiment was run and how much, if any, dioxin was found after the entire length of the experiment? Also, what were the soil conditions—wet or dry?

Reply: T. Sterling

I expect the experiment was run for 160 days. What was measured was the radioactivity remaining after that period of time. That this radioactivity was labeled dioxin which had been put into the soil to begin with is, of course, assumed. The soil condition was sandy but not dry.

Question: J. B. Neilands, Biochemistry, University of California, Berkeley

Is it not a fact that the increased incidence of birth defects in South Vietnam has more or less paralleled the deployment of herbicides, that is, starting in about 1961 and tapering off in the later years of the decade when the defoliation operation (but not the bombing and intensity of the war) was cut back? And is it not a fact that the highest levels of birth defects reported in South Vietnam were in Tay Ninh, the province most heavily defoliated?

Can you give us an estimate as to how many agencies of the U.S. government had access to the report of the Bionetics Laboratories and also the time when the report was available to the agencies? My motivation in asking this is the following. Some years ago the UN voted 80 to 3 to class herbicides as prohibited weapons under the 1925 Geneva Protocol—only Australia and Portugal joined the U.S. in voting "no"—and 95 nations have ratified this instrument. Even though the U.S. has not yet ratified the Protocol, we have admitted we are bound to it by customary international law and under the Constitution the latter is considered part of the body of U.S. law. Statistics aside, it is clearly a *war crime* to employ herbicidal sprays in military operations. Domestically, 2,4,5-T has also been used on a massive scale, again with assurances that it is innocuous to animal life. Yet we now know that a manufacturer was forced to close a 2,4,5-T plant because of chloracne, a dermatitis, in the workers. Thus if the Department of Agriculture knew of the toxicity of preparations of 2,4,5-T and delayed action against the herbicide, officials of that Department would seem to be irresponsible and derelict in their duty to protect the health of the American public. Can you comment on the legal status of 2,4,5-T?

Reply: T. Sterling

As both the Army and the HAC reports show, the questions on the correlation between birth defects and defoliation are not easy to answer. Certainly the analyses of the data conducted by Dr. Meselson (the HAC report) find some support for the increase in birth defects paralleling use of the defoliant. His work also uncovered that the Tay Ninh Province reported the highest level of birth defects.

Everything I have seen in my contact with 2,4,5-T would indicate that the actions of the government officials involved were extremely responsible. The Department of Agriculture, the Department of the Interior, and the Department of Health, Education, and Welfare took action immediately as soon as evidence emerged that 2,4,5-T had teratogenic potentials. While there are some scientists and industrialists who might even accuse government officials of acting in haste, I surely do not think we would find many individuals who would accuse officials of the Department of Agriculture (or of other departments) of having been derelict in their duty to protect the health of the American public. If I may add, a certain amount of fault rests on all of our shoulders. We have now been willing to tolerate sloppy experiments and data in the area of toxicology and public health to such an extent that rational and reasonable action becomes more and more difficult to take. Perhaps we, as scientists, ought to re-examine our procedures first before examining the actions of government officials who are much less well equipped to deal with the states of uncertainty created by inadequate scientific procedures.

ECOLOGICAL AND ENVIRONMENTAL PROBLEMS IN THE APPLICATION OF BIOMATHEMATICS

BURTON E. VAUGHAN
BATTELLE MEMORIAL INSTITUTE
PACIFIC NORTHWEST LABORATORIES
RICHLAND

1. Introduction

In view of the purpose of this symposium, I thought it appropriate to delineate some ecological aspects of pollution—aspects which require fairly sophisticated biomathematical approaches. If we are to have a plan to estimate health effects for any one of several suspected pollutants, be it DDT, a heavy metal, or a radionuclide, a good deal of descriptive information must be consolidated, quantized, and assigned priority. In some cases, the descriptive information still remains to be established, and in any event, we need a "road map" for the consolidation, quantizing, and assignment of priorities. We also need to be clear about objectives in such handling of the data, and I will have more to say on this at a later point.

During the past several days, considerable discussion about nuclear materials has taken place. It may be useful to take a brief look at the nuclear industry as regards other pollution problems. Let me say simply that the nuclear industry provides us with one of the few examples of a comprehensively planned technology. Operationally, it provided for the building of nuclear plants, their regulation, environmental monitoring, the setting of radiation exposure standards, and the support of studies on ecological and health problems. As a result, the assessment of risk to man and the underlying ecological pathways affected by discharge of radioactive wastes are probably more completely understood than any other kind of industrial risk. We have been taken by surprise with environmental deterioration (the poisoning of birds and fish by DDT) and with serious toxicity effects (the contamination of fish by mercury). Yet, the dissemination of many such pollutants, particularly persistent chlorinated hydrocarbon compounds, is analogous to the radiation situation. The attendant phenomena of dispersal, biological concentration, and concentration in feed webs were predictable.

2. Environmental deterioration as a concern

Let's examine our ecological objectives more closely. Emphasis, during the past few days, has been placed on narrow aspects of health—mortality, long term toxicity, and potential mutagenicity. I hope we can also agree that a good natural environment is a determinant of one's general well being. Environmental quality is much less well defined, at this point in time, than is physical health in the sense of one's freedom from disease. Yet, its over-riding importance has been clearly indicated by creation of the first Presidential Commission on Environmental Quality, by the report of that Commission [6], and by the creation of the Environmental Protection Agency, itself.

Most authorities agree that widespread environmental effects affecting our health could potentially overtake us, and rapidly, as a direct consequence of our very large technological growth rate (for example [6], [14]). This growth rate is typified by the projection of total energy demand in the United States shown in Figure 1. The growth rate shown is only about four per cent per year,

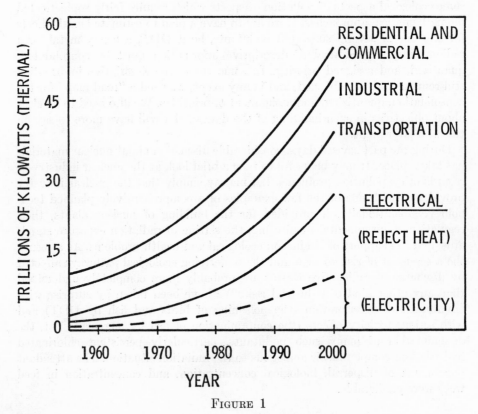

FIGURE 1

Total energy, projected demand to year 2000. Waste heat consequent on electrical generation in year 2000 will exceed present total electrical generation.

but the utilities companies frequently use larger rates for planning purposes. Note that waste heat ("reject" in Figure 1) will, in year 2000, exceed total electrical generation as of 1955. The problem of heavy pollution consequences— in this case waste heat—exists in a situation where there is little land reserve. Worldwide, land reserves amount to about ⅓ of global land surface and are principally in forest holdings [14]. The U.S. Picture is much tighter, with less than five per cent in any sense to be considered reserve [6].

How does pollution affect environmental quality? Are there important values other than personal esthetic values? These questions, in fact, have been answered well in several recent publications [5], [13], [14], so I will take time only to indicate one peculiarly ecological problem which the SCEP work group termed "pesticide addition" [14]. In this situation, regular use increases the need for and frequency of application of pesticide. Continued use creates new and some-times resistant pests. Also, new herbivorous insects find shelter among crops where their predator enemies cannot survive. To stop pesticide use in this sit-uation invites catastrophic crop damage. The problem shows another aspect. In Table I, if we compare crop yield with pesticide use, we see that a 10-fold

TABLE I

Pesticide Use Compared to Agricultural Yields,
in Selected World Areas

Source: FAO, *Production Yearbook*, 1963.

Area or nation	Pesticide use (grams per hectare)	Yield (kilograms per hectare)
Japan	10,790	5,480
Europe	1,870	3,430
United States	1,490	2,600
Latin America	220	1,970
India	149	820
Africa	127	1,210

increase in pesticide application led to a crop yield which increased only three-fold. Thus, we find ourselves locked into a new technology with highly polluting consequences and, perhaps, at an irreversible point.

3. The environmental approach to pollution effects

With a broader definition of health and pollution in mind, let us now approach the problem of planning an epidemiological study of pollution effects. To be systematic, we need to keep in mind several types of effort which, in fact, con-stitute determinants of any pollution process. Several of these factors have already been well recognized in nuclear energy development. They are:

(1) Environmental pathways; identifying the geographic distribution, of sources and magnitudes of pathways.

(2) Ecosystem structure; description of biotic systems likely to be affected: (a) food webs, food chains, and trophic levels; (b) physical factors, their interaction with different biota (climatic, meteorological, edaphic, and hydrological factors).

(3) Indicator species; monitoring their activity. Involves measurement of uptake, concentration, population levels, and loss for (a) organisms important to the functional integrity of ecosystems, and (b) organisms which are critical for certain food chain links.

(4) Biological effects and retention; measuring (a) toxicity data for extrapolation to man, and (b) most sensitive species for assessing ecological perturbation; for example, loss of eagles, insect predators, and fish.

(5) "Ultimate reservoir"; assessing potential size and possibilities for reentrainment of pollutant (may be air, organisms, soil, or water).

4. What pollutants need study?

A serious difficulty remains, in any of this work. For any pollutant of concern, we have, simultaneously, insufficient data in certain categories mentioned above and an overwhelming amount of secondary data in the other categories. In such a situation, we do well to start with as accurate as possible a case history for individual pollutants, then ask what kind of biomathematics is desirable and necessary to further a solution. Pesticides like DDT, petroleum, phosphorous, lead, and mercury, can be selected as good case studies. They are not necessarily the most important pollutants, but they differ significantly in their target, their effects, their sources, and their routes into living organisms. Considered together, they represent a broad spectrum of the considerations necessary to assess global pollution and its effects directly and indirectly on health.

Biocides used for the control of weeds have increased substantially during the past 20 years, particularly the organochlorine compounds. In 1964, 184 million pounds of herbicides were sold in the United States, and 97 million acres of agricultural land were treated with them [2]. In a major study [1], many substances were considered from an operational point of view. Such an analysis had to be ultimately limited to lead and to DDT, because of the scarcity of data regarding the environmental effects of the other biocides. DDT and lead were, of course, in widespread use, longer than any other material, and they had demonstrated wide interaction with all of the potential ecological and other transport pathways for man-made chemicals in the environment.

The relatively short time allotted this morning will not permit comprehensive discussion of lead, or the other pollutants mentioned above, so we will, this morning, confine our attention to a single case study, DDT. Partial information on the other pollutants can be found in the literature [1], [2], [14]. Basically, we approached each of these problems with the five-step approach, indicated above.

5. DDT case study

5.1. *Environmental pathways.*   Figure 2 is based on comprehensive studies of sources, production levels, and various reported concentrations in the environment [1], [16]. Geographic locations of DDT producers are less well known, however. One may note, also, that an operational analysis of this kind includes data from many workers, based on non-standardized technics. Thus, many of the reported water concentrations are too high, 10 or 1000-fold higher than the solubility limit of DDT. Undoubtedly, these samplings included microscopic organisms in which DDT was concentrated. It is of interest that of the DDT produced annually, about $\frac{1}{4}$ ends up in the ocean. This estimate is based on measured concentrations in rainfall [20] and estimated total oceanic precipitation [14]. There are, however, about 25-fold differences among some of the DDT samplings measured [14], [16], [17], and the discrepancies show a need for comprehensive world-wide monitoring. Since DDT binds to the surface of the soil clay particle, runoff concentrations as we have reported them here are

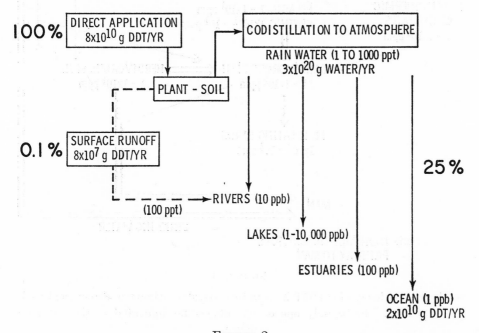

FIGURE 2

Environmental pathways for DDT, showing ranges of reported concentrations and estimated magnitudes of transport. Water concentrations except ocean probably represent mixed activity (see text).

probably overestimated for average conditions. Most of DDT finding its way into the oceans appears to be disseminated atmospherically after redistillation with water from the soil and water surfaces [14], [16]. An alternative explanation may be that soil particles carrying DDT are aerosolized by winds, since our (unpublished) data on co-distillation seem low in comparison to present estimates.

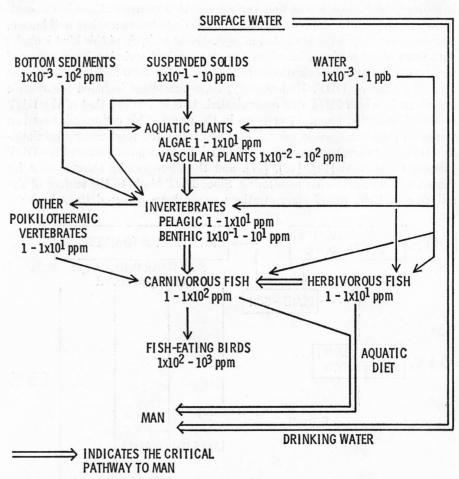

FIGURE 3

Ecological pathways for DDT in aquatic ecosystem, showing also critical pathways to man. This is only one of six subsystems indicated in the preceding figure.

5.2. *Ecosystem structure.* In Figure 3, we have summarized the ecological structure of a marine aquatic system, showing both typical classes of organisms according to their trophic (feeding) levels and their measured concentrations of DDT. The significant features here are (1) the 1000-fold concentration of DDT in aquatic plants as compared to water (shown in the preceding slide) and (2) another 1000-fold increase in concentration in the tissues of fish eating birds. At worst, a 10^6-fold concentration has taken place from water to bird. This representation is generalized, it also shows the set of pathways for DDT entering man through fish. I'll have more to say about man at a later point. Here we will temporarily ignore man and examine the rest of the ecological system from the standpoint of deterioration in environmental quality. The DDT levels shown represent approximately steady state magnitudes where world-wide concentrations are only slowly increasing. There are, however undoubtedly wide regional differences.

Comparable magnitudes also have been measured in experimental situations where radioactively labelled DDT was administered in a single aerial spraying. In such a study of a freshwater marsh [8], [9], we attempted to model a restricted aquatic ecosystem, in order to predict transfer rates between trophic levels. As shown in Figure 4, the maxima for tissue accumulation of DDT, for various organisms, were attained in about 30 days, with multiexponential loss curves thereafter. Some of difficulties of sampling are indicated in this study, in which, for example, had we rigidly adhered to a fixed time sampling program (desirable for reasons of economy) we would have missed important features about DDT turnover in sediments, tadpoles, and pondweeds.

The ecosystem shown in Figure 4 is in a dynamic state, with organisms in higher trophic levels dependent on next preceding organisms. With more complete data, one might be able to quantitatively estimate standing populations and energy transfer rates between trophic levels. However, most of the populations are changing rapidly with time, and in location, so that the model portrayed is not presently adequate as a quantitatively predictive model for system behavior. One may see how productivity of the whole system would be impaired if aquatic plants were to disappear. As it stands presently, this model is useful for revising sampling strategy, assessing organisms for their utility as biological concentrators of DDT, and for estimating transfer rates between trophic levels.

In the two preceding figures, we showed a marine and a marsh aquatic ecosystem. There are also several important terrestrial ecosystems to consider for DDT. Information on terrestrial food-webs is much more sparse, primarily as a consequence of the greater difficulty in sampling and monitoring (see [21]). We do not have adequate information about deterioration of terrestrial environments, although warning signs are up.

5.3. *Indicator species and biological effects.* Both of these pollution parameters are now fairly well established. Fish-eating birds, for example, petrel and pelican, are extremely sensitive because of their high concentrating capacity. Both the causative mechanism and the correlation between reproductive success and DDT levels are established for these birds and also for carnivorous fish, especially sea trout [16]. Robert Risebrough, earlier this week, documented the near extinction of the brown pelican as a consequence of DDT interference

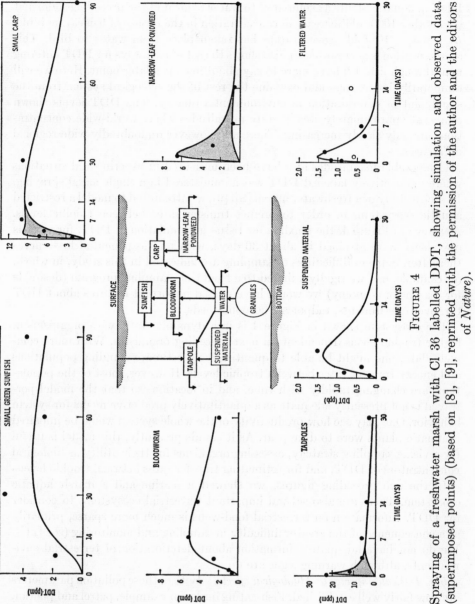

FIGURE 4

Spraying of a freshwater marsh with Cl 36 labelled DDT, showing simulation and observed data (superimposed points) (based on [8], [9], reprinted with the permission of the author and the editors of *Nature*).

with normal hatching. Laboratory toxicity determinations now show that *at least some marine organisms at every trophic level are exposed to lethal concentrations,* for DDT at the higher levels reported in Figure 3. This includes fish larvae, crab, shrimp, oysters, molluscs and other species [16]. Certain phylplankton are extremely sensitive to DDT, in the parts per billion range [15], [22]. There are, of course, myriad planktonic organisms, and it is not clear that the depth or range of species concerned are equally affected by water concentrations as presently measured. DDT and the polychlorinated biphenyls (PCB) seem to distribute together, and they have similar biological effects in birds [24]. It is not clear at present to what extent effects attributed to DDT are consequent on PCB. Analytic differentiation of the compounds is difficult, but it should be better established.

5.4. *Ultimate reservoir.* As to reservoir, we have only sketchy information, not readily quantifiable. Soil degradation rates are quite slow [12]. We know little about the long term cumulative effects on soil micro-organisms. The quantitative estimates I described earlier for codistillation suggest that the ocean is the ultimate accumulation site. However, we have little or no information on rainout patterns in relation to aquatic breeding sites—in estuaries and on continental shelves, as in Figure 5—nor do we know about depth distribution

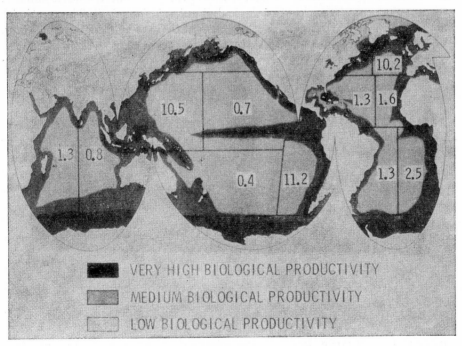

FIGURE 5

Productivity of marine fisheries in millions of metric tons (copied from [16]).

profile important to the behavior of organisms. DDT and other organochlorine residues are probably concentrated in the surface water by activity of the micro-organisms present [16].

6. Environmental factors affecting body burden of DDT in man

Study of environmental pathways is not only important from the standpoint of environmental deterioration. There are also less obvious routes by which man is affected, apart from ingesting food.

Human exposures can be divided into direct and indirect pathways [1]. The direct pathway deals with the uptake from primary sources, during manufacture and applications where human exposure is primarily by inhalation. Dermal uptake as well as ingestion are features of the direct pathway. The indirect pathway involves human exposure by translocation through the air, water, or food. Of these, the critical pathway to man is through the aquatic food chain, particularly fish. One should note, however, that conflicts exist in the data, as to indirect pathway. Worldwide, persistent pesticide content in man is remarkably constant. Within the United States there are, however, significant racial and geographical differences. Such differences are difficult to explain if food is the major transport pathway to man. Indirect evidence, from residues in animals, suggests that only 50 per cent of the body burden is from food. The remainder may come from inhalation of insecticide aerosols or dust laden with insecticides. Assuming the correctness of this deduction, then control of the human burden of pesticides by control of food residues, as is now practiced, is, at best, only partially effective.

7. Biomathematical aspects of pollution

The real problem, at this point in time, as exemplified by DDT, but holding equally for the other pollutants—oil, phosphate, mercury, 2-4-5-T, and lead— is the need for an adequate system of monitoring and sampling. This is needed in order to definitively establish the dissemination pathways and patterns. For DDT, operational considerations point to atmospheric dissemination and codistillation as key factors. We already know the sensitive species, with respect to environmental degradation and with respect to critical pathways to man. We probably have adequate work in progress on toxicity and related mechanisms; for example, inducible enzyme systems [4]. We also have reasonably adequate mechanisms for operational analysis, but inadequate data, in many cases.

7.1. *Monitoring needs.* Adequate monitoring would likely permit us to differentiate the environmentally critical pathways from secondary pathways for dissemination of DDT. For example, a synoptic series of maps, like Figure 5, is needed, showing DDT levels of water or biota in relation to precipitation patterns. An analogous series should be prepared showing DDT levels in relation to ocean currents. Such maps should also show locations and magnitude of manufacturing source and density of effluent. Similar analysis has already been done, with the necessary curve-fitting technics, for certain air pollutants, as Joseph Behar showed earlier this week. It is only in this way that operational analysis achieves its full potential as a useful tool.

7.2. *Sampling strategy.*   Monitoring is, however, costly, and optimum sampling strategy cannot usually be specified *a priori* without a good understanding of the descriptive aspects of the system.

The sampling problem has held our attention for several years [7] through [11]. We have generally assumed that the principles of first order kinetics provide a reasonable mathematical model of the actual processes governing pollutant movement through food chains. Our procedure generally involves (1) establishing "case histories," (2) constructing computer simulation models for individual case histories, (3) investigating linear least squares methods as tools for fitting models to data, (4) attempting to elucidate some general principles that may be appropriate for predicting rate constants for substances and species that have not been directly studied, and (5) formulating questions about efficient sampling procedures. At present, it seems necessary to limit an approach via simulation to some rather simple situations, inasmuch as there are a number of unresolved questions about descriptive aspects of the systems of concern. In Figure 6, and in Figure 4 shown earlier, we can examine two simulated ecosystems with a view toward optimizing sampling strategy. Basically, there are three elements to consider [8]: (1) sampling in time, (2) sampling the system, and (3) sampling the space.

Of these elements, sampling in time is basically a curve fitting process. Seasonal cycles in concentration are usual, and they will affect measurement levels. Also, an optimum sampling spacing for one species may be very inefficient for another.

Sampling the system involves assessing the potential pathways of accumulation for a pollutant. If we have extensive data available in the literature, the problem is simplified, but considerable experience and judgment are still required. Such judgment is evidently not yet generally available for efficient "systems analysis." If the pollutant is, in fact, doing serious damage to some species population, there is evidently need to bring in whole new areas of study, for example, population dynamics and dose response models. Field study of the effects of pollutants on natural populations is something that we see very much in prospect, but I do not at the moment perceive much more than that we somehow have to establish connections between "food chain kinetics" and population dynamics as the relevant fields of inquiry.

Sampling in space is complicated by seasonal cycles and by feedback effects. Of these two problems, potential recycling (re-entrainment) constitutes a formidable problem. Usually, only a small fraction of material is found in aggregate mass of biota compared to the "reservoir" locations. For this reason, a balance study must usually be coupled to the dynamic study. In this connection one

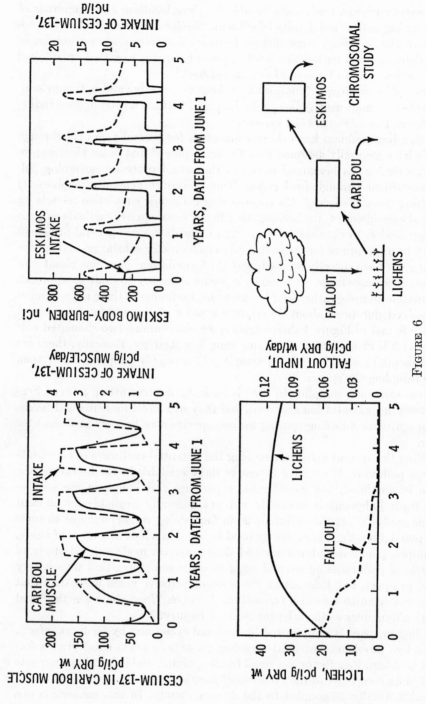

FIGURE 6

Simulation model of arctic ecosystem; showing effect of caribou feeding pattern on man's body burden of Cs 137. The model is based on extensive observational data (points omitted for clarity), see Figure 7.

should note the 1000-sample ocean baseline sampling program proposed in the SCEP report [14].

The Arctic ecosystem shown in Figure 6 lends itself particularly well to ecological modelling because the pathways are comparatively simple and because a very extensive body of data was accumulated on all aspects of the system. A remarkable feature of the system is how faithfully the seasonal increase in body burden of Cs 137, in those Eskimos following native ways, reflects the seasonal variation in Cs 137 in the caribou. The variation in caribou is a consequence of their winter dependence on lichens for forage (seasonal migration to feed on lichens). Lichens are good accumulators of Cs 137 from fallout. The Eskimo's body burden is related to that of the caribou herd and to its movement, thus, a complex consideration exists as to modelling parameters. In Figure 7, we can judge better the simulation in relation to observed data points for Cs 137 in muscle. The parameters here include time of the migration and the amount of lichen eaten, which then apply to the entire six season run of data. The

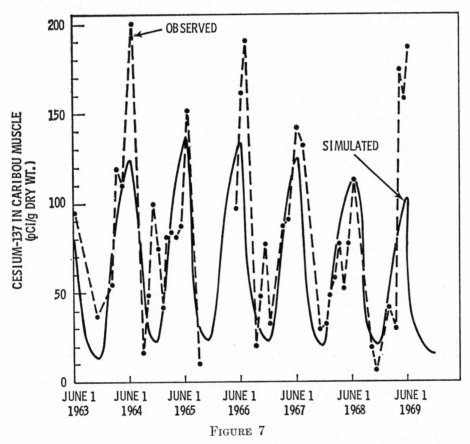

FIGURE 7

Comparison of simulation and observed points; Cs 137 in caribou muscle.

simulation is rather accurate in time phasing, but of note are the several high peaks, which we cannot yet model from *a priori* considerations.

Sample survey methods are a well established and essential feature of natural resource biometry. Since the ecosystems studied are dynamic, the problems encountered have many features in common with population studies. If objectives of a given study are taken to be the estimation of total quantities of a contaminant, it becomes necessary also to estimate the biomass of the species considered. So far, most studies have dealt with concentration rather than quantities, and interest in the immediate future appears likely to center around supplying parameter estimates for kinetic models expressed as concentrations. Beyond the evident difficulties of dealing with mobile forms of life (in the consumer and higher trophic levels) lie the prospects that concentrations may change rapidly in time. Perhaps the major departure from standard sample theory is that surveys are largely "analytic" in nature. That is, objectives are usually not so much to estimate totals as to discover and measure differences in time, space, and species. Brief treatments of analytic survey methods are available [3], [18], [23].

8. Conclusion

I have not attempted to describe in any detail our biostatistical procedures, but rather to delineate the broad range of ecological and environmental problems. Each problem has intensive biomathematical need. The needs encompass systematized operational analysis, statistical estimation, and, in a few cases, deterministic models of the ecological system.

I would like to thank my colleagues at Battelle Memorial Institute. My comments here have drawn freely on their studies.

REFERENCES

[1] BATTELLE MEMORIAL INSTITUTE, *Technical Intelligence and Project Information System for the Environmental Health Service*, Columbus, Battelle Memorial Institute, Vol. 3; see also, *ibid.* 29 June 1970, Vol. 4, pp. 20–38.
[2] BATTELLE MEMORIAL INSTITUTE, *Environmental Pollution* (background paper for the Atlantic Council and BMI conference on Goals & Strategy to Environmental Quality Improvement in the Seventies), Columbus, Battelle Memorial Institute, 1 December 1970, pp. 6–8.
[3] W. G. COCHRAN, *Sampling Techniques*, New York, Wiley, 1963 (2nd ed.).
[4] A. H. CONNEY, "Pharmacological implications of microsomal enzyme induction," *Pharmacol. Rev.*, Vol. 19 (1967), pp. 317–366.
[5] J. CONSTABLE and M. MESELSON, "The ecological impact of large-scale defoliation in Viet Nam," *Sierra Club Bull.*, Vol. 56 (1971), pp. 4–9.
[6] COUNCIL ON ENVIRONMENTAL QUALITY, *Environmental Quality* (first annual report of the President's council transmitted to Congress), Washington, D.C., U.S. Government Printing Office, 10 February 1970.
[7] L. L. EBERHARDT, "Modelling radionuclides and pesticides in food chains," *Proceedings*

of Third National Symposium on Radioecology, Washington, D.C., USAEC, 12 May 1971, in press.

[8] L. L. EBERHARDT, R. L. MEEKS, and T. J. PETERLE, "DDT in a freshwater marsh—a simulation study" (Document No. BNWL-1297), Richland, Battelle Memorial Institute, Pacific Northwest Laboratories, 1970.

[9] ———, "Food chain model for DDT kinetics in a freshwater marsh," *Nature*, Vol. 230 (1970), pp. 60–62.

[10] L. L. EBERHARDT and R. E. NAKATANI, "Modelling the behavior of radionuclides in some natural systems," *Symposium on Radioecology* (edited by D. J. Nelson and F. C. Evans), CONF-670503, Springfield, N.T.I.S., 1969.

[11] L. L. EBERHARDT and W. C. HANSON, "A simulation model for an arctic food chain," *Health Phys.*, Vol. 17 (1969), pp. 793–806.

[12] C. A. EDWARDS, "Insecticide residues in soils," *Residue Rev.*, Vol. 13 (1966), pp. 83–132.

[13] D. W. EHRENFELD, *Biological Conservation*, New York, Holt, Rinehart & Winston, 1970.

[14] MASSACHUSETTS INSTITUTE OF TECHNOLOGY, *Man's Impact on the Global Environment, The Study on Critical Environmental Problems* (SCEP Report), Cambridge, MIT Press, 1971.

[15] D. W. MENZEL, J. ANDERSON, and A. RANDTHE, "Marine phytoplankton vary in their response to chlorinated hydrocarbons," *Science*, Vol. 1967 (1970), pp. 1724–1726.

[16] NATIONAL ACADEMY OF SCIENCES, *Chlorinated Hydrocarbons in the Marine Environment* (Report of panel on monitoring persistent pesticides in the marine environment, Committee on Oceanography), Washington, D.C., National Academy of Science, 1971.

[17] T. J. PETERLE, "DDT in Antarctic snow," *Nature*, Vol. 224 (1969), p. 620.

[18] J. SEDRANSK, "An application of sequential sampling to analytic surveys," *Biometrika*, Vol. 53 (1966), pp. 85–97.

[19] H. V. SVERDRUP, M. W. JOHNSON, and R. H. FLEMING, *The Oceans*, New York, Prentice-Hall, 1942, p. 120.

[20] K. B. TARRANT and J. TATTON, "Organopesticides in rainwater in the British Isles," *Nature*, Vol. 219 (1968), pp. 725–727.

[21] U.S. DEPARTMENT OF AGRICULTURE, *Wastes in Relation to Agriculture and Forestry* (Miscellaneous publication no. 1065), Washington, D.C., U.S. Government Printing Office, March 1968.

[22] C. F. WURSTER, "DDT reduces photosynthesis by marine phytoplankton," *Science*, Vol. 159 (1968), pp. 1474–1475.

[23] F. YATES, *Sampling Methods for Censuses and Surveys*, New York, Hafner, 1960.

[24] R. W. RISEBROUGH, S. G. HERMAN, D. B. PEAKALL, and M. N. KIRVEN, "Polychlorinated biphenyls in the global ecosystem," *Nature*, Vol. 220 (1968), pp. 1098–1102.

Discussion

Question: E. L. Scott, Department of Statistics, University of California, Berkeley

The fit of the models shown by Dr. Vaughan is very impressive. From the slide alone, I do not know whether the deviations are important. They raise the question of the possible effects, with interactions, of other pollutants. In the Eskimo study might there not be, say, DDT from the atmosphere as well as cesium entering the food chain? Are there data towards questions like this and have they been considered?

Reply: B. E. Vaughan

Referring to the Arctic lichen-caribou-Eskimo food chain, the deviations from simulation curves, as shown for Cs 137 in caribou (Figure 7), are certainly important. We do not know why occasional groups of samples peak at higher than predicted values. I doubt that these peculiarities in the data represent interactions with other pollutants, because we measured Cs 137 specifically and because the concentrations measured are known to be without significant physiological consequences. It is much more probable that the early peak and unusually high values represent some as yet undescribed phenomena concerning the growth cycle of the lichens, or the migrating habits of the caribou, or both.

On the question of DDT as well as Cs 137 and other pollutants entering such food chains via the atmosphere or water, we try to get simultaneous data where possible for monitoring purposes. But analysis costs can be quite expensive. More importantly, there remain problems about suitable sampling strategy. Such data as we have are very few and not indicative of unexpected phenomena. It will be a very long time in the future before the question of possible synergistic effects can be considered experimentally.

Question: R. W. Gill, Department of Biology, University of California, Riverside

Would you comment on the relative importance of the different biological and physical factors as possible vectors for the transportation of DDT from ecosystem to ecosystem throughout the world?

Reply: B. E. Vaughan

Vectors in the sense of the carrier for DDT? Well, certainly atmospheric precipitation seems to be a prime factor in disseminating DDT worldwide, and it is governed chiefly by physical processes. This matter is not proven to rigorous standards, however, so we should beware of some of the discrepancies in the data.

Biological vectors seem to be unimportant. Bacteria and planktonic organisms, in surface water, which concentrate DDT thousands of times over its water solubility cannot really be considered a vector in the epidemiological sense of the term. They are ubiquitous and don't move far. Neither should carnivorous fish and sea birds be considered important as vectors, even though their feeding habits and the biochemical properties of DDT lead to high concentrations. Fish and birds are accumulators, not disseminators, and they are ultimately killed by DDT.

Question: Joel Swartz, Biophysics, University of California Berkeley

Is it feasible to obtain a lot more data points for the systems described? It seems that there are insufficient data to distinguish between models.

Reply: B. E. Vaughan

It is feasible but expensive. Fifteen years of effort and at least two people were involved in the Arctic food chain study. There is also nothing intrinsically more real about one or another mathematical model of the Arctic ecosystem. More than one approach can be designed; the important consideration is whether or not a given model allows a better evaluation of sampling or dynamical problems.

POSSIBLE MANIFESTATIONS OF WORSENING ENVIRONMENTAL POLLUTION

WILLIAM R. GAFFEY
CALIFORNIA STATE DEPARTMENT
OF PUBLIC HEALTH, BERKELEY

1. Introduction

On the assumption that many mathematical statisticians are not well acquainted with the kinds of data available to describe health, I propose to review those aspects of health which may be affected by the environment, and to make some comments on the measurement of the environment, as well as the statistical, or rather metastatistical, problems involved in establishing an association between environment and health.

2. Definition and measurement of environment

It is possible (and useful for some purposes) to define the "environment" of a human being as everything outside of his epidermis. Such a view would define the smoking habits of parents, for example, as part of their children's environment. It would also define the presence of an efficient ambulance service as part of the environment of the population of a city. In both these cases, the life expectancy of an individual may be effected by the factors mentioned, thus both are examples of environmental characteristics which affect health.

For our purposes, however, it is clearly expedient to take a narrower view, and to define "environment" as the sum total of the physical phenomena which an individual encounters: food, water, air, and other substances with which he comes in contact.

Those aspects of this environment which are thought to be harmful, or potentially harmful, are now often monitored more or less systematically. However, several qualifying comments must be made about the nature of the monitoring process.

First, the purpose of monitoring is usually to implement public policy regarding maximum levels of pollution. As a result, monitoring is often designed not to estimate the average level of pollution, but to detect violations of some maximum permissible level. It therefore tends to take the form of what might be called "suspicion" sampling. For example, a large proportion of our data on

DDT contamination of the environment comes from sampling of critical incidents, such as an overturned DDT tanktruck, or a suspiciously large die-off of birds in an area previously treated with the pesticide. The necessity for such sampling to implement safety regulations is clear, but the results are not terribly useful in estimating population exposures.

Second, the environmental characteristics which are monitored are those which are already known or suspected to be harmful, or those which public opinion has identified as an aesthetic hazard. Therefore, if one uses existing data to assess the effect of environment on health, one is limited to verifying the hypotheses which justified the establishment of the particular monitoring systems, or hypotheses about further health effects of those same pollutants. For example, one can investigate the effects of photochemical air pollution on cardiovascular disease because such pollution already causes subjective discomfort, and is regularly monitored in several areas for that reason.

Third, and somewhat paradoxically, the measurement of environmental effects may be less of a problem than the measurement of health effects in many situations. The reason is that suspected environmental pollution usually comes from an identifiable source, such as a nuclear power plant or a sewage outfall pipe. Therefore, if untoward health effects are found to occur near such sources, we are entitled to suspect environmental contamination even if we cannot immediately define and measure the precise pollutant involved.

The point here is that the possible spectrum of environmental pollutants is so large that most studies of the effect of particular environmental pollution will probably have to measure that pollution on an *ad hoc* basis. This will have to be done either by monitoring for specific pollutants, or by verifying the existence of probable pollution sources and accepting this as *prima facie* evidence of pollution.

3. Definition of health

The dependent variable in a study of the effects of environmental pollution is called for convenience "health," but when we get down to cases, we turn out in fact to be measuring lack of health, since this seems to be easier to talk about in specific terms.

In times past, when health hazards manifested themselves as severe communicable diseases or natural catastrophes, it was natural and easy to measure ill health simply by mortality, or possibly by the occurrence of illness, where by illness is meant a communicable disease diagnosed by a physician. It is apparent that these measures have become increasingly inadequate indices of ill health [2]. Recently we have attempted to measure disability and symptoms, and even more recently the presence or absence of factors such as immunization and medical services, which presumably affect the risk of future illness or death.

One is tempted to look for a single summary index which would adequately express these factors in one descriptive number, and indeed such indices have been devised for combining such aspects as occurrence and duration of disease, and mortality, [5], [6]. But the construction of such an index, however useful it would be for public information and certain aspects of health administration, might conceal phenomena of interest to the scientist. Consider the problems of such an index for mortality. It is clear that we can measure mortality by the number of deaths per 1000 population per year. If we wish to compare two populations whose age, sex and racial compositions are different, this crude mortality rate can be corrected in one of several ways to compensate for these differences. Using such a measure for California, we would find that the risk of death had decreased over the last several decades. However, the overall rate would conceal the fact that the decrease had been greatest in childhood and young adults, and least in older people. More important, both to the scientist and the planner of health programs, the causes of death have changed.

TABLE I

THE FIVE LEADING CAUSES OF DEATH BY SELECTED
AGE GROUPS, CALIFORNIA, 1950 AND 1968

Source: State of California, Department of Public Health, Bureau of Vital Statistics.

| Cause | \multicolumn{8}{c}{Age group} | | | | | | | |
| | \multicolumn{2}{c}{1–4 Rank} | \multicolumn{2}{c}{15–24 Rank} | \multicolumn{2}{c}{34–44 Rank} | \multicolumn{2}{c}{55–64 Rank} |
	1950	1968	1950	1968	1950	1968	1950	1968
Accidents, including motor vehicle	1	1	2	2	3	5		
Motor vehicle accidents	2	2	1	1	4			
Cancer	3	3	4	5	2	1	2	2
Pneumonia and influenza	4	5						
Suicides			5	3		3		
Homicides				4				
Tuberculosis			3		5		4	
Diseases of the heart					1	2	1	1
Cirrhosis of the liver						4	5	4
Stroke							3	3
Respiratory diseases commonly designated as obstructive								5

Table I shows the five leading causes of death in 1950 and 1968 in California for four large age groups. No changes have occurred in the group under age 5. However, in the 15–24 age group we see the disappearance of tuberculosis, the decline of cancer, the appearance of homicide, and the increase of suicide. In the older age groups we see the disappearance of tuberculosis, the appearance of cirrhosis of the liver, and the appearance of emphysema and similar respiratory diseases. One could say that the crude mortality rate has concealed the fact that behavioral disorders are becoming a more important cause of death, to the extent that alcoholism and smoking can be blamed for cirrhosis and emphysema.

If we broaden our consideration to other measures of ill health, the problem becomes more complex. Consider Table II, which lists the principal criteria for physical ill health which have been proposed in recent years [1], excluding certain measures of social function which seem to involve mental health.

TABLE II

POSSIBLE MEASURES OF REAL OR POTENTIAL ILL HEALTH

1. Mortality
2. Illness diagnosed by a physician
3. Disability
4. Other symptoms
5. Utilization of health services and facilities
6. Presence of metabolic or physiological abnormalities without illness, disability or symptoms
7. Lack of good health practices (proper immunization, and so on)
8. Absence of adequate health services and facilities
9. Presence of potentially threatening environmental conditions

Clearly, the content of any index calculated from these measures will depend on the use to which it is put. Health authorities would probably wish to consider all nine elements in planning programs to improve health. Those of us concerned with the health effects of environmental pollution will want to consider at most the first six, since the rest are clearly measures of education, public policy, economic status, or (in the case of item 9) redundant.

It seems clear that a descriptive vector, rather than a descriptive scalar number, is required to describe ill health.

4. Measurement of health

There are three possible mechanisms for recording or estimating the occurrence of the events in question: (1) compulsory legally sanctioned recording of all events in a given category as soon as they occur, or within some limited time thereafter; (2) sampling those persons (physicians) and/or facilities (hospitals) which provide medical care to a population; (3) sampling the population concerned.

Note that, by definition, some of these techniques are unusable for some measures of ill health. For example, mortality cannot be estimated from a population sample, and symptoms and disability cannot be estimated in their entirety from a physician sample.

More important, perhaps, compulsory registration and sampling of physicians and hospitals cannot in themselves provide measures of risk of ill health, since they do not count those persons to whom the events in question have not occurred. With these techniques, measures of the population at risk must be obtained for the place and time in question, which is frequently a difficult task, since definite population counts will in general be available once every decade through the U.S. Census. Estimates of small area populations in California are made annually by the State, but usually not for areas smaller than a county.

To the best of my knowledge, there is currently no regular sampling of U.S. physicians at this time, although this is being discussed both in California and at the Federal level. There are several systems in existence for obtaining data on hospital patients, but they do not represent either a probability sample or a complete census, so that the problem of determining the population at risk is not easily solvable. For all practical purposes, therefore, usable existing data are limited to those obtained by compulsory registration and population sampling.

Let us now consider the first six variables in Table II, and what is now being done to measure them.

4.1. *Mortality.* In all states, all deaths are recorded and tabulated by age, race, sex, cause and residence. The latter could theoretically be done for any area, no matter how small, but in fact deaths are tabulated by county and sometimes for large cities. Obtaining risk of death for smaller areas is difficult and unreliable because of the problem of estimating population. The exception is infant mortality, where the population at risk is the number of infants born, which is determined by a registration process similar to that for deaths. Tabulations of infant mortality also tend to be made on a county and large city basis, but most health jurisdictions are capable of producing rates for smaller areas if funds can be obtained to pay for the tabulation.

4.2. *Diagnosed illness.* Certain diseases must be reported to the authorities when diagnosed by a physician. The precise list of diseases varies from state to state, but they are generally the classical contagious diseases. Two things can be said of these reporting systems. They are incomplete, often grossly so, and the diseases are not usually the ones we think of in connection with environmental pollution.

A second source of data on diagnosed disease is the National Health Survey [3], which gives patients' reports on their own diagnosed illnesses. This is supplemented by the Health Examination Survey [4], which actually examines a small subsample of the NHS sample. Unfortunately, the NHS data come from a nationwide sample and give little detail for local geographic areas.

A third class of data for a rather specific class of diagnoses comes from birth certificates, which must record congenital malformations. As with all birth data, this information could be provided, given the resources, for very small areas. However, the completeness of reporting of congenital abnormalities is somewhat suspect, since minor abnormalities may escape notice, and others may not develop until some time after birth.

4.3. *Disability, symptoms, utilization of health resources.* Here again the only regular source of data is the National Health Survey, with the limitations noted above for small area calculations. Also, the particular variable of utilization, as measured by days in hospital, number of doctor visits, and so on, is not useful in most cases, since it is not clear that a low rate does not simply reflect the unavailability of resources, rather than a state of good health.

4.4. *Presence of abnormalities with no other symptoms.* The only data available in this area come from specific research studies of particular environmental pollutants. For example, groups exposed to certain pesticides may experience abnormal cholinesterase levels with no other symptoms, so that the abnormality itself is taken as evidence of exposure. However, there is no systematic surveillance of the vast range of possible abnormalities which might occur.

In summary, it seems clear that no regular health data system is usable in environmental studies. Either existing mortality data must be recalculated on a small area basis corresponding to the area of study, or an ad hoc data system set up, possibly involving a sample survey of the local population.

5. Relating health to environment

There are two ways, generally speaking, to establish that health is affected by environmental deterioration. The first is to observe health systematically under different environmental conditions; the second is to make health inferences on the basis of laboratory or animal experiments.

In the first case comparisons can be made either over time in one area, or between areas which differ in environmental pollutants. A practical problem will be the possibility that there is no immediate acute effect, but that cumulative exposure causes some chronic condition. In such a case, any examination of residents of a given area will have to involve some estimation of the length of exposure, and will be complicated by such situations as the individual who lives in a polluted area, but travels some distance to work in a nonpolluted area.

There is no need to discuss the specific statistical techniques which may be used, but I do wish to point out that we are here applying classical inference to a nonexperimental situation, since "treatment" is not randomized. Therefore our statistical decision making, which in the classic situation decides between chance and the experimental variable as an explanation of observed differences, here can only tell us that chance is or is not an acceptable explanation. If it is not, then the difference between, say, health in two areas is either due to the environmental difference or to some other difference, associated with environment, which we were not clever enough to detect.

Not much attention has been paid to untangling the latter problem, most likely because it is ill defined and makes us uneasy. The only systematic consideration of the problem of which I am aware is [7], in which Yerushalmy suggests the following rule of thumb: if a suspected "cause" is associated with a very large number of "effects," one should be suspicious of the reality of any of the cause-effect relationships. This is perhaps nothing more than a guide to one's intuition, but alternative suggestions do not appear to be forthcoming.

Laboratory experiments present no such philosophical problem of assessing effect, but they create equally knotty problems of generalization to human beings. A recent example is the relationship between DDT and liver cancer. Animal experiments indicate that massive doses of DDT will produce cancer-like liver tumors in rats and mice, but in no other laboratory animal. The problem of generalization consists then in deciding whether the human metabolism is closer to that of rats and mice than to, say, dogs.

Obviously there are many situations, especially those involving radiation, where we cannot conduct human experiments. Nevertheless the problem of generalization is there, and the generalizations we make are clearly going to be intuitive, dependent on the consequences of error as we see them, and hopefully, guided somewhat by rational statistical analysis.

In summary, it appears that (1) the routine measures of health as they are presently obtained cannot, generally speaking, be used to assess the impact of environmental pollutants; (2) routinely collected measures of environmental quality will not, as a rule, provide needed data except for a narrow range of investigations; (3) although prudence and necessity lead us to evaluate pollution through animal experiments, the generalization to man is a somewhat intuitive and nonstatistical procedure; and (4) even with all precautions, the results of systematic observations on a population must be further considered in the light of the fact that we do not operate in an experimental situation, and that little attention has been paid to plausible inference in such a situation.

REFERENCES

[1] S. B. GOLDSMITH, "Status of health status indicators," *New Eng. J. Med.*, in press.

[2] I. M. MORIYAMA, "Problems in the measurement of health status," *Indicators of Social Change*, New York, Russell Sage Foundation, 1968.

[3] NATIONAL CENTER FOR HEALTH STATISTICS, "Origin, program and operation of the U.S. National Health Survey," *Vital and Health Statistics* (PHS Pub. No. 1000—Series 1—No. 1), Washington, U.S. Gov't. Printing Office, August 1963.

[4] ———, "Plan and initial program of the Health Examination Survey," *Vital and Health Statistics* (PHS Pub. No. 1000—Series 1—No. 4), Washington, U.S. Gov't. Printing Office, July 1965.

[5] ———, "An index of health, mathematical models," by C. L. Chiang, *Vital and Health Statistics* (PHS Pub. No. 1000—Series 2—No. 5), Washington, U.S. Gov't. Printing Office, May 1965.

[6] ———, "Conceptual problems in developing an index for health," by D. F. Sullivan, *Vital and Health Statistics* (PHS Pub. No. 1000—Series 2—No. 17), U.S. Gov't. Printing Office, May 1966.

[7] J. YERUSHALMY, "On inferring causality from observed associations," *Controversy in Internal Medicine*, W. B. Saunders Co., 1966.

Discussion

R. J. Hickey, Institute for Environmental Studies, University of Pennsylvania, Philadelphia

I am uneasy about your "rule of thumb" comment to the effect that if a "cause" has "too many effects," the system, or "cause" (hypothesized) is "not real."

If one considers what is known about the biological effects of ionizing radiation, an effect on experimental animals (using relatively high dosages compared with background radiation levels) has been reported to be a shortening of survival, based on population studies. Perhaps the effect might be referred to as an increased rate of biological senescence, possibly based on cumulative somatic genetic degeneration. But if the experimental animals subjected to irradiation "age" more rapidly than the controls, an effect of such more rapid aging can be earlier occurrence of diseases of aging, which are many. Thus, the ionizing radiation "cause" presumably has many effects.

Comparably, if one examines the hypothesis that among nonradioactive atmospheric (and other environmental) chemicals are some which are radiomimetic, then one might expect certain effects in populations somewhat comparable to the effects of ionizing radiation, for example, positively correlated relationships between environmental concentration of a suspected chemical with risk in populations to mortality from certain diseases of senescence [1], [2]. Thus, such hypothesized "causes" might, perhaps, be expected to have many effects.

Regarding a "single index" of manifestations of worsening pollution, one might consider life expectancy. This, however, may not be easily estimated in, for example, different metropolitan populations. However, median age is re-reported by government agencies. But median age is determined in populations by birth, death, and migration. Median age has been "predicted" statistically from atmospheric concentrations of certain chemicals [1].

REFERENCES

[1] R. J. HICKEY, D. E. BOYCE, E. B. HARNER, and R. C. CLELLAND, "Ecological statistical studies concerning environmental pollution and chronic disease," *IEEE Trans. Geosci. Electron. GE-8* (1970), pp. 186–202.
[2] R. J. HICKEY, "Air pollution," Chapter 9, *Environment: Resources, Pollution, and Society*, Stanford, Sinauer Associates, Inc., 1971.

John R. Goldsmith, Environmental Epidemiology, California Department of Public Health

Professor Hickey has both emphasized the multiplicity of radiation reactions, and used the term "radiomimetic" to describe pollution effects. These notions are in conflict, since they obscure the specific effects of radiation on the one hand and of pollution on the other. As a basis for raising questions, the notion of radiomimeticity has merit, but as a basis for describing what is known, I feel it to be confusing.

Professor Gaffey has not mentioned the interrelationship of experimental results and epidemiological ones, but both are a basis for deciding what effect to study and how to interpret results on epidemiological studies. Experimental results are of crucial importance. This meeting offers an opportunity to statisticians, who have been extensively involved in experimental design and analysis, to help encourage the further interaction of the experiential and experimental branches of science.

Samuel W. Greenhouse, National Institute of Mental Health

You have raised very serious questions concerning the populations one studies and the thoroughness of the information elicited from each subject. Some discussants have raised questions about the need for studying interactions of different agents such as drugs, air pollutants and radiation, and so forth—questions which are of considerable importance to epidemiologists and biostatisticians. I wonder if you would react to the following thought which I have proposed at several meetings in the past, namely, borrowing the "panel" concept from the Census Bureau, and establishing in effect population laboratories which might be counties of size 100,000 to 200,000 residents to be observed carefully with respect to health, and to social behavior. In this body, populations could be well defined and information very detailed on each resident. The idea is obviously not new. The Johns Hopkins use of the Baltimore Eastern Health District is certainly in line with this suggestion.

Reply: W. Gaffey

I think Dr. Greenhouse's suggestion is an excellent one.

Up until now, our attempts to evaluate the kinds of relationships which Dr. Greenhouse mentioned have usually involved one of two strategies. The first is to make very simple measurements on extremely large groups of people, as typified by the usual calculations of mortality rates by residence, occupation, and so forth. The second is to make extremely careful and detailed measurements on small groups of people. An example is the intensive scrutiny of a wide range of physiological parameters which is often carried out on workers in pesticide plants.

The kinds of relationships in which we are interested fall between the cracks, so to speak. On the one hand, although the health effects in question concern the population at large, they are likely to manifest themselves in forms which are not measured by the routine data collection systems. On the other hand, the untoward events for which we are looking are likely to be rare, so that no observation of a small group, however carefully done, will turn up anything.

I see Dr. Greenhouse's suggestion as a feasible compromise, which recognizes that we must look, admittedly at some cost in money and effort, at a large enough group to find rare events, while avoiding the completely impractical alternative of making intensive measurements on the whole population.

EFFECTS OF TOXICITY
ON ECOSYSTEMS

ROBERT W. GILL

UNIVERSITY OF CALIFORNIA, RIVERSIDE

1. Definition of ecosystem

The purpose of this paper is to put the proposed epidemiological study of pollution effects into an ecological perspective. Before discussing effects of toxicity on ecosystems, I need to define a few terms. The term *ecosystem* was introduced by Tansley in 1935 [25] and its applicability to general ecological study was argued by Evans in 1956 [8]. An ecosystem is the sum of the organisms and the nonliving environment in a given area. A particular ecosystem can be as large as the whole earth or as small as the protozoa living in the gut of a termite; its actual size depends on the ecological questions being asked. In the study of an ecosystem as in the study of any other system, the functional pathways linking components and the interactions among the components are stressed. To a greater and greater extent, the methods and generalizations of systems analysis that have been developed in other fields are being brought to bear on the study of ecosystems [27], [28], [29].

A critical aspect of the description of an ecosystem is the delineation and study of its boundaries. All natural ecosystems are more or less permeable, that is, various substances cross the defined limits of the system and may have significant effects on its components. For example, a stream ecosystem receives water, dissolved substances, dead plant material, and numerous organisms from the land and radiant sunlight from above. Many organisms spend from a few moments to most of a life cycle within the stream, returning later to the land. There is extensive output from the system as it flows into the sea or some other body of water. Man removes water from the system for drinking and for various agricultural and industrial purposes and may return the water in a greatly altered form. He dumps various additional materials into the system at innumerable points. Thus, the study of what is going into and coming out of the system is almost as complicated as determining what is going on within the system itself.

2. Toxicity effects on the ecosystem

Toxicity effects on an ecosystem are chemical or radiation effects on a species or group of species that result in the reduction or elimination of these compo-

521

nents through death or sterility. Toxicity effects are to be distinguished from nontoxic, but perhaps equally disasterous, effects. Such nontoxic effects would include abnormal climatic patterns, man's mechanical removal of vegetation prior to cultivation, *eutrophication* (the addition of nitrates, phosphates, and other limiting substances to aquatic ecosystems producing increased growth of algae and other plant species), over harvesting of particular species, and so forth. The toxicity effects may generate a series of additional changes, and they may significantly interact with each other and with the nontoxic effects.

It should be stressed that not all toxic effects on ecosystems are caused by man's activities. Many plants release *phytotoxins* into the environment that give them an advantage over competing species, a phenomenon called *allelopathy* [17], [31]. Considerable interest has developed around the discovery that plants produce insecticides and other deterrent chemicals that protect them from herbivorous insects [6], [31]. Many potentially toxic substances, such as arsenic and other heavy metals, may enter terrestrial and aquatic ecosystems through natural weathering processes. Perhaps the most striking example of toxicity effects on an ecosystem in which man plays no apparent role is the red tide, in which one or more species of dynoflagellate algae become temporarily extremely abundant and produce such a high concentration of toxic chemicals that huge numbers of other organisms in the area are killed [4], [9]. Man himself may be poisoned by eating shellfish containing the dynoflagellate [3], [20].

3. Monitoring the ecosystem

Numerous problems are associated with the simple description of ecosystems. Adequate description of them requires monitoring all the essential components and all the relevant input and output of the system. Any ecological study, whether at the ecosystem level or at the population level, faces the basic problem of determining how many organisms of each kind are present in the environment. This problem is complicated by the fact that many organisms have a clumped distribution in nature, necessitating more extensive sampling than if the distribution were random or regular. Changes in the system may be associated with changes in the values of a number of abiotic factors, such as temperature, relative humidity, concentration of particular chemical substances, solar radiation, and so forth. Since any of the components may be changing rapidly, frequent monitoring of them is required. Even if human, technological, and financial requirements for complete monitoring of the ecosystem could be met, there is the real possibility that something approaching the uncertainty principle exists in ecology, namely, that by the very process of frequently measuring all of the components, the system will be altered. The samples removed and the other perturbations produced by monitoring may partly determine the particular output observed and the interactions discovered.

4. Ecosystem dynamics and model building

At this point, I want to draw a very strong distinction between the simple description of an ecosystem and the functional analysis of that ecosystem. The simple description would require enumerating the species present, determining who eats whom, and, for a particular toxin or set of toxins, determining their concentration in each species and in the various parts of the environment.

In contrast to this, a functional analysis of the ecosystem would require discovering the rates at which energy and various substances are flowing through the system. It would require determining the consequences of removing particular species or groups of species from the system, of altering the reproductive dynamics of particular species, of so altering the environment that particular exotic species move in, and so forth. In particular, a functional analysis would permit prediction of future ecosystem performance under a variety of possible management decisions; a descriptive analysis merely tells us what is there now and perhaps what appears to be going on now. A functional analysis requires model building and sensitivity analysis, followed by experimental manipulations to verify the predictions of the model and of the sensitivity analysis of the model.

Ideally, any experimental manipulation of an ecosystem, whether it is intentional or accidental, should have an associated control so that the effects of the particular manipulation can be distinguished from effects that would have occurred in its absence. But a control for any particular study is very difficult to find. No two large natural ecosystems are exactly alike; they differ with respect to the levels of abiotic components and species composition. Neither is it completely satisfactory to do "before and after" studies of the same ecosystem, since some of the ecosystems studied to date are capable of changes in structure and species composition in the absence of obvious external manipulation [23], [7]. Furthermore, most of the ecosystems of interest to us in any toxicity study have not been adequately studied prior to their alteration by an influx of toxic substances. For example, the decline in productivity of a coastal marine fishery has no control ecosystem for comparison, so it is possible that the decline is due to an influx of various toxins, or to erosion of the shoreline and silting of the bays, or to over exploitation of the fishery by man, or to natural cycles of abundance of predators and prey, or to some combination of these and/or other factors. Perhaps our only hope at present for such systems is to design statistical studies that will discover correlations between concentrations of particular toxins and unusual structural properties or states of components in the ecosystem. We can hope that future studies and experiments will demonstrate the causal relationships (if any) involved. But such purely statistical studies will not allow us to make firm predictions of the consequences of various management decisions.

One possible alternative to a control for our experiments is to develop a general model for the kind of ecosystem we are studying, and then see how our abused system differs from this general model. At present no sufficiently complete and detailed model for a natural ecosystem exists. However, one of the goals of the Analysis of Ecosystems section of the US International Biological Program (IBP) is the construction of an ecosystem model for each of six of the major biomes of the world. The models, if successfully constructed, would predict *primary productivity* (the rate at which plants produce material that is potentially available as food for other organisms) as a function of the values of the other components, and will functionally relate primary productivity to as many of the other factors and properties of the system as possible. As an integral part of the program, changes in all major components of the system will be monitored and correlated with changes in other components and parameters. Experimental manipulations of replicas of the systems are being performed, including watering, grazing, fertilizing, and so forth. A large team of researchers works on each project, with each researcher or group of researchers responsible for the study of one component or set of components of the system. It is hoped that the results obtained will be integrated into a coherent and robust model, using all the modern techniques of systems analysis. The model will then be simulated, and its more interesting and promising predictions will be tested by further experimentation. The models from the different ecosystems will be compared to discover their common properties and particular differences. An outline of a Canadian approach to the design and initiation of an IBP ecosystem study has been presented by Coupland and co-workers [5]. The major problems encountered in planning the study were location of an adequate study area, recruitment of competent researchers, and individual adherence to the group's research goals.

It will be at least a few years before these studies are complete enough to provide us with workable models for the study of toxicity and other effects on large ecosystems, although preliminary models are currently being circulated among participating researchers in the program. The progress of the US/IBP Analysis of Ecosystems section should be carefully followed. We should attempt to profit from their mistakes, and, where possible, should incorporate their results and progress into our experimental design.

5. Pollutant pathways in the food web

Given that our knowledge of ecosystems is very incomplete and is based largely on simplistic and incomplete models and theory, on causal inferences from observed correlations, and on studies of small subsets of ecosystems, what relevant generalizations are possible concerning ecosystems, and how do these generalizations relate to toxicity effects on ecosystems?

A common generalization is that the organisms in an ecosystem can be represented as a series of trophic levels, typically green plants, herbivores (which feed on the green plants), carnivores (which feed on the herbivores), secondary carnivores (which feed on the carnivores), and decomposers (which feed on the dead bodies, excretions, and other remains of the other organisms) [14]. In reality, the feeding relationships among the organisms are extremely complex, often vary during the life of the individual, and cannot be easily reduced to a simple trophic structure. For greater precision they should be represented as a food web or food net, in which all of the feeding relationships among species are shown. Whether trophic levels or food webs are used to represent feeding relationships among species in the system, the basic concept is extremely important for any study of direct toxic effects on man. Such a study would require following a toxic substance through the feeding relationships among the species and evaluating its residence time, concentration, and physiological effects in each species. The sampling methods and experimental design are relatively straightforward and have been applied to the study of dichloro diphenyl trichloroethane (DDT) and other chlorinated hydrocarbons and to radioactive substances in particular. For example, dichlorodiphenyl dichloroethane (DDD) (or tetrachlorodiphenylethane, TDE) was applied in 1949, 1954, and 1957 to Clear Lake, north of San Francisco, to kill the aquatic larvae of the midges, which are nuisance insects as adults [12]. Subsequent studies [16], [30] indicated that DDD had become extremely concentrated in numerous organisms in the lake, and the fatty tissues of fish contained 40 to 2500 ppm DDD. The concentration was its highest in the fat of the predatory fishes, which man prefers for sport and food. The flesh of the fish contained less DDD than the fatty tissues, but, for most fish, still exceeded the maximum tolerance level of 7 ppm set by the FDA for DDD residues in marketed foods [30]. A really thorough study of the toxicity effects on this ecosystem, including man as one of the consumers of fish, would involve periodic samples of all of the abundant species in the system to determine the concentration of DDD and its breakdown products, feeding observations and experiments to determine how DDD is flowing through the system, and detailed studies of its distribution in the bottom mud and other parts of the abiotic environment. Samples of the organisms should be tested periodically for sensitivity to DDD and these sensitivities compared with those of populations from areas less exposed to DDD to determine the degree of evolution of resistance to DDD. The third spraying of Clear Lake produced less kill of midges, implying evolution of resistance to DDD, at least on the part of the midges. The Clear Lake ecosystem would be an excellent one to study now, since there has been no deliberate input of DDD since 1957, and sampling of organisms for DDD concentration was stopped in 1965. A similar experimental design would be applied to the study of any other toxin in any other system.

A second generalization about ecosystems is that we know least about the decomposer portion of ecosystems. The bacteria and fungi are responsible for breaking dead plant and animal material down into simple substances that can be reused by the plants. The decomposers are assisted by numerous other organisms that also feed on dead organic matter and on the decomposers themselves. Very little is known about their abundance and detailed function in nature. Yet these are the very organisms that are basic to the detoxification of many toxins, and, in the case of elemental mercury, to the increase in its toxicity in an ecosystem [13]. The decomposers will pose as basic a problem to a study of the toxicity effects on an ecosystem as they do for any other ecological study: we just don't know enough about them and don't have the methodology to deal with the problem.

6. Factors affecting ecosystem stability

Another generalization about ecosystems is that diverse ecosystems are more stable than simple ecosystems. A diverse ecosystem is one composed of a large number of fairly abundant species. It is normally assumed that diverse ecosystems have more complex feeding relationships among the organisms and a greater number of pathways through which food energy can flow. From this it is assumed that more diverse systems are more stable. However, ecologists use the term stability in at least two different ways: (i) to mean constancy of numbers of individuals, and (ii) to mean constancy of species composition. A system in which all the species persist through time would be one in which no species becomes extinct and no species becomes so abundant that it competitively reduces some other species toward extinction. Such a system might be called a system with *protected diversity*, analogous to the term protected polymorphism recently introduced into the population genetics literature [19]. Increased diversity of itself does not always increase stability [10], [29]; it depends on where in the system the diversity is added. But, in the great majority of cases studied, increased diversity produces increased stability in one or both senses of the word, and greater simplification of the system produces greater fluctuations in numbers of individuals and greater probability of extinction. The simplest of man-made ecosystems are the mono-crop agricultural systems, which are very vulnerable to extensive defoliation by pests and to decimation by diseases.

Woodwell [32] has recently argued that ionizing radiation, persistent pesticides, and eutrophication each produce the same kind of simplifying effects on ecosystems. The nuclide, cesium 137, gamma radiation experiments conducted at Brookhaven indicated that at high doses of radiation the trees were eliminated, stronger radiation eliminated the tall shrubs, still stronger radiation eliminated the low shrubs and herbs, and the highest levels of radiation eliminated the lichens and mosses. This is the same order of susceptibility found in studying the effects of fire, exposure on mountains, salt spray, and water availability. The response of the vegetation to oxides of sulphur near the smelters in Sudbury, Ontario, was also similar: first the sensitive tree species were eliminated; then the whole tree canopy, leaving resistant shrubs and herbs widely recognized as characteristic of the development from open field to forest.

Extensive loss of nutrients from the system may accompany loss of the trees, as is illustrated by Bormann's study [2] in the northeastern U.S. in which he cut down a portion of a forest, left the dead vegetation in place, and followed the nutrient concentration and stream flow in the streams draining the area.

The greater simplification of ecosystems associated with toxicity effects has the potential for producing greater instability. This possibility should be thoroughly investigated in any study of toxicity effects on ecosystems, although the lack of an adequate control is a very serious limitation. The instability would presumably result from basic alterations in the structure of the system due to the elimination of a number of components and the possible introduction of additional components. In this sense, an ecologist does not care *how* something dies, but that it dies; and the consequences of its death for the ecosystem as a whole are what concern him most. The consequences may include greater fluctuations in abundance of species of interest to man, as in the case of a marine fishery; an increase in abundance of various species man considers undesirable; or the breakdown of a vital function of the system, such as the water-holding capability of a forest ecosystem. In any case, the loss of a species will probably produce an ecosystem that is less aesthetically pleasing to man.

There is considerable circumstantial evidence that extensive use of certain pesticides in agricultural systems can so alter the structure of the system that subsequently a more extensive use of pesticides is required, resulting in newer and more serious problems [21], [24]. The following pattern may develop: a pest is particularly abundant late in the growing season of a crop, so the farmer sprays to kill the pest. But at the same time, he kills a number of the parasites and predators of the pest, due to greater sensitivities, concentrating effects, or peculiarities of their life cycles. The system may then enter the next growing season with fewer predators and parasites, allowing the pests to achieve high densities earlier in the next season. The farmer sprays earlier and more often this next season, thereby setting up greater problems the following season. This positive feedback system may proceed until the predators of an organism that has never achieved pest densities before are killed, and this secondary pest emerges early in the season and causes extensive damage. If this secondary pest is one for which little chemical control is yet possible, as was the case for mites for a while [18], the farmer has no choice but to postpone his time of treating for a number of years until the system reestablishes more of a state of stability. The farmer suffers considerable economic loss in the process. Thus, in a really effective program of toxicity study it would be desirable to follow the actual structural changes in the ecosystem caused by the toxic substances so that the consequences of reducing toxic inputs could be foretold as well as the consequences of continued or increased toxic input. It may be that by trying to act on the basis of too little knowledge of ecosystem function and structure, we will cause greater problems than if we fail to act.

As an example of this latter possibility, considerable debate has developed around the feasibility and desirability of reducing the phosphate content of detergents [1], [11], [26]. Ryther and Dunstan [22] have studied this problem from the standpoint of the coastal marine phytoplankton, the single celled algae that are the base of the food chain of the ocean. They demonstrated, by some convincing observations and experiments, that nitrogen, not phosphorus, is the critical factor limiting coastal marine phytoplankton. About twice the amount of phosphate as can be used by the algae is normally present in polluted waters. This is a result of the nitrogen to phosphorus ratio in the input to marine environments and to the greater rate of recycling of phosphorus compared with nitrogen. They conclude that removal of phosphate from detergents is therefore not likely to slow the eutrophication process in costal marine waters, and its replacement with nitrogen containing nitrilotriacetic acid will only accelerate the problem.

Although eutrophication is not a toxicity effect on an ecosystem as I have defined the term, it may interact significantly with toxicity effects, as is suggested by Wurster [33] in his study of DDT sensitivity of algae from coastal marine waters. He found very significant reductions in photosynthesis for laboratory stocks of four very different species of marine algae at fairly low concentrations of DDT. Since eutrophication favors the development of certain species of algae and greatly alters the relative abundance of the trophic levels, its interaction with DDT inhibition could produce significant changes in the structure of the marine environment. But our knowledge of that ecosystem, as well as most others, is too incomplete to predict the form of the structural changes.

A fundamental problem in the study and description of changes in an ecosystem is to determine the amount of dimensionality necessary to describe the responses of the system. Lewontin [15] has recently discussed this problem and suggests that if the unaccounted for dimensionality is treated as a random variable, we may be able to generate a stochastic model that will adequately describe the state of the system. The only problem we may face in this connection is the possibility that the structure of the system is not very stable: it is possible that changes in values of parameters unaccounted for in the model will produce very different performances of the system. As Lewontin points out, this is very well illustrated by the classical predator-prey equations of Lotka and Volterra, in which the basic model predicts undamped oscillations, but the addition of density effects for each population produces damped oscillations.

7. Summary

In summary, any epidemiological study of pollution effects must concern itself with description of the concentrations in and pathways through the components of the ecosystems involved. But if the study is to have any predictive value and to form the basis for any intelligent management decisions with respect to environmental quality, the study must also include the functional analysis of these ecosystems. It must produce realistic and testable models that adequately represent the nature and consequences of the interrelationships of ecosystem components.

REFERENCES

[1] P. H. ABELSON, "Editorial: Excessive emotion about detergents," *Science*, Vol. 169 (1970), p. 1033.

[2] F. H. BORMANN and G. E. LIKENS, "The watershed-ecosystem concept and studies of nutrient cycles," *The Ecosystem Concept in Natural Resource Management* (edited by G. M. Van Dyne), New York, Academic Press, 1969, pp. 49–76.

[3] J. M. BURKE, *et al.*, "Analysis of the toxin produced by *Gonyaulax catenella* in axenic culture," *Ann. New York Acad. Sci.*, Vol. 90 (1960), pp. 837–842.

[4] A. COLLIER, "Some biochemical aspects of red tides and related oceanographic problems," *Limnology and Oceanography*, Vol. 3 (1958), pp. 33–39.

[5] R. T. COUPLAND, *et al.*, "Procedures for study of grassland ecosystems," *The Ecosystem Concept in Natural Resource Management* (edited by G. M. Van Dyne), New York, Academic Press, 1969, pp. 25–47.

[6] P. R. EHRLICH and P. H. RAVEN, "Butterflies and plants: a study in coevolution," *Evolution*, Vol. 18 (1964), pp. 586–608.

[7] M. D. ENGELMANN, "The role of soil arthropods in the energetics of an old field community," *Ecolog. Monogr.*, Vol. 31 (1961), pp. 221–238.

[8] F. C. EVANS, "Ecosystem as the basic unit in ecology," *Science*, Vol. 123 (1956), pp. 1127–1128.

[9] G. GUNTER, *et al.*, "Catastrophic mass mortality of marine animals and coincident phytoplankton bloom on the west coast of Florida, November, 1946, to August, 1947," *Ecolog. Monogr.*, Vol. 18 (1948), pp. 309–324.

[10] N. G. HAIRSTON, *et al.*, "The relationship between species diversity and stability: an experimental approach with Protozoa and bacteria," *Ecology*, Vol. 49 (1968), pp. 1091–1101.

[11] A. L. HAMMOND, "Phosphate replacements: problems with the washday miracle," *Science*, Vol. 172 (1971), pp. 361–363.

[12] E. G. HUNT and A. I. BISCHOFF, "Inimical effects on wildlife of periodic DDD applications to Clear Lake," *California Fish and Game*, Vol. 46 (1960), pp. 91–106.

[13] S. JENSEN and A. JERNELÖV, "Biological methylation of mercury in aquatic organisms," *Nature*, Vol. 223 (1969), pp. 753–754.

[14] E. J. KORMONDY, *Concepts of Ecology*, Englewood Cliffs, Prentice-Hall, 1969.

[15] R. C. LEWONTIN, "The meaning of stability," *Brookhaven Symp. Biol. No. 22: Divers. Stabil. Ecolog. Sys.* (1969), pp. 13–24.

[16] J. D. LINN and R. L. STANLEY, "TDE residues in Clear Lake animals," *California Fish and Game*, Vol. 55 (1969), pp. 164–178.

[17] C. H. MULLER, "The role of chemical inhibition (allelopathy) in vegetational composition," *Bull. Torrey Botan. Club*, Vol. 93 (1966), pp. 332–351.

[18] L. D. NEWSOM, "The end of an era and future prospect for insect control," *Proceedings Tall Timbers Conference on Ecological Animal Control by Habitat Management* (1970), pp. 117–136.

[19] T. PROUT, "Sufficient conditions for multiple niche polymorphism," *Amer. Natur.*, Vol. 102 (1968), pp. 493–496.

[20] E. F. RICKETTS and J. CALVIN, *Between Pacific Tides*, Stanford, Stanford University Press, 1968.

[21] W. E. RIPPER, "Effect of pesticides on balance of Arthropod populations," *Ann. Rev. Entomol.*, Vol. 1 (1956), pp. 403–438.

[22] J. H. RYTHER and W. M. DUNSTAN, "Nitrogen, phosphorus, and eutrophication in the coastal marine environment," *Science*, Vol. 171 (1971), pp. 1008–1013.

[23] A. M. SCHULTZ, "A study of an ecosystem: the arctic tundra," *The Ecosystem Concept in*

Natural Resource Management (edited by G. M. Van Dyne), New York, Academic Press, 1969, pp. 77–93.

[24] R. F. SMITH and R. VAN DEN BOSCH, "Integrated Control," *Pest Control* (edited by W. W. Kilgore and R. L. Doutt), New York, Academic Press, 1967, pp. 295–340.

[25] A. G. TANSLEY, "The use and abuse of vegetational concepts and terms," *Ecology,* Vol. 16 (1935), pp. 284–307.

[26] J. R. VALLENTYNE, "Letters: Eutrophication—key elements," *Science,* Vol. 170 (1970), p. 1154.

[27] G. M. VAN DYNE, "Implementing the ecosystem concept in training in the natural resource sciences," *The Ecosystem Concept in Natural Resource Management* (edited by G. M. Van Dyne), New York, Academic Press, 1969, pp. 327–367.

[28] K. E. F. WATT, *Systems Analysis in Ecology,* New York, Academic Press, 1966.

[29] K. E. F. WATT, *Ecology and Resource Management,* New York, McGraw-Hill, 1968.

[30] R. H. WHITTAKER, *Communities and Ecosystems,* Toronto, Macmillan, 1970.

[31] R. H. WHITTAKER and P. FEENY, "Allelochemics: chemical interactions between species," *Science,* Vol. 171 (1971), pp. 757–770.

[32] G. M. WOODWELL, "Effects of pollution on the structure and physiology of ecosystems," *Science,* Vol. 168 (1970), pp. 429–433.

[33] C. F. WURSTER, "DDT reduces photosynthesis by marine phytoplankton," *Science,* Vol. 159 (1968), pp. 1474–1475.

Discussion

Question: John R. Goldsmith, Environmental Epidemiology, California Department of Public Health

In considering the analogy of ecology and epidemiology there may be merit in considering how epidemiologists deal with the problem of the extremely large number of variables which may affect health. The most vital technique is the classification of variables, followed by choice of an index (or more than one) for each class.

By such methods, which should be applicable as well to ecology and epidemiology, one can then choose a specific hypothesis of association for testing. In a way, formulating such a hypothesis is the second vital step in epidemiology. Testing such a hypothesis is the third crucial step.

Reply: R. Gill

Your comment is well taken, in that ecologists are often overwhelmed by the complexity of the systems they study. Perhaps such an epidemiological approach to ecological systems would yield considerable understanding of them.

Question: B. E. Vaughan, Ecosystems Department, Battelle Memorial Institute, Richland, Wn.

Would you elaborate on monitoring and the statistical design needed for adequate monitoring? Are we at a point where we can formulate an adequate approach to sampling in yet poorly studied systems? Dr. Sterling yesterday described teratogenic considerations about 2,4,5-T. When 2,4,5-T is used for forest management purposes, does anyone here know, for example, (1) what concentrations in runoff water are typical? (For example, how do these concentrations compare to teratogenic levels?) (2) How tightly does 2,4,5-T bind to soil? Can you elaborate on this or analogous problems from the standpoint of better monitoring or sampling approaches?

Reply: R. Gill

To answer the last part of your question first, the work that has been done up to 1967 on ecological effects of herbicides and their movement through ecosystems has been summarized in W. B. House, *et al., Assessment of Ecological Effects of Extensive or Repeated Use of Herbicides: Final Report,* Midwest Research Institute, Kansas City, Missouri, 1967. This report contains discussion of a number of somewhat superficial studies that indicate 2,4,5-T is retained in the soil longer than 2,4-D is; from three months to perhaps a year, depending on climate and soil conditions. I don't know of any studies in which the distribution and concentration of 2,4,5-T in a whole ecosystem were adequately measured. Further, I don't think we know enough about 2,4,5-T's teratogenic properties in the organisms of the system to put the concentrations into a teratogenic perspective, once they are measured.

With respect to the rest of your question, monitoring within a given ecosystem would require following introduced toxic substances through the ecosystem until they were transported out of the system or rendered nontoxic. For widely used substances that are transported considerable distances, such as the chlorinated hydrocarbons, this would mean worldwide monitoring.

In general, I think we are at the point of formulating some, but not all, of the necessary aspects of the monitoring design for any given system. For the physical environment it must involve sampling of surface water, ground water (where possible), soil, and air (to test for possible codistillation, for example). It must involve enough stations to detect patterns of movement of the substances through the physical environment. Dr. Behar, for instance, worked with 20 sampling stations in his study of oxidants in the air of the Los Angeles basin, and this seems to me to be the bare minimum number of stations to get an effective picture of the formation and movement of oxidants in the air of the basin.

But in addition, there must be sampling of the organisms in the system that are exposed to the toxin, to determine their sensitivity to the toxin and the concentration of the toxin within them, if they retain it.

Particularly relevant to the problem of monitoring design is the possibility that monitoring from the standpoint of ecosystem structure and function may dictate a very different design from that dictated solely by consideration of human exposure to the pollutants. If pollutants are accumulating in the soil or in certain organisms, a monitoring system focusing on human exposure to pollutants will not detect this, but will produce strange and confusing results every time the pollutants are released from these ecological reservoirs. I have tried to point out that the presence of toxins in the ecosystem can have far reaching human effects due to alterations in ecosystem structure and function, and these effects might not be explicable from the results of a monitoring system built solely to consider human exposure problems.

Question: Unidentified discussant

Would you comment on the minimum number of components of an ecosystem that should be established to yield an adequate description of reality?

Reply: R. Gill

This varies with the system and the purpose of the study. If you are willing to establish a somewhat artificial set of black box categories, involving such concepts as trophic levels, 10 or 15 categories might produce some meaningful results. For more realistic categories, perhaps 50 components would be necessary. Dr. Vaughan, would you like to comment on this question?

Reply: B. E. Vaughan

I greatly disagree with generalized black box building. With due respect to the last questioner, the question (of how simple an ecosystem needs to be for adequate description of a real system) is badly framed. It is not a matter of how many boxes are strung together, but rather a judgmental matter of how representative a schema may be. It takes a great deal of ecological experience to judge the adequacy of an ecosystem model. You referred to the trophic level concept, which may be quite misleading for the multiply interconnected food web of an estuarine ecosystem. In such web systems, even a few interconnected boxes can be made to demonstrate some very remarkable properties.

DEMOGRAPHIC DATA
FOR LOCAL AREAS

HARLEY B. MESSINGER
INSTITUTE FOR HEALTH RESEARCH, BERKELEY

1. Data sources

1.1. *Census.* The 1970 Census offers much better access to local data than previous decennial censuses. Examples are "block-face" coding and the preparation of census tract directories. The latter permit coding street addresses to census tracts and were never before available for so many cities at one time, the former would permit obtaining data from housing along designated streets with several levels of auto traffic density for a study of air pollution effects on people living at the several levels.

1.2. *Local records.* Vital statistics records have long been popular for epidemiologic studies but recently the value of other routinely collected local items has become appreciated. Santa Clara County in the San Francisco Bay Area has linked files for different departments by coding census tract onto many county records.

1.3. *Surveys.* While much can be done with data gathered for other purposes, an epidemiologic study will usually have to collect information directly as well. Having an interviewer survey each individual household no longer is considered necessary to get valid data. Hochstim's Cervical Cytology Study in Alameda County showed how to increase efficiency of data collection by use of the telephone or the mails [2].

2. Data handling

2.1. *Data banking.* Larger studies may require huge data files meriting the label of "data bank." At one extreme, too little attention is paid to planning for this aspect and the study founders in a mass of doubtful data; at the other, the gathering of information becomes an end in itself and more is collected than can possibly be analyzed with funds available. With the growing realization that protection of confidentiality requires positive measures, even more care will be required to keep data processing budgets within bounds. In the "link file" system used by the American Council on Education for a longitudinal study [1] the identity of a questionnaire respondent is kept secret by having the file linking him to his particular set of answers stored in a foreign country.

2.2. *Record linkage.* Besides the aggregation of statistics by area, the linkage of records for individuals is also important, for example, births and early deaths. An early example of this was the British Columbia Population Study based on records of 114,000 marriages in that province over a ten-year period. Births resulting from these unions were linked with early deaths and also with records of handicapped children in a separate registry [4].

2.3. *Mapping.* Data for small areas may be displayed by mapping as an alternative to tabling. Patterns in census tract distributions of a disease, say, might resemble those of certain air pollutants. In the case of many chronic diseases, such a spatial correlation might not appear until age adjustment of the disease rates had been made. Trend-surface mapping [3], an application of the general linear model to map analysis, offers an approach to systematic use of this technique.

2.4. *Restructuring analytical units.* In census tracting an area, one tries to divide up larger political units such as counties into smaller homogeneous pieces using judgment to set boundaries. Even though this objective can be reasonably well achieved at one point in time, the passage of the years tends to remove any advantages of judiciously chosen partitions over randomly chosen ones. Despite this, census tracts remain more homogeneous than larger units simply because of their size and the fact that demographic characteristics show strong spatial correlation. Thus there is the opportunity to group similar census tracts into "pseudo-counties" as an alternative to the existing political units which may vary markedly in size and may be heterogeneous. An air pollution study in California would not be likely to use Los Angeles County as one unit nor would the census tract be the answer if reliable incidence figures for most chronic diseases were needed. Clustering of contiguous tracts to maximize within-cluster homogeneity would give a better areal unit.

3. Data problems

3.1. *Population mobility.* The high residential mobility of urban populations in the United States is well known. Much of the movement, however, is within the boundaries of political units such as counties or, at worst, standard metropolitan areas in the case of the larger urban agglomerations. Thus a study with coverage of a large city may be able to keep track of a population sample over a period of a year or two provided they have planned for the staff needed to do the tracing.

3.2. *Updating the census.* In an ongoing study based on census data, thought should also be given to detecting major changes in the physical characteristics of the local areas involved. A freeway, from the moment when it's first proposed can have a dramatic effect on the local units in its path. How could such complex changes be comprehended without costly periodic community surveys? And that is only one influence! If one tried as well to keep up with the less dramatic changes, the total effort would dominate all but the largest project. A good case can be made for replacing decennial censuses with continual sampling surveys even if it takes a constitutional amendment!

3.3. *Ecological correlations.* Criticisms have been raised as to the validity of correlations based on aggregated data and the tendency has been noted for their magnitude to increase as one moves up to higher levels as from census tracts to counties to states. They may even change sign as in Buechley's example of the negative correlation of income with cirrhosis of the liver mortality at the census tract level changing to a positive one at the state level [5]. Within a city, it's the "poor" tracts where the disease is common but it's the "rich" states that have the large urban centers with the "poor" tracts inside them. Both correlations are valid but it is easy to try to interpret these associations as if they were based on data for individuals.

4. Analytical methods

4.1. *Factor analysis.* Factor analysis provides an objective way to find patterns of association in large sets of demographic variables. Most studies have employed only data from the decennial censuses. To these demographic measures can be added ones relating specifically to health [6], [7] such as the liver cirrhosis mortality rates already mentioned. The association with economic status seen in states of the United States, incidentally, is not found in prefectures of Japan, cirrhosis in the latter country appearing in a different factor.

4.2. *Cluster analysis.* Just as multiple regression is much easier to interpret if the predictor variables are really stochastically independent, so clustering of areal units on the basis of a few orthogonal factor scores facilitates description of a cluster of census tracts. The late Professor Tryon pioneered in this technique not only in his primary field of psychology but also in the analysis of census tract data. For instance, comparisons of census tract "types" in San Francisco before and after World War II may be found in his book [8].

REFERENCES

[1] A. W. ASTIN and R. F. BOURCH, "A 'link' system for assuring confidentiality of research data in longitudinal studies," *Amer. Educ. Res. J.*, Vol. 7 (1970), pp. 615–624.

[2] J. R. HOCHSTIM, "A critical comparison of three strategies of collecting data from households," *J. Amer. Statist. Assoc.*, Vol. 62 (1967), pp. 976–989.

[3] W. C. KRUMBEIN and F. A. GRAYBILL, *An Introduction to Statistical Models in Geology*, New York, McGraw-Hill, 1965. (See Chapter 13.)

[4] H. B. NEWCOMBE, "Population genetics: population records," *Methodology in Human Genetics* (edited by W. J. Burdette), San Francisco, Holden-Day, pp. 92–113.

[5] A. PEARL, R. W. BUECHLEY, and W. R. LIPSCOMB, "Cirrhosis mortality in three large cities: implications for alcoholism and intercity comparisons," *Society, Culture, and Drinking Patterns* (edited by D. J. Pittman and C. R. Snyder), New York, Wiley, 1962. (See Chapter 19.)

[6] E. S. ROGERS and H. B. MESSINGER, "Human ecology: toward a holistic method," *Milbank Fund Quart.*, Vol. 45 (1967), pp. 25–42.

[7] E. S. ROGERS, M. YAMAMOTO, and H. B. MESSINGER, "Ecological associations of mortality in Japan and the United States: a factor analytic study," *Population Problems in the Pacific* (edited by M. Tachi and M. Muramatsu), Tokyo, 1971. (See Chapter 24.)

[8] R. C. TRYON and D. E. BAILEY, *Cluster Analysis*, New York, McGraw-Hill, 1970. (See references to the "Social Area" problem.)

Discussion

Question: John R. Goldsmith, Environmental Epidemiology, California Department of Public Health

The Current Population Survey and National Health Survey collect both demographic and health data for intercensus years. Data on a sample basis is valid for the country as a whole, but not for separate states, counties, or cities. With some additional resources, such data could be obtained. We obtained it for the Health Survey data for California from 1954 to 1960 and found this useful. The California Legislature is also considering support for a demographic service.

In response to the question as to whether census data are only of value for "fishing expeditions," my opinion is that such data are of great value in testing specific hypotheses. We have used them in health survey studies, in the two community mortality study, and in testing for the effect of demographic variables on other effects.

Reply: H. Messinger

We had thought about using National Health Survey data for a study with states as areal units but could not because comparable items would have been required for at least the conterminous states. The NHS questionnaire is a useful model, in any case, and the "five-foot shelf" of reports issued so far can serve as benchmarks for local surveys.

The term "fishing expedition" should no longer be one of opprobrium. In some areas, such as genetics, most experimental problems can be formulated as clear-cut tests of Hypothesis A against Hypothesis B. In other areas, especially those involving the behavior of humans as individuals and as groups, problems are not as easy to structure. At one stage, census data may be used as an aid in formulating hypotheses; at another, in testing hypotheses as Dr. Goldsmith points out.

Question: J. Neyman, Statistical Laboratory, University of California, Berkeley

Is it practicable to obtain local demographic data indicating, for example, that a given locality, originally comparable, is in the process of turning into a slum? (Or vice versa?)

Reply: H. Messinger

As of the date of a census, it's possible to identify a group of tracts as a slum (having defined this term) from the population and housing data collected, by computing various variables like family income, persons per room, percentage houses dilapidated, percentage unemployed, and also health data like mortality and morbidity rates. I would do this by methods mentioned in Section 4 of my remarks, but other ways could be devised. Then the problem is reduced to the one already touched on of updating the census. With change occurring at present rates, looking at the situation every ten years is inadequate if we are expecting to achieve any control over the process of urban decay.

Ways of extending the analytical methods of Section 4 to encompass change may be found in Harris [1], especially Tucker's article on three mode factor analysis [3]. A well documented study focusing on urban change is Murdie's study of Metropolitan Toronto from 1951 to 1961 [2].

REFERENCES

[1] C. W. Harris (editor), *Problems in Measuring Change*, Madison, University of Wisconsin Press, 1963. (See Chapters 6 to 10.)
[2] R. A. Murdie, "Factorial ecology of Metropolitan Toronto, 1951–1961," University of Chicago, Department of Geography Research Paper No. 116, 1969.
[3] L. R. Tucker, "Implications of factor analysis of three-way matrices for measurement of change," Chapter 7 of Harris, *Op. cit.*

UTILITY OR FUTILITY OF ORDINARY MORTALITY STATISTICS IN THE STUDY OF AIR POLLUTION EFFECTS

WARREN WINKELSTEIN, JR.
UNIVERSITY OF CALIFORNIA, BERKELEY

1. Introduction

Interest in the disease producing potential of air pollution in the United States was stimulated by the Donora episode which occurred in October 1948. This audience may be interested in a brief description of that episode which prefaces the exhaustive and now classic report [1] of the epidemic:

"This particular smog encompassed the Donora area on the morning of Wednesday, October 27. It was even then of sufficient density to evoke comments by the residents. It was reported that streamers of carbon appeared to hang motionless in the air and that visibility was so poor that even natives of the area became lost.

The smog continued through Thursday, but still no more attention was attracted than that of conversational comment.

On Friday, however, a marked increase in illness began to take place in the area. By Friday evening the physicians' telephone exchange was flooded with calls for medical aid, and the doctors were making calls increasingly to care for their patients. Many persons were sent to nearby hospitals, and the Donora Fire Department, the local chapter of the American Red Cross, and other organizations were asked to help with the many ill persons.

There was, nevertheless, no general alarm about the smog even then. On Friday evening the annual Donora Halloween parade was well attended, and on Saturday afternoon a football game between Donora and Monongahela High Schools was played on the gridiron of Donora High School before a large crowd.

The first death during the smog had already occurred, however, early Saturday morning at 2:00 a.m., to be precise. More followed in quick succession during the day and by nightfall word of these deaths was racing through the town. By 11:30 that night 17 persons were dead. Two more were to follow on Sunday, and still another who fell ill during the smog was to die a week later on November 8.

On Sunday afternoon rain came to clear away the smog. But hundreds were still ill, and the rest of the residents were still stunned by the number of deaths that had taken place during the preceding 36 hours . . ."

In retrospect it appears that despite the fact that almost 6000 people, 43 per cent of the population of the Donora area, were affected by the smog, the episode would probably have gone unnoticed except for the occurrence of 17 deaths in a single day. In Figure 1 the death rates by month for Donora are shown for the period 1945 through 1948. The logarithmic vertical scale tends to minimize the amplitude of the peaks but the episode is clearly reflected in the time series. However, there are other monthly peaks, one of which is particularly striking since it actually exceeds the episode peak. During this earlier month, April 1945, excess deaths were not concentrated into a few days nor could the Donora investigators find any evidence of an outbreak of illness. Nevertheless, Weather Bureau records indicate that meteorological conditions of low average daily wind velocity with high average early morning valley stability made the period from 7 to 14 April 1945 more favorable for nocturnal retention of smoke than for any other time from 1945 up to the October 1948 episode. Thus one is tempted to also attribute the excess mortality of April 1945 to the effects of air pollution.

While the problems inherent in the study of time series are undoubtedly well known to this audience, it might be worthwhile to mention briefly those most relevant to the study of the mortality effects of air pollution.

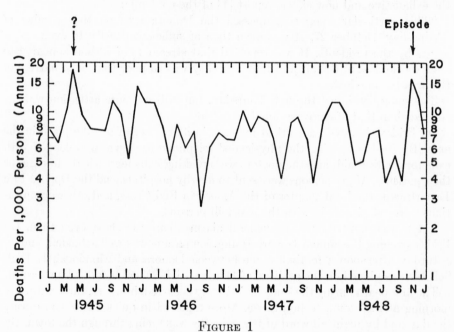

FIGURE 1

Death rates per 1,000 by month, Donora, Pennsylvania, and environs, 1945–1948 [1].

(1) As indicated in Figure 1, choice of an appropriate time interval is crucial if effects are to be detected. If the interval is too long the effect may be missed and if too short, it may be attenuated. Appropriate end points for intervals must be chosen.

(2) Allowance must be made for latent periods between exposure and mortality. For example, at Donora in 1948 the bulk of the deaths occurred on the fourth and last day of extreme smog.

(3) Other causes of excess mortality such as epidemics of infectious disease, accidents, and natural disasters must be removed from consideration.

(4) If specific causes of death are studied as effects, the usual cautions with respect to diagnostic criteria must be invoked.

(5) Since air pollution is a generic term, it is necessary to specify insofar as possible the particular component under consideration. Unfortunately, it is usually impossible to separate out particular components for study and whatever pollutant is measured becomes, in reality, the "index" of air pollution.

After the Donora episode, epidemiological research on the effects of air pollution in the United States took two general forms. On one hand there were studies of time series, notably those of Greenberg [2], McCarroll [3], and most recently of Hexter [4], and, on the other hand, geographical comparisons best exemplified by the Nashville [5] and Erie County [6] studies.

Some of the studies of time series demonstrated that acute episodes of high air pollution, measured by a variety of indices, were associated with detectable excess mortality. The geographic comparisons showed an association of high air pollution at place of residence with excess mortality, both general and from specific causes. However, these findings frequently have been questioned because of methodological problems.

In the remainder of this presentation I will discuss some of these methodological problems as they relate to the Erie County study.

2. The study plan

Buffalo and Erie County, New York, seemed to be a good place to examine the association of air pollution and mortality since the major contribution to air pollution here is derived from large industries located to the windward of the city. This produces wide bands of heavy pollution traversing areas of varied economic status, a factor of considerable importance. The design of the study was straightforward. Ambient air characteristics for the study area were determined by a network of air sampling stations designed to measure suspended particulates, settleable solids, and oxides of sulfur. Sampling was continued over a period of two years from July 1961 through June 1963. It was then possible to construct isopleths for particular pollutant levels for the study area. Those for suspended particulates are shown in Figure 2.

Using these isopleths, the approximately 125 census tracts making up the study area were classified into four groups according to two year average levels of suspended particulates. The areas are shown in Figure 3. Resident deaths for the pericensal period 1959–1961 were then utilized in conjunction with 1960 population data to construct age, sex, race, and cause specific death rates for each of the four census tract groups [7]. An exactly analogous approach was

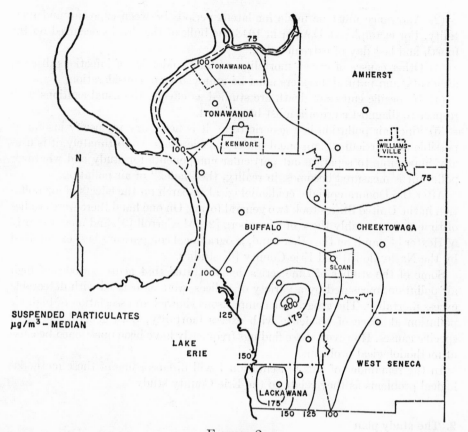

FIGURE 2

Suspended particulate isopleths, Buffalo, New York, and environs, 1961–1963.

used for oxides of sulfur [8]. Since settleable solids and suspended particulate levels were highly correlated, the latter was selected as the index of particulate pollution.

3. Problems to be overcome

There were five major problems to be overcome in the analysis and interpretation of the data generated in this study. They were (1) the strong association between low economic status and high air pollution level of census tracts; (2) the lack of information necessary to adjust the base populations and the deaths with respect to specific covariates known to affect death rates, namely, occupation, national origin, and tobacco smoking habits; (3) the mobility of the population; (4) the inability to separate the effects of various components of air pollution; and (5) the difficulty in making valid inferences from indirect associations.

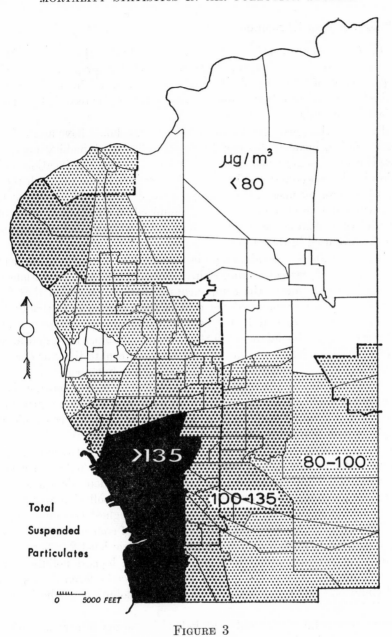

FIGURE 3

Suspended particulate air pollution areas, Erie County,
New York Air Pollution Study [7].

4. Solutions and partial solutions

4.1. *Controlling for economic status.* Since it was known that there would be a strong socioeconomic effect on mortality rates and that air pollution levels are not independent of living standards, the census tracts in the study area were divided into groups according to median family income as recorded in the 1960 published census data.

Ideally, all of the census tracts within each group should have approximately the same median income; this could be accomplished by making many small groups. However, many small groups would make cross classification with air pollution levels unproductive. The optimal procedure is one in which the variability of median incomes within groups is small compared to that between groups, while the number of groups is minimized. A visual inspection of the distribution of median incomes of the 125 census tracts revealed the five natural groups that are used in the subsequent analyses.

In order to evaluate any advantage in this method of classification over the more conventional percentile approach, a one-way analysis of variance was performed. Examination of the components of variance revealed that for the five visually separated groups, 85 per cent of the variance was attributable to between group variance and 15 per cent to within group variance, while for quintiles the comparable figures were 80 per cent and 20 per cent, and for quartiles, 76 per cent and 24 per cent. It is of interest to note that for deciles the intergroup component of variance was increased to only 89 per cent.

The cross tabulation of the total death rates for white men is shown in Table I. The empty cells indicate census tract groups with no population representation and confirm the hypothesis of a strong association between low economic status and high air pollution level.

The marginal totals show large differences of mortality between areas of low and high economic status and between high and low air pollution exposure. However, the rows and columns of the table indicate independent associations. It is perhaps interesting to note that in the lowest air pollution area the effect of the economic gradient is minimal while in the highest economic area, the effect of the air pollution gradient is minimal. This, of course, suggests an interaction which needs further examination. In economic level 2, where each air pollution level is represented, the death rate in the highest pollution area is 50 per cent greater than in the lowest. The difference between the rate in the highest pollution—lowest economic area and lowest pollution—highest economic area is threefold.

Since economic status was used as an index of a variety of personal and social characteristics, it is of interest to examine the homogeneity of the economic levels based on family income with respect to other characteristics enumerated by the census. Data for median years of school completed, per cent of laborers in the labor force, and per cent of sound housing in each of the economic levels are presented in Table II. It is apparent that family income is remarkably stable

TABLE I

Average Annual Death Rates from All Causes per 1000 Population
According to Economic and Air Pollution Levels of Census Tract
of Residence, White Men 50–69 Years of Age, Buffalo, New York
and Environs, 1959–1961 [7]

Economic level is based on median family income for each census tract: 1 = $3,005–$5,007;
2 = $5,175–$6,004; 3 = $6,013–$6,614; 4 = $6,618–$7,347; 5 = $7,431–$11,792.
Air pollution level is based on average suspended particulate levels: 1 = <80 micrograms
per cubic meter per 24 hours; 2 = 80–100 μg/cu m/24 hr; 3 = 100–135 μg/cu m/24 hr; and
4 = >135 μg/cu m/24 hr.

Economic level of census tract of residence	Air pollution level of census tract of residence				
	1 (low)	2	3	4 (high)	Total
1 (low)	—	36	41	52	43
2	24	27	30	36	29
3	—	24	26	33	25
4	20	22	27	—	22
5 (high)	17	21	20	—	19
Total	20	24	31	40	26

over all four air pollution levels in each economic grouping. However, for
median number of years of school completed there is a downward trend with
increasing air pollution levels in each of the economic groupings, most marked
in the two lower levels. For both per cent of laborers in the labor force and
per cent of sound housing, air pollution level 4 seems consistently different from
the other three in the direction of higher proportions of laborers and lower
proportions of sound housing in the two lower economic groupings.

Despite this evidence that air pollution level 4 may differ from levels 1, 2,
and 3 with respect to certain socioeconomic factors, it seems unlikely that these
are responsible for the apparent air pollution effects. This is demonstrated by
the fact that all of the gradients persist when air pollution level 4 is left out of
the comparisons.

4.2. *Accounting and adjusting for the effects of covariates.* The punched cards
from which the mortality data were obtained did not carry all of the data
available from the death certificates. Of particular concern in this study was
the lack of information on usual occupation. Since residence in an area of high
air pollution may be highly correlated with occupation in dusty or dangerous
industries, and since such occupations have been associated with excess mortality, it would be useful to study the mortality pattern of a group without such
exposures. Thus, one would expect the air pollution—mortality relationship
revealed in Table I to disappear for women if occupation was the causal factor.
While some women might have such occupational exposure, this would be a
very small proportion of the population. The cross tabulation of death rates for
white women 50–69 years of age according to economic status and air pollution
level of place of residence is shown in Table III. Since the pattern of association

TABLE II

SELECTED POPULATION CHARACTERISTICS ACCORDING TO AIR POLLUTION LEVEL,
BUFFALO AND ENVIRONS, 1960 [7]

Population characteristics are family income and years school completed: median of median
values for census tracts.
Per cent laborers in labor force and per cent homes sound: median per cent for census tracts.

Economic level	Population characteristics	Air pollution levels				Total
		1 (low)	2	3	4 (high)	
1 (low)	family income	—	4,323	4,274	4,248	4,274
	yrs. school comp.	—	9.1	8.7	8.3	8.4
	% laborers in L.F.	—	19	15	24	19
	% homes sound	—	62	67	59	62
2	family income	5,734	5,852	5,749	5,618	5,764
	yrs. school comp.	9.5	9.4	9.0	8.6	9.1
	% laborers in L.F.	8	7	9	15	8
	% homes sound	86	87	90	65	86
3	family income	—	6,411	6,378	6,178	6,378
	yrs. school comp.	—	10.0	9.4	9.8	10.0
	% laborers in L.F.	—	5	7	15	7
	% homes sound	—	97	93	92	94
4	family income	6,833	6,997	6,789	—	6,906
	yrs. school comp.	11.4	11.6	10.9	—	11.4
	% laborers in L.F.	5	4	8	—	5
	% homes sound	97	97	97	—	97
5 (high)	family income	8,024	8,094	7,431	—	8,000
	yrs. school comp.	12.4	12.5	11.5	—	12.4
	% laborers in L.F.	2	2	5	—	2
	% homes sound	98	99	98	—	99

TABLE III

AVERAGE ANNUAL DEATH RATES FROM ALL CAUSES PER 1000 POPULATION
ACCORDING TO ECONOMIC AND AIR POLLUTION LEVELS OF CENSUS TRACT
OF RESIDENCE, WHITE WOMEN 50–69 YEARS OF AGE, BUFFALO,
NEW YORK AND ENVIRONS, 1959–1961 [7]

Economic level of census tract of residence	Air pollution level of census tract of residence				Total
	1 (low)	2	3	4	
1 (low)	—	16	19	32	22
2	12	13	17	20	15
3	—	13	16	18	14
4	13	11	15	—	12
5 (high)	11	12	9	—	11
Total	12	12	17	22	14

is similar to that for men, we can be reasonably confident that the association is not due primarily to the confounding effect of occupation.

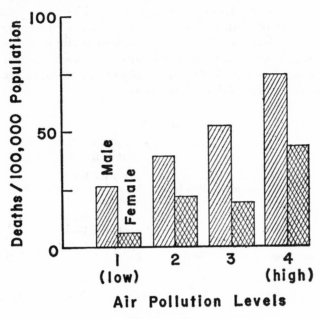

FIGURE 4

Death rates per 100,000 population for gastric cancer in white men and women 50–69 years of age for economic levels 2–4 combined, Buffalo, New York, and environs, 1959–1961 [9].

Because death certificates contain only a limited amount of information, it was necessary to use an adjustment procedure to deal with national origin. In Figure 4 I have summarized the association between suspended particulate air pollution and cancer of the stomach [9]. Haenszel had previously shown an association between foreign birth, particularly Polish, and stomach cancer [10], and since Buffalo has relatively one of the largest Polish-born populations of any American City, it was imperative to rule out the possibility that the association of mortality and air pollution was the result of confounding by an ethnic factor. In fact, when we examined the composition of the study population, as shown in Table IV, both Polish-born and American-born of Polish parents showed increasing proportions in areas of increasing air pollution. To adjust for this, we assumed a constant Polish mortality differential of $2\frac{1}{3}$ times the average death rate for cancer of the stomach observed among white men and women 50–69 years of age. Selection of the factor $2\frac{1}{3}$ was based on the data presented by Haenszel. This rate (70/100,000) was then applied to the estimated population of foreign-born and foreign-born Polish plus native-born of Polish or mixed

Polish parentage in the various cells of the air pollution—economic level matrix. The resulting numbers of deaths were then removed from the total deaths. The ethnic populations were estimated by applying the appropriate percentages derived from the 1960 census tabulations to the combined populations of white men and women 50–69 years of age in each cell of the air pollution—economic level matrix. After the population of each ethnic subgroup was subtracted from the total populations, new rates were computed. In effect, the two ethnic subgroups were treated as though they were subject to a constant excess risk but not subject to an additive risk from air pollution.

TABLE IV

Ethnic Composition of Study Population
According to Air Pollution Level

Study population is at economic levels 2–4.
Polish stock means Polish-born plus native-born of Polish
or mixed Polish parentage.

Group designation	Air pollution level			
	1 (low)	2	3	4 (high)
Polish stock	4%	8%	19%	22%
Foreign-born	8%	11%	9%	13%

The recomputed rates are shown in Table V. Since the associations between suspended particulate air pollution and stomach cancer are essentially unchanged after making adjustments for the possible competing risk of high ethnic susceptibility, the results do not indicate that the association can be explained by the ethnic composition of the populations at risk.

TABLE V

Average Annual Death Rate per 100,000 Population According
to Air Pollution Levels Assuming an Ethnic Effect:
White Males and Females, Economic Levels 2–4,
Buffalo and Environs, 1959–1961 [9]

Ethnic effect assumed to be $2\frac{1}{3}$ times overall age specific rate.
Polish stock means Polish-born plus native-born of Polish
or mixed Polish parentage.
Numbers in parentheses indicate sample sizes.

Group designation	Air pollution level			
	1 (low)	2	3	4 (high)
Total	16 (10)	30 (50)	34 (29)	58 (16)
Less Polish stock	15 (9)	27 (41)	24 (17)	51 (11)
Less foreign-born	10 (6)	25 (37)	31 (24)	54 (13)

Since cigarette smoking has been associated with so many different causes of death, it seemed important to account for the distribution of smoking patterns in the study population. We were, of course, unable to determine the smoking habits of the dead and, unfortunately, the census did not collect information on this characteristic in the 1960 census. However, a population survey had been conducted in Buffalo in 1963 [11] for other purposes and we were able to obtain information from this survey with respect to smoking habits. The distribution of 50–69 year old white male smokers of more than one pack per day is shown in Table VI. While sample sizes are rather small in a number of cells, only in economic levels 2 and 4 is there any suggestion of a positive relationship between heavy smoking and air pollution. Since economic level 2 contained the largest population and had representation in each of the four air pollution levels, it was important to evaluate this trend. The total chi squared value for this row was 3.55 which is not sufficient to permit partitioning to test for regression in the manner suggested by Cochrane [12]. It is of interest to note that the lung cancer rates in economic level 2 followed the pattern of the cigarette smoking habits of the sample [7]. Incidentally lung cancer was not associated with suspended particulate air pollution in this study.

TABLE VI

Per Cent of Sample of 50–69 Year Old White Men Smoking More than One Pack of Cigarettes per Day According to Economic and Air Pollution Levels, Buffalo Only, 1963

Numbers in parentheses indicate sample sizes.

Economic level of census tract of residence	Air pollution level of census tract of residence			
	1 (low)	2	3	4 (high)
1 (low)	—	100 (4)	32 (22)	75 (4)
2	18 (17)	35 (48)	27 (33)	45 (20)
3	—	47 (34)	55 (9)	37 (8)
4	22 (18)	42 (19)	55 (9)	—
5 (high)	33 (3)	38 (13)	33 (6)	—

4.3. *Mobility.* There are several kinds of mobility that ideally need to be accounted for. These include short term mobility in relation to duration of exposure at place of residence, occupation, and recreation, and long term mobility with respect to change of residence and occupation within the study area or outside of it. Unfortunately, we were unable to devise any very good way of getting at this. The census gave information regarding residence at present address for more than five years and when this was examined we found that the proportion of the population living longer than five years at present residence was lower in the lower economic groups than in the higher, as expected, but there was no relationship to air pollution level.

4.4. *Separating component effects.* Frequently all measurable components of air pollution are highly correlated in their areal distribution. However, in the Erie County Air Pollution Study, sulfation and suspended particulates had sufficiently different distributions to permit an analysis of their independent and compound effects [8]. The data are shown in Table VII. Unfortunately, the population distributed somewhat unevenly among the various areas. However, it is quite apparent that oxides of sulfur are not contributing to the mortality effect either independently or in combination with suspended particulates.

TABLE VII

AVERAGE ANNUAL MORTALITY PER 1000 POPULATION FROM ALL CAUSES
IN WHITE MEN, 50 TO 69 (1959–1961) [8]

Suspended particulates: High is >100 μg/cu m/24 hr; low, <100 μg/cu m/24 hr.
Oxides of sulfur: high, >0.45 mg/sq cm/30 days; low, <0.45 mg/sq cm/30 days.

| | Air pollution level (oxides of sulfur and suspended particulates) | | | | | | | |
| | Susp. part. low SO$_2$ low | | Susp. part. low SO$_2$ high | | Susp. part. high SO$_2$ low | | Susp. part. high SO$_2$ high | |
Economic level	Pop.	Rate	Pop.	Rate	Pop.	Rate	Pop.	Rate
1 (low)	530	36	—	—	4,413	46	2,822	41
2	13,383	26	—	—	2,245	33	7,908	32
3	7,684	24	—	—	4,189	28	1,063	26
4	13,771	21	735	19	2,639	27	—	—
5 (high)	11,428	19	1,301	16	574	20	—	—
Total	46,796	23	2,036	17	14,060	34	11,793	33

It is of some interest in this regard to note that in London, where suspended particulate air pollution has been drastically reduced, as a result of the enforcement of a "Clean Air Act," while sulfation has remained high, severe fogs of recent years have not been accompanied by nearly the degree of excess mortality characterizing the fogs of earlier years.

4.5. *Inferences from indirect associations.* Epidemiologists and biostatisticians are well aware of the problems involved in the analysis of incomplete cross tabulations based on classifications which are indirect measurements of the characteristics under study, in this case, air pollution level of place of residence controlled for economic status. Because of this, my colleagues and I have refrained from applying ordinary statistical tests to the data from the Erie County Air Pollution Study and have, instead, evaluated the results by inspection of the rates for trends and consistency. This has aroused some comment [13], but I believe that this approach is less risky than the application of inappropriate statistical tests. Further, we have tried to be cautious and guarded in our interpretations of observed associations.

The fundamental problem in interpreting these types of data is the avoidance of what Karl Pearson termed "spurious correlation" and what Professor Neyman has more cogently described as inappropriate methods of studying correlation [14]. In the present study the problem revolves around the possibility that some intervening factor associated with both risk of death and air pollution level has produced the observed associations. In Sections 4.1 through 4.3 I have discussed a number of such possibilities. Professor Neyman suggested that I look further into the question particularly with respect to population size. Specifically, he was concerned with the possibility that the populations of the census tracts and consequently their density might be positively correlated with both air pollution and risk of death. Thus, in Table VIII I have subdivided the 40 census tracts making up economic level 2 into three groups according to their size. When the total death rates for white men 50–69 were computed for each of the four air pollution levels in each size group, the trend revealed in Table I was confirmed in each group. From this evidence, it appears unlikely that the observed association is simply a function of population size. However, it does not dispose of Professor Neyman's basic concern. Perhaps my colleagues and I have overlooked the potential value of more sophisticated correlation techniques because of our belief that the indirect nature of all of our measurements except the counts of deaths and population precluded their use.

In making inferences from the mortality data in the Erie County Study we have given special attention to two characteristics, first, to biological plausibility and, second, to confirmatory evidence from other studies. Thus, of the seven diseases or groups of diseases (all causes [7], chronic respiratory disease [7], cancer of the stomach [9] and prostate gland [15], arteriosclerotic heart disease, cerebrovascular disease [16], and cirrhosis of the liver [17]) for which mortality has been reported associated with suspended particulate air pollution in Erie County, only cirrhosis and cerebrovascular disease have not been reported in association with air pollution in the Nashville Study. With respect to cirrhosis, there is a rational explanation for the association. It is hypothesized that toxic agents in industrial air pollution act individually and synergistically with alcohol to produce mortality. For cerebrovascular disease, greater caution is indicated. While it could be postulated that air pollution in Erie County contains agents toxic to the human vasculature, the data are not as consistent over all economic levels and between the sexes as one would like.

It is interesting to note that even after mortality from the six associated diseases or groups of diseases are removed from the total, a substantial gradient remains. This may mean that the association is spurious or it may suggest that particulate air pollution is a general environmental stress analogous at our present state of knowledge to poverty. Nevertheless, there were a number of diseases which were definitely not related to suspended particulate air pollution while revealing strong inverse associations with economic status. Notable among these were lung cancer and infant mortality.

TABLE VIII

DEATHS AND DEATH RATES FROM ALL CAUSES PER 1000 POPULATION
ACCORDING TO AIR POLLUTION LEVEL OF CENSUS TRACT OF RESIDENCE
AND SIZE OF TRACT, WHITE MEN 50–69 YEARS OF AGE,
BUFFALO, NEW YORK AND ENVIRONS, 1959–1961

Population of tract	Number of tracts	Air pollutiou level				
		1 low	2	3	4 high	Total
		Population				
99–555	23	470	3,550	1,434	2,181	7,635
592–899	10	2,217	3,963	1,607	—	7,787
939–1925	7	976	2,207	3,927	1,004	8,114
Total	40	3,663	9,720	6,968	3,185	23,536
		Deaths (3 yrs.)				
99–555	23	35	306	109	242	692
592–899	10	167	326	157	—	650
939–1925	7	65	167	364	98	694
Total	40	267	799	630	340	2,036
		Average annual death rate				
99–555	23	25	29	25	37	30
592–899	10	25	27	33	—	28
939–1925	7	22	25	31	33	29
Total	40	24	27	30	36	29

5. Conclusion

There is no particular mystery about the disease producing potential of air pollution. If toxic agents are discharged into the air, whether they be in the form of ionized particles, elemental metals, or complex organic molecules, they will produce disease if they are delivered to a susceptible host in sufficient dosage. It seems to me that this potential has now been demonstrated repeatedly in differently designed studies. The fact that a number of diseases of unknown etiology appear to be related to air pollution suggests that important causal relationships may be revealed through air pollution studies. Such knowledge can, of course, be utilized for the betterment of the public health.

REFERENCES

[1] H. H. Schrenk, et al., "Air pollution in Donora, Pennsylvania," *Epidemiology of the Unusual Smog Episode of October, 1948: Preliminary Report*, Bull. 306, Federal Security Agency, Public Health Service, Bureau of State Services, Division of Industrial Hygiene, 1949.

[2] L. Greenburg, et al., "Report of an air pollution incident in New York City, November, 1953," *Pub. Health Rep.*, Vol. 77 (1962), pp. 7–16.

[3] J. McCarroll and W. Bradley, "Excess mortality as an indicator of health effects of air pollution," *Amer. J. Pub. Health*, Vol. 56 (1966), pp. 1933–1942.

[4] A. C. Hexter and J. R. Goldsmith, "Carbon monoxide: association of community air pollution with mortality," *Science*, Vol. 172 (1971), pp. 265–267.

[5] L. D. Zeidberg, J. J. Schuenemann, P. A. Humphrey, and R. A. Prindle, "Air pollution and health: general description of a study in Nashville, Tennessee," *J. Air Poll. Cont. Assoc.*, Vol. 11 (1961), pp. 289–297.

[6] W. Winkelstein, Jr., "The Erie County air pollution—respiratory function study," *J. Air Poll. Cont. Assoc.*, Vol. 12 (1962), pp. 221–222.

[7] W. Winkelstein, Jr., S. Kantor, et al., "The relationship of air pollution and economic status to total mortality and selected respiratory system mortality in men. I. Suspended particulates," *Arch. Environ. Health*, Vol. 14 (1967), pp. 162–171.

[8] W. Winkelstein, Jr., S. Kantor, et al., "The relationship of air pollution and economic status to total mortality and selected respiratory system mortality in men, II. Oxides of sulfur," *Arch. Environ. Health*, Vol. 16 (1968), pp. 401–405.

[9] W. Winkelstein, Jr. and S. Kantor, "Stomach cancer, positive association with suspended particulate air pollution," *Arch. Environ. Health*, Vol. 18 (1969), pp. 544–547.

[10] W. Haenszel, "Cancer mortality among the foreign-born in the United States," *J. Nat. Cancer Inst.*, Vol. 26 (1961), pp. 37–132.

[11] W. Winkelstein, Jr., "Study of blood pressure in Buffalo, N.Y.," *Ann. N.Y. Acad. Sci.*, Vol. 107 (1963), pp. 570–575.

[12] W. G. Cochran, "Some methods for strengthening the common χ^2 tests," *Biometrics*, Vol. 10 (1954), pp. 417–451.

[13] S. C. Morris and M. A. Shapiro, "A statistical note on the association of air pollution and cirrhosis of the liver," *Arch. Environ. Health*, in press.

[14] J. Neyman, "On a most powerful method of discovering statistical regularities," *Lectures and Conferences on Mathematical Statistics and Probability*, Washington, D.C., Graduate School of the U.S. Department of Agriculture, 1952, pp. 143–154 (2d ed.).

[15] W. Winkelstein, Jr. and S. Kantor, "Prostatic cancer: relationship to suspended particulate air pollution," *Amer. J. Pub. Health*, Vol. 59 (1969), pp. 1134–1138.

[16] W. Winkelstein, Jr. and M. L. Gay, "Arteriosclerotic heart disease and cerebrovascular disease: further observations on the relationship of suspended particulate air pollution and mortality in the Erie County Air Pollution Study," *Proc. 16th Ann. Meet., Inst. Environ. Sci., Boston, 1970*, pp. 441–447.

[17] W. Winkelstein, Jr. and M. L. Gay, "Suspended particulate air pollution, relationship to mortality from cirrhosis of the liver," *Arch. Environ. Health*, Vol. 22 (1971), pp. 174–177.

[18] E. Sawicki, W. E. Elbert, T. R. Hauser, F. T. Fox, and T. W. Stanley, "Benzo(a)-pyrene content of the air of American communities," *Amer. Ind. Hyg. Assoc.*, Vol. 21 (1960), pp. 443–451.

[19] E. C. Hammond, I. J. Selikoff, and P. J. Lawther, "Inhalation of benzopyrene and cancer in man," First Fall Scientific Assembly of American College of Chest Physicians, Chicago, 1969.

Discussion

Question: R. J. Hickey, Institute for Environmental Studies, University of Pennsylvania, Philadelphia

You commented on environmental polycyclic hydrocarbons which are known to have carcinogenic properties, such as 3,4-benzopyrene. Do you recommend that people cease the great American custom of favoring charcoal broiled steaks, smoked meats, overly grilled sandwiches, and other foods which are known to, or may, contain these polycyclic hydrocarbons?

As you probably know, Sawicki, Elbert, Hauser, Fox, and Stanley [18] investigated the 3,4-benzopyrene content in the atmosphere of a number of cities in the United States. They also estimated the number of micrograms of this compound inhaled per capita in a year. They observed, in these early studies, that 3,4-benzopyrene data and mortality data for lung cancer in various cities "failed to reveal a significant relationship." In this connection it should be recognized that a mammalian enzyme (or enzyme system) exists, called benzopyrene hydroxylase, which is involved in metabolizing 3,4-benzopyrene and perhaps other polycyclic compounds.

Further, if atmospheric 3,4-benzopyrene is in fact dangerous, through posing a risk of contracting, for example, lung cancer, then it might be expected that high level occupational exposure, such as in certain tarring and roofing occupations, could lead to high risk of lung cancer in people in these occupations. Interestingly, Hammond, Selikoff, and Lawther [19] conducted an investigation on this question and, as I recall, were unable to demonstrate a high frequency of occurrence of lung cancer in these people as compared with comparable people in the average population. Curiously, so far as I know, the paper has not yet been published.

Reply: W. Winkelstein

I would certainly recommend the cessation of cigarette smoking, but I am not sure that total austerity is worth the slight improvement in risk.

Lung cancer was not associated with particulate air pollution in either the Buffalo or Nashville studies.

Question: E. Tompkins, Human Studies Branch, Environmental Protection Agency

Have you looked at the distribution of causes of death in Buffalo as compared to the distribution in other similar cities without comparable air pollution (within economic groups)?

Reply: W. Winkelstein

Except for comparing the results with the Nashville Air Pollution Study, no other comparisons have been made.

SKELETAL PLAN FOR A COMPREHENSIVE EPIDEMIOLOGIC STUDY OF POLLUTION: EFFECTS OF EXPOSURE ON GROWTH AND DEVELOPMENT OF CHILDREN

JOHN R. GOLDSMITH

CALIFORNIA STATE DEPARTMENT OF PUBLIC HEALTH

I. Basis for choosing this association to study

A. An "effect" which could be associated with atmospheric pollution, radiation, nutrition, occult lead or heavy metal poisoning or infectious or parasitic diseases.

B. An "effect" which is not currently under active study, but which is of general interest in that many populations will want to avoid a negative association.

C. An "effect" with some interesting statistical properties, but also possessing the full range of epidemiological complexity.

D. Practical, or apparently so, on an adequate scale.

E. Likely to be sensitive to existing exposure levels. (? sensitive for radiation exposures.)

F. Symposium participants are likely to be similarly familiar and informed relative to, say, chronic respiratory disease morbidity and mortality, or infant mortality, for which some participants are better informed than others.

G. Such a study is likely to attract support from available funds.

II. Background

Wetzel grid used to describe the growth function, based on height, height-weight ratio, age, and "body type."

Kapalin distribution diagram describes "location" of population, variance, and homogeneity. Fiducial limits were not applied. Kapalin has demonstrated apparent association of height, with nutrition, economic status of family, family size, pollution exposure, stature of parents. Similar analyses of RBC count, and hemoglobin. Reversibility suggested with change of location or with nutritional program.

Availability of preliminary data should be good—height and age of children in different schools, with census data, for schools with different climatological and pollution exposures.

Temporo-spatial strategy is likely to be appropriate.

III. Two study plans

Study plan A

Neonatal growth in relation to areas affected by pollution which can be defined for high levels of air pollution and with nitrate pollution of drinking water. Control areas, similar in socioeconomic factors and in climatology can be defined. Other variables are family size, smoking of parents, housing quality, breast feeding—sources of formula or bottle feeding, ethnicity, time of year.

Hypothesis to test: That during the months with relatively high levels of pollution, there is a greater unfavorable difference between polluted and control areas. (Judged by position of distribution, variance or deviant sub-population.)

Questions for discussion: Study size, suitability of various statistical criteria, how to treat replicate observation within same individual. Value of blood tests. Developmental yardsticks. How to treat illness episodes. Duration and nature of follow-up. Treatment of missing data (or data from different sources).

Study plan B

Skeletal growth, bone mineral, exercise capacity, lung function, and hematological levels in relation to pollution among eight to ten year old school children.

Measurements to be made at least four times yearly for two years and with hypotheses like those in Plan A. Height to be measured monthly, exercise capacity, lung function four times yearly and hematological variables and bone mineral (I^{125} scan, less than 10 mr per test) at randomized-within-individual times—yearly per individual, but clustered at 4 times a year.

Pollution estimates to be based on extension of existing air and water monitoring programs. Diet samples may be suitable.

Questions for discussion: Compare absolute values (or difference in central tendency estimate) or rate of change. Statistical test suitable to individual variables and or their combination. Statistical yardsticks for maximal performance test repetitions (lung function, exercise capacity) standardizing motivation. Inter-laboratory and inter-instrument comparability. Meteorological or climatological factors as missing variables. Experimental models for such effects. Relevance of growth and development for health as adult or for aging.

It is of great importance that there be at least one center which demonstrates the utility of a comprehensive approach. It is more important than three or four centers starting at once. It takes a tremendous amount of time for the different disciplines involved to work out their mutual problems. By the time one adds the dimension of geography to disciplinary interactions the problem becomes unreasonably large.

REFERENCE

[1] V. KAPALIN, "Evaluation of the level of development in children and adolescents," *Children and Adolescents* in *Collection of Scientific Reports of the Institute of Hygiene, Prague,* Prague, Institute of Hygiene, 1970, pp. 7-82.

SKELETAL PLAN FOR A STUDY
OF DAILY MORTALITY

ALFRED C. HEXTER

CALIFORNIA STATE DEPARTMENT OF PUBLIC HEALTH

The study reported in *Science*, Vol. 172 (1971), pp. 265–267 used multiple regression to study the association between air pollutants and mortality in Los Angeles county. Of the three regressions reported therein, the second provides a method for predicting the total number of deaths each day using a typical multiple regression model with terms for cyclic variation and trend and a quadratic formulation for temperature with lags of up to three days.

The approach should be applicable to any reasonably homogeneous area with sufficient population. By obvious extensions it can be used to examine specific causes or other variables such as (dropping temperature) "proportion of births with congenital malformations."

The essential features are: (1) daily occurrences are examined; and (2) allowance is made for cyclic variation (such as season of year), secular changes (trend) and the most important confounding variable, temperature.

Two applications to the study of specific pollutants (carbon monoxide and total oxidants) are given in the paper. The method could be applied to any pollutant for which comparable data are available.

Another extension, not included in that paper, is application of standard industrial quality control methods to compare the actual number of deaths with the predicted number. This could be used to detect unusual increases in mortality such as might be caused by an epidemic or by an episode of air pollution. Only three inputs are required for each day: date, maximum temperature and a count of deaths. In principle this could be operated on very nearly a "real-time" basis.

SKELETAL PLAN OF A COMPREHENSIVE STATISTICAL HEALTH-POLLUTION STUDY

JERZY NEYMAN

STATISTICAL LABORATORY, UNIVERSITY OF CALIFORNIA, BERKELEY

1. *Objective.* The objective of the proposed comprehensive statistical health-pollution (CSHP) study is to estimate the relationship between selected characteristics of health conditions and the proliferating pollutants as they appear in the actual environment.

2. *Selection of health characteristics and of the pollutants to study.* The selection of health characteristics and of pollutants to be studied falls within the field of competence of specialists in biology, in health sciences, in chemistry and in physics.

Two aspects of the problem appear to require separate consideration. First, there are suspected deleterious effects of pollutants on health of "normal" humans now living. Second, it is presumed that certain pollutants are mutagens which affect adversely future generations. The subject of study could have been simplified if these two different aspects would be treated separately. As things stand now, there is a substantial overlap: mutagenic effects seem to parallel carcinogenic effects, which manifest themselves in the now living generations. Also [1], mutagens are being suspected as causes of abnormalities at birth. These points, as well as difficulties in monitoring, will be discussed at the conference during the Thursday morning session, July 22nd.

3. *Necessity of simultaneous treatment of all the suspected deleterious pollutants with reference to a number of localities.* Even though many current studies refer to just one pollutant (frequently radiation), it must be clear that, in order to be able to evaluate the effect of a single pollutant, it is unavoidable to evaluate, perhaps only summarily, the effects of all others. The point is that all the deleterious pollutants "compete" with each other for human health and lives. The number of victims claimed by a particular pollutant A, in a given locality and during a particular year, can be small or large depending on whether another pollutant B kills many or only a few people (respectively), preventing them from succumbing to A. This remark applies to cases where the cause of death (or other condition) is unambiguously defined as, for example, death from cancer. The quantities discussed in some current pollution-health studies are what is technically called "crude rates," for example, of deaths from cancer. The proper

measure is the "net rates" computable taking into account deaths from other causes [2].

In most pollution-health studies the causes of deaths (or other health conditions) are not identified and the purpose of such studies is precisely to find out, tentatively, whether the actual causes might have been the suspected pollutants. An unambiguous answer could be obtained in ideal conditions of having two localities, say L_1 and L_2, inhabited by identical populations and affected by identical pollutants, with one exception: in addition to pollutants affecting L_1, the locality L_2 is affected by a single extra pollutant A. If such two localities could be found, then the differences in health conditions in L_1 and L_2 would be ascribable to A operating with all other pollutants in the background.

Clearly, no such two localities exist in the real world and the results of an actual CSHP study must be only tentative, subject to indirect confirmation by experiments with animals, etc. However, in order to be able to obtain even such tentative evaluations of particular pollutants it is unavoidable to study simultaneously a number of localities with different patterns of pollution. A study of this kind, involving 38 cities in the United States, was recently performed [3] and the authors, Hickey et al., deserve recognition for their effort and for the initiation of what might be described as multivariate-multilocality health-pollution studies.

In a sense, the proposed CSHP study might parallel the study of Hickey et al., with a few modifications. One is that radioactive pollution, omitted by Hickey et al., should be included in the CSHP study. The other modifications refer to statistical methodology and, probably, to selection of localities to be studied. One example is as follows.

The statistical methodology used by Hickey et al. is based on multivariate linear regression techniques. Contrary to this, a substantial biological literature indicates that the regression of unsatisfactory health conditions on two or more pollutants might well be far from linear. The literature in question, to be discussed in the Thursday afternoon session, is concerned with the term "co-carcinogens." Apparently, certain agents exist, say A and B, which, by themselves, are only mild carcinogens if at all. A small exposure to either A or B produces only a few cancer cases in experimental animals. Thus, the regression of cancer incidence in A alone may be linear but with only a small regression coefficient, and similarly for B.

On the other hand, if a small dose of A is followed by exposures to increased doses of B, the incidence of cancer grows fast. In these conditions, the regression of cancer incidence on both A and B simultaneously will be far from linear.

The above applies to what might be labeled "positive" co-carcinogens. Professor F. N. David tells me that there are also "negative" co-carcinogens. Another point to consider: if co-carcinogens exist, referring specifically to cancer, is it not plausible that some other ailments, perhaps respiratory diseases, are affected by some co-factors or co-pollutants? The a priori adoption of a linear repression scheme may bring out misleading conclusions.

The nonlinearity of regression is just one of the points of divergence between the methodology adopted by Hickey et al. and what would be my own preference. There are other such points.

4. *Localities to be included in the proposed study.* Because of the public concern about deleterious effects of radiation, it is proposed that the CSHP study cover (a) all the localities of the operating nuclear power facilities (Messrs. Patterson and Thomas promised to include the list of such facilities in their paper to be given during the afternoon session of July 21st). (b) For comparison, it is proposed to include in the study the sites of a comparable number of large factories using mineral fuels. These factories should be randomly selected from a reasonably complete, appropriately stratified list. Possibly Dr. Waksberg from the Bureau of the Census could help to secure such a list. (c) The same public concern with radioactivity suggests the desirability of inclusion in the study of localities surrounding the nuclear weapons test sites in Nevada. Here again Drs. Patterson and Thomas will be helpful with information about monitoring facilities. (d) In order to provide a comprehensive picture of the health-pollution relationships in the country, the CSHP study should include representative samples of major cities and of countryside localities, particularly those exposed to chemical pollutants. Here the suggestions of Dr. G. Morgan are likely to be very useful. Also I expect important information from Drs. R. Risebrough and T. Sterling.

5. *Difficulties with the data.* It is to be anticipated that the CSHP study will encounter considerable difficulties in securing reliable data. The prospects in this respect may be judged by the already quoted article of Dr. Hook [1], describing a special two-day conference held last year in Albany, N.Y. The specific purpose of this conference was the finding of means to improve the monitoring of human birth defects. It appears that, in a particular state, a change in the method of monitoring malformations resulted in an increase in the records in the ratio of 1 to 3.5! While this applies to malformations of the new-born, it appears plausible that the accuracy of other health records must also vary from one state to another and, probably be a greater extent, from one county to the next. (Inspection of actual data and also conversations with some knowledgeable persons convinced the present writer that this supposition is very plausible.)

If the proposed CSHP study is to bring out reliable information, special care must be taken to see that differences in the health data reflect true differences in health conditions rather than differences in routines of data collection. How to achieve this? What existing organizations can be helpful? It is possible that some agency of the Federal Public Health Services conducts routinely some spot checking of the precision of data collection. Hopefully, information on this point will be forthcoming in the discussions at the conference.

The incompleteness of records is not the only possible source of uncertainties regarding health data. For instance, the incidence of various diseases may depend very much on the stratum of the population. Many diseases rather rare in conditions of relative comfort are likely to be quite frequent in the slums.

Thus, in order to be able to assign deteriorations in health conditions to pollutants it is essential to eliminate biases resulting from equal treatment of data referring to slums and to comfortable suburbs. It follows that the proposed CSHP study will need data on particular racial and economic strata of the population in the various localities. The difficulties in this respect are enhanced by the fact of apparently widespread uneasiness about proximity of a nuclear facility. It is not implausible that, when such a plant is constructed in a given locality, the economically comfortable inhabitants move out, prices of property decrease and the immediate vicinity rapidly turns into a slum.

Clearly, all the above is just speculation which can be easily countered by other speculations. For example, it may be argued that the construction of a nuclear power facility leads to closing down of some smoky mineral fuel power plants, to cleaning of the atmosphere and the consequent increase in the general standard of living. However, if the CSHP study is to yield reliable information, it must be based not on speculations of one kind or another, but on verifiable data. Here questions arise: how and where could one secure reliable data on the stratification of inhabitants of localities selected for the study? Special sample surveys? What agency could and would conduct them? How much would such surveys cost? The participation of Dr. Waksberg in the discussion of these and some other similar problems will be greatly appreciated.

In addition to uncertainties of health data, there are also uncertainties about monitoring of pollutants. The already mentioned paper by Hickey *et al.* deals with an impressive array of pollutants including SO_2, NO_2, Cu, Ti, *etc.*, *etc.* In addition, however, one must also think of residues of pesticides and of defoliants found in milk and other foods. In an optimistic mood one is inclined to take it for granted that all the relevant data are reliable, even though possibly scarce. The disquieting detail in this connection is the paper by Fred S. Goulding concerned with "an improved analytical tool for trace element studies." The title of the paper suggests the possibility that the monitoring of, say, Cu and Ti in the atmosphere is conducted by not just one method in the United States, but by a variety of different methods, using different tools. If this be so, then the data on pollutants must be subject to the same kind of uncertainties as the data on health conditions: the records of trace metals, as well as the records of particulates, and so forth, coming from different localities may reflect not only the real differences in the degree of pollution but also the differences in the method of ascertainment. If the proposed CSHP study is to be reliable, this particular point requires serious attention. How uniform are the techniques of monitoring pollutants? Is it at all feasible to reduce the observations made in different localities by different methods to some kind of a common denominator? Here, Dr. George Morgan may provide very important information.

6. *Recommended structure of the CSHP study.* The authoritativeness of the proposed CSHP study will be greatly increased if it is to be performed by not just one but by at least two or three academic statistical groups, perhaps one in the East, one in the Midwest and one on the West Coast. It would be essential

to arrange that the three groups could work in frequent contact with each other and in cooperation with governmental agencies, but with a guaranteed independence from these agencies.

One of the reasons for such multiplicity of effort is the fact that statistical study of a given complex phenomenon allows a number of different approaches (or "models") which, at first sight at any rate, may appear equally plausible. The choice between such possibilities must depend upon earlier experiences of the individuals concerned and, undoubtedly, on the ubiquitous preconceived ideas that affect all of us.

The interactions between cooperating groups, combined with unlimited access to the same data and to all other sources of information are likely to result in an increase in objectivity and reliability of the findings. It may be hoped that the cooperative study will result in a single joint factual report. In addition there will be interpretations and/or conclusions. These are expected to be somewhat different, depending upon personal attitudes of the authors. However, the public and the Government are likely to gain from having at their disposal the presentations from several different points of view.

7. *Practical steps.* The above discussion suggests that, in order that the proposed comprehensive statistical health-pollution study be reliable, it must be preceded by substantial interdisciplinary preparation.

(i) A preliminary list of health parameters of pollutants and of calendar years to be studied must be established.

(ii) A preliminary list of localities must be compiled.

(iii) A group of knowledgeable persons must investigate the reliability of health data available for the chosen localities with possible deletions in lists (i) and (ii).

(iv) Presumably, another group of knowledgeable persons must investigate the availability and the comparability of data on contemplated pollutants. This is likely to result in further deletions in the two tentative lists under (i) and (ii).

(v) The results of efforts under the above four points must be somehow collated, presumably in a substantial conference.

(vi) Ordinarily, desirable things do not just happen. Also, ordinarily, they cost money. If the CSHP study is to be performed, whether conforming with the above skeleton plan or in some other way, a small group of interested persons must pick up and carry the ball, including budgeting, search for funds and all the innumerable organizational details.

In the end, the conclusion may be that the CSHP study is now impossible or that it is too messy to attempt!

A Postscript. In reply to a question, I wish to make it clear that my generally favorable reference to the paper by Hickey *et al.* does not imply my endorsement of the conclusions they reached. Specifically, I question the statistical methodology leading to the conclusion that, in cases specified, some of the pollutants, like SO_2 and NO_2, tend to increase mortality and some other pollutants, like Cu, tend to decrease it. The conclusions may or may not be true but the statistical analysis which led to them is in my opinion invalid.

REFERENCES

[1] E. B. HOOK, "Monitoring human birth defects and mutations to detect environmental effects," *Science*, Vol. 172 (1971), pp. 1363–1366.
[2] C. L. CHIANG, *Introduction to Stochastic Processes in Biostatistics*, New York, Wiley, 1968.
[3] R. J. HICKEY, D. E. BOYCE, E. B. HARNER, and R. C. CLELLAND, "Ecological statistical studies concerning environmental pollution and chronic diseases," *IEEE Trans. Geosci. Electron.*, Vol. GE-8 (1970), pp. 186–202.

SOME SKELETAL PLANS FOR STUDYING HEALTH EFFECTS OF AIR POLLUTION

HANS K. URY

PERMANENTE MEDICAL GROUP. OAKLAND, CALIFORNIA

1. Filtered air studies on populations at risk

One of the difficulties in evaluating the health effects of photochemical pollution lies in the fact that these tend to be long term effects, in contrast to immediate air pollution disasters, such as the London "Killer fog" of 1952. A method of obtaining short term photochemical pollution effects consists of considering the results of lung function tests obtained under different pollution conditions on subjects suffering from respiratory diseases.

In Ury and Hexter [5], a number of univariate and multivariate statistical procedures are discussed for a study in which a series of lung function tests and other physiological measurements were obtained for 15 emphysematous subjects, both under ambient and under filtered air conditions, in a Los Angeles hospital room with controllable air supply. A significant association was found to exist between airway resistance and oxidant levels. (Subsequently, a lesser but still significant association was found to exist between airway resistance and NO_2.)

This type of study can obviously be extended to subjects suffering from any respiratory or circulatory disease or to any other population at risk, and to any pollutant which can be effectively filtered out. In order to isolate the effects of specific pollutants, partial correlation or step-wise regression should presumably be used.

2. Simple and robust "preliminary" methods, after blocking

In Ury [3] and Ury et al. [4], a statistically significant association is shown to exist between the frequency of automobile accidents and oxidant levels in Los Angeles, while no association is found between such accidents and CO levels.

The statistical method used in these reports is extremely simple. Accident frequencies and pollutant levels are compared for the same hour of the day on the same day of the week at time intervals of exactly one week, in order to equalize as many covariables as possible. If the hour with the higher pollutant level has the higher accident frequency, this is scored $+1$, and the opposite case, -1. Ties in either variable are scored 0. Thus, a concordance sign test, or Kendall's tau with $n = 2$, is used to investigate the possible association between accidents and pollutant levels.

This almost embarrassingly naive method has all the earmarks of a preliminary look-in. Before proceeding to more sophisticated techniques one should, however, note the following. (a) Even nine pollutant monitoring stations can scarcely cover every area of Los Angeles County precisely. (b) Pollutant measurements are far from reliable; errors of 50 per cent or more are quite common. (c) For the many other potential pitfalls see [4]. In short, any study dealing with area wide pollution rather than with pollutants measured at a specific place is likely to run into a number of problems, and the unreliability of pollutant measurements can affect even studies of the type given in Section 1.

While the concordance sign test procedure is not especially applicable to health studies (since such factors as hospital admissions are not too meaningful on an hourly basis), it is applicable to crime studies, for example, in any area that has monitoring stations that provide hourly averages, or maximum values, for various pollutants.

3. Simple follow-up methods for daily mortality studies

A skeletal plan for a study of daily mortality is given elsewhere by Hexter, using the multiple regression technique outlined in Hexter and Goldsmith [1]. After making allowance for cyclic variation, trend and temperature, one can obtain at least a preliminary idea of the effects of various pollutants during short periods of heavy "smog" by subjecting the residuals to simple sign tests of the type outlined in Section 2. This is covered in detail in [2].

4. Multiple location studies of physiological measurements with different pollutant levels

As a specific example, consider the residence effect on blood lead level in four locations: a rural town or village in California, the Los Angeles beach area, the Oakland area, and downtown Los Angeles. A randomly chosen sample of n subjects from each area, matched with regard to sex, age, socioeconomic status, occupational exposure and length of residence, can presumably give a fair idea of the pollution effect on blood lead levels. If an *a priori* ordering of the locations with regard to pollutant levels is feasible, regression or trend tests can be used.

This type of study can be applied to any pollutant and physiological measurement and to any set of communities sufficiently close to monitoring stations that measure this pollutant. Once again, simple and robust procedures should be used, in view of measurement unreliability.

REFERENCES

[1] A. C. HEXTER and J. G. GOLDSMITH, "Carbon monoxide: association of community air pollution with mortality," *Science*, Vol. 172 (1971), pp. 265–267.

[2] H. K. URY, "Statistical procedures for testing the relationship between daily mortality and air pollution," California State Department of Public Health technical report, December 1967.

[3] H. K. URY, "Photochemical air pollution and automobile accidents in Los Angeles," *Arch. Env. Hlth.*, Vol. 17 (1968), pp. 334–342.

[4] H. K. URY, J. G. GOLDSMITH, and N. M. PERKINS, "Possible association of motor vehicle accidents with pollutant levels in Los Angeles," Research Project S-23, Project Clean Air Research Reports, University of California, Vol. 2, September 1970.

[5] H. K. URY and A. C. HEXTER, "Relating photochemical pollution to human physiological reactions under controlled conditions," *Arch. Env. Hlth.*, Vol. 18 (1969), pp. 473–480.

SUMMARY OF PANEL DISCUSSION PLANNING A COMPREHENSIVE STUDY OF EFFECTS OF POLLUTION ON HEALTH

E. L. SCOTT

UNIVERSITY OF CALIFORNIA, BERKELEY

1. Introduction

The invited panel, chaired by Professor Herbert A. David, consisted of 12 individuals: P. Armitage, C. L. Chiang, F. N. David, J. R. Goldsmith, B. G. Greenberg, R. J. Hickey, E. B. Hook, F. J. Massey, G. B. Morgan, J. Neyman, H. W. Patterson and C. A. Tobias. Also active in the discussion were S. W. Greenhouse, E. Landau, H. L. Rosenthal, E. J. Sternglass, W. Winkelstein, Jr., and B. E. Vaughan.

After opening the conference, Chairman David invited the authors of the four skeletal plans for a comprehensive statistical study of the health-pollution problem to outline their ideas briefly. Following the presentation of the plans by Goldsmith, Hexter, Neyman and Ury, the twelve panelists offered their remarks. Then there were comments from the floor.

The four skeletal plans differ considerably in their underlying ideas. One extreme contemplates what amounts to experimentation in hospitals on patients suffering from respiratory and heart diseases. A small number of patients, such as 15, can be divided into groups placed in separate rooms with different controlled degrees of air pollution. Frequent physiological observations could then provide information on the effects of each pollutant separately and of their combinations. Apparently experiments of this kind have been performed in Los Angeles. An intermediate point of view is represented by the recommendation of observational research typified by a study also performed in Los Angeles. Here, a measure of health conditions or a health parameter, such as number of deaths, is correlated with a variety of observational data, such as the cyclically varying temperatures and the concentration of selected air pollutants. Hope is expressed that, if such studies could be repeated in many urban areas in different countries, then the degree of agreement of the conclusions reached might be indicative of the real effects of the pollutants considered.

The other extreme proposal advocates a study involving simultaneously all the pollutants suspected to exercise major effects on health. The proposal is based on the observation that each particular locality is affected not by just one pollutant but by a combination of several of them. Hence, in order to be able to assign to each of the pollutants its own effect on the health parameter chosen for investigation, it is unavoidable to include in the study a number of localities with different patterns of pollutants. According to this point of view, a really informative study must be both a multipollutant and a multilocality study. The proposal emphasizes the desirability that the suggested study be conducted by several independent statistical groups having access to the same data and enjoying the opportunity of frequent consultations.

The discussion that followed was as varied as the skeletal plans. It is summarized under the following headings: (1) subjects for study, (2) organizational difficulties, (3) availability of data and (4) statistical methodology.

2. Discussion of specific subjects for investigation

There seems to have been a general agreement that the effects of the various pollutants on health of small animals, such as *Drosophila*, are easier and perhaps preferable to study than the effects on human health. Because of the possibility of combining experiments with observational investigations, studies of animals might lead to a better understanding of the phenomenon in the large. Studies of this kind might reveal chains, particularly food chains leading to accumulation of noxious chemicals in the body of man. Also, studies of food chains might show the existence of important pathways from environment to man other than food chains. For example, according to a study of the Battelle Memorial Institute, over half of the body burden of DDT in man comes from other than food sources. Inferentially, this other half may come via inhalation of pesticide-laden soil particles or the like.

In spite of the greater ease of studying lower animals rather than humans, it is primarily human health that must be the ultimate subject of contemplated studies. While this general point of view was vigorously emphasized by several speakers, there was considerable variation in detail.

Some speakers felt it desirable to investigate health hazards from the use of thousands of new chemicals that appear on the market year after year, some as food additives, some others for use in commercial laundries, and so forth. The effects of these chemicals may be carcinogenic or teratogenic and they may be cocarcinogenic or coteratogenic. Also these effects may be long range. In order to be able to detect them, frequent and detailed observations of the exposed individuals are necessary. Even with generous financial support such observations could be made only on a limited number of individuals. But then, if the sample studied is small, there will be little hope of detecting the effects that one wishes to detect. On the other hand, if the study is to be based on a large sample, only a few health parameters could be included.

Two foreign studies were quoted as examples of what might be done in this country. One of them, performed in the United Kingdom, is a multilocality study on the general lines of the research by Hickey and co-workers (*IEEE*

Transactions on Geoscience Electronics, Vol. GE-8 (1970), pp. 186–202) but apparently less extensive. The other study, by V. L. Kapalin, at least in its part described at the panel discussion, gives tabulations of measurements on several groups of children, living in localities with different levels of air pollution and belonging to different income strata of Czechoslovakia. Among other things, it appears that the growth of children in polluted localities is slower than in localities with cleaner air.

With reference to possible health effects of radiation in the vicinity of nuclear power facilities, there seemed to be general agreement that, if this subject is studied, then the investigation should also indicate the localities with large plants using mineral fuels.

The culminating point of this part of discussion was an appeal for forming some kind of consensus on what is the question that has to be answered. If this question is settled then the discussion of the means of obtaining the answer could be fruitful.

3. Organizational difficulties

Several speakers expressed opposition to the idea that, if a comprehensive statistical study is attempted, it be conducted by several statistical groups working in parallel. The chief argument against a "multicenter" organization is the difficulty of reaching consensus on what should be done and how. Several speakers proposed that some kind of statistical consulting center be established where substantive research workers could request and receive authoritative advice.

4. Availability of reliable data

While results of monitoring various pollutants are published by appropriate agencies, concern was expressed as to the adequacy of monitoring stations to produce measurements to characterize substantial areas which these stations are supposed to represent. It was suggested that, in order to characterize the air pollution of an area like Los Angeles so that the measurements of pollution could be meaningfully correlated with health parameters relating to the same area, more stations than there are now would be needed. Also, it was suggested that these stations be randomly distributed over the area. The importance of expert design and surveillance of the monitoring equipment was emphasized.

There was extensive discussion of the desirability of using special surveys of health parameters and of the heavy cost and possible incompleteness of health data.

5. Statistical methodology

The following problems were discussed.

(i) *Problem of competing risks.* Ordinarily, an individual in a population is exposed to the risk of death within a unit period of time not just from one particular cause C_1, but also from a number of other causes, say C_2. The probability of death from C_1 in the presence of "competition" from C_2 is labeled the crude rate of death from C_1. Clearly, the value of the crude rate of death from C_1 depends not only on C_1, but also on the severity of all other causes symbolized by C_2. The concept characterizing C_1 alone is labelled the net risk of C_1. This is the conditional probability of death from C_1 computed on the assumption that all other risks C_2 are inoperative.

The rates of death from various causes ordinarily published by various institutions are estimates of the crude rates. It was suggested that for pollution-health studies net rates of risks would be more relevant.

(ii) *Spurious correlations.* Several speakers warned against direct juxtaposition of morbidity data published for different geographical localities with the corresponding levels of pollutants. One possibility is that both the morbidity and the level of pollution are closely related to the socioeconomic status of the inhabitants in the different localities. Hence the unconditional correlation between pollution and morbidity may be very high even if the pollutants do not affect the health at all. In particular, a thorough statistical study may well reveal that the much discussed pollution with radioactive wastes is not really correlated with any damage to health.

Many present studies compute correlations between rates which have correlated (even identical) denominators. The researchers tend to infer that the observed correlation of rates implies correlation between the numerators. However, this inference may be entirely spurious, even in the wrong direction as has been illustrated by classical examples.

6. Recommendations

There was not complete unanimity on any plans for a comprehensive study of effects of pollution on health. However, there was general agreement that the results from this Symposium should be studied and that small working agencies should be set up to organize and recommend the details of the study.

EPILOGUE OF THE HEALTH-POLLUTION CONFERENCE

JERZY NEYMAN

STATISTICAL LABORATORY, UNIVERSITY OF CALIFORNIA, BERKELEY

1. *General remarks.* The papers published in this volume represent the result of the effort to compile a realistic cross section of the contemporary statistical thinking on the problems of pollution and health. The first seven papers stem from public institutions and, with unavoidable differences as to the amount of detail, reflect these institutions' interests. The next seven papers illustrate the sharp dispute about health effects of radioactive pollutants. The subjects of the remaining papers are varied, each representing a different "case history" connected with the problem of pollution. Thus far the problem of health and pollution has not attracted the attention of many mathematical statisticians and this volume contains just one paper, by Richard E. Barlow, that contains a theorem.

Four papers of the first group, one by Totter and the other three by Finklea, by Riggan, and by Nelson, with collaborators, describe in detail the very impressive programs of activities of the Biology-Medicine Division of the Atomic Energy Commission and of the Division of Health Research, Environmental Protection Agency. To a considerable extent, this includes not only the work of the two important agencies of the Federal Government, but also that conducted by the various contractors. As a result, these four articles do give a firsthand account of a large section of the contemporary statistical work on pollution. The paper by Sirken illustrates the commendable concern with the reliability of data collected by the National Center for Health Statistics.

While this coverage of institutional research is gratifyingly broad, it is regretted that the information in this volume on the impressive amount of work (with 468 papers published up to June 1969!) performed under the aegis of the National Academy of Sciences–National Research Council is only secondhand, being fragmentarily reported by several speakers. In particular this applies to the NAS-NRC biology-health studies of the atomic bomb casualties in Hiroshima and Nagasaki.

Even though the announced ultimate goal of the conference was to discuss and to plan a comprehensive statistical study of the relationship between human health and the various pollutants, and even though four skeletal plans have been submitted and published above, other material in the volume shows little enthusiasm for the project. In fact, Gofman and Tamplin are explicit in opposing

This investigation was partially supported by PHS Grant No. GM-10525-08, National Institute of Health, Public Health Service.

the idea. Most other authors showed indifference. Particularly this is the case with regard to the controversial pollutants such as radioactivity: a hot potato effect, perhaps. Yet, the papers presented do suggest that the current information on health effects of the various pollutants is fragmentary and that firm data needed for the formulation of national policy might be obtained only through a difficult, large, broadly multipollutant and multilocality statistical study. In particular, this applies to radiation because, apparently, the biological effects of radiation are not uniquely induced by radiation, but are also caused by other environmental agents. Among the other agents that may compete with radioactivity, various papers emphasize DDT, lead, mercury, and a defoliant 2,4,5-T. The programs of the Environmental Protection Agency and, particularly, of the Biology-Medicine Division of the Atomic Energy Commission are so extensive that the proposed comprehensive statistical health-pollution study might be included within the sphere of these institutions' activities. The drawback is that both the AEC and the EPA appear to be parties to the sharp controversy which the proposed comprehensive statistical study could help to resolve.

The purpose of the present *Epilogue* is to emphasize a few arguments in favor of a comprehensive study, perhaps on the lines of the "Skeletal Plan . . ." published earlier in this volume, and also to bring out some methodological-statistical difficulties which seem to have escaped the attention of a number of authors.

2. *Fragmentary character of information available on health effects of radio-activity*. Figure 2 in the paper by Totter refers to an experiment with mice irradiated with a high rate of X-ray. It contains a tentative extrapolation to humans. The figure seems to imply that, if a fetus is irradiated *in utero* at any time some two to five weeks after conception, then it is almost certain to develop abnormalities; also, a very substantial proportion of live births, perhaps 40 to 50%, are followed by neonatal deaths. All this applies to a high dose of X-ray radiation administered at a high rate, not in conditions of real life. Figure 4 in the same paper of Totter indicates that the effect of a fixed dose of gamma ray irradiation depends substantially on the rate at which this dose is administered. Specifically, the decrease of the rate from 6.7 rads/min to some 0.003 rads/min decreases the shortening of life of the experimental animals in the ratio of, roughly, 4 to 1.

All the above are results of experiments performed on animals and one is naturally interested in the effects on humans of such rates of radiation as one is likely to find in actual life. Here the paper by Vaughan provides some relevant information. This paper gives a description of an interesting study conducted in Alaska. The study is concerned with the food chain: fallout → lichens → caribou → Eskimos. The study seems to have covered the period 1963–1969 and involved direct observations and measurements of cesium 137 in the fallout, in lichens, in caribou, and in humans. The findings are summarized in Figure 6. Briefly and roughly, they are as follows. The fallout content of cesium 137 diminished steadily and reached essentially zero by the end of 1966. The lichens' content of cesium 137 had a maximum in 1965. Then it showed a gentle decline, ending at a level about double that at the beginning of the study period. The curve representing the body-burden of cesium 137 in the Eskimos is somewhat complicated, showing a yearly cyclic variation. The yearly minima grow from about 160 nCi in 1964 to about 300 nCi in 1968. At the same time, the maxima increase, roughly, from 575 nCi to 700 nCi. No indications are given of the health effects that the accumulation of the radioactive cesium may have caused either to adult Eskimos or to their progeny. However, a detail in Figure 6 indicates a chromosomal study as the next step in continuing research. This may or may not mean an effort to answer the question suggested by Totter's Figure 2 about the effects on human fetuses of radiation administered at now observable levels.

It must be obvious that, while highly interesting, the study described by Vaughan can provide only a fragmentary answer to the general question of radioactive pollution and health. This will continue to be the case even when the indicated chromosomal study is completed and even if it is accompanied by similar studies of other pollutants mentioned by Vaughan: DDT, lead, mercury, and the defoliant 2,4,5-T. The reasons are that (a) malformations at birth and postnatal deaths need not be reflected in the observable changes in the chromosomes and (b) that whatever may be found for Eskimos in Alaska will refer to the whole local Alaskan pattern of pollutants and not to any one of them in particular.

Clearly, in order to be able to evaluate n unknowns (these would be separate effects on a particular health parameter of n different pollutants or their combinations), one needs to have n independent equations. Over a unit of time, each locality, such as Alaska, can produce only one such equation. To estimate n effects, one needs at least n localities with different patterns of pollutants; in order to allow for random variation, a multiple of n localities is needed.

One other paper in this volume is concerned with the body-burden of a radioactive chemical. This is the paper by Rosenthal giving measurements of strontium 90 in teeth and in bones of children, apparently accumulated through milk consumed by expectant mothers. The information given is obviously important but, by itself, it is only a fragment of the general picture that seems important to have.

As described by Totter, the effect of the accumulating strontium 90 body-burden in man, resulting from the world wide fallout from nuclear tests, is a continuing concern of the AEC. A number of experiments with various animals have been conducted and are still in progress. There seems to be no doubt that, at sufficient rates of intake, strontium 90 is deadly to animals through a number of forms of cancer. As far as humans are concerned, there is the omnipresent problem of extrapolation.

Figure 6 of Vaughan suggests the intriguing question about the source of cesium 137 in the fallout over Alaska, quite large in June 1963 and then declining to essentially zero in 1966. Here, several figures of Sternglass, beginning with

Figure 4, come to one's mind. Among other things, Figure 4 indicates that 1961–1963 were three years of the latest large H-bomb tests. These tests were conducted at distances from Alaska measured in thousands of miles and one is inclined to skepticism at the suggestion that just these tests could have resulted in measurable quantities of cesium 137 in the fallout, not to speak of the measurable body-burden of this chemical in the Eskimos that persisted at least up to 1968. Is there any other imaginable source of cesium 137?

The diagrams produced by Sternglass are intended to support his opinion that the observable deceleration in the decline of infant mortality is likely to have been caused by tests of nuclear weapons. While the covariation asserted by Sternglass seems to be there, the arguments for causality based on this covariation alone appear tenuous. However, if no imaginable source of cesium 137, in the fallout over Alaska, can be found other than the 1961–1963 H-bomb tests, then the findings described by Vaughan contribute to the credibility of the Sternglass hypothesis.

Remark. In what follows the words cause, causality, and the like are occasionally used. The reader will realize that these words are used without any metaphysical connotation. With reference to empirical facts, the statements to the effect that "A is caused by B" mean simply that in the past, the appearance of B was always followed by A. With reference to theoretical speculations "causality" means the hypothesis that not only in the past but also in the future the appearance of B will be followed by A, possibly, through a particular hypothetical mechanism.

No observational study can ever establish causality. The best one can hope for is that an observational study will suggest real causal relations. Thus, the proposed multipollutant and multilocality statistical health-pollution study cannot resolve all the controversies. However, if conducted with all due care, it can help to unify the existing fragmentary information to form a coherent general picture of actual happenings. Among other things, such a study could answer the question whether the body-burden of radioactive chemicals (cesium 137, strontium 90, etc.) in people living close to electric power generators using mineral fuels is less or is larger than among those living *in comparable conditions* next door to nuclear power plants. Another illustrative question that the proposed study might answer is that about the frequency of malformations at birth in *comparable localities* in which, however, the body-burdens of radioactive chemicals are different. Both these questions seem interesting and important. There are many others of the same kind.

The two questions just mentioned illustrate some of the difficulties of the proposed study. For one thing, the monitoring of malformations at birth must be made reliable. As illustrated by the conference organized last year by Professor Hook, this is a difficult problem. For another thing, the program of monitoring pollutants organized by the EPA should be enlarged to include a new kind of "environment"—the bodies of the people: the body-burdens of radioactive chemicals will be necessary for the study.

Whether all of these and a host of similar other problems can be authoritatively solved, and by what uncommitted organization, is debatable. With reference to the defoliants, Professor Sterling mentions the American Statistical Association. An alternative possibility might be the International Association for Statistics in Physical Sciences which exhibited an independent interest in the matter. On the initiative of this organization an international symposium on pollution was recently held at Harvard, organized by Professor John W. Pratt.

The phrase "in comparable conditions" is easy to write. However, because of all the complexities of monitoring physical factors, and because of the relevance of race and of socioeconomic status, as discussed by Landau and by Winkelstein, the difficulties in reaching a reasonable level of comparability are enormous.

3. *Pitfalls of competing risks.* The much debated question of reliability or otherwise of the safety standards for man-made radioactivity hinges considerably on extrapolations of the results of studies conducted on several groups of individuals, who underwent heavy exposures to radiation administered at high rates. This includes the survivors of atomic bomb attacks on Hiroshima and Nagasaki. The studies concerned were conducted at least partly under the aegis of the National Academy of Sciences–National Research Council, in part through the Atomic Bomb Casualty Commission. No direct report on these studies is published in this volume, but a few fragmentary references to them indicate some of the difficulties involved. For example, the dosimetry proved difficult: on occasion it was difficult to estimate the exact dose of radiation to which a survivor at Hiroshima or Nagasaki was exposed. However, there was another great difficulty in the same studies which, with the apparently single exception, did not seem to attract the attention of the various authors. The difficulty in question is that of allowing for the presence of so-called "competing risks." [1]

The directly computable death rate from a cause C_1 is what is called the "crude rate," the value of which depends not only on the intensity of C_1 but also on that of all other competing causes, say C_2. The rate that characterizes C_1 alone, is the "net rate" which is the rate of interest. If C_2 is only mild, then the difference between the net and the crude rates of C_1 can be trivial. On the other hand, if the combination C_2 of all other causes competing with C_1 is intense, then the net rate of death from cause C_1 can be a large multiple of the crude rate.

It is quite plausible that the radiation induced deaths of infants in Hiroshima and Nagasaki had very intense competitors in, say, starvation, lack of maternal care, and the like. Again, with reference to some other studies in this country and abroad, of groups of individuals subjected to X-ray treatment against some disease D, deaths from that same disease, and from the various complications thereof, must have competed with deaths from the radiation induced cancer. How heavy were these cases of competition? What was done, and how, to elicit the all important estimates of net rates of deaths directly caused by radiation? The competence and the authority of the National Academy of Sciences are and should be great. Therefore, it is very regrettable that the methodology used to solve the problem of competing risks is not described in this volume.

Even though the term competing risk is not mentioned by Gofman and

Tamplin, their discussion indicates that they are fully aware of the issues involved, complete with the complication that the radiation induced leukemia tends to develop earlier than other radiation induced cancers.

4. *Pitfall of incomplete comprehensiveness of a multipollutant, multilocality study.* In a multipollutant, multilocality statistical investigation one of the threatening pitfalls is incompleteness of the set of pollutants studied: if the study involves a certain number s of pollutants, say $P_1, P_2, \cdots, P_s$, but neglects another pollutant P_0 that happens to be important, then the conclusions regarding $P_1, P_2, \cdots, P_s$, suggested by even very highly significant findings, may be completely misleading.

A case in point is a recently published multipollutant and multilocality study of a substantial number of pollutants, which omitted radioactivity. One of the findings was that an increase of copper content of the air diminishes significantly the experienced frequency of death from a certain disease D. The correct, but somewhat lengthy interpretation of the result is, roughly, as follows:

Among the so many localities studied (and there were quire a few of them), the average frequency of deaths from disease D in localities with high copper pollution is less than in other localities in which copper pollution is low.

This interpretation is just a statement of facts and, apart from being cumbersome and somewhat incomplete, is not objectionable. However, after obtaining a result like this, one is tempted to go just a little farther and conclude that an increase in copper pollution of the air tends to diminish the frequency of deaths from D.

Radioactivity in the air, or in food or in water is frequently a suspected cause of premature death. Also, the levels of radioactive pollutants vary considerably from one locality to the next as, undoubtedly, do the levels of pollution with copper. *A priori* it seems possible that the dust of copper containing chemicals in the air is deleterious to health (contrary to conclusions suggested by the actual study) but that it is much less deleterious than radioactivity. Finally, it is possible that the localities with high air pollution levels of copper (possibly, localities with strip mining of copper ore) have relatively little pollution with radioactivity. In other words, among the localities studied in the particular investigation, there may have been a strong negative correlation between levels of radioactivity and of copper pollution. If this was so, then, on the average, the localities with plenty of copper in the air were also localities with little radioactivity and hence with relatively low mortality from D, even though the presence of copper in air caused some increases in deaths.

The reader will realize that all the above is purely hypothetical and that there is no intention to suggest that copper is deleterious or that the radioactivity is substantially more so. The purpose of the discussion is to indicate the danger of incomplete inclusiveness of pollutants in observational studies. Various considerations of convenience, and others, may (and do) suggest the formulation of the policy, such as:

We are interested in health effects of pollutants P_1, P_2, $\cdots$, P_s but, provisionally at least, not in others; therefore, even though we are aware of claims that some other pollutants P_{s+1}, P_{s+2}, $\cdots$, P_n are deleterious to health, our own multipollutant and multilocality study shall be limited to P_1, P_2, $\cdots$, P_s.

This is a dangerous policy.

While practical considerations must impose limitations on the number of pollutants to be included in an investigation, it is important to be aware of what the omission of a particular pollutant may entail. In particular, the reader will have no difficulty in visualizing how the omission from the study of a particular pollutant, say P_{s+1}, may result in the appearance that the pollutant studied P_1 is deleterious to health while, in actual fact, it is beneficial.

5. *Pitfalls of "spurious correlations."* Spurious correlations have been ruining empirical statistical research from times immemorial. Apparently the first publicly discussed incident is recorded in three contributions published next to each other in 1897. The credit for identifying the noxious phenomenon belongs to Karl Pearson [2]. The victim whose spurious correlation mishap stimulated the discussion was W. F. R. Weldon [3]. The third contribution, intended to make Pearson's developments more clear intuitively, is due to Francis Galton [4].

Even when contrasted with "organic correlation," the term "spurious correlation" seems to be a misnomer. As used by Karl Pearson, the term refers to a very real and easily computable correlation, say R_1. What is spurious is the interpretation of R_1 as having something to do with another correlation, say R_2, termed "organic" which happens to be of primary interest but is not easy to compute. As rightly noted by Karl Pearson and fully understood by Galton, R_1 may have no relation to R_2.

Pearson's own awareness of difficulties connected with spurious correlations stemmed from studies of errors committed independently by several observers. Later he noticed similar difficulties in biology and economics. Weldon's studies were concerned with shrimp. In more modern times, spurious methods of studying correlations were involved in a great variety of empirical research: in astronomy, in farm economics, in biology, in the study of elasticity of demand, in the problems of drunkenness and crime, of railroad traffic, and of racial segregation. On occasion, they were used in arguments about public policy matters. This applies to the health-pollution literature, including some papers in this volume, which is the justification for the present somewhat long section of this article. In general terms referring to public policy matters, the situation is as follows.

Consider a not directly controllable phenomenon P as it develops in some units of observation U (perhaps different localities in a country in a given year, or over several consecutive years in the same locality). The phenomenon P manifests itself in some variable Y which is of public concern: the currently observed values of Y appear unacceptably high (or low). It is suspected that Y is somehow connected with another variable X which is subject to at least partial control. (Y may be the number of deaths from a disease D in a specified section of the population, perhaps in a particular age group; X may stand for the level of a pollutant). A public measure, perhaps legislation, is contemplated to enforce a change in the values of X with the hope that this would result in desirable changes in Y, perhaps only on the average.

Authoritative information as to whether changes in X will cause changes in Y at least on the average, can be obtained through a well designed experiment. However, in the circumstances considered experiments are impossible and one is compelled to conduct an observational study. While no causal relations between X and Y can be expected from such investigation, it can reveal how the average values of Y in units of observation where X is large differ from those where X is small, which may be valuable information in deciding on the public measure contemplated.

The situation would be relatively simple if enough observational units could be found identical in all respects, except for values of X and Y. In real life such favorable conditions cannot be expected and one must be prepared to find that, generally, the observational units vary not only in values of X and Y but also in many other respects. In particular some variables $Z_1, Z_2, \cdots, Z_s$ come under consideration, the variation of which is likely to influence either X alone or Y alone or both. For example, in a health-pollution study with X and Y denoting, respectively, the level of a particular pollutant and Y the number of deaths from D, the several "nuisance variables" Z may be numbers of the exposed to risk from disease D who belong to particular racial and socioeconomic categories of the population. Again, some other nuisance Z may refer to pollutants other than that under study. Possibly Z_i may mean the average body-burden of strontium 90 and Z_j that of cesium 137 and the like. Obviously, the practical problem of estimating the changes in Y, that may result from the contemplated intentional changes in X, calls for the study of conditional regression, say $Y(x|z)$, of Y on X, with all the Z's maintaining some fixed values symbolized by the latter z. With a moderate number of the nuisance variables Z, and in favorable conditions of linearity of regression, etc., such an investigation is not very difficult. However, in actual life various complications occur that suggest looking for some shortcuts. One kind of shortcut may mean the replacement of several measures of qualitatively different radioactive pollutants by a single measure of all such pollutants combined. Another kind of shortcut may result from an effort to correct the number of deaths for the variation from one unit of observation to the next in the total number of exposed to risk and also in the numbers of those who belong to different racial and socioeconomic groups. Frequently, the temptation to take a shortcut is strong and one is involved in studying spurious correlations.

In the simplest case, for each unit U of observation one computes the supposedly corrected value, say V, of X and/or the supposedly corrected value W of Y which these variables X and Y would have had if all the nuisance Z did not vary from one unit of observation to the next, but maintained some typical values. In effect V and W are certain functions f_1 and f_2 of the directly observed X, Y and Z,

$$\begin{aligned} V &= f_1(X, Z), \\ W &= f_2(Y, Z). \end{aligned}$$

(1)

Then the correlation between V and W is taken to represent that of X and Y, with the influence of the nuisance Z being eliminated.

Emphatically, the convincingness of the correcting functions f_1 and f_2 notwithstanding, THE CORRELATION OF V AND W NEED NOT BE INDICATIVE OF THE PARTIAL CORRELATION BETWEEN X AND Y WHEN THE NUISANCE VARIABLES Z ARE FIXED.

Some theoretical considerations relating to this problem will be found on pp. 143–154 of reference [5]. I am indebted to Robert Traxler for the following numerical example illustrating, with somewhat exaggerated precision, what may result from a "shortcut" in the case where, given the nuisance Z, the variables X and Y are strictly independent.

The example refers to the simplest case of spurious correlation studies, involving just one nuisance variable Z, which represents the number of individuals exposed to the risk of death from a disease D, and where only one of the variables studied, namely Y, is corrected for the variation of Z. Furthermore, the method of correcting is so convincing that it appears incredible that it may involve some pitfalls.

Traxler considers 54 localities with the necessary data given in Table 1. This table is divided into six panels, each panel referring to 9 localities, all characterized by the same number Z of exposed to risk measured in some convenient units, such as 10,000 or the like. In the first panel $Z = 5$, in the second $Z = 6$, etc., up to $Z = 10$ in the last panel.

While the value of Z in each panel is constant, there is a variation in X, the level of the pollutant studied. Measured in some units and from a conventional zero point, X has three different values in each of the six panels: 1, 2, 3 in the first, 2, 3, 4 in the second and so forth. This change in the values of X from one panel to the next reflects Traxler's idea that the increase in the number Z of the exposed to risk may mean an increase in the total population, with the more populated localities being more intensely polluted than those with smaller populations. Of course, this need not be always the case and the reader may find it interesting to investigate the situation in which the relation between X and Z is contrary to that of Traxler.

The detail of Table 1 which deserves particular attention is the arrangement of columns giving X and Y. In each panel, to each of the three different values of X there correspond three localities with varying numbers of deaths Y. The important point is that to all the three values of X there corresponds *the same* triplet of values of Y. In the first panel this triplet is 1070, 1100 and 1130 deaths from D. In the second panel, with somewhat more people at risk, the triplet of the numbers of deaths is 1270, 1300, and 1330, and so on.

TABLE I

SIX GROUPS OF LOCALITIES WITH POLLUTANT X, NO. OF DEATHS Y AND NO. AT RISK Z

Locality no.	No. at risk Z	Pollutant level X	No. of deaths Y	Locality no.	No. at risk Z	Pollutant level X	No. of deaths Y	Locality no.	No. at risk Z	Pollutant level X	No. of deaths Y
1		1	1070	19		3	1465	37		5	1860
2	5		1100	20	7		1500	38	9		1900
3			1130	21			1535	39			1940
4		2	1070	22		4	1465	40		6	1860
5			1100	23			1500	41			1900
6			1130	24			1535	42			1940
7		3	1070	25		5	1465	43		7	1860
8			1100	26			1500	44			1900
9			1130	27			1535	45			1940
10		2	1270	28		4	1665	46		6	2060
11	6		1300	29	8		1700	47	10		2100
12			1330	30			1735	48			2140
13		3	1270	31		5	1665	49		7	2060
14			1300	32			1700	50			2100
15			1330	33			1735	51			2140
16		4	1270	34		6	1665	52		8	2060
17			1300	35			1700	53			2100
18			1330	36			1735	54			2140

Table 1 produced by Traxler shows many facets of regularity which cannot be expected in any real study. Rather than have six sets of localities with exactly the same number Z of individuals at risk within each set, in a real study it would not be surprising to find that no two localities have the same value of Z. Here, then, there will be a real problem of dealing with the variation in Z and the method that the practicing statistician is likely to use is easy to perceive. The expected reasoning would be somewhat as follows. My primary objective is to find out the changes in the *frequency* of deaths from D that are to be expected if the level X of the pollutant is intentionally changed. Thus, the most direct way of studying the problem is by computing, for each locality, the death rate $W = Y/Z$, by classifying all the localities according to the level X of the pollutant and, finally, by computing the mean death rate $\overline{W}(X)$ that corresponds to any given value of X.

What such an analysis may lead to is illustrated in Table II. Table II is arranged so as to simplify the calculation of the $\overline{W}(X)$ somewhat. The simplification is based on the fact that all the 54 localities considered are divided into triplets, each triplet being characterized by a combination of values of X and Z. Thus, there is no point in calculating the death rate W separately for each locality. It is sufficient to compute the average rate per triplet, then to classify the triplets according to (X, Z) and to average over values of Z so as to obtain the desired $\overline{W}(X)$. Accordingly, the 8 lines of Table II correspond to the eight values of X and the six columns to the six values of Z. The last column gives the desired average death rates from D corresponding to increasing values of X.

TABLE II

ANALYSIS OF THE EFFECT OF POLLUTANT BASED ON DEATH RATES

Pollution level X	Mean death rates $\overline{W}(X, Z)$ in triplets of localities cross classified according to X and Z = no. at risk						Mean death rate $\overline{W}(X)$ in localities with pollution X
	$Z = 5$	6	7	8	9	10	
1	220.0						220.0
2	220.0	216.7					218.3
3	220.0	216.7	214.3				217.0
4		216.7	214.3	212.5			214.5
5			214.3	212.5	211.1		212.6
6				212.5	211.1	210.0	211.2
7					211.1	210.0	210.6
8						210.0	210.0

If any real study exhibited the correspondence between the average death rates $\overline{W}(X)$ and the level X of the pollutant studied, anywhere comparable to that in Table II (however, no comparable regularity can be expected!), the interpretation would be somewhat as follows:

(i) The pollutant studied *does* influence the death rates from D.

(ii) At least as far as the disease D is concerned, the pollutant studied is beneficial: if the level of the pollutant is low, say if it is lower than $X = 6$, then an increase in the level of the pollutant decreases noticeably the death rate from D.

(iii) For the above reasons, the adoption of a public measure should be considered to increase the level of the pollutant, perhaps by spraying the countryside, at least in those localities in which the current level of the pollutant is low.

In order to see what would be the result of adopting any such measure as suggested in (iii), we must return to Table I. Each of the six panels refers to localities with three different levels of the pollutant: 1, 2, 3, or 2, 3, 4, etc. It is seen that if, through spraying or otherwise, the two lower levels of the pollutant are replaced by the highest, the effect on the numbers of deaths would be exactly nothing. The numbers of deaths implied by the data would be unchanged. Thus, the contrary conclusion suggested by Table II is not inherent in the data. It is an artifact produced by dealing not with the triplets of values of (X, Y, Z) as given directly by the observations, but by values of X and $W = Y/Z$ computed for each locality. The anatomy of the phenomenon is interesting and the reader is urged to examine it both theoretically as in reference [5] and numerically. In particular, it may be interesting to see how Table II would be modified if the six values of Z in Table I ranged not from 5 to 10 but, say, from 4 to 9 or from 7 to 12, all other details of Table I remaining without change. While such changes in the range of Z may seem of little consequence, their effect on the appearance of Table II is likely to appear dramatic. Another question that may be interesting to answer is whether the range of values of Z can be so adjusted as to force Table II to yield an answer to the basic question which is at least approximately true.

Finally, the "anatomical" study of Tables I and II can help to meet the possible objection that, while Table I is unambiguous about the apparent effect of an increase in the pollution X in each of the six categories of localities by one or two units, it does not say anything about the possible effect of bringing X to its highest value $X = 8$, uniformly in all the 54 localities studied. Obviously, in order that Table I provides this kind of information, it must include data for more than 9 localities in each of its six panels, with the consequent increase in its complexity. However, it is likely to be interesting to consider how the objective could be attained.

Still another detail of Table II is worth noticing. This is that, while the computed $\overline{W}(X)$ decrease when X grows, the decrease is not linear.

The general principle that Traxler's example is intended to suggest may be heuristically formulated as follows. The object of the empirical study is to estimate the effect on a variable Y of an intentional change in the level of another variable X (or variables X). Here, the term "effect" is understood to refer to not any single unit of observation (locality) but to a population of such units. The information available for the study consists of values of not only X and Y but also of some s other variables $Z_1, Z_2, \cdots, Z_s$ which are suspected of being somehow involved in the mechanism that connects X and Y. The safe method of studying the population effect on values of Y of an intentional change in X, while the values of the Z's are left to vary as they will, is through an investigation of the *joint* variability of all the $s + 2$ variables involved $(X, Y, Z_1, Z_2, \cdots, Z_s)$. It is this simultaneous variation that characterizes the complex mechanism involved, of which we are interested in a single detail: what will happen to the values of Y (number of deaths) if the values of X are modified in a specified manner. Admittedly, the direct investigation of the variability of $(X, Y, Z_1, Z_2, \cdots, Z_s)$ is cumbersome and the tendency to reduce the number of the nuisance variables Z is understandable. However, any such reduction is equivalent to the injection into the mechanism studied of some elements that are extraneous to it. Traxler's example illustrates the pernicious effect of replacing the triplet (X, Y, Z) by the pair $(X, W = Y/Z)$ which looks very natural. If the reader investigates the suggested modifications in the range of Z, he will find that the effect of substituting the study of (X, W) for the study of the triplet (X, Y, Z) depends considerably on the properties of the joint distribution of the three variables, a detail of which is precisely the subject of investigation and *a priori* is unknown. The relationship between reality and the outcome of a real study in which one attempts, for example, to summarize in just one variable such directly observable quantities as, say, the body-burden of cesium-137, the body burden of strontium 90 and the radiation from walls of buildings, is a subject for speculation.

I am grateful to Professor F. N. David for providing the references for Francis Galton and Karl Pearson.

REFERENCES

[1] C. L. CHIANG, *Introduction to Stochastic Processes in Biostatistics*, New York, Wiley, 1968.
[2] KARL PEARSON, "Mathematical contributions to the theory of evolution—On a form of spurious correlation which may arise when indices are used in the measurement of organs," *Proceedings, Royal Society of London*, Vol. 60 (1897), pp. 489–498.
[3] W. F. R. WELDON, "Note, January 13, 1897," *Proceedings, Royal Society of London*, Vol. 60 (1897), p. 498.
[4] FRANCIS GALTON, "Note on the memoir by Professor Karl Pearson, F.R.S., on spurious correlation," *Proceedings, Royal Society of London*, Vol. 60 (1897), pp. 498–502.
[5] J. NEYMAN, *Lectures and Conferences on Mathematical Statistics and Probability*, Graduate School, U.S. Department of Agriculture, Washington, D.C., 1952, pp. 143–154 (2nd ed.).